Effects of a Meteoroid Impact on Steel and Aluminum in Space

By

R. L. Bjork[1]

(With 10 Figures)

(Received June 15, 1959)

Abstract — Zusammenfassung — Résumé

Effects of a Meteoroid Impact on Steel and Aluminum in Space. In this paper the formation of craters by hypervelocity particles is calculated in detail using a hydrodynamic model. The range of velocity considered is from 5.5 to 72 km/sec, which encompasses the meteoric velocity regime.

The assumption is made that the strength of the materials involved does not affect the dynamics of the process, so that even steel appears to act as an inviscid, compressible fluid. The validity of this assumption may be qualitatively deduced *a priori* by examining the pressures and densities created in a one-dimensional collision between two plates. This case may be trivially solved analytically if the equation of state of the materials is known. The solutions are exhibited for steel and aluminum.

Under this assumption, the compressible hydrodynamic equations govern the dynamics of crater formation. The calculations of this paper consist of solving these equations subject to the appropriate boundary conditions. A meaningful impact problem requires at least three independent variables, two for space and one for time, and the restriction to only two space variables demands the use of cylindrical symmetry. The solution is severely complicated by the presence of shocks, which represent moving surfaces across which discontinuities of the dependent variables occur. To solve the problem, it was necessary to resort to numerical techniques. The numerical method devised is restricted to problems wherein the target and projectile are of the same material.

The results for aluminum and steel indicate that the craters formed are hemispherical with radii that increase with about the one-third power of impact velocity, and are proportional to the characteristic dimension of the projectile.

The hydrodynamic assumption is verified *a posteriori* by the fact that the calculated craters agree in both shape and size with experimental ones created at 6.3 and 6.7 km sec for aluminum and steel, respectively. The agreement is especially satisfying inasmuch as the hydrodynamic assumption becomes better at higher velocities.

These calculations lead to craters considerably smaller than previously predicted by extrapolations from experimental data. Using a reasonable estimate of the meteoroid flux as a function of mass leads to the conclusion that space structures will survive about two orders of magnitude longer in time than had previously been supposed.

---

[1] Aero-Astronautics Department, The RAND Corporation, Santa Monica, California, U.S.A. (The views expressed in this paper are not necessarily those of the Corporation.)

Wirkungen eines Meteoriten-Treffers auf Stahl und Aluminium im Weltraum. In der vorliegenden Arbeit wird die Bildung von Kratern durch hyperschnelle Teilchen unter Verwendung eines hydrodynamischen Modells berechnet. Angenommen wird ein Geschwindigkeitsbereich von 5,5 bis 72 km/sec, was den Geschwindigkeiten der Meteoriten entspricht.

Es wird angenommen, daß die Festigkeit des untersuchten Werkstoffes die Dynamik des Prozesses nicht beeinflußt, so daß auch Stahl sich wie eine kompressible Flüssigkeit zu verhalten scheint. Die Gültigkeit dieser Annahme kann aus der Untersuchung der Drucke und Dichten, die bei der eindimensionalen Kollision zwischen zwei Platten entstehen, abgeleitet werden. Bei einer solchen Kollision entsteht ein einfaches Stoß-system, dessen Lösung nur die RANKINE-HUGONIOT-Gleichungen und die Zustands-gleichung erfordert. Bei Stahl erzeugt eine relative Aufschlag-Geschwindigkeit von 5,5 km/sec einen Druck von etwa 2 megabar. Verglichen mit der Zugfestigkeit von gutem Stahl mit 0,0068 megabar (entsprechend 70 kg/mm²) ist diese tatsächlich ver-nachlässigbar. Darüber hinaus werden in dem verdichteten Werkstoff hohe Tempera-turen erzeugt, die die Festigkeitswerte vermindern. Eine relative Auftreff-Geschwindig-keit von 72 km/sec erzeugt einen Druck von 144 megabar. Bei 5,5 km/sec wird der Stahl anfänglich auf das 1,5fache seiner ursprünglichen Dichte verdichtet und bei 72 km/sec beträgt die Stoßverdichtung das 3,5fache des Normalzustandes. Die Verdichtungswirkungen sind daher wesentlich. Man nimmt an, daß die kompressibel-hydrodynamischen Gleichungen die Dynamik der Kraterbildung bei hyperschnellen Teilchen richtig wiedergeben, und zwar je besser, je höher die betrachtete Geschwindig-keit ist. Die Berechnungen dieser Arbeit bestehen in der Lösung dieser Gleichungen unter entsprechenden Grenzschicht-Bedingungen.

Die Behandlung eines Aufschlagproblems erfordert mindestens drei unabhängige Veränderliche, zwei für den Raum und eine für die Zeit; bei der ersten Annahme ist dabei bereits eine gewisse Symmetrie vorausgesetzt. Es wurde axiale Symmetrie gewählt. Die Lösung wird relativ kompliziert durch das Auftreten von Stößen, was die abhängigen Veränderlichen diskontinuierlich macht. Um das Problem zu lösen, war es nötig, numerische Methoden zu verwenden. Ferner wird vorausgesetzt, daß das auftreffende Teilchen und der getroffene Körper aus demselben Werkstoff bestehen.

Die Zustandsgleichungen der Berechnungen basieren bei hohen Drucken und Dichten auf einer theoretischen FERMI-THOMAS-Berechnung. Bei niedrigen Drucken und Dichten werden die FERMI-THOMAS-Ergebnisse korrigiert, um den richtigen Elastizitätsmodul unter Normalbedingungen und Vergleichswerte für die explosive Kompression von Metallen zu erhalten. Die experimentellen Daten gehen bis etwa 0,5 megabar. Die erhaltene Zustandsgleichung ergibt Werte, die zwischen den experi-mentellen Daten mit niedrigen Drucken und den FERMI-THOMAS-Berechnungen für hohe Drucke liegen.

Die Ergebnisse für Stahl und Aluminium zeigen, daß halbkugelförmige Krater entstehen, deren Radien mit etwa ¹/₃ der kinetischen Energie wachsen. Bei einer gegebenen Geschwindigkeit ist der Radius des Kraters proportional zur charakteristi-schen Dimension des auftreffenden Teilchens. Die Berechnungen zeigen eine hervor-ragende Übereinstimmung mit einem experimentellen Punkt bei 6,7 km für Stahl und 6,3 km/sec für Aluminium sowohl bezüglich der Kraterform wie -größe. Das Ergebnis ist besonders befriedigend, weil die Rechnung direkt ohne die Hilfe anzu-passender Konstanten die Geometrie ergibt.

Die beiden genannten experimentellen Punkte wurden erst kürzlich erhalten und stellen die bisher schnellsten kontrollierten Schüsse einzelner Partikel dar. Um Krater-Effekte über diesen experimentellen Bereich voraussagen zu können, fanden es die meisten Bearbeiter nötig, in irgendeiner Form zu extrapolieren. Die meisten herkömmlichen Extrapolationen basierten auf der Annahme, daß die Kratergröße mit ²/₃ der kinetischen Energie wächst. Die Wirkung hoher Geschwindigkeiten wurde dadurch überschätzt. Da die zu erwartende Lebensdauer einer Konstruktion im Weltraum stark von der Wirkung einzelner Meteoritentreffer abhängt, führte die erwähnte Überschätzung zu einer vorausgesagten Lebensdauer, die etwa um zwei Größenordnungen zu gering ist.

Xth INTERNATIONAL ASTRONAUTICAL CONGRESS
LONDON 1959

X. INTERNATIONALER ASTRONAUTISCHER KONGRESS

Xe CONGRES INTERNATIONAL D'ASTRONAUTIQUE

PROCEEDINGS

BERICHT   COMPTES RENDUS

HERAUSGEGEBEN VON   EDITORIAL BOARD   COMITÉ DES RÉDACTEURS

W. von BRAUN · A. EULA · B. FRAEIJS de VEUBEKE · F. HECHT
W. B. KLEMPERER · J. M. J. KOOY · F. I. ORDWAY III · E. SÄNGER
K. SCHÜTTE · L. I. SEDOV · L. R. SHEPHERD · J. STEMMER

SCHRIFTLEITUNG   EDITOR-IN-CHIEF   RÉDACTEUR EN CHEF
F. HECHT

MIT 464 FIGUREN   WITH 464 FIGURES   AVEC 464 FIGURES

II
⟨P. 505—946⟩

SPRINGER-VERLAG WIEN GMBH 1960

ISBN 978-3-662-38961-4   ISBN 978-3-662-39914-9 (eBook)
DOI 10.1007/978-3-662-39914-9
Softcover reprint of the hardcover 1st edition 1960

Effets d'un impact météoritique sur l'acier et l'aluminium. La formation de cratères par des particules à très grande vitesse est calculée en détail sur la base d'un modèle hydrodynamique. Le domaine de vitesse considéré s'étend de 5.5 à 72 km/sec et englobe le régime des vitesses météoritiques.

Par hypothèse la résistance des matériaux considérés n'affecte pas la dynamique du phénomène, de sorte que même l'acier se comporte comme un fluide parfait compressible. Une justification qualitative *a priori* de cette hypothèse s'obtient en examinant les pressions et densités créées lors d'un impact uni-dimensionnel entre deux plaques. Ce cas peut être résolu analytiquement de façon élémentaire à partir de l'équation d'état des matériaux. Les solutions sont données pour l'acier et l'aluminium.

Les équations des fluides parfaits compressibles gouvernent donc la dynamique de la formation des cratères. Les calculs consistent à résoudre ces équations avec les conditions aux limites appropriées. Pour avoir une signification le problème d'impact requiert au moins deux variables spatiales en dehors de la variable temps. La restriction à deux variables spatiales demande la symétrie cylindrique. La solution est compliquée par la présence de chocs, surfaces mouvantes à travers lesquelles se présentent des discontinuités des variables dépendantes. Il a fallu appliquer des techniques numériques. Celle qui a été imaginée est restreinte au cas où la cible et le projectile sont fait dans le même matériau.

Les résultats obtenus pour l'acier et l'aluminium indiquent que les cratères sont hémisphériques avec des rayons croissants à peu près comme la puissance 1/3 de la vitesse d'impact et proportionnels à la dimension caractéristique du projectile.

L'hypothèse hydrodynamique est vérifiée *a posteriori* par le fait que les cratères calculés sont conformes en grandeur et forme aux cratères expérimentaux obtenus pour l'acier et l'aluminium aux vitesses respectives de 6.3 et 6.7 km/sec. L'accord est d'autant plus satisfaisant que l'hypothèse hydrodynamique devient plus exacte avec l'augmentation de la vitesse.

Ces calculs conduisent à des cratères beaucoup plus petits que ceux antérieurement prédits par extrapolation de données expérimentales. Une estimation raisonnable du flux météoritique en fonction de la masse conduit à la conclusion que la durée de vie des structures spatiales est supérieure de deux ordres de grandeur aux valeurs antérieurement supposées.

I. Introduction

The hazard posed by meteoroids to space vehicles is too well-known to merit an extensive introduction. The problem of assessing the hazard divides itself naturally into two parts:

1. Estimating the effects of a collision between an individual meteoroid and some component of the vehicle.

2. Estimating the flux of meteoroids in space as a function of their mass and velocity.

This paper is essentially confined to the first part of the problem.

Many estimates regarding the effects of such collisions have been made in the past, and these differ widely among themselves. The estimates are featured in the main by *ad hoc* assumptions which are difficult to justify. Many involve extrapolations in some form from data taken at low impact velocities.

The following is an attempt to calculate the phenomenology of an impact at meteoric velocities from fundamental principles, making only simple assumptions which may be justified by order-of-magnitude arguments. The validity of the assumptions receives support in the end by the agreement with experiment of the calculated shape and dimensions of the craters.

II. Shocks in One-Dimensional Collisions

In the high-speed impact process, shock waves induced in the solid projectile and target are of primary importance. A feeling for the order of magnitude of the physical quantities associated with such shocks may be gained by examining the one-dimensional case, for which a solution may be obtained in closed form. This case is illustrated in Fig. 1.

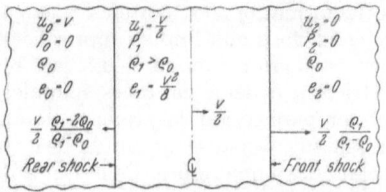

Fig. 1a illustrates the type of collision under consideration, wherein the

Fig. 1. Center of mass coordinate system after impact. *a* Laboratory coordinate system. *b* Center of mass coordinate system before impact

Fig. 2. Laboratory system after impact. Rear shock moves to left if $\varrho_1 < 2\,\varrho_0$. Rear shock moves to right if $\varrho_1 > 2\,\varrho_0$

colliding bodies are both semi-infinite and are bounded by planes. Both are of the same material. To take full advantage of the symmetry of the problem, one must view the process from the "center of mass" of the system, which is a point

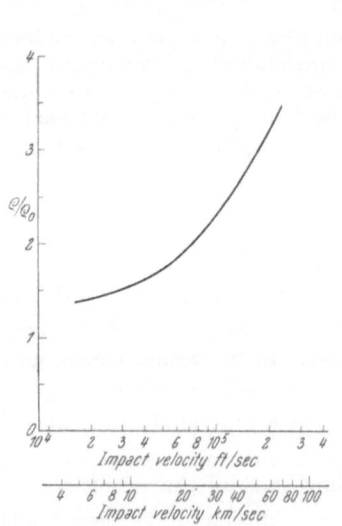

Fig. 3. ϱ/ϱ_0 versus impact velocity behind a one-dimensional shock for iron

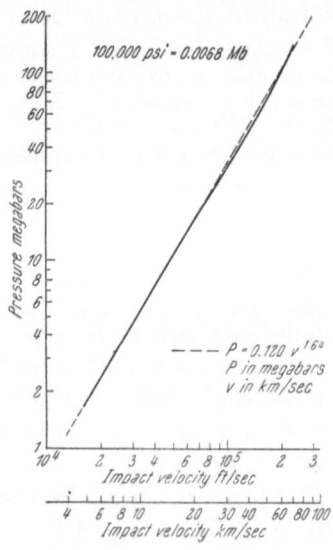

Fig. 4. Pressure vs impact velocity for one-dimensional collisions of iron

initially midway between the two bodies, and moving with one-half of the impact velocity. The situation as viewed from this frame of reference is shown before impact in Fig. 1b, and after impact in Fig. 1c.

By symmetry, the interface of the two bodies must be brought to rest upon impact, and must remain at rest at all future times, since there is no preferred direction for it to move. As shown in Fig. 1c, two shock fronts race outward from the stationary interface, encompassing more and more material as time progresses. The material in region 1 between the two shocks is at rest, is compressed to greater than normal density, and is under high pressure. The material in regions 0 and 2 moves inward toward region 1, is of normal density, and is under no pressure. In some instances, it will be possible for elastic waves to precede the shock fronts into regions 0 and 2, but the stresses and particle velocities carried by these waves will always be insignificant in comparison with those occasioned by the shock, so that they will be neglected in this treatment. Thus the material outside the shock fronts is essentially not affected by the impact, and will remain in its initial state until a shock front reaches it.

In Fig. 2 the situation is shown from the point of view of the observer fixed in the laboratory, who sees the initial conditions exemplified in Fig. 1a. This case is derivable from Fig. 1c by merely adding a velocity $v/2$ to the right to everything in Fig. 1c. Depending on the degree to which the material is compressed, the rear shock may move either to the right or to the left. If $\varrho_1 < 2\,\varrho_0$, it moves to the left, otherwise it moves to the right.

For the case of iron, the degree to which the material is compressed in region 1 is shown in Fig. 3 as a function of impact velocity. At 5.5 km/sec, ϱ/ϱ_0 is about 1.4, and becomes larger at higher velocities. At 72 km/sec, the speed of the fastest meteor, ϱ/ϱ_0 is about 3.5, which means that the iron occupies only about $^2/_7$ of its original volume. These figures apply equally well to iron or steel, because the small amount of carbon in the steel is of negligible importance in this regime.

In Fig. 4, the pressures engendered in region 1 are shown as a function of impact velocity. The ultimate yield stress of a very good steel is on the order of 100,000 psi, which is equal to 0.0068 megabars. (1 megabar $= 10^{12}$ dynes/cm^2, or about one million atmospheres.) All of the pressures encountered in impacts above 5.5 km/sec are very much in excess of this figure. As the pressures so greatly exceed the material's strength, it is taken to be a good approximation that the strength may be neglected in calculating the motion.

III. Description of Problem Solved

Under such an approximation, the correct equations of motion for the process are the compressible, inviscid, adiabatic hydrodynamic ones exhibited in Table I. The validity of neglecting the viscosity and heat conduction may also be justified *a posteriori*.

Table I. *Hydrodynamic Equations in the Inviscid, Adiabatic Approximation*

$$\varrho\,\frac{\partial \bar{u}}{\partial t} + \varrho\bar{u} \cdot \operatorname{grad} \bar{u} + \operatorname{grad} P = 0 \qquad (1)$$

$$\frac{\partial \varrho}{\partial t} + \bar{u} \cdot \operatorname{grad} \varrho + \varrho\,\operatorname{div} \bar{u} = 0 \qquad (2)$$

$$\varrho\,\frac{\partial e}{\partial t} + \varrho\bar{u} \cdot \operatorname{grad} e + P\,\operatorname{div} \bar{u} = 0 \qquad (3)$$

$$P = f\,(\varrho, e) \qquad (4)$$

$\bar{u}$ fluid velocity
P pressure
e specific internal energy
ϱ density
$\bar{x}$ space coordinates
t time

The solution of this set of equations is severely complicated by the presence of shocks, which represent moving surfaces on which discontinuities of the dependent variables occur. To solve an interesting impact problem, at least two space dimensions must be used. No one has yet been able to solve such a problem analytically, so that it is necessary to resort to numerical techniques.

The type of problem upon which attention has been focused is shown in Fig. 5. These problems have an axis of symmetry to reduce to two the number of space variables required to describe the process. This necessitates using a projectile of cylindrical symmetry which is directed normally to a cylindrically symmetric target. For simplicity, the shape of the projectile was chosen to be a right circular cylinder whose height equals its diameter, and the target selected was a semi-infinite solid bounded by a plane. The numerical procedure used requires that the projectile and target be of the same material.

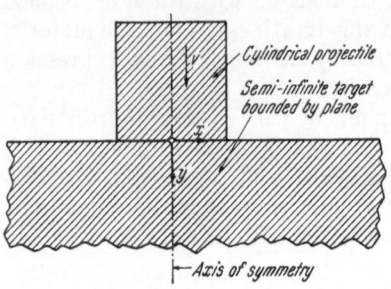

Fig. 5. Initial conditions. $P = e = 0$ everywhere. $\varrho = \varrho$ everywhere. $t = 0$. Velocity $= 0$ in target. Velocity $= v$ in y direction in projectile. Projectile has just contacted target

The origin of coordinates was chosen arbitrarily to be the intersection of the axis of symmetry and the plane bounding the target. The positive y-axis is directed downward along the axis of symmetry, and the positive x-axis to the right. Time is counted from the instant the projectile makes contact with the target surface. Initial conditions are prescribed as: pressure and specific internal energy zero everywhere, zero velocity in the target, and a constant y-velocity throughout the projectile.

IV. Discussion of Results

Using the initial conditions described, a series of problems was run wherein projectiles 10 cm in both height and diameter strike targets of the same material at various velocities. Two materials, iron and aluminium, were considered. Fig. 6 illustrates the pressure and velocity field for an iron collision at the early time of 3.5 microseconds where the impact velocity was 5.5 km/sec (18,000 ft/sec). The velocity vector is drawn for each mesh point, whose location is at the tail of the vector. It may be seen that at this instant of time all of the velocity vectors near the axis are parallel to it. Where this is the case, the material has not yet "received news" that the projectile is not of infinite radius, and therefore behaves as it would in the case of the one-dimensional collision previously described.

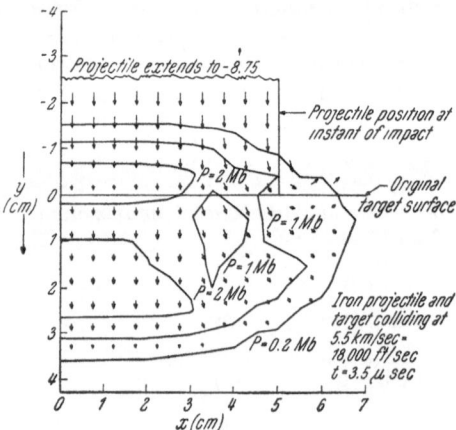

Fig. 6. Pressure contours and velocity field

The "news" is carried inward in the form of a rarefaction wave which starts at the outer circumference of the projectile, and propagates with the velocity of sound in the compressed material. In the figure, one may identify the material encompassed by the rarefaction wave from the fact that it has acquired an x-component of velocity by being accelerated toward the region of lower pressure.

The one-dimensional behavior of the material near the axis permits a check on whether the numerical method is correctly describing the behavior of shock waves. Ideally, the shock would represent a discontinuous jump of the pressure, density, specific internal energy. and velocity. The numerical method "smears" the jump out over a few mesh spaces, so that the change is rapid but continuous. An examination of the results showed that for the variables mentioned, the method gives correctly the jumps across the shock. Moreover, if one picks the center of the jump as the position of the shock front, the method gives correctly the veloc-

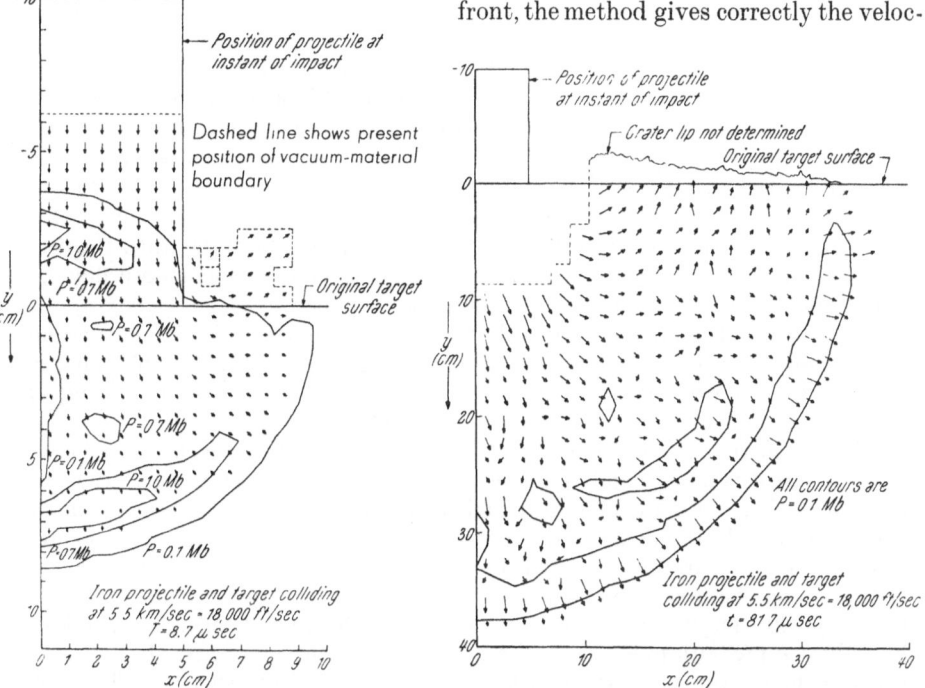

Fig. 7. Pressure contours and velocity field   Fig. 8. Velocity and pressure contours

ities of the shock fronts themselves. As the density ratio, ϱ/ϱ_0, is 1.4 in this case, the rear shock moves upward into the projectile, rising above the level of the original target surface, just as predicted by the one-dimensional analysis.

In the case of higher velocity impacts, where ϱ/ϱ_0 is greater than 2, the rear shock never rises above the level of the original target surface, but is carried down below it from the outset.

The material which has not yet been reached by a shock wave has no information that a collision has occurred, or more explicitly, no forces have been exerted on it. Thus, the material in the target remains at rest and the material in the projectile moves downward with unimpeded velocity until a shock is encountered. For high velocity impacts. this means that the projectile material may proceed far beneath the level of the original target surface before it even suffers its first deceleration, a fact which is contrary to the supposition that the penetration might go to zero in the limit of infinite velocity.

A further check on the method was made by calculating the pressure, velocity, density, and specific internal energy fields for a nuclear blast. A uniform atmosphere, zero gravitational force, and the absence of the ground were assumed, so that the problem treated actually had spherical symmetry. Such a problem had been previously calculated in detail by BRODE using a one-dimensional code, [1]. The method under discussion gave a two-dimensional solution, which, if accurate, should check BRODE's solution as one proceeds outward along any radius. Three radii were picked, $x = 0$, $y = 0$, and $x = y$, and the solution on each actually checked BRODE's solution very well. One may conclude that the method at hand correctly calculates the hydrodynamic problem in the presence of shocks.

Proceeding through the gamut of impact velocities, the pressures and densities engendered in the initial shock increase smoothly, as illustrated in Figs. 3 and 4. There is no evidence of a critical velocity which has been postulated by some authors. It has been proposed that at impacts above the velocity of sound in the target, the energy is unable to escape as it cannot propagate with a velocity greater than that of sound, so that it is concentrated in a small volume, resulting in an explosion. Based on the results of the present calculations, this supposition is easily seen to be incorrect. The velocity of the initial shock wave increases smoothly with the impact velocity and reaches speeds far in excess of the velocity of sound in the target. Indeed, for an impact at 72 km/sec. the initial speed of the shock which moves down into the target is 50.4 km/sec, about ten times the velocity of sound in iron. At all velocities investigated, one finds a smooth flow field quite similar to that illustrated in the present paper.

Fig. 7 shows the pressure and velocity fields at the time of 8.7 microseconds. At this time the rarefaction wave has progressed inward to the axis of symmetry, separating the single original high pressure region into two regions. As may be seen, the flow field is very regular at this time.

Fig. 8 illustrates conditions at 81.7 microseconds. The shock wave moving into the target is now roughly a hemisphere of 38 cm radius. Immediately behind this shock there exists a region of high pressure, but as one progresses farther behind the shock, the pressure diminishes rather rapidly and becomes negligible. This type of pressure behavior will be called a pressure pulse. In the lower part of the figure, the first pressure pulse is closely followed by a second. These two pulses quickly merge as time progresses and the leading shock is then followed by only one pressure pulse, as in the upper part of the figure. At the tail of the pressure pulse, the particle velocity and pressure simultaneously fall to zero, so that as the pulse moves into the target, it leaves behind a stationary region where the pressure is zero. The numerical method approximates this region with small velocities, randomly oriented, so that the average velocity of any particle is zero. The shock at the head of the pulse always moves with a velocity greater than that of sound in undisturbed target material, whereas the pressure pulse tail moves at just the velocity of sound. Consequently the width of the pulse increases with time. The amount of material contained in the pulse increases both for this reason, and also because of the inverse square law. As the pulse contains a fixed amount of momentum and energy, the average velocity and specific internal energy in the pulse quickly decrease as the pulse progresses.

In all cases investigated to date, pressure pulses of this type break away from the region of flow immediately surrounding the crater. These pulses would cause spalling if they encountered a free surface in a target of finite thickness, and this explains how the rear surface can be damaged by spalling and the front surface by cratering, while there may be a region of undamaged material between the two.

In the pulse, the material has greater than normal density, so that as the pulse moves, material is displaced away from the crater. When the pulse encounters

free surfaces and there is no spalling the excess material will appear as bulges in
the surfaces.

Near the forming crater one still has a smooth flow field, and as the calculation proceeds to later times, some of the material flows away, being ejected
from the crater, and the rest is brought to rest at the crater walls. Fig. 9 shows
the craters deduced from the calculations for iron impacts at various velocities.
The quantities, Δ, shown in the figure are estimates of the possible error in the
radius of the craters. The errors arise from the fact that there exists some residual

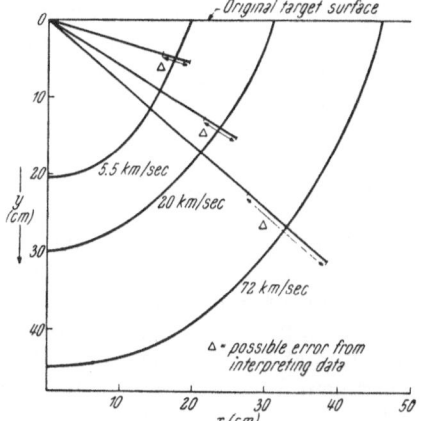

Fig. 9. Crater cross-sections

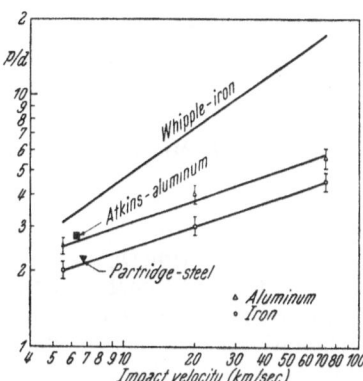

Fig. 10. Penetration as a function of impact
velocity

motion over a few mesh spaces, so that the crater boundary delineated by the
method is not perfectly sharp. The possible error amounts to about 10 per cent
of the radius in each case. Within the limits of this error, the craters found in
each case for both iron and aluminum were hemispherical.

The velocity dependence of the craters' dimensions is illustrated in Fig. 10,
where the depth of the crater, p, divided by the projectile's characteristic dimension, d, is shown as a function of velocity. Since the diameter equals the height
for the projectile under consideration, that dimension is used as the characteristic
one. The iron points are fitted very well by a straight line of slope one-third on this
log-log plot, and the aluminum points are fitted to a reasonable degree of approximation by a straight line of the same slope.

V. Comparison with Experiment

Experimental data is sparse in the hypervelocity region wherein the model is
most applicable. Indeed, the two points in Fig. 10 have only recently been achieved
by the cited experimentalists. The point labeled "Atkins-aluminum" represents
a shot made by W. W. ATKINS of the Naval Research Laboratory, Washington,
D.C., at 20,700 feet per second. The projectile in this shot was a right circular
cylinder of diameter, $D = .348$ in., and length, $L = .301$ in., and the crater was
roughly hemispherical with a depth of 2.33 cm. Both the target and projectile
were of aluminum. The characteristic dimension to choose for such a projectile
is not obvious. The one selected was $(LD^2)^{1/3}$, which results in the point shown
on the graph.

The point labeled "Partridge-steel" was taken by W. S. PARTRIDGE of the Utah Research and Development Company, Salt Lake City, Utah, and represents a shot at 22,200 feet per second of a steel ball bearing into a steel target. The diameter of the projectile was .125 in., and the depth of the hemispherical crater was .276 in. Using the diameter as the characteristic dimension results in the point shown on the graph.

The agreement obtained for both the shape and dimensions of the craters is especially satisfying in view of the fact that no adjustable constants have been used to fit the data. The only numbers put into the model are those which specify the equation of state and the initial conditions. The equation of state at high pressures and densities is calculated from fundamental principles by the THOMAS-FERMI method. At low pressures, it is derived from experiments involving the explosive compression of metals carried out at Los Alamos [2]. A numerical fit is made which fairs between the two regions.

VI. Meteor Hazard Smaller than Previously Supposed

The crater dimensions computed on this model are considerably smaller at high impact velocities than the estimates previously made by most workers in this field. This is primarily because these previous estimates have involved extrapolations of the experimental data taken at low impact speeds rather than an attempt to calculate directly what the penetration at a given velocity should be. In the low velocity regime where most experimental data are taken, the penetration increases with a power of velocity which varies between about two-thirds and unity, and extrapolating this behavior leads to predictions of large craters at high velocities. However, the calculations of this paper show that at the high velocities where the hydrodynamic model adequately describes the process, the penetration actually increases with about the one-third power of velocity, indicating that such extrapolations are inaccurate.

As an example to illustrate the difference, the results of WHIPPLE [3] are also given in Fig. 10. At the average meteoroid velocity of 30 kilometers per second, WHIPPLE's prediction for the penetration of a steel skin by an iron meteoroid exceeds the present one by about a factor of three, while at 72 kilometers per second his results are larger by about a factor of four. In fairness, it should be stated that WHIPPLE recognized the crudity of his approximations, and cited the need for a new theory of high-speed penetration.

As the flux of meteors of mass greater than m is commonly accepted to vary as $1/m$ [3], and the penetration to vary as the characteristic dimension of the projectile, or as $m^{1/3}$, the factor of three or four in the predicted penetration makes a large difference in the frequency with which a metal shell in space may be expected to be penetrated. Indeed, one concludes from these two laws that the average time between penetrations to a depth, p, or greater varies as p^3. Therefore the factor of three in predicted penetration means that metal shells may be expected to have a useful life about two orders of magnitude longer than previously supposed.

References

1. H. L. BRODE, Numerical Solutions of Spherical Blast Waves. J. Appl. Physics 26, 766 (1955).
2. J. M. WALSH et al., Shock-Wave Compressions of Twenty-Seven Metals. Equations of States of Metals. Physic. Rev. 108, 196 (1957).
3. F. L. WHIPPLE, The Meteoric Risk to Space Vehicles. Vistas in Astronautics, pp. 115—124. New York: Pergamon Press, 1958.

La sostentazione getto-orbitale nei trasporti ultraveloci

C. E. Cremona[1, 2] e D. Cunsolo[1, 3]

(Con 7 Figure)

(Ricevuto il 24 giugno 1959)

Abstract — Zusammenfassung — Résumé

Jet-Orbital Lift in Ultra-Fast Transports. The orbital lift, beyond the atmosphere, enables a body to travel over an elliptical trajectory arc, until it returns into the atmosphere, at an extremely high speed, and therefore in a time that is much lower than that of any other land, sea or air-vehicle, for the same geographical distance.

Besides the incapability of said vehicles to achieve such high speeds, their energetic cost would be immensely higher than that of the orbital lift velocity, especially when considering that the latter is not (or nearly) dependent on the geographical range, and therefore it is highly profitable in the case of long distances.

Moreover, by taking into account—as it has been done in the present paper—the possibility of a partial jet lift, the most convenient reduced velocity of the orbital lift and its most suitable direction in space can be found.

The lifting jet could contribute to solve efficiently the problem of re-entering the atmosphere, if a jet orientation device would be reliable; furthermore, in the last part of the return flight it allows to attain the most suitable and profitable combination of air and jet sustenance.

Orbitaler Auftrieb durch Strahlantrieb bei überschnellen Transporten. Oberhalb der Atmosphäre gestattet der Auftrieb infolge der Krümmung der Erde („orbitaler Auftrieb") einem Körper, mit äußerst großer Geschwindigkeit einen elliptischen Bahnbogen zu beschreiben, bis er in die Atmosphäre zurückkehrt. Die Flugzeit ist viel kleiner als diejenige eines jeden Land-, See- oder Luftfahrzeuges über die gleiche geographische Distanz.

Wegen der Unfähigkeit der genannten Fahrzeuge, so große Geschwindigkeiten zu erlangen, würde ihr Energieverbrauch außerordentlich größer sein als der bei der Geschwindigkeit des „orbitalen Auftriebs", insbesondere wenn man bedenkt, daß dieser nicht oder fast nicht von der geographischen Entfernung abhängt. Deshalb ist er im Fall weiter Entfernungen sehr sparsam.

Wenn man ferner wie in der vorliegenden Arbeit die Möglichkeit eines partiellen Strahlantrieb-Auftriebes annimmt, kann die günstigste verminderte Geschwindigkeit des orbitalen Auftriebes und seine geeignetste Richtung im Raum gefunden werden.

Der Auftrieb beim Strahlantrieb könnte wirksam zur Lösung des Problems des Wiedereintritts in die Atmosphäre beitragen, wenn ein Gerät zur Orientierung des Strahlantriebes verfügbar wäre; schließlich erlaubt er im letzten Teil des Fluges bei der Rückkehr, die geeignetste und günstigste Kombination der Stützung durch die Luft und den Strahlantrieb zu erreichen.

Portance orbitale par réaction dans les transports ultra-rapides. Au-delà de l'atmosphère la portance due à la courbure de la terre permet à un objet de parcourir

---

[1] Associazione Italiana Razzi, Piazza S. Bernardo, 101, Roma, Italia.
[2] Professore della Accademia Militare Aeronautica, Nisida (Napoli), Italia.
[3] Università di Roma, Roma, Italia.

avec une très grande vitesse un arc elliptique de trajectoire jusqu'à son retour dans l'atmosphère. Le temps du trajet est inférieur à celui de tout autre véhicule terrestre, marin ou aérien pour la même distance géographique.

De plus la dépense d'énergie de ces autres véhicules serait beaucoup supérieure, spécialement dans le cas des grandes distances. En fin quand on tient compte, comme dans cet article, d'une portance partielle par jet, la meilleure vitesse réduite et la meilleure direction de la portance orbitale peuvent être trouvées.

La portance par jet peut contribuer à résoudre efficacement le problème de la rentrée, si un dispositif d'orientation du jet peut être utilisé avec confiance. Dans la dernière partie du trajet de rentrée il permet d'obtenir la meilleure combinaison de sustentation par jet et surface portante.

I. Premessa

Il volo di un mobile lungo una traiettoria orbitale — sollecitato dalla sola forza centrale secondo la legge di Newton — è caratterizzato dall'equilibrio dinamico fra il vettore rappresentativo della accelerazione centrale e quello risultante delle accelerazioni secondo due assi — per esempio quello centrifugo e quello tangenziale — e si svolge secondo una curva piana.

Si suppone, per comodità, in questa indagine, che il moto avvenga al di fuori dell'atmosfera — resistenza all'avanzamento nulla — e che la curva descritta dal mobile sia un arco di ellissi compreso fra due punti I ed F simmetrici rispetto all'asse maggiore e quindi situati alla stessa quota h (Fig. 1) cioè alla distanza r_0 dal centro della Terra.

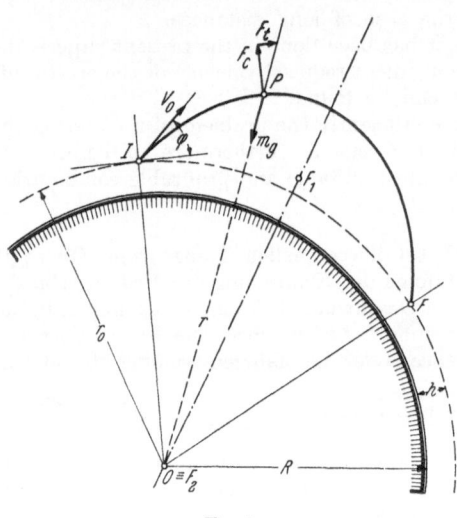

In una posizione generica P del mobile lungo l'arco dell'ellissi dovrà essere, quindi, indicando con g il valore della accelerazione di gravità locale:

$$g = [V^4 c^2 + \dot{V}^2]^{\frac{1}{2}} \qquad (1.1)$$

essendo c la curvatura e V la velocità tangenziale.

Fig. 1

Questa condizione impone valori della velocità — e di conseguenza quella $V_I = V_F$ dei punti I di *lancio* ed F di *rientro* — molto elevata e ciò rende molto costoso, dal punto di vista energetico il moto e tecnologicamente complicato il problema del lancio e quello del rientro nella atmosfera.

Si propone in questa nota un metodo tendente a ridurre sia la velocità di sostentazione orbitale — e quindi anche quella di lancio e di rientro — sia, di conseguenza, la complicazione del problema del rientro.

II. La sostentazione getto-orbitale

Se al mobile si considera applicata una forza aggiuntiva, costantemente diretta verso il centro della Terra, la (1.1) diventa:

$$a_r = g - \frac{T}{m} = [V_1^4 c_1^2 + \dot{V}_1^2]^{\frac{1}{2}} \qquad (2.1)$$

la quale mette in evidenza i vantaggi conseguibili per effetto di una riduzione del valore della velocità tangenziale e di una riduzione della curvatura istantanee (Fig. 2).

La forza T può essere ottenuta, però, solo attraverso l'impiego di un getto; di conseguenza la massa m che compare nella (2.1) decrescerà nel tempo, in modo tanto maggiore quanto più si avvicini il valore iniziale di $\dfrac{T_0}{m_0}$ a quello dell'accele-

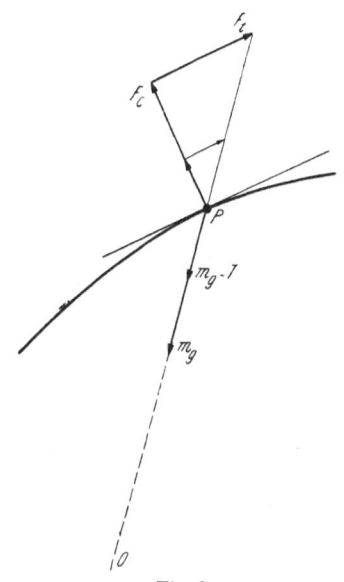

Fig. 2

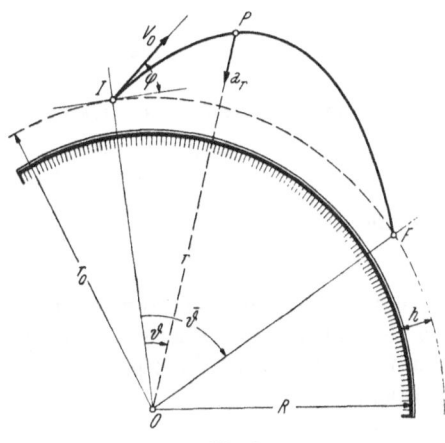

Fig. 3

razione di gravità g_0 alla quota iniziale h e sempre nel senso di rendere minore la differenza $g - \dfrac{T}{m}$ nello sviluppo del volo.

La (2.1) assumerà, in tal caso, supposto costante il valore di T, l'aspetto:

$$a_r = g - \frac{T}{m_0\left(1 - \zeta \dfrac{t}{t_c}\right)} = [V_1{}^4 c_1{}^2 + \dot{V}^2]^{\frac{1}{2}} \tag{2.2}$$

con

$$g > \frac{T}{m_0\left(1 - \zeta \dfrac{t}{t_c}\right)}$$

nella quale t_c è la durata della combustione e

$$\zeta = \frac{m_p}{m_0}$$

è il rapporto fra la massa del propellente e la massa del mobile al lancio trascurando, cioè, come d'uso, i transienti all'inizio ed alla fine del volo orbitale.

La (3) mette in evidenza il fatto che la velocità tangenziale e la curvatura andranno diminuendo nel tempo e la traiettoria del mobile assumerà un suo caratteristico andamento, come verrà precisato nel caso più semplice riportato nel prossimo paragrafo.

III. Traiettoria con getto-sostentazione verticale costante

La traiettoria descritta dal mobile soggetto all'azione concomitante della sostentazione orbitale e della getto-sostentazione, dovrà sottostare, come si è detto, alla legge delle aree:

$$r^2 \dot{\vartheta} = A \qquad (3.1)$$

essendo A il doppio della velocità areolare.

Esprimendo la equazione di equilibrio, lungo la congiungente la posizione generica del mobile P con il centro della Terra, in coordinate polari, si potrà scrivere (Fig. 3):

$$a_r = \ddot{r} - r \dot{\vartheta}^2 \qquad (3.2)$$

la quale, per la (3.1) diventa (formula di Binet):

$$a_r = -\frac{A^2}{r^2} \left(\frac{1}{r} + \frac{d^2}{d\vartheta^2} \frac{1}{r} \right). \qquad (3.3)$$

L'equazione (3.3) del moto diventa, dunque, nel caso in esame:

$$\frac{A^2}{r^2} \left(\frac{1}{r} + \frac{d^2}{d\vartheta^2} \frac{1}{r} \right) = g_0 \frac{r_0^2}{r^2} - \frac{T}{m_0 + \dot{m}\, t} \qquad (3.4)$$

nella quale

$$T = -v_e \dot{m} \qquad (3.5)$$

è la spinta radiale di getto-sostentazione; $-\dot{m}$ è la portata, in massa, dell'effusore; v_e è la velocità di efflusso dei gas dall'ugello; r_0 è la distanza iniziale del mobile appena fuori della atmosfera; g_0 è la accelerazione di gravità iniziale alla quota h e t il tempo.

Ponendo

$$A^2 = r_0^3 g_0 B^{-2} \qquad (3.6)$$

$$\frac{r_0}{r} = \varrho \qquad (3.7)$$

$$T = m_0 g_0 b \qquad (3.8)$$

$$\dot{m} = -\frac{T}{v_e} = -\frac{m_0 g_0}{v_e} b \qquad (3.9)$$

$$t = \tau \sqrt{\frac{r_0}{g_0}} \qquad (3.10)$$

la (3.4) assume la forma adimensionale:

$$B^{-2} \varrho^2 \left(\varrho + \frac{d^2 \varrho}{d\vartheta^2} \right) = \varrho^2 - \frac{b}{1 - \dfrac{\sqrt{r_0 g_0}}{v_e} b \tau}. \qquad (3.11)$$

In questa espressione si può ora inserire la (3.1) la quale, per la (3.6) diventa

$$\frac{d\vartheta}{dt} \frac{r^2}{r_0^2} = B^{-1} \sqrt{\frac{g_0}{r_0}}$$

ossia, tenendo conto della (3.7) e della (3.10),

$$\frac{d\vartheta}{d\tau} = \varrho^2 B^{-1} \qquad (3.12)$$

che, integrata, fornisce

$$\tau = B \int_0^\vartheta \frac{d\vartheta}{\varrho^2} \qquad (3.13)$$

ottenendo, infine:

$$\varrho + \frac{d^2 \varrho}{d\vartheta^2} = B^2 \left[1 - \frac{\dfrac{b}{\varrho^2}}{1 - K B \displaystyle\int_0^{\vartheta} \frac{d\vartheta}{\varrho^2}} \right] \tag{3.14}$$

nella quale si è posto:

$$K = \frac{\sqrt{r_0\, g_0}}{v_e}\, b\, .$$

La (3.14) per $b=0$ è la nota equazione dei moti kepleriani. Il significato fisico del parametro b è, come risulta dalla (3.8), quello di una *accelerazione iniziale di sostentamento misurata in g*.

IV. Osservazioni

Da quanto esposto nel paragrafo II i valori di b che possono interessare il problema in esame sono solo quelli:

$$0 < b < 1\, .$$

La (3.14) potrà, quindi, essere integrata, con i soliti procedimenti, tra due valori dell'angolo ϑ definiti da, per esempio,:

$$\vartheta = \vartheta_0 \quad \text{nel punto } I$$
$$\vartheta = \overline{\vartheta} \quad \text{nel punto } F$$

con le condizioni:

$$\varrho\,(\vartheta_0) = \varrho\,(\overline{\vartheta}) = 1$$

oltrechè:

$$\varrho'(\vartheta_0) = \varrho_0'$$

essendo ϱ_0' il valore relativo al moto kepleriano.

Si potranno, così, ottenere famiglie di traiettorie caratterizzate dal fatto di essere tutte tangenti nel punto di lancio alla velocità di lancio V_0 in funzione del parametro b e, di conseguenza più facilmente confrontabili.

In tal modo si potranno mettere a confronto, come indicato in Fig. 4, famiglie di traiettorie aventi uguali direzioni di lancio φ e diverse energie cinetiche iniziali — tutte minori di quelle occorrenti per l'orbita ellittica — valutate per unità di massa del carico utile.

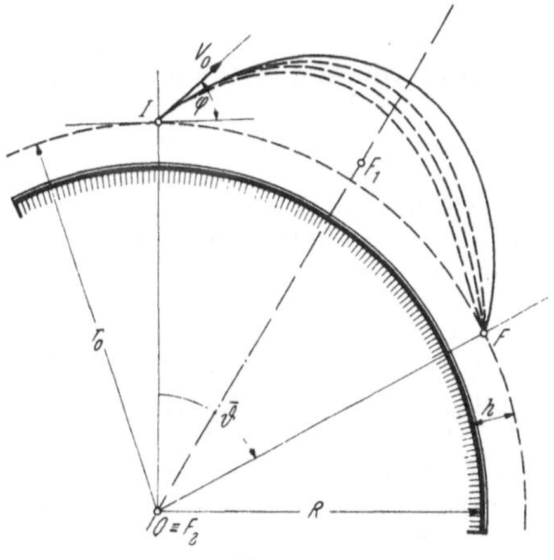

Altre famiglie di traiettorie possono essere, poi, individuate per valori diversi della direzione di lancio φ a parità di energia cinetica iniziale.

Si potrà determinare, quindi, dal loro confronto l'energie minima occorrente a coprire una data gittata in funzione dell'angolo φ.

Fig. 4

In effetti si tratta di ricercare il minimo della funzione $V_0\,(\varphi; b)$ nell'intervallo considerato.

Si osservi, inoltre, che assumendosi per esempio per r_0 la distanza di 6.500 chilometri dal centro della Terra ed una corrispondente accelerazione in quota di

9,36 m/sec², la $\sqrt{r_0/g_0}$ risulta di circa 800 secondi ed il fattore $\dfrac{1}{v_e}\sqrt{r_0 g_0}$ dell'ordine di qualche unità (~ 4).

Partendo quindi dalla condizione di minima velocità necessaria a coprire una voluta gittata geodetica in un moto kepleriano, sarà possibile controllare, al variare del parametro b, di quanto si possa ridurre tale velocità con l'ausilio di una getto-sostentazione.

Questa indagine consente non solo di determinare le condizioni migliori all'inizio del moto ma altresì quelle relative al raccordo della traiettoria nel vuoto (*cenonautica*) con quella di rientro nell'atmosfera (*aeronautica*).

La getto-sostentazione, infatti, può non arrestarsi nel punto F ma, opportunamente orientata anche durante il volo, fornire un prezioso contributo al problema della riduzione della velocità nella fase del rientro ed a quello di creare il migliore raccordo fra i due rami di traiettoria.

Questa analisi appartiene alla serie dei problemi di optimalizzazione e sarà sviluppata nel seguito.

Per ora si riportano in appendice i risultati numerici delle prime indagini eseguite.

V. Conclusioni

E' infatti indispensabile procedere alla impostazione del confronto dei bilanci energetici in quanto mentre nel moto kepleriano l'energia cinetica necessaria è quella corrispondente alla forza viva da imprimere al mobile, alla quota prescelta, nel moto in esame alla minore energia cinetica necessaria all'inizio occorre aggiungere quella consumata durante il volo spaziale.

Si dovrà però, sempre tenere nel debito conto il triplice sensibile vantaggio di un minor peso della macchina alla partenza dalla superficie terrestre (comunque essa avvenga); della minore velocità al rientro e dei minori valori delle accelerazioni inerziali sia al lancio che al rientro, vantaggi che concorrono, senza dubbio, a rendere più agevole ed immediato il problema di consentire anche all'Uomo l'impiego di questi mobili ultraveloci.

Bibliografia

1. LEVI-CIVITA ed U. AMALDI, Lezioni di Meccanica Razionale, Bologna: Ed. Zanichelli.
2. C. E. CREMONA, Sulle possibilità di impiego della navigazione extra-atmosferica nelle comunicazioni mercantili. VI. Convegno Internazionale delle Comunicazioni — Ottobre 1957 — Genova.
3. E. CREMONA, Lezioni di Dinamica dei Missili. Università di Roma — Scuola di Ingegneria Aeronautica. Siargraph N. 15 marzo 1958 — Roma.
4. C. E. CREMONA, Dall' Aeronautica alla Balistica Aeronautica. Rassegna tecnica A.N.I.A.I. Anno XII N. 5 — maggio 1958, Napoli.
5. C. E. CREMONA, Il tramonto dell'ala e l'avvento del getto. Associazione Meccanica Italiana, Sezione di Napoli — Conferenza tenuta il 16 febbraio 1959 nella Sede Sociale.
6. C. E. CREMONA, La Celonautica. Conferenza tenuta presso il Corso di alta Informazione Scientifica del Civico Istituto Colombiano il 14 marzo 1959. — Genova.

Appendice

Nella integrazione della (3.14), che, in base alle (3.13) diventa:

$$\varrho + \frac{d^2\varrho}{d\vartheta^2} = B^2\left[1 - \frac{b/\varrho}{1 - k\tau}\right] \tag{A.1}$$

si sono imposte le seguenti condizioni:

a) la traiettoria ha i punti iniziale e finale fisseti ad una quota determinata:

$$h_0 = h_j = 130 \text{ km};$$ (A. 2)

b) il lancio sia di un sestante

$$\vartheta_0 = 0 \qquad \vartheta_j = 60° $$ (A. 3)

che corrisponde ad una gittata di

$$\frac{2\pi}{6} 6.500 = 6.800 \text{ km}$$

risulterà, per le (3.7), in base ad (A. 2)

$$r_0 = r_j = 6.500 \text{ km}.$$

Se V_r e V_ϑ sono le componenti radiale e trasversale della velocità si ha

$$V_r = \frac{dr}{d\vartheta}\,\dot\vartheta = -\frac{r_0}{\varrho^2}\frac{d\varrho}{d\vartheta}\frac{A}{r^2} = -\frac{\sqrt{r_0 g_0}}{B}\frac{d\varrho}{d\vartheta}$$ (A. 4)

avendo fatto uso delle (3.7), (3.6) e (3.1).

Inoltre

$$V_\vartheta = r\,\dot\vartheta = \frac{A}{r} = \frac{\sqrt{r_0 g_0}}{B}\,\varrho.$$ (A. 5)

Combinando allore le (A. 4) (A. 5):

$$V = \frac{\sqrt{r_0 g_0}}{B}\sqrt{\varrho^2 + \left(\frac{d\varrho}{d\vartheta}\right)^2}.$$ (A. 6)

La (A. 6) mette in luce il legame tra la constante B che figura in (A. 1) e la velocità iniziale.

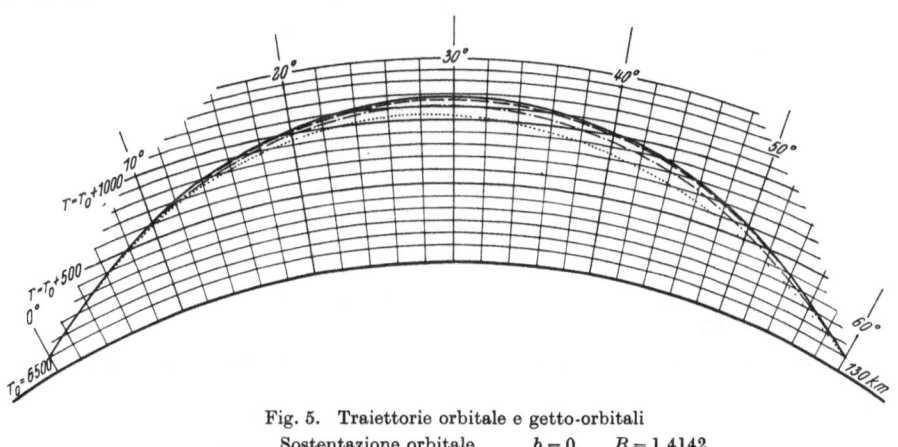

Fig. 5. Traiettorie orbitale e getto-orbitali

———— Sostentazione orbitale $b = 0,$ $B = 1,4142$
– – – – Sostentazione getto-orbitale $b = 0,04,$ $B = 1,4561$
–·–·– Sostentazione getto-orbitale $b = 0,08,$ $B = 1,5187$
········ Sostentazione getto-orbitale $b = 0,10,$ $B = 1,6065$

$$b = \frac{T}{m_0 g_0} \qquad B = \frac{(r_0 g_0)^{1/2}}{A} \qquad a = g\,\frac{T}{m_0\left(1 - \zeta\frac{t}{t_c}\right)} \qquad \zeta = \frac{m_p}{m_0}$$

$$\frac{A}{r^2}\left(\frac{1}{r} + \frac{d^2}{d\vartheta^2}\frac{1}{r}\right) = g_0\frac{r_0^2}{r^2} - \frac{T}{m_0 + \dot m\, t}$$

Il problema che ci si propone è di trovare le velocità iniziali e finali date le posizioni iniziale e finale del missile e la direzione iniziale della velocità. Questo occorre per ogni valore di b cioè del rapporto tra spinta e peso iniziale alla quota h_0.

Si è supposto, infine, la spinta costante per tutta la durata della traiettoria.

La (A. 1) integrata per $b=0$ dà:

$$\varrho = B^2 - 2\,\frac{B^2-1}{\sqrt{3}}\,\mathrm{sen}\left(\vartheta + \frac{\pi}{3}\right) \qquad (A.\ 7)$$

avendo imposto

$$\varrho_0 = \varrho_f = \varrho_{\pi/3} = 1 \qquad (A.\ 8)$$

e, volendo imporre la minima velocità iniziale per la prescelta gittata, si ha da (A. 6) ed (A. 7) il valore di:

$$B = \sqrt{2} = 1{,}4142 \qquad (A.\ 9)$$

corrispondente quindi a $b=0$.

La (A. 7) diventa allora

$$\varrho = 2 - \frac{2}{\sqrt{3}}\,\mathrm{sen}\left(\vartheta + \frac{\pi}{3}\right) \qquad (A.\ 10)$$

mentre

$$\varrho'(\vartheta) = \varrho_0' = -\frac{1}{\sqrt{3}}\,. \qquad (A.\ 11)$$

Il metodo di integrazione usato per la (A. 1) nei casi nei quali $1>b>0$, che sono quelli che interessano, è consistito nel considerare l'ultimo termine a secondo membro della (A. 1) come un termine perturbatore.

La (A. 1) viene scritta come

$$\frac{d^2\varrho}{d\theta^2} = \lambda(\vartheta) \qquad (A.\ 12)$$

con:

$$\lambda(\vartheta) = B^2\left(1 - \frac{b/\varrho}{1-k\tau}\right) - \varrho \qquad (A.\ 13)$$

nella quale il secondo membro è calcolato lungo la traiettoria imperturbata.

Il valore di $\varrho(\vartheta)$ che risulta della (A. 12) viene posto in (A. 11) ed (A. 13) con metodo iterativo.

Ad ogni iterazione si modifica il valore di B in modo da ottenere infine

$$\varrho_{\pi/3} = 1.$$

I valori di b usati sono stati: 0,04; 0,08; 0,10; 0,11; 0,119; per ogni b è stata considerata come traiettoria inturbata, per iniziare il calcolo, quella risultante del valore precedente di b.

Nella Tab. 1 sono riportati i valori di ϱ per alcuni valori di ϑ e nella Tab. 2 le quote h raggiunte per gli stessi valori di ϑ al variare del parametro b. Nella Fig. 5 sono riportate, a confronto, le traiettorie ottenute al variare di b e quella orbitale ($b=0$).

Tabella 1. *Valori di* $\varrho = \dfrac{r}{r_0}$

ϑ_0	$b = 0$	$b = 0,04$	$b = 0,08$	$b = 0,10$	$b = 0,11$	$b = 0,119$
0	1.00	1.00	1.00	1.00	1.00	1.00
4	1.03933	1.03920	1.03907	1.03879	1.03858	1.03832
8	1.07599	1.07556	1.07532	1.07382	1.07294	1.07186
12	1.10888	1.10806	1.10677	1.10403	1.10200	1.09952
16	1.13688	1.13557	1.13290	1.12847	1.12485	1.12042
20	1.15896	1.15712	1.15284	1.14640	1.14082	1.13399
24	1.17423	1.17188	1.16594	1.15733	1.14955	1.14004
28	1.18203	1.17928	1.17181	1.16107	1.15105	1.13877
32	1.18203	1.17903	1.17037	1.15779	1.14565	1.13078
36	1.17423	1.17117	1.16183	1.14792	1.13403	1.11696
40	1.15896	1.15603	1.14668	1.13218	1.11711	1.09860
44	1.13688	1.13427	1.12563	1.11147	1.09602	1.07708
48	1.10888	1.10671	1.09953	1.08681	1.07199	1.05398
52	1.07599	1.07436	1.06933	1.05927	1.04639	1.03125
56	1.03933	1.03826	1.03597	1.02993	1.02070	1.01162
60	1.00	1.00	1.00	1.00	1.00	1.00

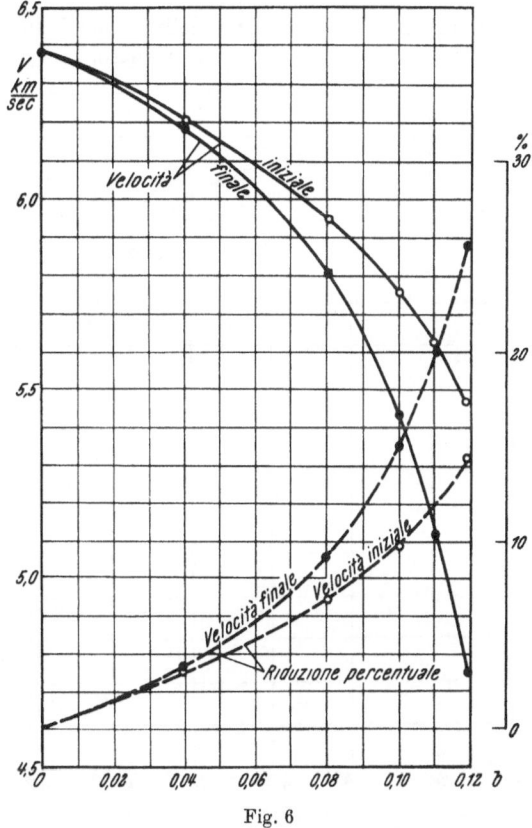

Fig. 6

La (A. 6) dà infine, in base oi valori di B trovati di volta in volta, i valori della velocità iniziale e finale della traiettoria sostentata.

Tabella 2. *Valori di h in chilometri*

ϑ_0	$b = 0$	$b = 0,04$	$b = 0,08$	$b = 0,10$	$b = 0,11$	$b = 0,119$
0	130	130	130	130	130	130
4	385,62	384,80	383,92	382,14	380,76	379,10
8	623,91	621,16	619,55	609,83	604,10	597,07
12	937,69	832,37	824,03	806,18	793,02	776,88
16	1112,73	1011,21	993,86	965,07	941,55	912,74
20	1263,25	1151,28	1123,46	1081,61	1045,34	1000,96
24	1362,47	1247,24	1208,59	1152,62	1102,10	1040,27
28	1413,19	1295,34	1246,75	1176,96	1111,79	1032,02
32	1413,19	1293,71	1237,39	1155,61	1076,71	980,05
36	1362,47	1242,57	1181,90	1091,48	1001,20	890,25
40	1263,25	1144,22	1083,45	989,19	891,25	770,88
44	1112,73	1002,74	946,60	854,58	754,12	631,05
48	937,69	823,64	776,95	694,29	597,94	480,87
52	623,91	613,34	580,61	515,27	431,51	333,15
56	385,62	378,72	363,83	324,55	264,56	205,50
60	130	130	130	130	130	130

La Fig. 6 mostra la congrua riduzione delle velocità iniziali occorrenti e la ancor più grande riduzione delle velocità al rientro.

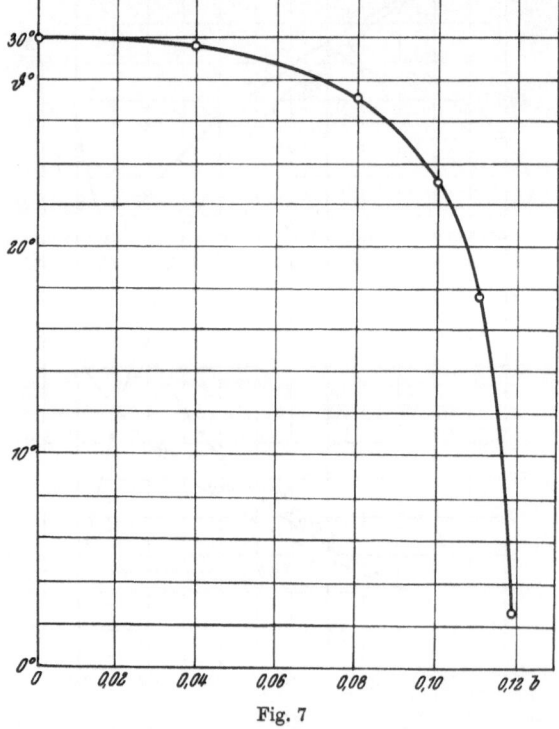

Fig. 7

Infine la Fig. 7 mostra la sensibile diminuzione degli angoli di impatto con l'atmosfera, al rientro, al crescere del parametro b.

E' attualmente in corso l'analisi del problema dal punto di vista variazionale nelle ipotesi:

1° — della variabilità temporale della spinta

2° — della variazione direzionale della spinta

3° — della variazione contemporanea della entità e della direzione della spinta.

Simboli

a	accelerazione centrale	t	durate parziale della combustione
g	accelerazione di gravità	t_c	durata della combustione
c	curvatura istantanea della traiettoria	ζ	rapporto di massa del missile
		m_p	massa dei propellenti
V	velocità del mobile	m_0	massa del mobile pronto al lancio
T	spinte del getto	r	distanza dal centro della Terra
h	quota iniziale sul livello del mare	ϑ	angolo girato da r.

Re-entry Paths for Manned Satellites[1]

By

W. F. Hilton[2]

(With 8 Figures)

(Received August 27, 1959)

Abstract — Zusammenfassung — Résumé

Re-entry Paths for Manned Satellites. Tables of values are given for the rocket braking required at apogee to decrease perigee height by a fixed amount. It is found that a small amount of rocket fuel (1 % of vehicle weight) will produce a 15 mile (24 Km) decrease of perigee height, and cause re-entry to take place.

It is desirable in the case of manned flight to have a perigee height of about 80 miles (128 Km) to ensure ultimate re-entry even if the retarding rocket cannot be operated. A digital computer programme has been developed to compute orbits under the force of gravity, while allowing for passage through a standard non-rotating atmosphere at any value, of lift drag. Some typical results are plotted in Figs. 1 to 3.

It is found that at orbital speed aerodynamic drag is essential in initiating re-entry, aerodynamic lift does not help to sustain the vehicle, but causes the axis of the elliptic orbit to rotate about the earth.

At sub-orbital speeds, however, lift plays a vital part in sustaining the vehicle and increases the time of flight and time available for dissipation of heat by radiation, at the same time decreasing the deceleration to less than 2 g.

It is shown that high wave drag is necessary to heat the surrounding air and to decrease the heating of the vehicle. The author is of the opinion that "high drag plus high lift" vehicles will be used for manned re-entry, and that "high drag with zero lift" will be reserved for simple unmanned re-entry, or very early manned flights.

The most important advantage of a lifting re-entry vehicle is the ability of the pilot to land anywhere over a large area of the earth's surface. The point of impact of a non-lifting re-entry vehicle is entirely determined once it plunges into the atmosphere at sub-orbital speeds.

A number of re-entry paths have been calculated for perfect and imperfect entry into the upper atmosphere.

Wiedereintrittswege für bemannte Satelliten. In der vorliegenden Arbeit werden Tabellen von Werten für die Bremsung von Raketen angegeben, wie sie beim Apogäum für die Verringerung der Perigäumshöhe um eine bestimmte Größe erforderlich sind. Es wurde gefunden, daß eine kleine Menge Treibstoff (1% des Gewichts des Fahrzeugs) eine Erniedrigung der Perigäumshöhe um 24 km und das Stattfinden des Wiedereintritts bewirkt.

---

[1] This work forms an integral part of a comprehensive study of manned and unmanned spaceflight at A.W.A., a member Company of Hawker Siddeley Aviation Division.

[2] D.Sc., D.I.C., Ph. D., F.R.Ae.S., A.R.C.S., Head of Astronautics Division, Advanced Projects Group, Hawker Siddeley Aviation Ltd., Richmond Road, Kingston-upon-Thames, Surrey, United Kingdom.

Im Fall des Fluges einer bemannten Rakete ist eine Perigäumshöhe von ungefähr 128 km wünschenswert, um einen Wiedereintritt letzten Endes auch dann zu sichern, wenn die Bremsrakete nicht zum Arbeiten gebracht werden kann. Ein Digitalrechnerprogramm wurde zur Berechnung von Bahnen unter Einwirkung der Gravitation entwickelt, wobei Durchgänge durch eine nicht rotierende Standardatmosphäre bei jedem Wert des Verhältnisses Auftrieb zu Widerstand mit in die Rechnung einbezogen werden. Einige typische Ergebnisse werden in den Abb. 1 bis 3 dargestellt.

Es zeigte sich, daß für die Einleitung des Wiedereintritts bei Kreisbahngeschwindigkeit aerodynamischer Widerstand wesentlich ist. Aerodynamischer Auftrieb trägt nicht dazu bei, das Fahrzeug zu stützen, bewirkt jedoch, daß die Achse der elliptischen Bahn um die Erde rotiert.

Bei Unter-Kreisbahngeschwindigkeiten aber spielt der Auftrieb eine notwendige Rolle, um das Fahrzeug zu stützen, und vergrößert sowohl die Zeit des Fluges wie auch die Zeit, die für die Zerstreuung der Wärme durch Strahlung verfügbar ist, während die Verzögerung auf weniger als 2 g vermindert wird.

Es wird ferner gezeigt, daß hoher Wellenwiderstand nötig ist, um die umgebende Luft zu erhitzen und die Erhitzung des Fahrzeugs zu verringern. Der Verfasser ist der Ansicht, daß großer Widerstand plus großer Auftrieb für den Wiedereintritt bemannter Fahrzeuge benützt werden wird, während großer Widerstand mit Auftrieb null dem einfachen Wiedereintritt unbemannter Fahrzeuge oder dem frühen Anfangsstadium bemannter Fahrzeuge vorbehalten bleiben wird.

Der wichtigste Vorteil eines Fahrzeugs mit Auftrieb beim Wiedereintritt ist die Möglichkeit für den Piloten, irgendwo innerhalb eines weiten Gebietes der Erdoberfläche zu landen. Der Einfallspunkt eines Fahrzeuges für auftrieblosen Widerstand ist völlig bestimmt, sobald es einmal mit Unter-Kreisbahngeschwindigkeit in die Atmosphäre eintaucht.

Eine Anzahl von Wiedereintrittswegen wurde für die vollkommenen und unvollkommenen Bedingungen des Wiedereintritts in die Hohe Atmosphäre berechnet.

Trajectoires de ré-entrée pour satellites habités. Le freinage par fusée requis à l'apogée pour réduire le périgée d'une valeur constante est calculé numériquement et tabulé. Une faible quantité d'ergol, de l'ordre de 1% du poids du véhicule, produit une diminution du périgée de 24 km. et provoque l'initiation de la ré-entrée.

Dans le cas du vol avec équipage il est désirable d'avoir un périgée d'environ 128 km. pour assurer le retour même si la fusée de freinage était inopérante. Un programme de calcul digital a été dressé pour évaluer les orbites décrites sous l'effet de la gravitation dans une atmosphère type sans rotation pour toutes valeurs de la finesse. Quelques résultats typiques sont illustrés sur les figs. 1—3.

Aux vitesses orbitales la traînée aérodynamique joue un rôle essentiel dans la ré-entrée. La portance n'est pas utilisée pour la sustentation mais provoque une rotation autour de la Terre de l'axe de l'orbite elliptique.

Aux vitesses suborbitales la portance joue un rôle sustentateur essentiel; en outre elle accroît le temps de vol et le temps disponible pour évacuer la chaleur par rayonnement, tandis qu'elle réduit la décélération à une valeur inférieure à 2 g.

On montre qu'une forte traînée est nécessaire pour échauffer l'air et réduire l'échauffement du véhicule. L'opinion de l'auteur est que la ré-entrée de satellites habités se fera avec portance et traînée élevées, les véhicules à forte traînée et portance nulle étant réservés aux ré-entrées sans équipage ou aux premiers essais avec équipage.

Le plus grand avantage de la portance est la possibilité de choisir une aire d'atterrissage; l'impact d'un véhicule sans portance étant fixé dès l'instant où il plonge dans l'atmosphère avec une vitesse suborbitale.

Un certain nombre de trajectoires de ré-entrée ont été calculées pour des conditions parfaites et imparfaites de pénétration dans l'atmosphère.

I. Lifting Versus Non-Lifting Re-entry Vehicles

In order to bring a vehicle safely back to earth we must either use rocket retardation, or atmospheric retardation by means of aerodynamic drag.

Full rocket retardation is at present out of the question and atmospheric retardation must be used. At circling speed each pound of mass has 4 k.w.h. of stored kinetic energy (8Kwh/Kg). This is sufficient to vaporise any known substance. Atmospheric retardation must convert the whole of this kinetic energy into heat energy, if the vehicle is to be brought safely to rest on the earth's surface (Joule's law). In general, drag due to shock waves will heat the air through which the vehicle is travelling; skin friction will heat the body and air in the boundary layer in equal proportions.

Thus, in order to minimise heat input to the vehicle, we require a shape of vehicle having a high shock-drag and a low friction drag. This is quite unlike any current aircraft or missile design.

Here we come to an unsolved question—drag only, or drag plus lift? A "drag only" shape, such as a cone, can be made very stable aerodynamically, and come down to earth without the need for a pilot. For "drag plus lift" vehicles, even if they are stable, the direction of the lift must be controlled by some sort of human or automatic pilot.

Thus "drag plus lift" demands more skill from the pilot than "drag only". However, this report shows that all other advantages are with the "drag plus lift" vehicle.

Given a perfect re-entry in both cases, the force to which the vehicle is subjected in the "drag plus lift" case is only a little more than that of gravity. The retardation of a "drag only" vehicle depends on the air density ϱ, the speed V and the mass per unit area (the mass loading) and is given by

$$f = \frac{1}{2} \varrho V^2 / (\text{mass loading}) \tag{1}$$

and is found to have values of 10 g or more in actual designs of manned re-entry vehicles.

Of course it can be argued that the pilot and vehicle must be able to survive this sort of acceleration on the way up, so that no saving of weight can result from limiting "g" to lower values on re-entry.

However, the lifting vehicle flies for a longer period, perhaps 30 minutes, instead of 5 minutes. This spreads the heat input over a longer period, and enables the beneficial effects of thermal radiation to take place.

It is useful to reduce the stresses when the structure is at high temperature, giving another advantage to the lifting re-entry vehicle.

What is perhaps more important still, the pilot will have a wide choice of landing site with a lifting vehicle. Entering the earth's atmosphere at orbital speed, this may just about cover the entire surface of the earth. The radius of action decreases of course, as the vehicle loses speed and altitude.

II. Description of Proposed Manned Flight

A design study has been carried out at Sir W. G. Armstrong Whitworth Aircraft Ltd. (part of Hawker Siddeley Aviation Ltd.) for a complete project to put two men into satellite orbit and for them to land safely back on the earth. The vehicle has been described by Mr. H. R. Watson (Technical Director) in [1]. The present paper examines the dynamics of the orbit and recovery trajectory.

In planning an orbit it was decided to keep below the J. Van Allen radiation belt, by limiting height at apogee to 680 miles (1100 Km). The perigee height

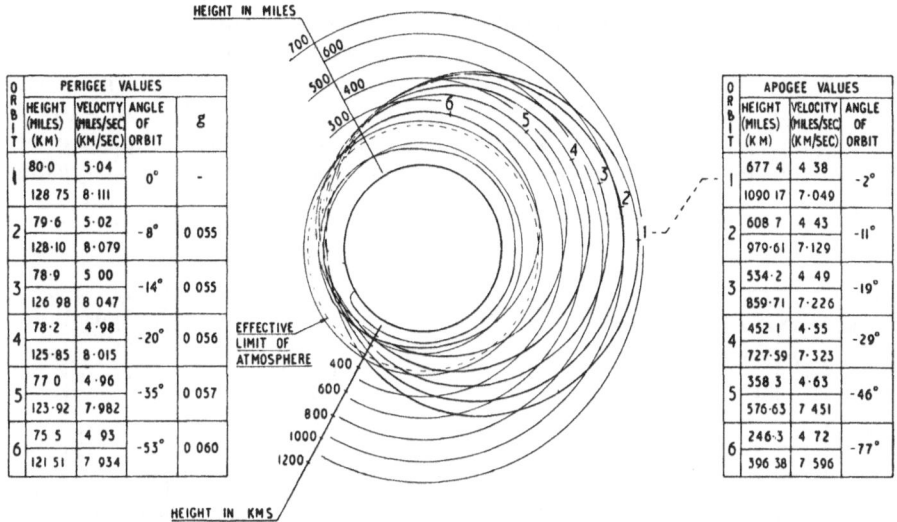

Fig. 1. Satellite orbits and re-entry paths. Loading = 12.4 lbs/ft² = 60.6 Kg/M²; lift/drag = 2.0; initially, height = 80 miles = 128 Km, velocity = 5.04 miles/sec. = 8 Km/sec., angle of ascent = 0°. Total time of flight = 11 hours 0 mins.

was chosen at 80 miles (129 Km) so as to ensure ultimate re-entry into the atmosphere, even if the braking rocket did not function.

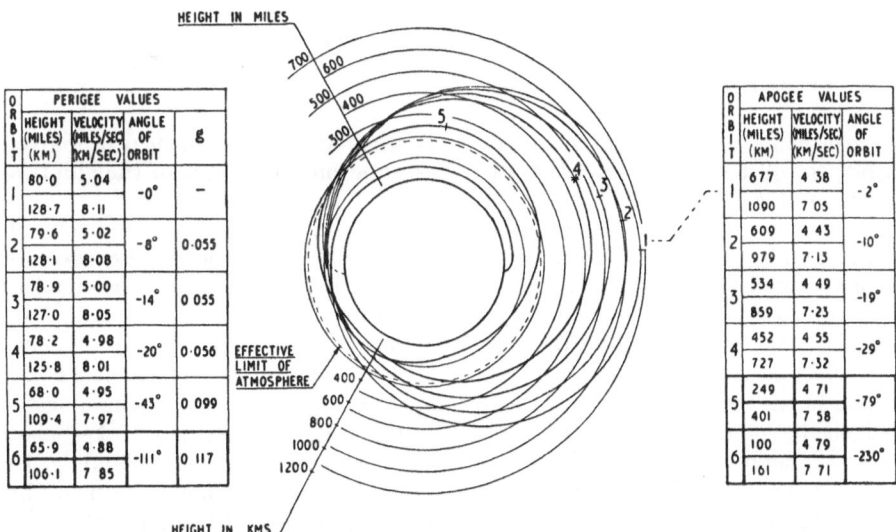

Fig. 2. Satellite orbits and re-entry paths. Loading = 12.4 lbs/ft² = 60.6 Kg/M²; lift/drag = 2.0; initially, height = 80 miles = 128 Km, velocity = 5.04 miles/sec. = 8 Km/sec., angle of ascent = 0°. Total time of flight = 8 hours 54 mins. Retro-rocket fired at apogee No. 4 (g/20 for 15 secs.)

The proposed orbit is shown in Fig. 1 together with its decay and re-entry after six revolutions. It should be noted that Figs. 1 to 3 have exaggerated height scales. Heights above the earth's surface are shown ten times bigger than the scale to which the earth is drawn. This distortion is necessary if we are to illustrate small differences in these near-circular orbits.

The calculation of these orbits is discussed in Section IV of this paper. It will be noticed that atmospheric braking at each perigee reduces height at apogee by some 12 % but only reduces the perigee height by about $1/_2$ %.

A value of aerodynamic lift/drag $= 2$ was assumed for Fig. 1. This would result in somewhat higher heating rates on the final descent, than for $L/D = 1$, but it prolongs the duration of the orbiting flight from two complete orbits in 3.8 hours to six orbits in 11 hours. This gives an idea of the aerodynamic

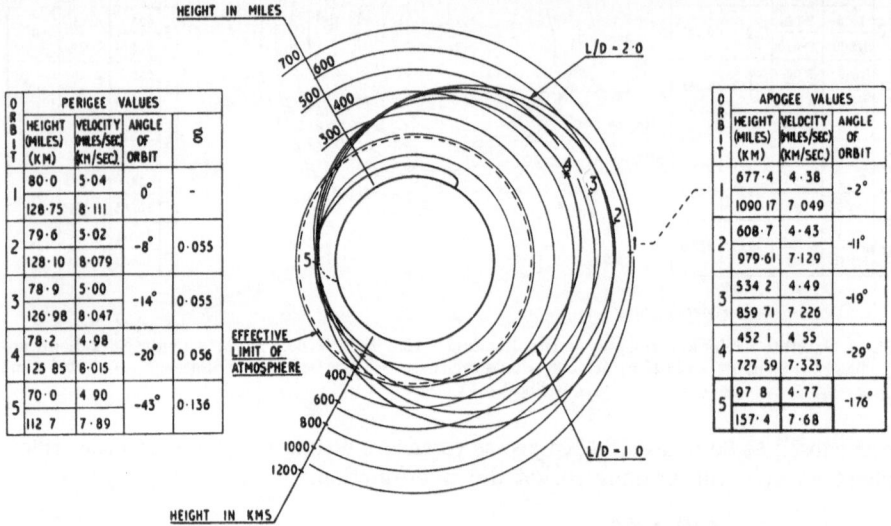

PERIGEE VALUES

ORBIT	HEIGHT (MILES) (KM)	VELOCITY (MILES/SEC) (KM/SEC)	ANGLE OF ORBIT	g
1	80·0 128·75	5·04 8·111	0°	—
2	79·6 128·10	5·02 8·079	-8°	0·055
3	78·9 126·98	5·00 8·047	-14°	0·055
4	78·2 125·85	4·98 8·015	-20°	0·056
5	70·0 112·7	4·90 7·89	-43°	0·136

APOGEE VALUES

ORBIT	HEIGHT (MILES) (KM)	VELOCITY (MILES/SEC) (KM/SEC)	ANGLE OF ORBIT
1	677·4 1090·17	4·38 7·049	-2°
2	608·7 979·61	4·43 7·129	-11°
3	534·2 859·71	4·49 7·226	-19°
4	452·1 727·59	4·55 7·323	-29°
5	97·8 157·4	4·77 7·68	-176°

Fig. 3. Satellite orbits and re-entry paths. Loading $= 12.4 \text{ lbs/ft}^2 = 60.6 \text{ Kg/M}^2$; lift/drag $= 1.0$ and 2.0; initially, height $= 80$ miles $= 128$ Km, velocity $= 5.04$ miles/sec. $= 8$ Km/sec., angle of ascent $= 0°$. Total time of flight $= 7$ hours 0 mins. Retro-rocket fired at apogee No. 4 ($g/20$ for 15 secs.)

control which an intelligent pilot can exert when already established in a given orbit. It will be noticed that the lift causes rotation of the axis of the elliptical orbits in Fig. 1.

However, it is not planned that the pilot should depend completely on aerodynamic forces to control the vehicle. Section III describes the effect of firing a small rocket while in orbit. It is found that a deceleration as small as $g/20$ for 15 seconds (or $0.75 g$ for 1 second) will lower the next perigee height by 15 miles and measure immediate re-entry into the atmosphere. Fig. 2 shows the effect of retaining the L/D ratio of 2, which causes the vehicle to "float" round in the upper atmosphere. Fig. 3 shows the result of the pilot trimming the vehicle to an incidence of 45° and using $L/D = 1$. Re-entry now takes place more suddenly and the vehicle does not quite make one orbit before landing. By varying the angle of incidence it is apparent that a skilful pilot can control his point of landing very considerably. This is particularly true at transorbital speeds, where the vehicle is "space-borne" rather than "air-borne", and aerodynamics is freed from the predominant necessity of providing an average lift equal to the weight of vehicle.

We are thus free to direct the lift downwards to retain contact with the atmosphere, or to roll the vehicle so as to produce a yawing force to permit steering in a horizontal plane.

If the vehicle is sufficiently low in the atmosphere to produce significant aerodynamic forces (say below 80 miles or 129 Km), it will be essential for the pilot to fly at incidences greater than 20° to avoid burning the upper part of the vehicle, as described in [2].

At great altitudes, where the aerodynamic forces are very small, the heating effects will be correspondingly small and "flying" at low incidence will be quite harmless, although nearly 50 % of the retarding drag force will be converted into heat on the surface of the vehicle. 50 % of a very small quantity is less important than say 5 % of the maximum retardation, converted to heat.

As was also described in [2] the greatest heating rates will occur at 0.6 of orbital speed. These will occur mainly on the underside of the vehicle, which will be of special construction to resist heat.

The choice of landing ground will be reduced as the vehicle loses speed and altitude. A very useful flight plan might be to fly banked at 45° producing 2 g resultant force; this would enable the pilot to circle the desired continent at say 0.5 orbital speed while keeping it in view over the lower wing-tip.

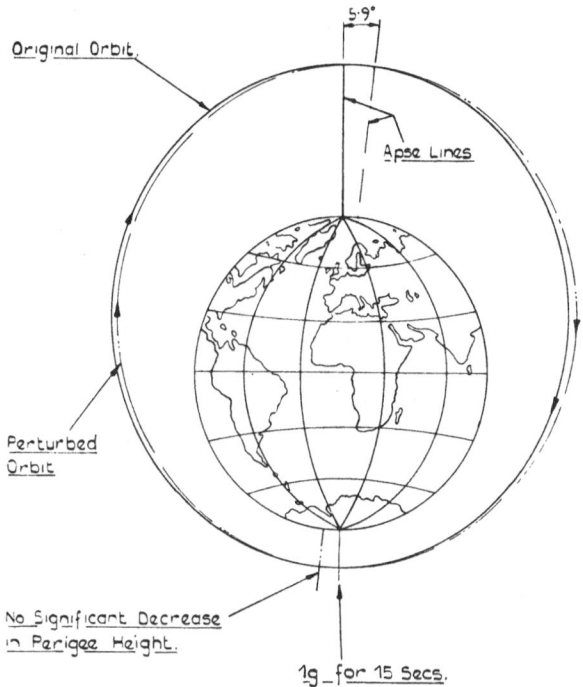

Fig. 4. Effect of "vertical" rocket burst applied at perigee on the orbit of an earth satellite

Final landing on the earth's surface at subsonic speeds would take place still using the same aerodynamic controls as at high speeds. No unusual effects are anticipated with the delta planform in this respect. Due to the low wing loading necessary to reduce heat transfer at higher speeds, little trouble is expected due to rough landing. Structural studies indicate a possible wing loading of 12.4 lbs/ft² 60.6 Kg/M²).

III. Perturbation of Satellite Orbits by Firing a Rocket

a) Normal Force

In order to change an orbit already established in space, it is necessary to apply a perturbing force. This is conveniently done by means of a rocket built in to the satellite. Any force can be resolved into a normal and a tangential component, and we will examine these in turn. The question arises as to what is the most economical method of achieving a given change of orbit. This was calculated by Mr. J. F. SWALE [3], working under my direction. His report showed that accelerating towards the earth at perigee does not bring the satellite any nearer to the earth, but results in rotation of the axis of the ellipse. The underlying physical cause is the fact that no mechanical work is done by a force acting perpendicular to the direction of motion. The case of an acceleration normal to an orbit of low eccentricity ε was considered in [2]. The position θ from perigee in the orbit at which the eccentricity is most greatly affected is given by

$$\cos \theta = -2\,\varepsilon + 3\,\varepsilon^3 \; - 8\,\varepsilon^5 + 0\,(\varepsilon^7)\,. \tag{2}$$

No change whatever can be produced in the magnitude of the major axis, which depends only on the energy of the orbit.

Fig. 4 shows the effect of firing a 1 g rocket for 15 seconds, directed towards the earth at perigee. This figure is plotted for a larger orbit than Figs. 1 to 3, in order to represent it in true scale. Again the large rotation of apse line due to a small rocket burst will be apparent.

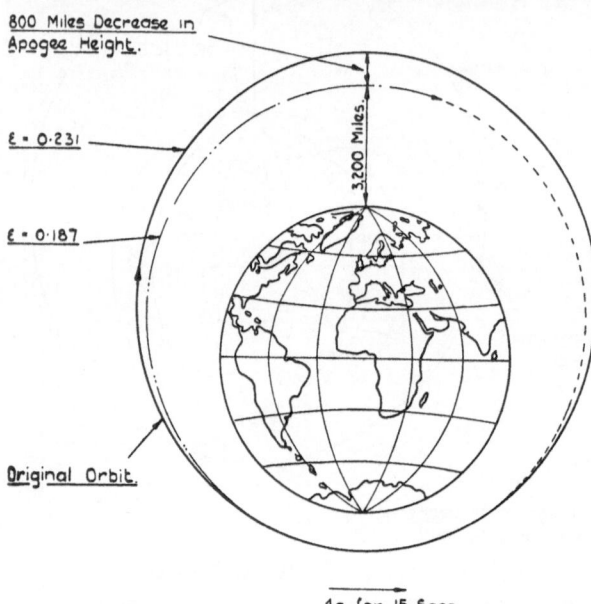

Fig. 5. Effect of forward-facing rocket burst applied at perigee on the orbit of an earth satellite. Orbit after application of thrust at perigee. Scale: 1 cm = 2000 miles

b) Tangential Force

The most effective method of burning fuel is to apply a force tangential to the orbit, so that the fuel does most work. The work done will be proportional to the velocity, so that a given rocket burst will be more effective at perigee than at apogee. However, retardation at perigee reduces height at apogee as shown in Fig. 5. Therefore we must tolerate the slightly less efficient retardation at apogee in order to lower the height

at perigee, and get down into the atmosphere again, as shown in Fig. 6.

The appropriate relationship using the notation of [3] is

$$\frac{1}{4f}\frac{dr_0}{dt} = \sqrt{\frac{a^3}{\mu}}\left[\frac{1-\varepsilon}{1+\varepsilon}\right]^{1/2} = \frac{a}{V_p} \tag{3}$$

and some values of the decrease in perigee height per unit change in velocity are given in Table I below. If velocities are in feet/second, heights will be in feet, and if in metres/second, heights will be in metres. Thus a 4000/250 mile ellipse will be lowered 4462 feet at perigee by 1 ft/sec^2 for 1 sec. applied at apogee.

The same ellipse will be lowered 4462 metres by a 1 metre/second decrease in apogee velocity.

From Table I it is apparent that a small rocket producing 1 g for 1 second at apogee will lower perigee height by the significant amount of about 20 miles (32 Km). If we now assume a specific impulse of 200 seconds for the fuel, we must burn 1/200 lb. of fuel per lb. of vehicle to produce this effect, which is only $1/2$% of the vehicle weight.

Table I. *Decrease in Perigee Height per Unit Change in Velocity at Apogee*

Maximum altitude (apogee)		Minimum altitude (perigee)					
		250	200	150	100	50	miles
miles	Km	400	320	240	160	80	Km
30,000	48,000	12,003	11,908	11,813	11,718	11,622	
25,000	40,000	10,529	10,443	10,356	10,270	10,183	
20,000	32,000	9,059	8,981	8,903	8,825	8,747	
15,000	24,000	7,596	7,527	7,457	7,387	7,318	
10,000	16,000	6,147	6,085	6,024	5,962	5,901	
5,000	8,000	4,734	4,679	4,625	4,570	4,516	
4,000	6,400	4,462	4,408	4,354	4,301	4,248	
3,500	5,600	4,328	4,274	4,221	4,168	4,116	
3,000	4,800	4,196	4,142	4,090	4,037	3,985	
2,500	4,000	4,066	4,013	3,961	3,909	3,857	
2,000	3,200	3,939	3,886	3,834	3,783	3,731	
1,500	2,400	3,815	3,763	3,711	3,660	3,608	
1,000	1,600	3,696	3,645	3,593	3,541	3,490	
750	1,200	3,639	3,587	3,535	3,484	3,433	
500	800	3,584	3,531	3,480	3,428	3,377	
250	400	3,530	3,478	3,426	3,374	3,323	

$R = 3,957$ miles $\mu = 95,600$ miles^3 sec^{-2}

Thus the amount of fuel necessary to initiate aerodynamic braking for re-entry is altogether smaller than the fuel needed to brake the rocket to zero velocity

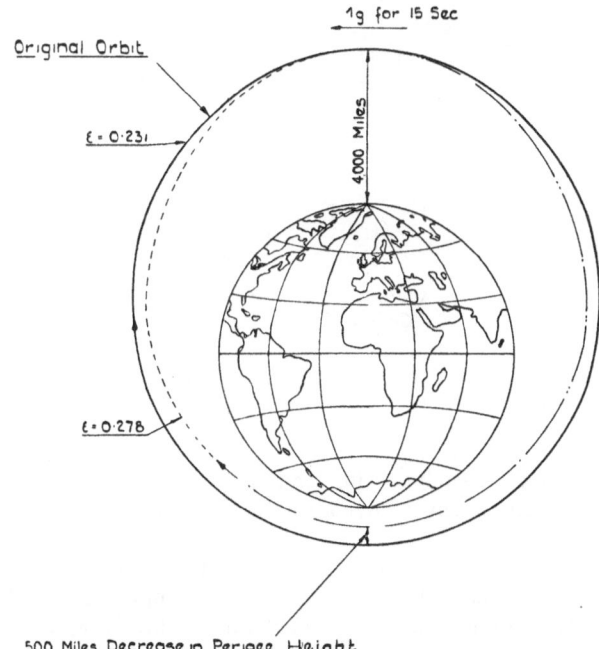

Fig. 6. Effect of forward-facing rocket burst applied at apogee on the orbit of an earth satellite. Orbit after application of thrust at apogee. Scale: 1 cm = 2000 miles

to avoid aerodynamic heating effects. Complete rocket braking would require a thrust period some 800 times as long as that needed to initiate positive re-entry.

This analysis indicates that, if it were required to change the orbit of any satellite, whether for scientific reasons or to evade an enemy, direct rocket action should be avoided. Rocket braking at apogee should cause entry into the atmosphere at perigee; aerodynamic forces should be used to change the orbit into the required new plane and apse line. Rocket power may then be used to establish the new orbit clear of the atmosphere.

Values of L/D as high as 5 should be employed, and the extra heating which will result should be minimised by flying at very high altitudes.

IV. Re-entry Path in Atmosphere

Various workers have described flight paths through the atmosphere under re-entry conditions, but we have been employing the Pegasus digital computer at A.W.A. to study a variety of re-entry paths on and off-design, to ascertain

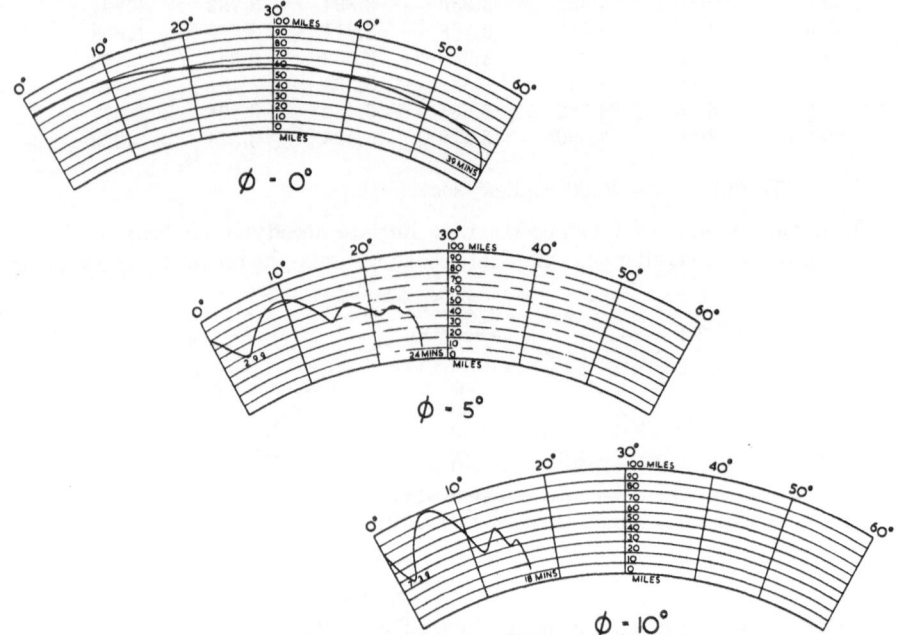

Fig. 7. Typical re-entry trajectories. Height exaggerated 10 times. Initial height 81 miles, initial speed 4.8 miles/sec., loading 12.4 lbs/sq. ft., initial angle of descent ϕ

the effects of design parameters such as wing loading, and of that elusive quantity termed "pilot error". Clearly we must have a reasonable tolerance on the piloting skill required.

A computer programme was developed by Mr. A. TRICE and is described in [4]. Basically this assumes a plane shock wave inclined at angle α to the direction of motion. Effectively this implies a predetermined value of lift/drag, where higher values of L/D are obtained from a slightly reduced lift divided by a much reduced drag, due to a lower angle of incidence δ of the wing. For $L/D=1$, $\delta=45°$ and for $L/D=2$, $\delta=26°\ 36'$, neglecting skin friction drag.

The equations used were the approximations for high MACH numbers given in [2]. In particular

$$\frac{L}{D} = \tan \delta = \frac{\sin 2a}{\gamma + \cos 2a}. \tag{4}$$

Assuming a spherical earth, a step by step solution of the equations was found for a small satellite moving under the combined influence of gravity, aerodynamic lift and drag.

This programme enabled "flights" to be undertaken on the computer, and the resulting trajectories to be plotted in detail. A wing loading of 12.4 lbs/ft² was used in the later work given here, in order to conform with the requirements of the A.W.A. manned satellite system described in [1].

The orbit considered passed through the atmosphere at a perigee height of only 80 miles, to ensure safety as mentioned above; the most striking result was the profound effect of L/D ratio at high velocities and high altitudes. A flight

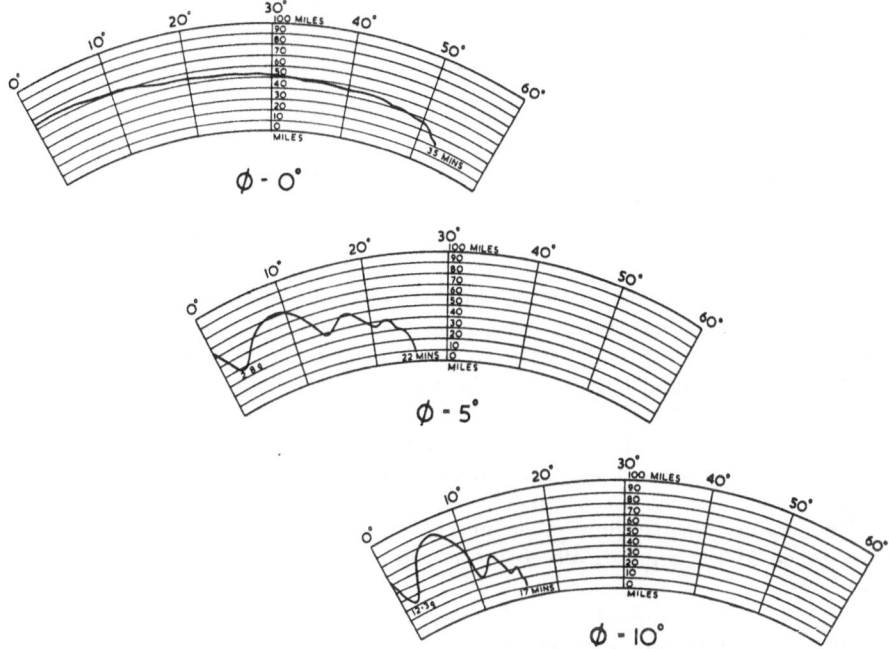

Fig. 8. Typical re-entry trajectories. Height exaggerated 10 times. Initial height 65 miles, initial speed 4.8 miles/sec., loading 100 lbs/sq. ft., initial angle of descent ϕ

of 6 orbits in 11 hours at $L/D = 2$ was reduced to 2 orbits in 3.3/4 hours when $L/D = 1$. It is not thought that the heating rates would be excessive in either case. This effect is, of course, due to the variation in drag force with L/D. The final descent should be made at $L/D = 1$ approx. to minimise heating rates.

The effect of levelling off at the right or wrong altitudes, resulting in "skip" trajectories can readily be studied by this technique. Fig. 7 shows a re-entry, initially tangential ($\varphi = 0$), from 80 miles altitude at 4.8 miles/sec. initial velocity, for a wing loading of 12.4 lbs/ft². Excessive rocket braking could cause re-entry at a greater inclination φ to the earth's surface; trajectories are shown in Fig. 7 for $\varphi = 5°$ and $\varphi = 10°$.

In an appendix to [3], J. F. SWALE showed that a velocity decrement of 555 ft/sec. at apogee would be required to produce $\varphi = 10°$ and hence a force of 7.9 g. This compares with the decrement of 24 ft/sec. recommended for perfect re-entry. Clearly as we can allow 20 times the rocket braking, the duration of firing is unimportant, provided we have a pilot aboard.

Turning now to Fig. 8, we consider the high wing loading of 100 lbs/ft² from 65 miles altitude. This results in very similar trajectories, though at a lower altitude due to the increased vehicle density. Heating rates would be increased in the same ratio as density.

Since the computer programme is a step by step one, it can be stopped at any step, in particular at apogee, the velocity decreased by a certain amount to correspond to firing a braking rocket, and the calculation re-started. The rocket chosen was $1/20\,g$ for 15 seconds, i.e. 24 ft/sec. velocity decrement, fired at the fourth apogee. Others gave results as below:

Retro-rocket fired for 15 secs. Number of "g"	Total time of flight	New perigee height	Deceleration at perigee
0	11.00 hrs.	79 miles	—
1/20	8.90 hrs.	79 miles	0.14 g
1/10	8.10 hrs.	63 miles	0.18 g
1/5	7.50 hrs.	59 miles	0.40 g

It was found that for a given height there existed a unique tangential velocity V_e to give a re-entry path free from dives and climbs. With higher values of V_e an oscillatory path results. However, a 10 % deviation of 0.5 miles/sec. in velocity results in a pull-out of only 2 g. Again, the pilot has a large margin of error in braking.

These oscillations will be seen to reduce the range of the satellite considerably.

The same computer programme was used to solve the case of failure to ignite various stages during the ascent.

Thus we find the computer a useful tool in simulating flights, and a pilot would have much to learn practically from a study of such programmes on a digital computer.

The tolerances available to a skilled pilot in a vehicle of proven ruggedness, would appear to be ample, if not generous, in spite of zero power, and to permit an aerodynamic landing at any chosen spot. An extra circuit of the landing field, might be possible by using any unused reserves of rocket fuel.

References

1. H. R. Watson, Astronautics at Armstrong Whitworths. British Commonwealth Astronautics Symposium. Aug. 1959.
2. W. F. Hilton, Recovery after Re-entry, by the Use of Aerodynamic Lift. R.Ae.S. — B.I.S. Symposium, Cranfield, July 1957.
3. J. F. Swale, Perturbations of Gravitational Orbits by Natural Forces or Rockets. A.R.L. Report No. 58/34—Armstrong Whitworth Aircraft and Appendix to above (58/34/1).
4. A. Trice, Investigation of Possible Orbits and Re-entry Paths of a Manned Satellite of Given Aerodynamic Design. A.R.L. Report No. 59/24—Armstrong Whitworth Aircraft.

Launching Conditions and the Geometry of Orbits in a Central Gravity Field

By

Fang-Toh Sun[1]

(With 6 Figures)

(Received June 26, 1959)

Abstract — Zusammenfassung — Résumé

Launching Conditions and the Geometry of Orbits in a Central Gravity Field. An analysis of the influence of the launching conditions on the geometry of the orbit of a projectile launched in a central gravity field is presented. Appropriate formulas relating the principal geometrical parameters of the orbit to the launching parameters defined at the final burnout point are developed. A unified treatment is given to all three basic types of the unperturbed orbit, namely, the elliptic, the parabolic, and the hyperbolic. The problem of burnout precision is also briefly discussed. Finally, an energy-momentum diagram is presented to show the essential geometrical aspects of the possible orbits in a single chart.

Startbedingungen und die Geometrie von Bahnen in einem zentralen Schwerefeld. Es wird eine Analyse des Einflusses der Startbedingungen auf die Geometrie der Bahn eines Geschosses in einem zentralen Schwerefeld angegeben. Geeignete Formeln werden entwickelt, welche die grundlegenden geometrischen Parameter der Bahn mit den Startparametern bei endgültigem Brennschluß in Beziehung setzen. Einheitlich behandelt werden alle drei Grundtypen der ungestörten Bahn, nämlich die Ellipse, die Parabel und die Hyperbel. Auch das Problem der Präzision bei Brennschluß wird kurz erörtert. Schließlich wird ein Energie-Moment-Diagramm angegeben, um die wesentlichen geometrischen Gesichtspunkte der möglichen Bahnen in einer einzigen Darstellung zu zeigen.

Conditions de lancement et géométrie des orbites dans un champ de gravitation central. L'influence des conditions de lancement sur la géométrie de l'orbite est analysée. Des formules sont établies pour relier les paramètres de lancement définis en fin de combustion et les paramètres géométriques principaux de l'orbite, que celle-ci soit elliptique, parabolique ou hyperbolique. Le problème de la précision en fin de combustion est aussi brièvement discuté. Enfin un diagramme énergie-quantité de mouvement est présenté, qui met en évidence les aspects géométriques essentiels d'une orbite.

Nomenclature

a	semi-major axis of elliptic orbit, or semi-transverse axis of hyperbolic orbit	b semi-minor axis of elliptic orbit, or semi-conjugate axis of hyperbolic orbit

---

[1] Member of the Astronautical Society of the Republic of China; Professor of Mechanical Engineering, Taiwan Provincial Cheng Kung University, Tainan, Taiwan, China.

c	center-to-focus distance of conic orbit	α	direction angle of asymptote to hyperbolic orbit, with respect to local vertical at burnout point
G	universal gravitational constant		
K	orbital energy per unit orbiting mass		
J	angular momentum per unit orbiting mass	γ	direction angle of asymptote to hyperbolic orbit, with respect to transverse axis of the orbit
M	gravitating mass		
r	radial distance from the center of the gravity field	ε	orbit eccentricity
$\bar{r}$	semi-latus rectum	θ	polar angle
r_A	apocenter radius	$\dot{\theta} = d\theta/dt$	
r_P	pericenter radius	$\ddot{\theta} = d^2\theta/dt^2$	
$\dot{r} = dr/dt$		θ_0	polar angle of the apsidal axis
$\ddot{r} = d^2r/dt^2$		λ	speed ratio $= V/V*$
V	orbital velocity	μ	gravitational constant $= GM$
V_r	r-component of orbital velocity		
V_θ	θ-component of orbital velocity	$\psi_0 = \dfrac{\pi}{2} - \varphi_0$	
$V*$	critical speed $= \sqrt{2\mu/r}$		
V_s	circular speed $= \sqrt{\mu/r}$	φ_0	angle of departure, the angle between the velocity vector and the local horizon at final burnout
X	dimensionless orbital energy $= r_0 K/\mu$		
Y	dimensionless angular momentum $= J/\sqrt{\mu r_0}$		*Subscript*
		0	condition at the final burnout (except θ_0)

I. Introduction

The geometry of the orbit of a projectile in a central force field has been well developed in theoretical and celestial mechanics, but its relations with the launching conditions did not arrest serious attention until the ballistic missiles and artificial satellites came into being. Fragmentary discussions on this subject are scattered in the literature from the early work of Cranz [1] down to the many current papers made by various authors [3—9], most of them dealing with elliptical orbits; yet no thorough treatment is available. The purpose of the present paper is to present a systematic study on the influence of the launching conditions defined at the final burnout of a rocket projectile on the geometry of its subsequent orbit. Appropriate formulas relating the principal geometrical parameters of the orbit to the burnout parameters are developed, and a unified treatment is given for the three basic types of the un-perturbed orbit; namely, the elliptic, the parabolic, and the hyperbolic. The following analysis is based mainly on Cranz's work on vacuum trajectory [1], and extends it toward a broader comprehension of the subject.

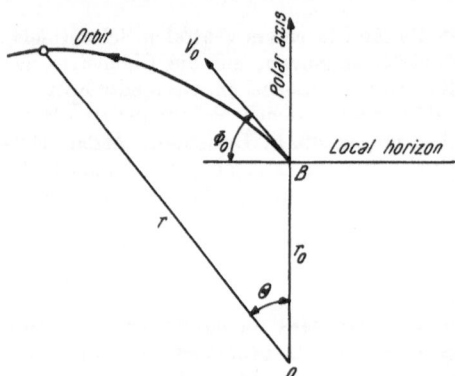

Fig. 1. Polar coordinates for the orbit

II. Equation of the Orbit

The motion of a projectile which is subjected to no external forces except the attraction of a central gravity field will lie in the plane determined by the velocity vector and the radius vector at the final burnout. For such motion a planar description of Fig. 1 is sufficient. The treatment of such motion is found in most texts on mechanics. To facilitate the subsequent development, a brief account is given below.

In an inverse-square field, the equations of motion of the projectile in polar coordinates in the orbital plane, with the origin placed at the center of the field, are

$$\left| \ddot{r} - r\,\dot{\theta}^2 = -\frac{\mu}{r^2} \right. \tag{1}$$

$$\left. r\ddot{\theta} + 2\dot{r}\dot{\theta} = 0 \right. \tag{2}$$

where $\mu = GM$, G is the universal gravitational constant, and M is the gravitating mass. Integration of eqs. (1) and (2) gives

$$\left| \frac{1}{2}(\dot{r}^2 + r^2\dot{\theta}^2) - \frac{\mu}{r} = K \right. \tag{3}$$

$$\left. r^2\,\dot{\theta} = J \right. \tag{4}$$

where the constants of integration K and J are recognized as the orbital energy and the angular momentum per unit mass of the projectile respectively. Upon eliminating dt from eqs. (3) and (4) and integrating, a polar equation of the orbit is obtained,

$$\frac{1}{r} = \frac{\mu}{J^2}\left[1 + \sqrt{1 + \frac{2KJ^2}{\mu^2}}\ \cos(\theta - \theta_0) \right]. \tag{5}$$

This is known as the equation of a conic having one focus at the origin and θ as the polar angle of its apsidal axis. Writing in the standard form,

$$\frac{1}{r} = \frac{1}{\bar{r}}[1 + \varepsilon \cos(\theta - \theta_0)] \tag{5a}$$

and comparing it with eq. (5) yields

$$\bar{r} = J^2/\mu \tag{6}$$

$$\varepsilon = \sqrt{1 + \frac{2KJ^2}{\mu^2}} \tag{7}$$

where $\bar{r}$ is the semi-latus rectum, and ε the eccentricity.

It is a well known fact that eq. (5) gives three different types of orbits: elliptic, parabolic, and hyperbolic, according as the orbital energy K is negative, zero, or positive. Further evaluation of the constants K, J, and θ, from the burnout condition is given in the following section.

III. Relations between the Burnout Parameters and the Geometrical Parameters of the Orbit

Let the polar axis be fixed in the inertia space and pass through the point B at which the final burnout occurs. At the instant of the final burnout, the velocity components of the projectile are

$$\begin{aligned}(V_r)_0 &= \dot{r}_0 = V_0 \sin \Phi_0 \\ (V_\theta)_0 &= r_0\,\dot{\theta}_0 = V_0 \cos \Phi_0 \, .\end{aligned} \tag{8}$$

Thus the orbital energy and the angular momentum may be evaluated as

$$K = \frac{1}{2}V_0^2 - \frac{\mu}{r_0} \tag{9}$$

$$J = r_0 V_0 \cos \Phi_0 \, . \tag{10}$$

Introducing the dimensionless speed parameter λ, which is defined as

$$\lambda^2 = \left(\frac{V}{V*}\right)^2 = \frac{rV^2}{2\mu} \tag{11}$$

eqs. (9) and (10) may be written as

$$K = \frac{\mu}{r_0}(\lambda_0^2 - 1) \tag{9a}$$

$$J = \sqrt{2\mu r_0}\, \lambda_0 \cos \Phi_0 . \tag{10a}$$

Thus the principal parameters of the conic orbit, $\bar{r}$ and ε may now be expressed in terms of the burnout parameters λ_0, Φ_0 and r_0 as

$$\bar{r} = 2r_0\, \lambda^2_0 \cos^2 \Phi_0 \tag{12}$$

$$\varepsilon = \sqrt{1 - 4\lambda_0^2(1 - \lambda_0^2)\cos^2 \Phi_0}. \tag{13}$$

The polar angle of the apsidal axis may be determined by applying the burnout condition, $r = r_0$ at $\theta = 0$, to the orbital eq. (5); thus

$$\cos \theta_0 = \frac{2\lambda_0^2 \cos^2 \Phi_0 - 1}{\sqrt{1 - 4\lambda_0^2(1 - \lambda_0^2)\cos^2 \Phi_0}} . \tag{14}$$

Furthermore, the orbit must be tangent to the velocity vector at the burnout point, that is

$$r_0\left(\frac{d\theta}{dr}\right)_0 = \cot \Phi_0$$

or

$$\sin \theta_0 = -\frac{\lambda_0^2 \sin 2\Phi_0}{\sqrt{1 - 4\lambda_0^2(1 - \lambda_0^2)\cos^2 \Phi_0}} . \tag{15}$$

Combination of eqs. (14) and (15) gives

$$\cot \theta_0 = \frac{1}{\lambda_0^2}\csc 2\Phi_0 - \cot \Phi_0 . \tag{16}$$

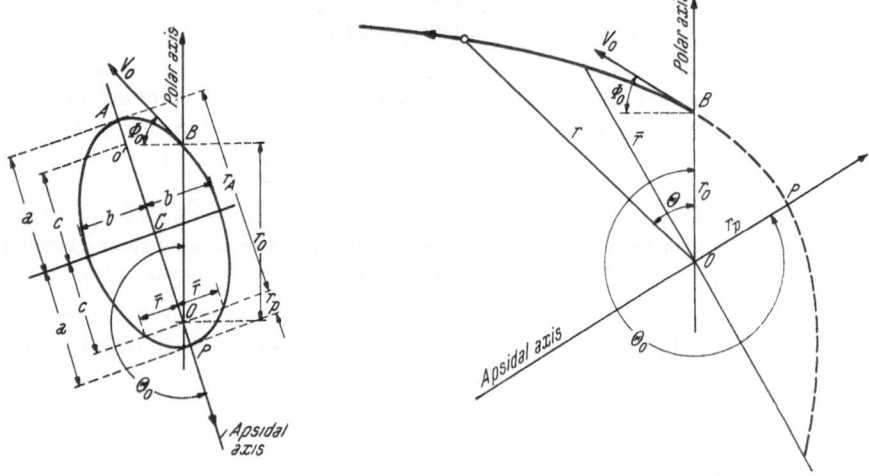

Fig. 2a. The elliptical orbit   Fig. 2b. The parabolic orbit

Any two of the eqs. (14) to (16) are sufficient to determine a unique value of θ_0 within 2π radians.

The fact about the type of orbit mentioned in the last section may now be stated as follows:

$\lambda_0 < 1$ ($\varepsilon < 1$): elliptic orbit
$\lambda_0 = 1$ ($\varepsilon = 1$): parabolic orbit
$\lambda_0 > 1$ ($\varepsilon > 1$): hyperbolic orbit.

It is to be noted that in the case of vertical launching, $\cos \Phi_0 = 0$, ε is unity for all three types of orbits. In such limiting case, the elliptic orbit degenerates into a line segment, and the parabolic and hyperbolic orbits both degenerate into a straight line extending to infinity.

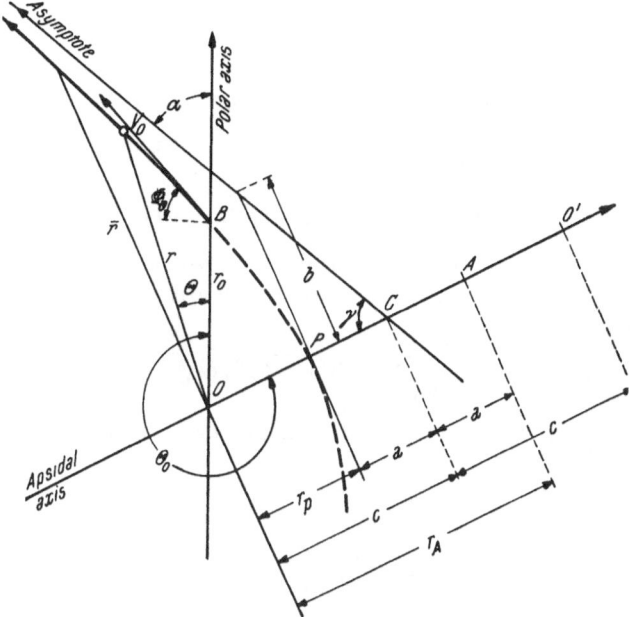

Fig. 2c. The hyperbolic orbit

Formulas for other principal geometrical parameters will be given separately for each of the three types of the orbit (for nomenclature, see Fig. 2a, b and c).

Elliptic Orbit ($\lambda_0 < 1$)

The peri-center P and the apo-center A may be located by setting $dr/d\theta = 0$ in eq. (5a), which yields $\theta = \theta_0$ or $\theta_0 + \pi$. At P: $\theta = \theta_0$, the peri-center radius is

$$r_P = \frac{\bar{r}}{1+\varepsilon} = \frac{r_0}{2(1-\lambda_0^2)} \left(1 - \sqrt{1 - 4\lambda^2_0(1-\lambda^2_0)\cos^2\Phi_0}\right). \tag{17a}$$

At A: $\theta = \theta_0 + \pi$, the apo-center radius is

$$r_A = \frac{\bar{r}}{1-\varepsilon} = \frac{r_0}{2(1-\lambda_0^2)} \left(1 + \sqrt{1 - 4\lambda_0^2(1-\lambda_0^2)\cos^2\Phi_0}\right). \tag{18a}$$

The pericenter and the apocenter are also the points of the farthest and closest approach of the projectile from the center of the field respectively.

Besides, the semi-major axis a and the semi-minor axis b may be expressed as

$$a = \frac{1}{2}(r_A + r_P) = \frac{r_0}{2(1-\lambda_0^2)} \tag{19}$$

$$b = a\sqrt{1-\varepsilon^2} = \frac{r_0 \lambda_0 \cos \Phi_0}{\sqrt{1-\lambda^2}}. \tag{20a}$$

And the distance between the two foci is given by

$$2C = r_A - r_P = \frac{r_0}{1-\lambda_0{}^2}\sqrt{1-4\lambda_0{}^2(1-\lambda_0{}^2)\cos^2\Phi_0} \qquad (21\,a)$$

where c is the radial distance of the center of the conic from the origin.

From eq. (5a) it is seen that the elliptic orbit will become circular if $\varepsilon = 0$, that is,

$$4\lambda_0{}^2(1-\lambda_0{}^2)\cos^2\Phi_0 = 1$$

or

$$\lambda_0{}^2 = \frac{1}{2}(1 \pm \sqrt{-\tan^2\Phi_0})\,.$$

Evidently λ_0 has a real solution only if $\Phi_0 = 0$, giving $\lambda_0 = \frac{1}{\sqrt{2}}$. Thus a circular orbit will result only from horizontal launching at the circular speed.

Hyperbolic Orbit ($\lambda_0 > 1$)

Same formulas for the elliptical orbit may also be used for the hyperbolic orbit[1]; however, in order to avoid confusion they are written in slightly different forms as follows:

$$r_P = \frac{\bar{r}}{\varepsilon+1} = \frac{r_0}{2(\lambda_0{}^2-1)}(\sqrt{1+4\lambda{}^2{}_0(\lambda_0{}^2-1)\cos^2\Phi_0} - 1) \qquad (17\,b)$$

$$r_A = \frac{\bar{r}}{\varepsilon-1} = \frac{r_0}{2(\lambda_0{}^2-1)}(\sqrt{1+4\lambda_0{}^2(\lambda_0{}^2-1)\cos^2\Phi_0} + 1) \qquad (18\,b)$$

$$a = \frac{1}{2}(r_A - r_P) = \frac{r_0}{2(\lambda_0{}^2-1)} \qquad (19\,b)$$

$$b = a\sqrt{\varepsilon^2-1} = \frac{r_0\lambda_0\cos\Phi_0}{\sqrt{\lambda_0{}^2-1}} \qquad (20\,b)$$

$$2C = r_A + r_P = \frac{r_0}{\lambda_0{}^2-1}\sqrt{1+4\lambda_0{}^2(\lambda_0{}^2-1)\cos^2\Phi_0}\,. \qquad (21\,b)$$

It is to be noted that, while the pericenter here is also the point of closest approach of the projectile to the center of the field just as in the case of elliptic orbits, the apocenter is not the point of the farthest approach, nor does it lie in the trajectory (but rather in the other branch of the hyperbola). Besides, the projectile will not be able to come to the pericenter P if it is launched in a direction above the horizon, since in this case P is prior to the burnout point B. With regard to the semi-axes, customarily a is called the transverse axis and b the conjugate axis of the hyperbola.

In addition, the direction of the asymptote of the hyperbolic orbit is given by

$$\cos\gamma = \frac{1}{\sqrt{1+\left(\frac{b}{a}\right)^2}} = \frac{1}{\varepsilon} \qquad \gamma < \frac{\pi}{2} \qquad (22)$$

where γ is the direction angle made with the transverse axis of the orbit. With respect to the polar axis the direction angle is given by

$$\alpha = \theta_0 - \gamma - \pi$$

or

$$\cos\alpha = \frac{1+2\lambda_0{}^2\cos^2\Phi_0\left(2\lambda_0\sqrt{\lambda_0{}^2-1}\sin\Phi_0-1\right)}{1+4\lambda_0{}^2(\lambda_0{}^2-1)\cos^2\Phi_0}\,. \qquad (23)$$

[1] Identical formulas may be used if r_A, a, b^2, and c are considered as negative.

Parabolic Orbit ($\lambda_0 = 1$)

Setting $\lambda_0 = 1$ in either the formulas for elliptic orbits or those for hyperbolic orbits gives the principal parameters of a parabolic orbit as follows:

$$\bar{r} = 2r_0 \cos^2 \Phi_0 \tag{12c}$$

$$r_p = r_0 \cos^2 \Phi_0 \tag{17c}$$

$$\theta_0 = 2(\pi - \Phi_0). \tag{14c}$$

The eccentricity is unity, and the apocenter lies in infinity. Just as in the case of hyperbolic orbits, the projectile may or may not be able to come to the pericenter according to the launching direction is below or above the local horizon.

IV. Influence of the Launching Parameters on the Geometry of Orbits

In the light of eqs. (12) to (23) and their graphical representations Fig. 3a to h, the influence of the launching parameters on the geometry of the orbit may now be examined.

Eccentricity

At a given angle of departure, the eccentricity of an elliptic orbit tends to decrease as the burnout speed increases from zero to circular velocity, and then

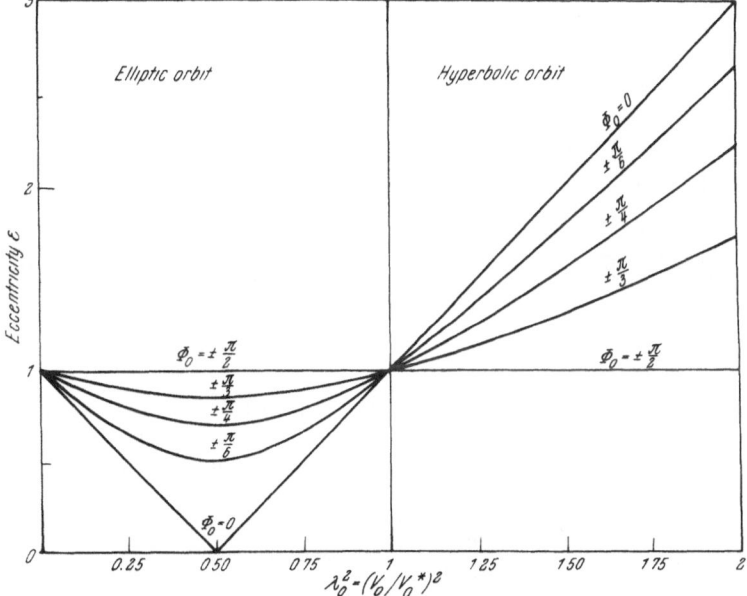

Fig. 3a. Eccentricity vs speed ratio

to increase as the speed further increases (Fig. 3a). A minimum eccentricity is attained at circular launching speed $\left(\lambda_0^2 = \dfrac{1}{2}\right)$, the value of which is given by

$$\varepsilon_{min} = |\sin \Phi_0| \tag{24}$$

which can be easily obtained from eq. (13). The lowest eccentricity will be obtained in horizontal launching, for which $\Phi_0 = 0$ and $\varepsilon = 0$, and a circular orbit results.

In hyperbolic orbit, the eccentricity will increase with the burnout speed without limit.

At a given burnout speed, the higher the deviation from horizontal launching, the higher will be the eccentricity if the orbit is elliptic; and the opposite is true if the orbit is hyperbolic, that is, the higher the deviation, the less will be the eccentricity.

In vertical launching eccentricity is unity for all values of the burnout speed.

Pericenter and Apocenter

The effect of the burnout speed or the angle of departure on the pericenter is the same for both elliptic and hyperbolic orbits.

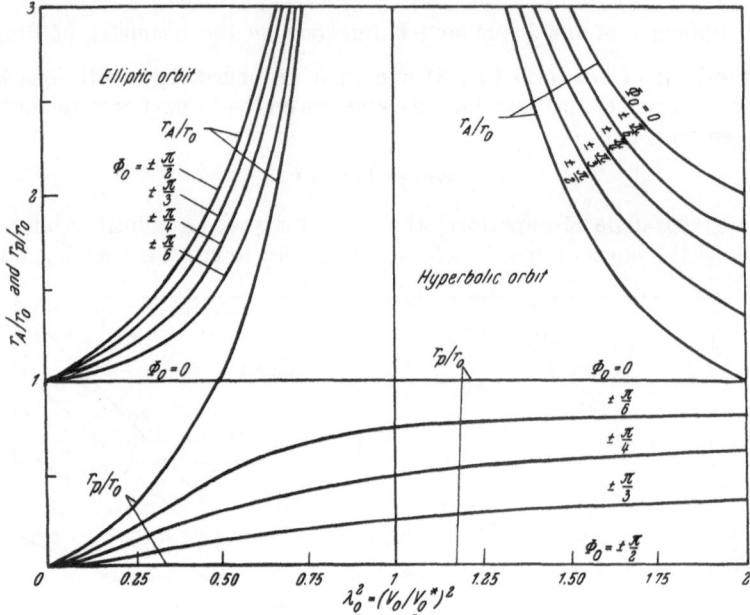

Fig. 3b. Apocenter and pericenter radii vs speed ratio

At a constant angle of departure, the increase in burnout speed tends to increase the pericenter radius (Fig. 3b). As the speed increases indefinitely, the pericenter radius approaches a definite limit determined by the angle of departure according to eq. (17),

$$\frac{r_P}{r_0}\bigg|_{\lambda_0 \to \infty} \to \left|\cos \Phi_0\right|. \tag{25}$$

In inclined launching the pericenter radius remains less than the burnout radius for all values of burnout speed. Same is true in horizontal launching if the speed is less than the circular speed; however, when the circular speed is reached, the pericenter coincides with the burnout point, and the further increase in speed will not change its position at all.

At a constant burnout speed the higher the deviation from horizontal launching, the closer will be the pericenter to the origin.

The effect of the burnout speed or the angle of departure on the apocenter is different in elliptic and hyperbolic orbits. At a constant angle of departure, the higher the burnout speed, the farther will be the apocenter from the origin if the orbit is elliptic; as the burnout speed increases from zero to critical, the apocenter moves from the burnout point to infinity. However, in hyperbolic orbit, the higher the burnout speed, the closer will be the apocenter to the origin; as the speed increases indefinitely, the apocenter radius approaches the same limit as the pericenter, thus the two points will coincide at infinite burnout speed.

In inclined launching the apocenter of the elliptic orbit remains greater than the burnout radius for all values of elliptical speed. However, in horizontal launching the apocenter will remain at the burnout point for all speeds less than the circular speed, and it moves away from the burnout point when the circular speed is exceeded.

At constant burnout speed, the higher the deviation from the horizontal launching, the farther will be the apocenter from the origin if the orbit is elliptic; and the closer will it be if the orbit is hyperbolic.

Semi-Latus Rectum

The semi-latus rectum $\bar{r}$ is proportional to the square of the burnout speed at a given angle of departure; and at a given burnout speed, the less the deviation

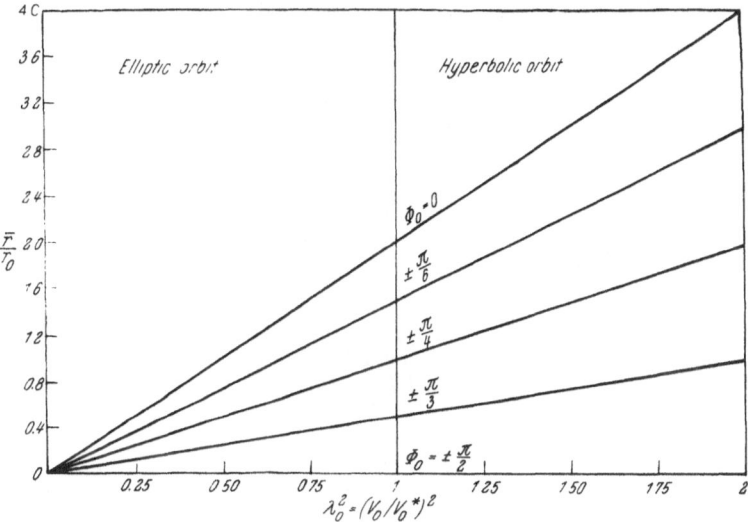

Fig. 3c. Semi-latus rectum vs speed ratio

from the horizontal direction, the greater will be the value of $\bar{r}$ (Fig. 3c). Maximum $\bar{r}$ occurs at horizontal launching, and is given by

$$\bar{r}_{max} = 2r_0 \lambda^2{}_0 . \tag{26}$$

The foregoing statement is true for both elliptic and hyperbolic orbits. It is also to be noted that in parabolic orbit, $\bar{r}$ is always twice the pericenter radius.

Semi-Axes

In either the elliptical orbit or the hyperbolic orbit the semi-axis a depends on the burnout speed only, and is independent of the angle of departure; while the semi-axis b depends on both (Fig. 3d, e).

In elliptic orbit both a and b tend to increase with the burnout speed; besides, b also increases as the deviation from horizontal direction decreases.

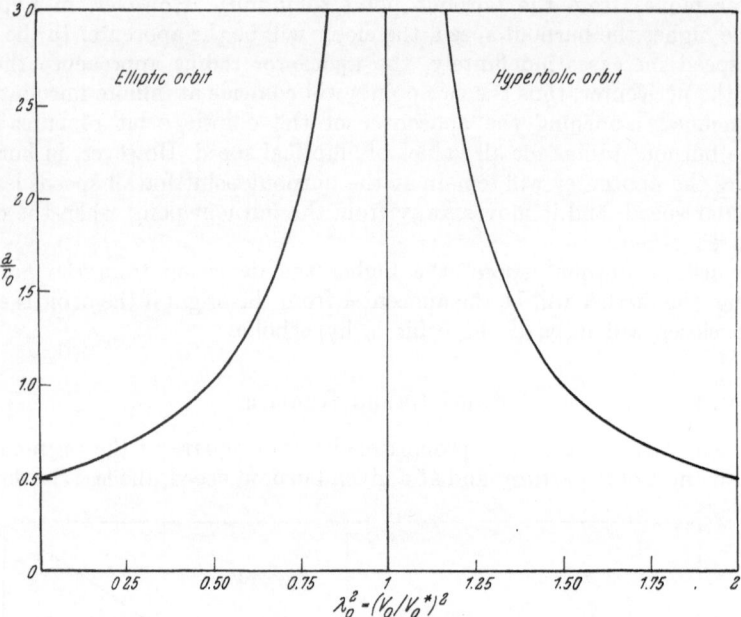

Fig. 3d. Semi-major axis and semi-transverse axis vs speed ratio

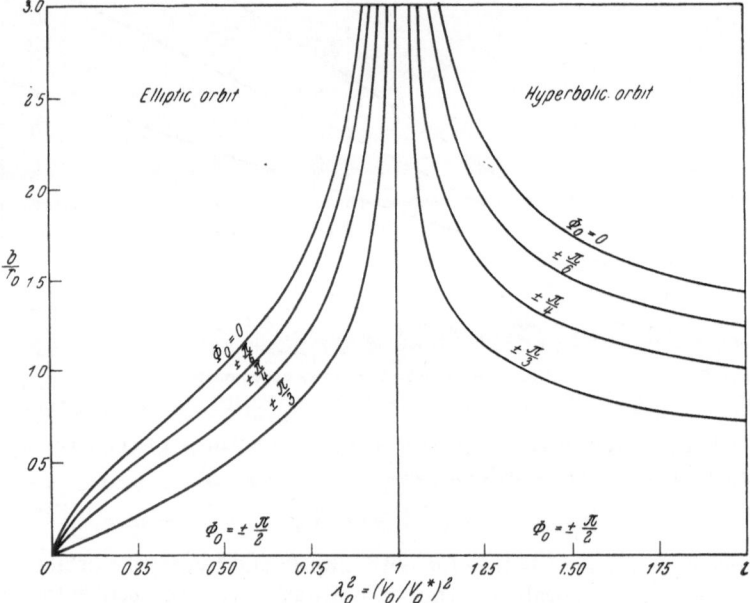

Fig. 3e. Semi-minor axis and semi-conjugate axis vs speed ratio

In hyperbolic orbit, a and b both tend to decrease as the burnout speed increases; besides, b also increases as the deviation from horizontal launching decreases.

In parabolic orbit, both a and b extend to infinity.

Second Focus and Center of the Conic

In elliptical orbit, if the burnout speed increases from zero to critical, the radius of the second focus first decreases, and then increases if the launching

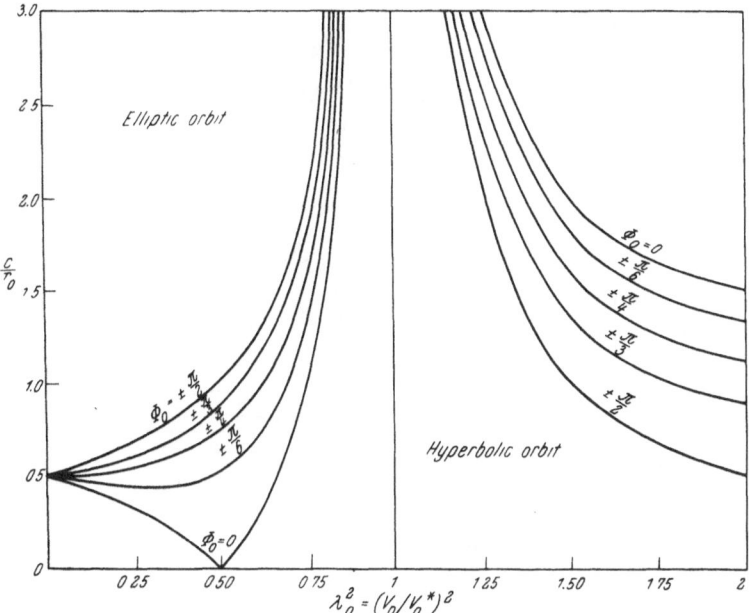

Fig. 3f. Center distance vs speed ratio

is made within $\pm 45°$ with the horizontal (Fig. 3f). The closest approach of the foci will be reached at the speed ratio given by

$$\lambda_0{}^2 = 1 - \frac{1}{2}\sec^2 \Phi_0 \tag{27}$$

which can be obtained by setting $dc/d\lambda^2 = 0$ in eq. (21a). The minimum distance between the foci is then

$$(2C)_{min} = r_0 \sin 2\Phi_0 . \tag{28}$$

A similar situation exists for the center of the ellipse since it is always situated midway between the two foci. If the launching direction is beyond $\pm 45°$ with the horizontal, the second focus and the center of the ellipse will always move away from the origin with increasing distance as the speed increases. At critical speed, both will be at infinity, and the orbit becomes a parabola.

In hyperbolic orbit as the burnout speed increases indefinitely at a constant angle of departure, the second focus and the center of the hyperbola move from infinity toward the origin. In the limiting case of infinite burnout speed the center of the hyperbola will coincide with the pericenter, and the second focus will be at its closest approach to the origin.

At a given burnout speed, the higher the deviation from horizontal launching, the farther will be the second focus and the center of the conic orbit from the origin if it is elliptic; and the closer will they be if it is hyperbolic.

Apsidal Axis

The apsidal angle θ_0 determines the angular displacement of the pericenter and apocenter from the polar axis. In general, the value of θ_0 depends on both the burnout speed and the angle of departure. If the burnout speed is less than

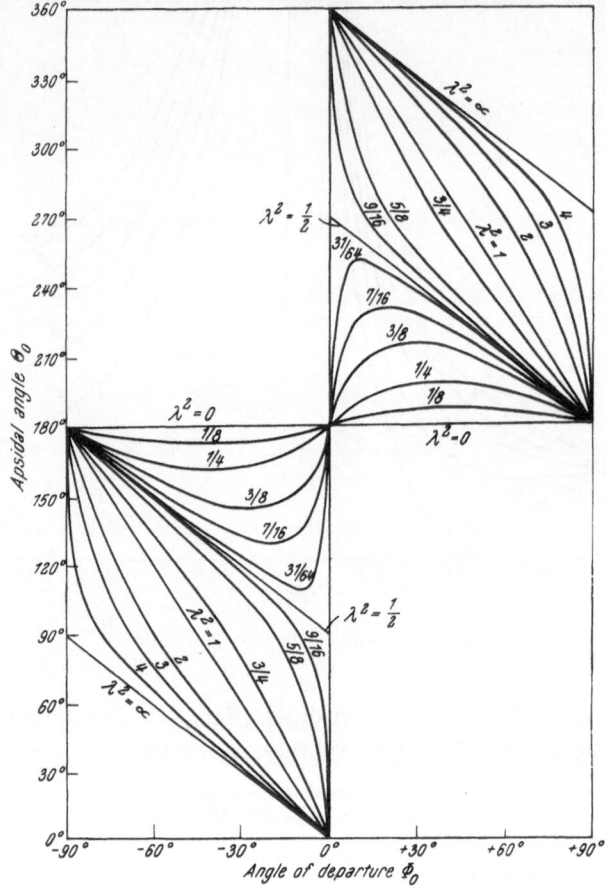

Fig. 3g. Apsidal angle vs angle of departure

the circular speed, as Φ_0 varies from $-\dfrac{\pi}{2}$ to $+\dfrac{\pi}{2}$, the apsidal axis will swing about the polar axis in a limited range according to the following equation

$$(\theta_0)_{max} = \pi + \sin^{-1} \frac{\lambda_0^2}{1 - \lambda_0^2}$$

$$(\theta_0)_{min} = \pi - \sin^{-1} \frac{\lambda_0^2}{1 - \lambda_0^2}. \tag{29}$$

The particular angles of departure giving the extremal values of θ_0 are

$$\Phi_0 = \pm \sec^{-1} \sqrt{2(1 - \lambda^2_0)} \tag{30}$$

where the plus sign corresponds to maximum θ_0, and the minus sign to minimum θ_0. Eqs. (29) and (30) can easily be obtained by setting $\dfrac{d\theta_0}{d\Phi_0} = 0$ in eq. (16). If circular speed is exceeded, then the apsidal axis may rotate about the polar axis in the full range of 360°.

As shown in Fig. 3g, at a given angle of departure, the smaller the burnout speed, the closer will be the apsidal axis to the polar axis; and at a constant burnout speed exceeding the circular speed, the less the deviation from horizontal launching, the farther apart will they be. If the burnout speed is less than the

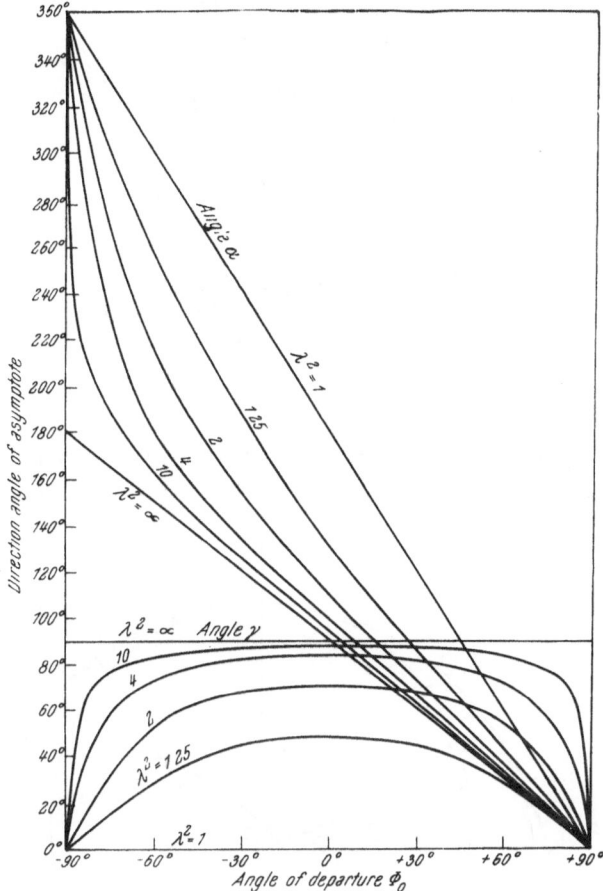

Fig. 3h. Direction angle of asymptote vs angle of departure

circular speed, then as the deviation from horizontal direction increases, the apsidal axis first swings away from the polar axis, and then after the extremal polar angle [eq. (29)] is reached, it swings back toward the polar axis again. In horizontal launching the apsidal axis will remain coincident with the polar axis for all non-circular speeds.

It is to be noted here that for some particular values of the burnout speed the apsidal angle θ_0 bears linear relations with the angle of departure Φ_0 as follows:

	$-\dfrac{\pi}{2} \leqslant \Phi_0 \leqslant 0$	$0 \leqslant \Phi_0 \leqslant \dfrac{\pi}{2}$
$\lambda_0 = 0$	$\theta_0 = \pi$	$\theta_0 = \pi$
$\lambda_0 = \dfrac{1}{\sqrt{2}}$	$\theta_0 = \dfrac{\pi}{2} - \Phi_0$	$\theta_0 = \dfrac{3\pi}{2} - \Phi_0$
$\lambda_0 = 1$	$\theta_0 = -2\Phi_0$	$\theta_0 = 2\pi - 2\Phi_0$
$\lambda_0 = \infty$	$\theta_0 = -\Phi_0$	$\theta_0 = 2\pi - \Phi_0$

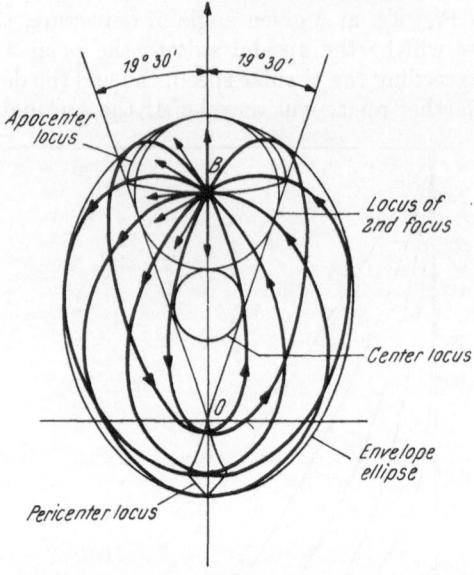

Fig. 4a. Elliptic orbits under varying angle of departure, $\lambda_0^2 = \dfrac{1}{4}$

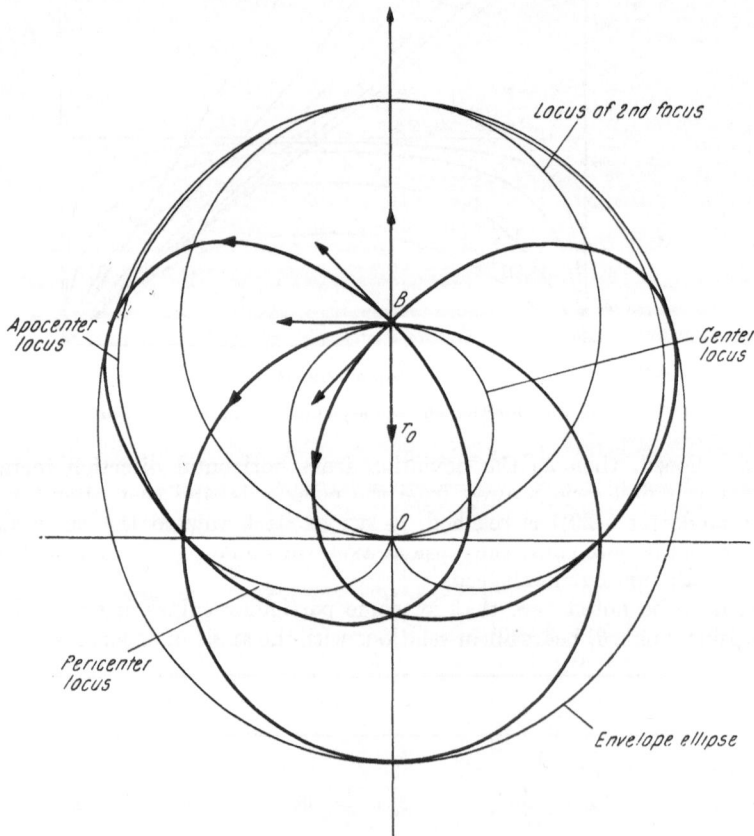

Fig. 4b. Elliptic orbits under varying angle of departure, $\lambda_0^2 = \dfrac{1}{2}$

Asymptote Direction of Hyperbolic Orbit

As shown in Fig. 3h, at a given speed ratio, the polar angle of the asymptote to the orbit decreases monotonically as the angle of departure varies from $-\frac{\pi}{2}$ to $+\frac{\pi}{2}$. At a given angle of departure, the higher the speed, the smaller will be the polar angle. In the limiting case of infinite speed, the hyperbolic orbit becomes a straight line, and the asymptote direction coincides with the launching direction.

V. Orbits under Varying Launching Conditions: Loci and Envelopes

Constant Burnout Speed, Varying Angle of Departure

Under constant burnout speed but varying angle of departure, the orbits constitute a system of conics all having the same length of semi-axis a and one

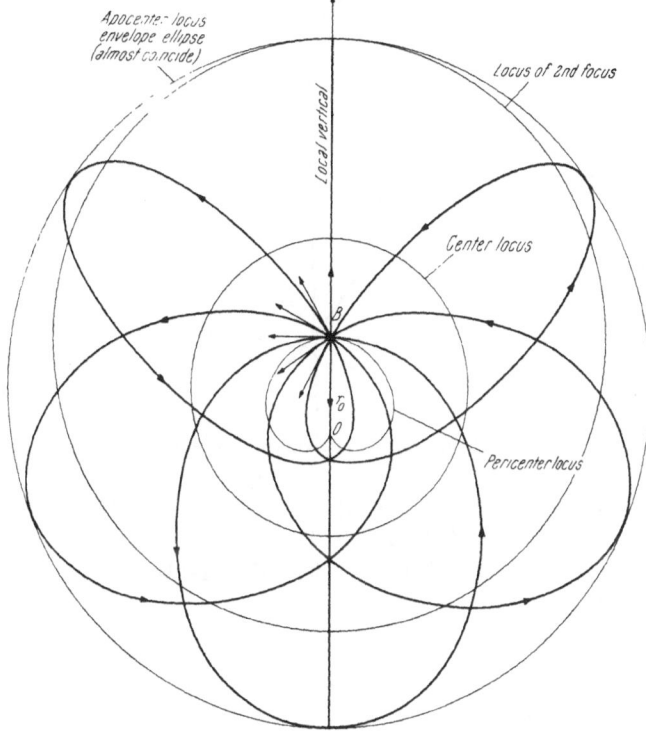

Fig. 4c. Elliptic orbits under varying angle of departure, $\lambda_0{}^2 = \dfrac{3}{4}$

common focus at the origin. The conics will be ellipses, parabolas or hyperbolas according as λ_0 is less than, equal to, or greater than 1 as previously mentioned.

It is interesting to note that as $\varPhi_0$ varies from $-\frac{\pi}{2}$ to $+\frac{\pi}{2}$, the second focus, the center of the conic, and the vertices all move around closed curves respectively.

It can be easily proved that the locus of the second focus is a circle with the center at the burnout point, since, according to the property of the conic, the sum or the difference of the focal radii is constant; that is,

$$O'B \pm OB = 2a$$

where the upper sign is for the elliptic orbit, and the lower sign for the hyperbolic orbit. It follows that the radius of the circle is

$$O'B = 2a \mp r_0 = \frac{r_0 \lambda_0{}^2}{|1 - \lambda_0{}^2|}. \tag{31}$$

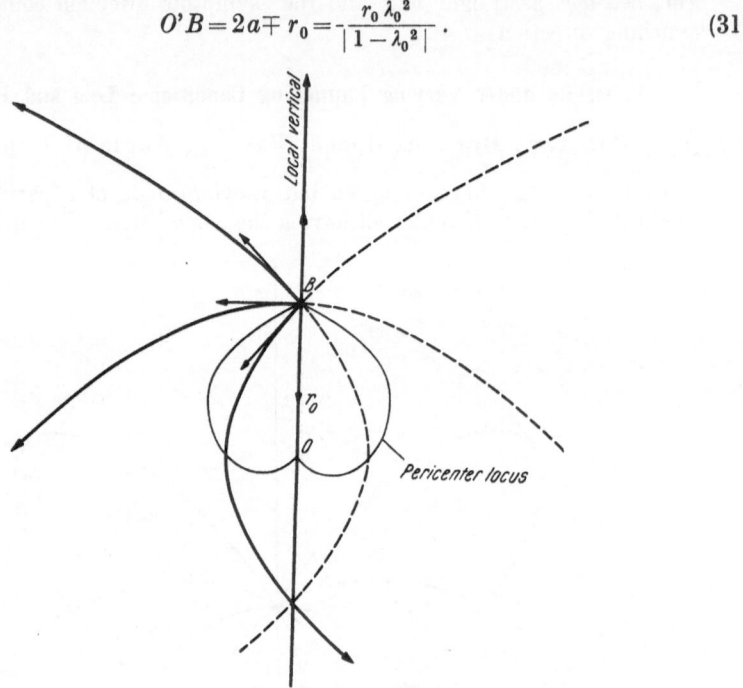

Fig. 4d. Parabolic orbits under varying angle of departure, $\lambda_0{}^2 = 1$

Similarly the center of the conic is also a circle with center at the middle point of OB, and radius equal to half of $O'B$. The locus of the vertices may be obtained analytically from eqs. (14), (17), etc. and its equation is found to be

$$\cos \theta = \frac{2(1 - \lambda_0{}^2)\left(\dfrac{r}{r_0}\right)^2 - 2\dfrac{r}{r_0} + 1}{2(1 - \lambda_0{}^2)\dfrac{r}{r_0} - 1}. \tag{32}$$

The curve given by eq. (32) in general consists of two closed portions, one for the apocenter A, and one for the pericenter P. The form of the curve depends on the value of λ_0. When $\lambda^2{}_0 < \frac{1}{2}$, the two portions are external to each other, both are bounded by the straight lines of $\theta = \pm \sin^{-1} \frac{\lambda_0{}^2}{1 - \lambda_0{}^2}$, conforming to eqs. (29) given in the previous section. When $\lambda^2{}_0 > \frac{1}{2}$, the apocenter locus encloses the pericenter locus. When $\lambda_0{}^2 = \frac{1}{2}$, the two curves unit to become a cardioid given by

$$r = r_0 (1 + \cos \theta). \tag{33}$$

At critical speed, $\lambda_0 = 1$, the orbit becomes a parabola, the second focus, the center of the conic, and the apocenter all lie in infinity, and the pericenter locus, according to eq. (32), becomes also a cardioid given by

$$r = \frac{1}{2} r_0 (1 + \cos \theta) . \qquad (34)$$

The various loci at some typical values of λ_0 are shown in Figs. 4a to e.

Finally in the case of elliptical orbits an envelope is found to enclose all the

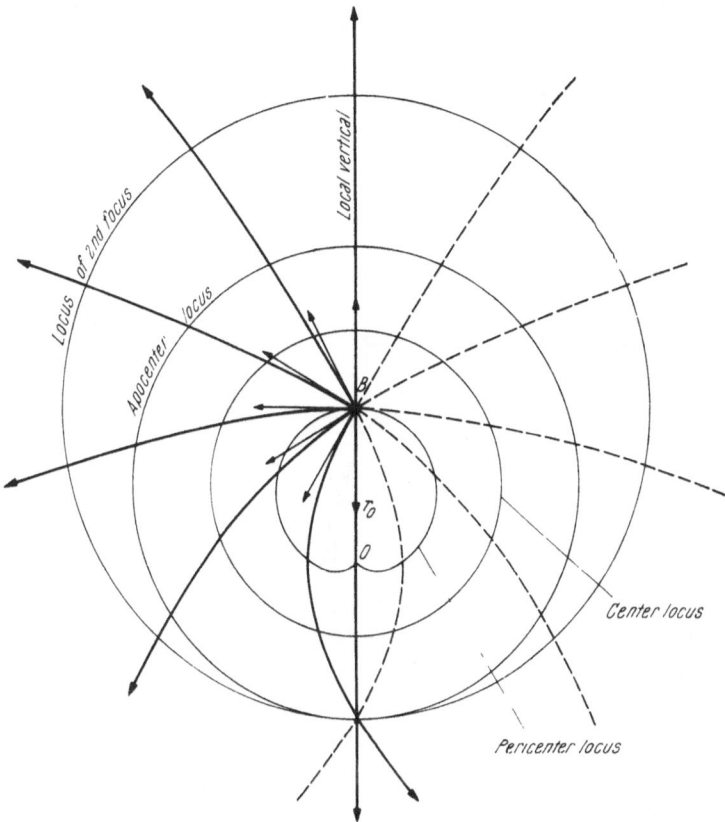

Fig. 4e. Hyperbolic orbits under varying angle of departure, $\lambda_0^2 = 2$

orbits of the family. To find the equation of the envelope, it is more convenient to write eq. (5a) with the aid of eqs. (14) and (15) in the form

$$f (r, \theta, \Phi_0) = \frac{r_0}{r} - \frac{1 - \cos \theta}{2 \lambda_0^2 \cos^2 \Phi_0} - \frac{\cos (\Phi_0 + \theta)}{\cos \Phi_0} = 0 . \qquad (35)$$

Setting $\dfrac{\partial f}{\partial \Phi_0} = 0$, and eliminating Φ_0 from the equations yields

$$\frac{r}{r_0} = \frac{2 \lambda_0^2/(1 - \lambda_0^4)}{1 - \dfrac{1 - \lambda_0^2}{1 + \lambda_0^2} \cos \theta} . \qquad (36)$$

Thus the envelope is also an ellipse with its foci situated at the burnout point and the origin, and its eccentricity equal to $(1 - \lambda_0^2)/(1 + \lambda_0^2)$. However, the envelope itself does not belong to the family of the orbits.

Constant Angle of Departure, Varying Burnout Speed

When the projectile is launched at a given angle of departure, with increasing speed, the orbit will first be elliptic for $\lambda_0 < 1$, it becomes parabolic at $\lambda_0 = 1$, and turns out to be hyperbolic for $\lambda_0 > 1$. Since the tangent vector of a conic always bisects the angles formed by the focal radii, simple geometry will show that the second focus of the orbit will make a constant angle with the polar axis, equal to twice the polar angle of the launching direction. Thus as Φ_0 varies, the second focus moves along a straight line passing through the burnout point at the constant polar angle of $2\,\Psi_0$, where Ψ_0 is the angle the launching direction

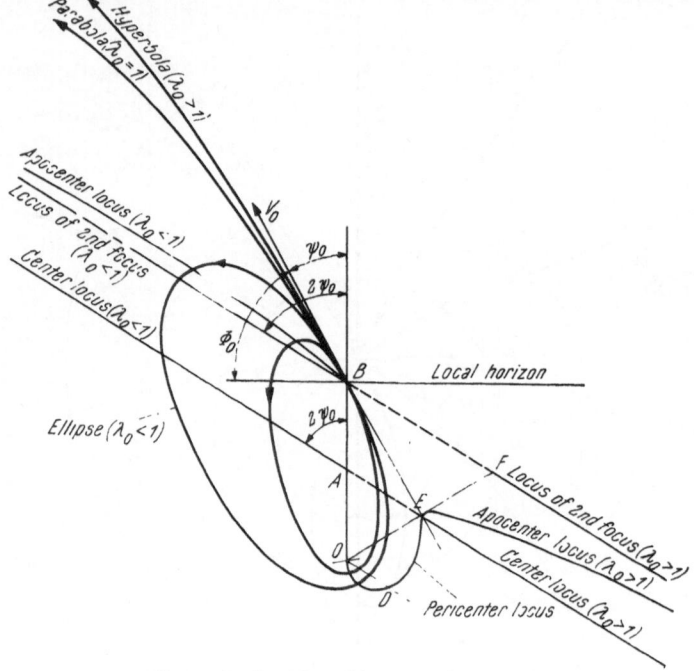

Fig. 5a. Inclined launching at various speeds

made with the local vertical, an angle complementary to Φ_0 (see Fig. 5a). Since the center of the conic is midway between the foci, it also moves along a straight line passing through the mid-point of OB and parallel to the locus of the second focus.

The locus of the vertices of the orbit may be obtained analytically from eqs. (14), (17) etc. Its polar equation is found to be

$$\cos \theta = \frac{(\sec^2 \Phi_0 - 2)\left(\dfrac{r}{r_0}\right)^2 + 2\,\dfrac{r}{r_0} - 1}{\left(\dfrac{r}{r_0}\right)^2 \sec^2 \Phi_0 - 2\,\dfrac{r}{r_0} + 1}. \tag{37}$$

As shown in Fig. 5a, the locus consists of two branches. When λ_0 increases from zero to 1, the apocenter moves along the upper branch from the burnout point toward infinity; while the pericenter moves along the lower branch from the origin to D, which is the vertex of the parabolic orbit. As λ_0 further increases, the orbit changes to hyperbola, and the pericenter continues to move from D to E, while the apocenter moves along the same branch from infinity also toward E. At infinite speed both coincide at E, which is the foot of the perpendicular drawn from the origin to the straight orbit.

In horizontal launching all these loci become a straight line coinciding with the polar axis (Fig. 5b).

VI. Burnout Precision in Satellite Launching

In the light of foregoing analyses the burnout precision in satellite launching may now be considered.

The most important thing in satellite launching is to keep the vehicle not too close to the mother planet; thus the pericenter radius is the most critical

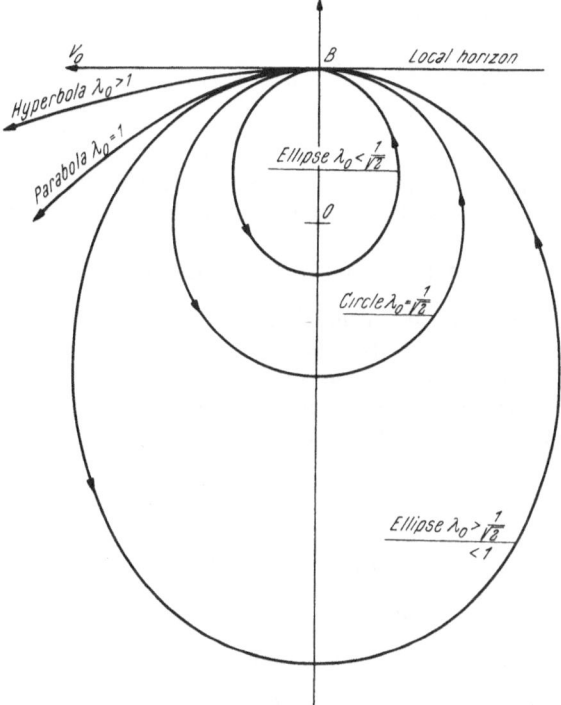

Fig. 5b. Horizontal launching at various speeds

	$\lambda = 0$ to 1	$\lambda_0 = 1$ to ∞
2nd Focus: st. line	B to infinity	Infinity to F
Center: st. line	A to infinity	Infinity to E
Apocenter: curve	B to infinity	Infinity to E
Pericenter: curve	O to D	D to E

parameter to be watched. Normally a satellite is to be launched at the circular velocity in the horizontal direction. Suppose the direction is correct, the deviation of r_p from r_0 due to the deviation in burnout speed may be found as follows. Since in horizontal launching the excess in speed will not change the pericenter radius, only the deficiency in speed needs to be considered. For horizontal launching at speed less than the circular speed, eq. (17a) gives

$$\frac{r_P}{r_0} = \frac{\lambda_0{}^2}{1 - \lambda_0{}^2} \qquad \lambda_0{}^2 < \frac{1}{2} \, . \tag{17a—1}$$

Thus

$$\frac{d\left(\frac{r_P}{r_0}\right)}{d\lambda_0} = \frac{2\lambda_0}{(1-\lambda_0{}^2)^2} \, . \tag{38}$$

When λ_0 is close to $\dfrac{1}{\sqrt{2}}$, eq. (38) gives

$$\frac{\varDelta r_P}{r_0} \cong 4\,\frac{\varDelta\lambda_0}{\lambda_0} = 4\,\frac{\varDelta V_0}{V_0} \qquad \begin{matrix} \varDelta\lambda_0 < 0, \\ \varDelta V_0 < 0. \end{matrix} \tag{39}$$

Thus 1 % reduction in speed will cause a 4 % reduction in pericenter radius[1]. In launching from an altitude of 400 km from the earth surface, this amounts to a reduction of 271 km in perigee altitude (earth radius $= 6370$ km).

Furthermore, since $V_0 \cong \sqrt{\mu/r_0}$ eq. (39) may be written as

$$\varDelta r_p \cong \frac{4}{\sqrt{\mu}}\, r_0{}^{\frac{3}{2}}\, \varDelta V \qquad \varDelta V < 0. \tag{39a}$$

Thus, for a given velocity deficiency, the shortening of r_P is approximately proportional to the 3/2 power of the burnout radius. In the gravitational field of earth, μ may be taken as 4×10^5 km^3/sec^2. At the burnout altitude of 400 km, eq. (39a) gives

$$\varDelta r_P \text{ (in km)} \cong 0.975\,\varDelta V \text{ (in km/hr)}.$$

Thus a deficiency of 1 km/hr in burnout speed corresponds to a shortening of approximately 1 km in perigee altitude.

Suppose the burnout speed is correct, the deviation of pericenter radius due to the deviation from horizontal direction may be found as follows. For launching at the circular speed, eq. (17a) gives

$$\frac{r_P}{r_0} = 1 - |\sin\varPhi_0|. \tag{17a—2}$$

It follows that

$$\frac{d\left(\frac{r_P}{r_0}\right)}{d\varPhi_0} = \pm\cos\varPhi_0. \tag{40}$$

The plus sign is used for $-\dfrac{\pi}{2} < \varPhi_0 < 0$ (launching direction inclined downwards), and the minus sign is used for $0 < \varPhi_0 < \dfrac{\pi}{2}$ (launching direction inclined upwards). For small deviation, $\varPhi_0$ is close to 0, eq. (41) gives[1]

$$\varDelta r_P \cong -r_0\,|\varDelta\varPhi_0|. \tag{41}$$

Thus a deviation from horizontal launching will always cause the shortening of the pericenter radius no matter it is inclined upwards or downwards. In launching at 400 km altitude from the earth surface, a $\pm 1°$ in direction will shorten the perigee altitude by 120 km.

Sometimes the determination of the error in apocenter radius due to the error in velocity or direction may also be desired. Since in horizontal launching the deficiency in velocity will not change the apocenter radius, only excess in velocity needs to be considered. For horizontal launching at the speed greater than the circular speed the apocenter radius, according to eq. (17a), is

---

[1] Same result was previously found by Blitzer, Weisfeld and Wheelon with a little different approach (see [4]).

$$\frac{r_A}{r_0} = \frac{\lambda_0^2}{1 - \lambda_0^2} \qquad \lambda_0^2 > \frac{1}{2} \tag{18a—1}$$

which is identical to r_P/r_0 for the case of $\lambda_0^2 < \frac{1}{2}$. Thus a differentiation of this equation leads to the similar results:

$$\frac{\Delta r_A}{r_0} \simeq 4\frac{\Delta\lambda_0}{\lambda_0} = 4\frac{\Delta V}{V} \qquad \begin{matrix} \Delta\lambda_0 > 0 \\ \Delta V_0 > 0 \end{matrix} \tag{42}$$

or

$$\Delta r_A \simeq \frac{4}{\sqrt{\mu}} r_0^{\frac{3}{2}}\Delta V \qquad \Delta V > 0. \tag{42a}$$

Thus in launching from 400 km altitude, a 1 % excess of velocity will lengthen the apogee altitude by 271 km, or 1 km/hr excess of velocity will lengthen it by 1 km.

Furthermore, at the correct speed but incorrect direction, eq. (42) applies equally to the apocenter radius but with the opposite sign, since the semi-major axis a is unchanged by directional variation. Thus

$$\Delta r_A = +r_0|\Delta\Phi_0|. \tag{43}$$

This shows the deviation from horizontal launching will always lengthen the apocenter radius by the same amount the pericenter radius is shortened.

VII. The Energy-Momentum Diagram

An examination of eq. (5) shows that the orbit in a central gravity field is uniquely determined by the two parameters: the orbital energy K and the angular momentum J. Thus a pair of values of K and J are sufficient to determine the geometry of the orbit. In other words, all the geometrical aspects may be shown in a single Energy-Momentum diagram. In the following such a diagram is presented. For convenience the dimensionless energy X and the dimensionless angular momentum Y are used instead of K and J; they are defined by

$$\begin{cases} X = \dfrac{r_0 K}{\mu} & (44) \\[2mm] Y = \dfrac{J}{\sqrt{r_0\mu}}. & (45) \end{cases}$$

Through eqs. (6), (7) etc. all the principal geometrical parameters may be expressed in terms of X and Y, such as

$$\varepsilon = \sqrt{1 + 2XY^2} \tag{46}$$

$$\frac{\bar{r}}{r_0} = Y^2 \tag{47}$$

$$\frac{r_P}{r_0} = \frac{1}{2X}(\sqrt{1 + 2XY^2} - 1) \tag{48}$$

$$\frac{r_A}{r_0} = \frac{1}{2|X|}(\sqrt{1 + 2XY^2} + 1) \tag{49}$$

$$\frac{a}{r_0} = \frac{1}{2|X|} \tag{50}$$

$$\frac{b}{r_0} = \frac{Y}{\sqrt{2|X|}} \tag{51}$$

$$\frac{c}{r_0} = \frac{\sqrt{1 + 2XY^2}}{2|X|} \tag{52}$$

and so on. Besides, the launching parameters are connected to X and Y through eqs. (9a) and (10) as follows:

$$\lambda_0^2 = X + 1 \tag{53}$$

$$\cos \Phi_0 = \frac{Y}{\sqrt{2(X+1)}} \qquad \left(-\frac{\pi}{2} \leqslant \Phi_0 \leqslant \frac{\pi}{2}\right). \tag{54}$$

With the aid of these equations, a momentum-energy diagram is constructed as shown in Fig. 6. In order to simplify the diagram, the curves are plotted

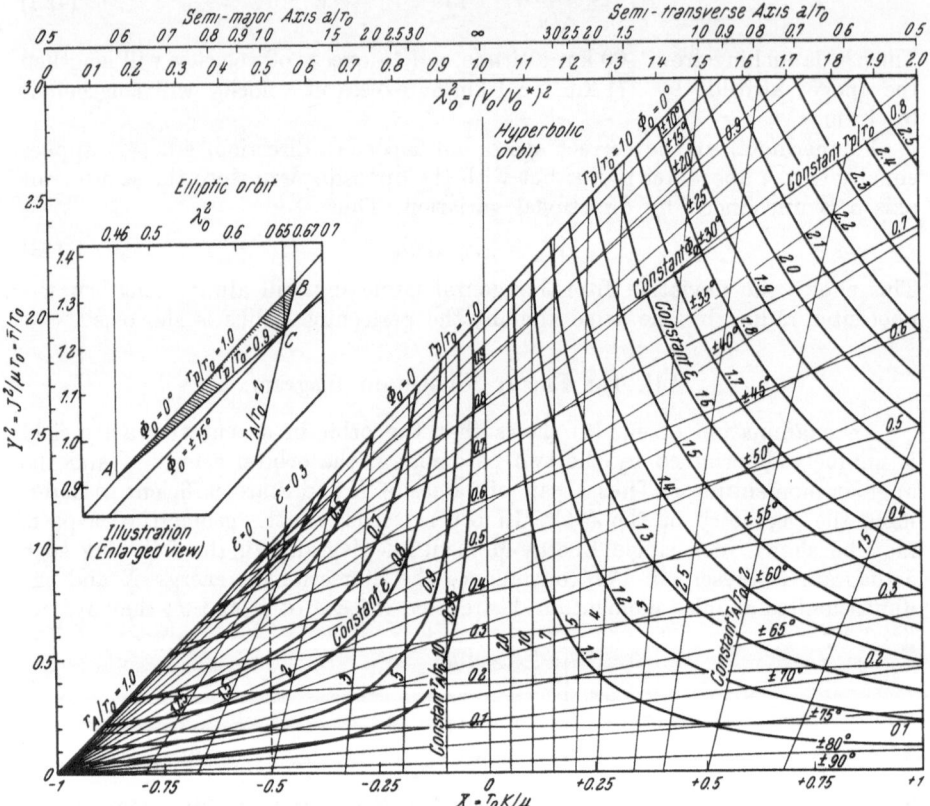

Fig. 6. Energy-momentum diagram

with X versus Y^2, and only the most essential geometrical parameters are shown. It is to be noted that eqs. (53) and (54) are subjected to the restrictions: $\lambda_0 \geqslant 0$ and $0 \leqslant \cos \Phi_0 \leqslant 1$, in other words,

$$X \geqslant -1, \quad 0 \leqslant Y \leqslant \sqrt{2(X+1)}.$$

Thus only the portion of the XY-plane bounded by the lines $X = -1$, $Y = 0$ and $Y^2 = 2(X+1)$ are physically possible. As the diagram is constructed without referring to earth in particular or any other gravity field, it is universal as long as the field is of the inverse-square type.

As an illustration, suppose a projectile is launched from the earth at the burnout speed of 9.6 km/sec and an angle of 15° from the horizon at an altitude of 850 km, then, taking the earth radius = 6370 km, we have

$$\lambda_0^2 = r_0 V_0^2/2\mu = (6370 + 850)(9.6)^2/8(10)^5 = 0.833.$$

As shown in Fig. 6, the orbit will be an ellipse with the following characteristics:

Eccentricity $\qquad \varepsilon \quad = 0.69$

Apogee radius $\qquad r_A/r_0 = 5,$ $\qquad r_A = 36{,}100$ km or 29,730 km altitude

Perigee radius $\qquad r_p/r_0 = 0.92,$ $\qquad r_p = 6{,}640$ km or 270 km altitude

Semi-major axis $\qquad a/r_0 = 2.96,$ $\qquad a = 21{,}370$ km

Semi-latus rectum $\qquad \bar{r}/r_0 = 1.55,$ $\qquad \bar{r} = 11{,}200$ km.

As another example, suppose an earth satellite is to have its r_p/r_0 not to be less than 0.9, and its r_A/r_0 not to exceed 2, then, from Fig. 6 (left corner) the allowed maximum variation in λ_0^2 is from 0.46 to 0.67 at horizontal launching, corresponding to a speed variation from 7.1 to 8.6 km/sec at an burnout altitude of 850 km. The maximum directional deviation allowed is about $\pm 15°$ from the local horizon (at $\lambda_0^2 = 0.65$). The combined variation in burnout speed and angle of departure allowable is represented by the points in the shaded triangular area ABC in Fig. 6. To determine the margins in speed and angle for orbits of low eccentricity will require more accurate plottings than the elementary form illustrated here.

References

1. C. Cranz, Lehrbuch der Ballistik, pp. 22-28. Berlin: Springer, 1925.
2. R. A. Becker, Introduction to Theoretical Mechanics, pp. 223-236. New York: McGraw-Hill, 1954.
3. C. D. Baker and J. J. Hart, Maximum Range of a Projectile in a Vacuum. Amer. J. Physics 23, 253 (1955).
4. L. Blitzer, M. Weisfeld and A. D. Wheelon, Perturbations of a Satellite's Orbit Due to the Earth's Oblateness. J. Appl. Physics 27, 1141 (1955).
5. S. F. Singer and R. C. Wentworth, A Method for Calculating Impact Points of Ballistic Rockets. Jet Propulsion 27, 407 (1957).
6. R. E. Roberson, Orbital Behavior of Earth Satellites. J. Franklin Institute 264, No. 3, 181 (1957).
7. L. Gold, Aspect of High Energy Ballistics. J. Franklin Institute 264, No. 4, 301 (1957).
8. S. F. Singer and R. C. Wentworth, A Method for Calculating Impact Points of Ballistic-Rockets: Convenient Representations. Jet Propulsion 28, 684 (1958).
9. G. S. Gedeon and R. E. Dawley, The Influence of the Launching Conditions on the Orbital Characteristics. Jet Propulsion 28, No. 11 (1958).

Design Compromises in Space Power Systems

By

M. A. Zipkin[1, 2] and **E. Schnetzer**[1, 3]

(With 16 Figures)

(Received August 27, 1959)

Abstract — Zusammenfassung — Résumé

Design Compromises in Space Power Systems. An adequate supply of power is essential to electrical space propulsion systems. Since this power supply also represents a major portion of the total propulsion system's weight, its design can control the accelerating capability and utility of the propulsion system and space vehicle.

The design of a space power supply, like that of any earthbound engine, represents a series of compromises among many interacting system parameters and components. The purpose of this paper is to investigate the power conversion equipment of the space power supply and to indicate some of the design compromises required in selecting specific designs. Static conversion and turbomachinery systems are compared. The type of cycle and the primary cycle variables of working fluid, temperature and pressure are analyzed and their effects on major system components (i.e., turbine and radiator) are indicated. The effects of the unusual space design requirements of hermetically sealing, meteorite interception, heat rejection by radiation, and zero gravitational field on the system design are also shown.

The report concludes with the design of a nuclear electric one megawatt, two-phase, liquid metal system. This power supply is compact, compatible with space environment, and light-weight.

Kompromisse im Entwurf elektrischer Krafterzeugungsanlagen für Raumschiffe. Hinreichend große Mengen elektrischer Energie bilden die Voraussetzung für elektrische Raumantriebssysteme. Da die Krafterzeugung den Großteil des gesamten Antriebgewichtes darstellt, wird ihr Entwurf bestimmend für die Beschleunigungsfähigkeit und Wirtschaftlichkeit des Antriebssystems und des damit ausgerüsteten Raumschiffes.

Der Entwurf einer Raumkraftzentrale stellt ebenso wie der irgendeiner erdgebundenen Maschinenanlage einen Kompromiß dar zwischen vielen einander gegenseitig beeinflussenden Faktoren und Komponenten. Der Zweck dieser Abhandlung ist, die Kraftumwandlungsausrüstung des Raumkraftsystems zu untersuchen und einige Entwurfskompromisse, die bei der Auswahl spezifischer Entwürfe nötig sind, aufzuzeigen. Statische Umwandlung und Turbogeneratoren werden miteinander verglichen. Der geeignetste Wärmekreisprozeß und seine wichtigsten Größen, wie Temperatur- und Druck-Niveau, sowie die Wahl der Arbeitsflüssigkeit werden besprochen und ihr Einfluß auf wichtige Bauelemente wie Turbine und Kühler nachgewiesen. Der Einfluß der ungewöhnlichen Konstruktionsbedingungen für ein Wärmekraftwerk im

---

[1] Flight Propulsion Laboratory Department, General Electric Company, Cincinnati 15, Ohio, U.S.A.

[2] Advanced Propulsion Systems, FPLD.

[3] Space Power Systems, FPLD.

Weltraum, wie hermetisches Abschließen, Schutzmaßnahmen gegen Meteoriten, Wärmeabgabe durch Strahlung und Schwerelosigkeit wird ebenfalls gezeigt.

Der Bericht schließt mit dem Entwurf einer typischen Atomkraftanlage, mit einem durch Kaliumdampf getriebenen Turbogenerator für 1 MW (= 1000 KW) elektrische Leistung. Diese Stromquelle ist klein, gedrängt und leicht und den Bedingungen des Weltraumes angepaßt.

Les compromis dans les systèmes de propulsion dans l'espace. Une source de puissance suffisante est essentielle aux systèmes de propulsion électriques dans l'espace. Puisque cette source représente aussi une portion majeure du poids total du système de propulsion, sa conception peut contrôler les possibilités d'accélération et l'utilité du propulseur ainsi que celle du véhicule spatial.

Le dessin d'un propulseur spatial, comme celui de n'importe quel moteur terrestre, représente une série du compromis entre plusieurs systèmes et organes en interaction. L'objet de cet article est d'examiner les appareils possibles pour la conversion de puissance de la source et d'indiquer certains des compromis requis pour des conceptions spécifiques. Les convertisseurs statiques et les turbomachines sont comparés. Le type de cycle et la nature des fluides du cycle primaire, leur température et pression sont analysés et leurs effets sur les organes principaux du système (tel que turbine et radiateur) sont indiqués. Les effets sur le système des exigences spéciales telles que la fermeture hérmetique, l'interception de météorites, l'élimination de chaleur par radiation et le champ gravitationnel nul sont aussi montrés.

Une présentation du plan d'un système nucléaire électrique à métal liquide d'un mégawatt à deux phases termine le rapport. Le système est compact, compatible avec le milieu spatial et de poids réduit.

I. Summary

The design of a space power supply, like that of any earth-bound engine, represents a series of compromises among many interacting system parameters and components. The purpose of this paper has been to investigate the turbogenerator power conversion equipment of the space power system and to indicate some of the compromises required in selecting specific designs. The most significant results of these studies, carried out in the General Electric Flight Propulsion Laboratory, are indicated below.

1. In addition to the conventional design requirements and limitations, space power systems must be compatible with the unique limitations of the space environment. The most critical limitation imposed by this environment is the rejection of all unavailable energy by radiation and the attendant large radiator areas vulnerable to meteoroid penetrations.

2. The resultant need to minimize radiator area favors cycles that can reject heat isothermally at the highest practical temperature. Consequently, the Rankine cycle, although more complex mechanically, is better suited to space environment than is the Brayton cycle.

3. The radiator for waste heat rejection should be as small and as light in weight as possible, consistent with an acceptable probability of achieving system design life in a meteoroid environment. Minimum size can be obtained by raising the condensing temperatures in the cycle to the highest practical value at the cost of some loss in cycle efficiency. Significant reductions in radiator vulnerability for a given weight can be achieved by employing finned tube radiator configurations.

4. Selection of the cycle working fluid is dependent upon the state-of-the-art in allowable temperatures and pressures. For a liquid metal system, where near-future temperatures are anticipated to approach 2100° R with refractory metal

construction, potassium appears to be the best compromise among the potentially available working fluids.

5. The turbogenerator design is a compromise among the conflicting requirements of high efficiency, low generator weight, low stress levels and low blade erosion rates. These requirements result in a multi-stage turbine design of moderate rotational and pitchline speeds.

6. Utilizing the compromises indicated for the principal system components and parameters, a 1 MW power supply has been designed. With the selection of potassium as a working fluid, a four-stage molybdenum turbogenerator operating at 12,000 RPM has been designed. The vapor is generated in a boiler heated by hot liquid metal coming from the heat source. An appropriate pump with necessary valves and piping would interconnect the components. This system has the over-all dimensions of 12 ft. in diameter and 70 ft. in length. The largest component is the radiator, having an effective area of 2000 sq. ft. The over-all weight of the system, including the heat source, is 13,000 lbs.

The system appears technically feasible and the weight estimates conservatively based on a large number of design studies. It should be emphasized however, that much development work must be done before the system becomes a reality.

II. Introduction

Space power supplies which can be visualized for the near future will consist of an energy source and a conversion mechanism which converts part of this energy into electrical output and rejects the unavailable heat generated by the source into space. The purpose of this paper is to investigate the power conversion equipment of the space power supply and to indicate some of the compromises required in selecting specific designs. Although discussions of the conversion equipment cannot be entirely divorced from the energy sources with which they may be coupled in space, considerable insight can still be obtained by considering the problems of the conversion system separately. A complete discussion of the possible combinations of energy sources and conversion systems is a paper in itself and is beyond the time allowed for this presentation.

The most promising systems at the present state of technology employ conversion of heat to mechanical energy and subsequently, by means of a generator, into electrical energy. Although static conversion systems for direct conversion of heat to electricity, such as thermoelectric and thermionic devices, are under development, present devices are low in efficiency, high in weight and further in the future than turbogenerator systems. Considerable development must still be done before thermoelectric and thermionic devices will be practical, particularly in larger power sizes. Consequently, this presentation is further limited to turbogenerator systems and their design compromises when built to operate in a space environment. In this paper the unique design requirements of the space power system are established, the principal types of conversion systems are reviewed and the major design considerations are indicated. A typical 1 MW alkali metal vapor system is then described.

III. Unique Design Requirements

In addition to all of the conventional design requirements and limitations, space power systems must be designed to be compatible with the new restrictions imposed by launch considerations and by the space environment. The major additional requirements are listed below. Their effect on the over all system and specific components will become apparent in the later sections of the paper.

A. Launching

Compact

Power supply must be compact in order to be compatible with reasonable size launching vehicles.

Light Weight

Launching rocket size and system cost are determined by the payload. Consequently, light weight power supplies can reduce launching costs very significantly.

High Acceleration

Launching of the power supply may impose acceleration forces in the order of fifteen times the acceleration of gravity on the system.

B. Space Environment

Hermetic Sealing

Inasmuch as the power supply operates in essentially a vacuum, it must be completely hermetically sealed to prevent loss of the working fluid.

Heat Rejection by Radiation

The lack of an atmosphere also precludes the possibility of convective heat transfer and requires transfer of all unavailable energy by radiation.

Meteorite Interception

Heat rejection by radiation will require large exposed areas that will be vulnerable to penetration by high-speed meteorites. Unless these radiating surfaces can be made compatible with the space environment, the power supply will have insufficient operating life to warrant launching it into space.

Liquid/Vapor Separation

The lack of a gravitational force causes difficulty whenever two phases of the working medium exist simultaneously. During either boiling or condensing, artificial means must be provided to separate the two phases.

Long Life Without Maintenance

Because of the initial time and effort required to launch the space vehicle, the power supply must be capable of a long, reliable life. This long life must be achieved in a system in which there is no opportunity for maintenance or repair.

IV. Principal Types of Conversion Systems

The closed cycle turbogenerator systems discussed in this paper may be classified under two distinct groupings according to the thermodynamic cycle employed in the respective power plants. In one cycle an alkali metal vapor is used as the working fluid and the power plant operates on a RANKINE Cycle in much the same manner as present-day steam turbogenerators. In the other power plant an inert gas is employed and the BRAYTON Cycle common to gas turbine conversion systems is used. Both of the power plants employ a turbine as the prime mover and in both systems waste heat is rejected to space by radiating surfaces.

A. The Brayton Cycle

Fig. 1 shows an inert gas cycle diagram and a schematic arrangement of the power plant. A single flow loop with helium gas as the working fluid is used in

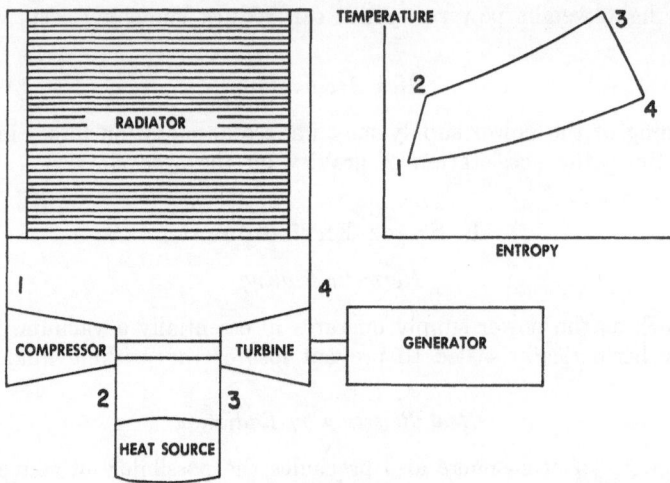

Fig. 1. Cycle diagram and schematic arrangement of inert gas power supply

this cycle. Compression is accomplished by an axial flow compressor (1—2); heat is then added at constant pressure from a suitable heat source (2—3). Gases are then expanded through a turbogenerator (3—4) and the waste heat is rejected in a radiator (4—1) at a continuously decreasing temperature.

The design of an illustrative 1 MW inert gas conversion system is shown in Fig. 2. The turbomachinery is on a single shaft, and rotates at 24,000 RPM. The whole assembly is contained in a single casing, with the heat source and generator attached at either end. Ducting losses are thus minimized and hermetic sealing is made possible. The hermetically sealed shell, containing the reactor, generator and turbomachinery, is 17 ft. long and 1 ft. in diameter at the narrowest section.

The use of helium as the working fluid results in a large number of compressor and turbine stages because of its low atomic weight. The compressor requires 41 stages, while 12 are needed for the turbine. The turbomachinery is complex, but the sizes and weights are small and no insurmountable development problems are foreseen. The compressor design is rendered more practical by the fact that all stages are nearly identical.

All of the rotating machinery runs on gas bearings lubricated by compressor discharge helium. The turbine rotor is the most critical component of the assembly. Although it operates at slightly lower temperatures than the heat source, it is under much higher stress levels due to its high tip speed. For 10,000 hours of operating life, creep of the turbine buckets becomes the limiting factor in the design.

The alternator shown is a brushless induction generator, which produces three-phase alternating current at 6,000 volts and 400 cycles.

The radiator, not shown, is comprised of three rectangular sections 57 ft. long and 49 ft. wide attached tangentially at one edge to an 8 ft. diameter cylinder and equally spaced about its periphery. The total effective radiating area is 14.500 sq. ft. Each section of the radiator is a tube sheet consisting of a large number of small tubes. Entrance and exit headers for the tube sheets are located inside the 8 ft. cylinder. The radiator is wrapped around the central cylinder for

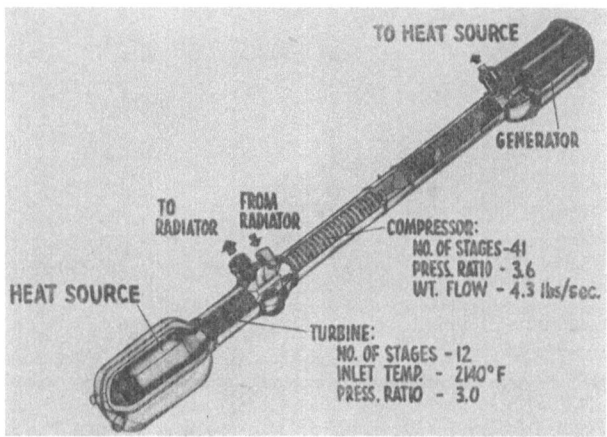

Fig. 2. Inert gas one megawatt power supply. Rotative speed, 24,000 RPM; working fluid, helium

launching and, when the constraints are released, unwraps due to its elasticity.

The principal attractive features of the BRAYTON Cycle are:

1. The inherent simplicity of a single loop.

2. The corrosion-free atmosphere provided by the inert gas will allow use of uncoated high temperature refractory alloys without fear of corrosion or oxidation.

3. Similarly, the inert gas system should also be erosion-free because there will be no solid particles in the working fluid.

4. Starting, controlling and restarting of the system will be relatively simple.

5. Gas bearings, actuators and valves for this system can be developed without major advances in technology.

The major disadvantages of the BRAYTON Cycle are:

1. Since most of the heat of the cycle is not added at the highest temperature or rejected at the lowest temperature of the cycle, as indicated on Fig. 1, efficiency is considerably lower than for cycles more closely approaching the CARNOT Cycle.

2. Considerable pumping power is required in the compression process. While this can be minimized by reducing the radiator outlet temperature, a significant amount of the turbine power must be used for the compression process.

3. The continuously decreasing temperature in the radiator and the low radiator outlet temperature increase the heat rejection problem. Since the only means of heat rejection in space is by radiation, these factors result in very large requirements for radiating area.

B. The Rankine Cycle

The alkali metal two-phase system is shown schematically in Fig. 3. It comprises two separate flow loops, the first of which serves to transfer heat from the heat source to the boiler. The alkali metal in this loop remains in the liquid phase.

The fluid in the secondary loop receives heat from the boiler and is alternately vaporized and condensed through its flow path. Referring to the cycle diagram shown in Fig. 3, heat is added at constant pressure in the boiler until the condition of saturated liquid (2) is reached and is then vaporized at constant temperature

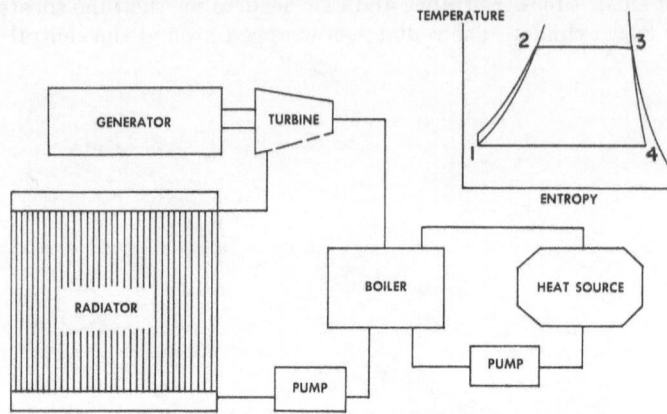

Fig. 3. Cycle diagram and schematic arrangement of liquid metal vapor power supply

and pressure (2—3). Expansion through the turbine occurs from 3—4 and the wet vapor is then condensed at essentially constant pressure and temperature in the radiator (4—1).

Fig. 4 shows an illustrative 1 MW two-phase system consisting of a heat source, boiler, turbine and generator. The boiler is a tube and shell type heat

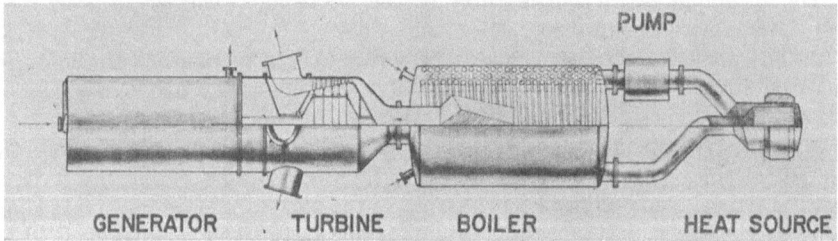

Fig. 4. Alkali-metal-vapor one megawatt power supply. Rotative speed, 12,000 RPM; working fluid, potassium

exchanger. The turbine is a 5-stage impulse type built of molybdenum. The alternator is of the brushless inductor type and generates a 2000 cycle per second current at 3000 volts. The turbogenerator and pumps have hydrodynamic-type bearings lubricated with liquid metal.

After the working fluid has left the turbine it is condensed in the radiator, not shown. The radiator is of the fin and tube type. The total effective radiating area is 2000 sq. ft. The condensate is returned to the boiler by means of a traction-type pump driven by a partial admission Curtiss turbine.

The principal advantages of the Rankine Cycle are:

1. Since most of the heat is added isothermally and all of the waste heat is rejected isothermally, the efficiency of the Rankine Cycle approaches that of the Carnot Cycle much more closely than does the Brayton Cycle.

2. The isothermal rejection of waste heat is particularly desirable in a radiation environment. The heat rejection temperatures can also be considerably higher in this cycle than for the Brayton Cycle.

The principal disadvantages of the RANKINE Cycle are:

1. The cycle is mechanically complex. In actual cycles two or more loops, possibly using different working fluids, may be required.

2. The corrosion and erosion problems associated with metal vapors will limit allowable turbine inlet temperature to lower values than for inert gas cycles and may also adversely affect system life.

3. The requirements for starting, controlling and possibly restarting in space are considerable and may require complex auxiliary systems.

In comparing the two systems for space applications, it is apparent that neither one is perfect. It will therefore be necessary to consider the design problems associated with each of the systems, as well as the design compromises within each of the systems, in order to determine which system is most desirable and what configuration it should take.

V. Major Design Considerations

The major decisions that must be made in the design of the conversion equipment for a space power system are the type of cycle, the radiator design, the working fluid and the turbine configuration. Cycle selection is primarily dependent upon radiator size and weight, working fluids by the allowable temperatures and pressures, and the turbogenerator configuration by compromising among efficiency, low stress, long life and minimum weight. Examples of these design compromises, as applied to a 1 MW system, will be discussed in the following sections.

A. Radiator

The radiator, when designed, to be compatible with the space environment of heat rejection by radiation and low vulnerability to meteoroid penetration, becomes a major portion of the size and weight of the power conversion system. The primary design objectives for the radiator are: small area, small volume, light weight and continuous operation without loss of working fluid, due primarily to meteoroid penetrations.

1. Area

Radiator area is a function of cycle efficiency, which determines the amount of heat to be rejected; and radiator temperature, inasmuch as this heat must be rejected in accordance with the fourth power radiant heat transfer law. The desire for high cycle efficiency results in a requirement for low radiator temperatures in order to minimize the amount of heat to be rejected. Since the heat rejected, however, is by the process of radiation, it would be desirable to maintain the highest possible temperature in the radiator. The result of a conflict between the desires for high efficiency and high radiator temperature results in optimization study that minimizes radiator area for any given set of cycle conditions.

Shown in Fig. 5 is the variation of the required radiator area as a function of radiator outlet temperature for several turbine inlet temperatures. On the left side of the figure are shown the variation of radiator area with radiator outlet temperature for inert gas systems. Inert gas systems are limited to relatively low turbine outlet temperatures because of the high pumping power requirements. On the right side of the figure are shown similar variations in area for metal vapor systems. The characteristic shape of the curves at constant turbine inlet temperature indicates that a minimum area corresponding to a particular radiator outlet temperature occurs for each turbine inlet temperature. It is apparent from

Fig. 5 that at the same turbine inlet temperature the minimum radiator area required for an inert gas system is at least an order of magnitude greater than for the metal vapor system. Although the values of the maximum turbine inlet tem-

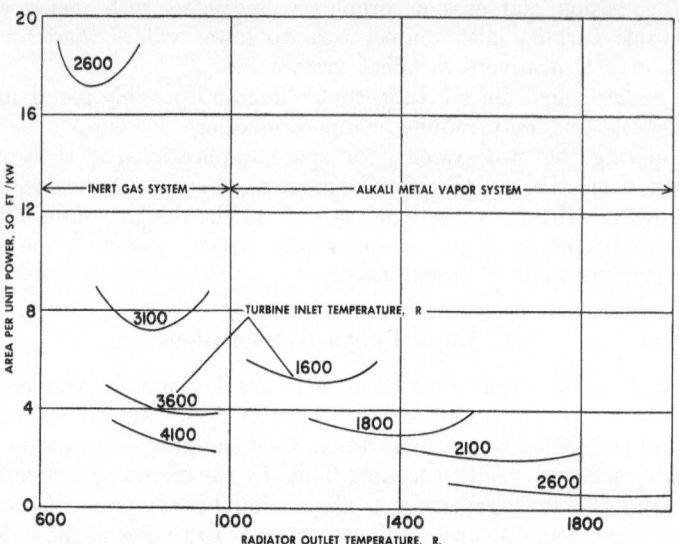

Fig. 5. Radiator area per unit power versus radiator outlet temperature. Sink temperature, 550° R

peratures of inert gas and metal vapor systems obtainable in the near future are not necessarily equal, it is apparent that substantially less radiator area per KW of power is required for the metal vapor systems than for the inert gas systems.

For example, assume near future technology will allow maximum cycle temperatures of 2600° R and 2100° R for the inert gas and liquid metal systems respectively. Then the required radiator area would be 17 sq. ft./KW for the inert gas systems and 2 sq. ft./KW for the liquid metal systems—a ratio of 8.5:1 in favor of the liquid metal systems.

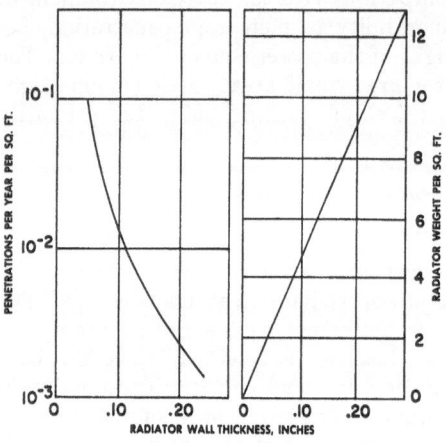

Fig. 6. Effect of radiator wall thickness on radiator vulnerability and radiator weight. Tube material, steel

2. Weight and Vulnerability

In addition to radiator area, radiator weight consistent with an acceptable vulnerability in a space environment is another factor that must be evaluated. Since radiator areas for useful power levels are quite large, the radiator must be designed to prevent loss of working fluid by means of meteoroid penetrations for an acceptable period of time. Using the data on meteoroid penetration of F. W. Whipple, the probable number of penetrations in 10,000 hours per sq. ft. of radiator area in a steel tubular radiator is plotted as a function of the radiator wall thickness in Fig. 6. Radiator weight as a function of tube wall thickness is also shown in this figure.

Although very thin wall thicknesses are adequate from stress considerations in the radiator, it is apparent from the figure that relatively heavy walls must be used if the number of punctures is to be kept within tolerable limits. For example, a 1000 sq. ft. radiator designed for a probability of two punctures in 10,000 hours would require a wall thickness of 0.21 inches and would weigh 8600 lbs., an enormous weight for such a small radiator. Fortunately, there are methods available to the designer that give somewhat more complicated radiator designs, but greatly reduce radiator weight. A technique useful for designing light weight radiators that are still relatively invulnerable utilizes finning, such as shown in Fig. 7. Here relatively few tubes are used and sheet metal fins between the tubes transfer most of the heat. Cycle working fluid passes from the large hot header through the tubes to the cold header. Heat is transferred from the fluid to the tube wall and to the fin, where it is simultaneously conducted down the fin and radiated to space from the fin surface. A meteoroid punc-

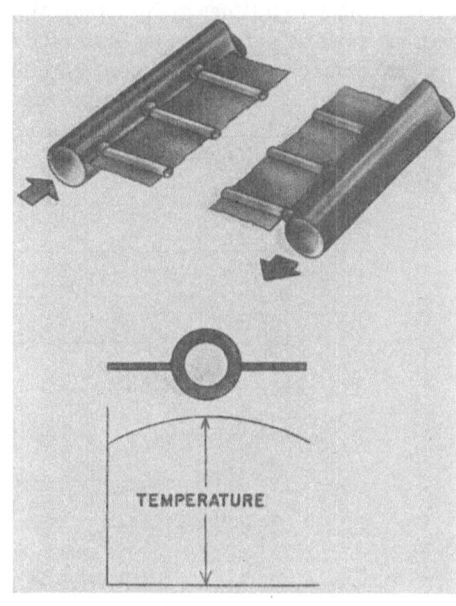

Fig. 7. Schematic of finned tube radiator design

turing the fin does not disturb its heat transfer. Tube diameters of a size sufficient to pass the required flow with a low pressure drop are used; the tube walls themselves are relatively heavy to withstand striking meteorites.

A comparison between variation of radiator weight per unit effective area of tube and tube-fin radiators is shown in Fig. 8 as a function of meteoroid penetrations in 10,000 hours. A fin-tube radiator of 1000 sq. ft. effective area designed for the probability of two penetrations in 10,000 hours weighs approximately 2300 lbs., one-fourth of the unfinned radiator. These radiator weights are based on the use of steel tubes and steel clad with copper fins.

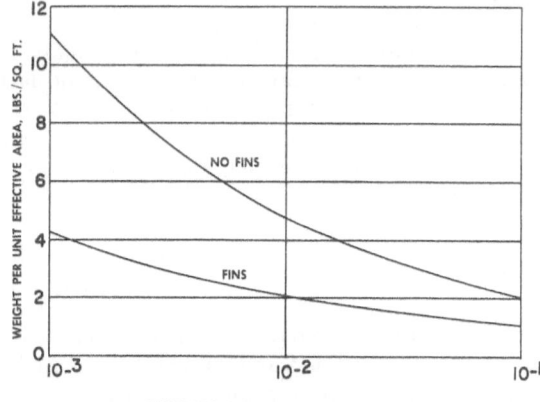

Fig. 8. Effect of radiator vulnerability on radiator weight with and without fins. Tube material, steel; fin material, copper with steel cladding

The use of materials that have a high ratio of thermoconductivity to mass density, such as beryllium, would reduce the weight of a fin-tube radiator substantially below the values shown in the chart.

If the data on area per unit power output of Fig. 5 and weight per unit of area

of Fig. 8 are combined, the specific weight of the radiator as a function of vulnerability can be obtained and is shown in Fig. 9 for inert gas and metal vapor systems.
Turbine temperatures consistent with near future technology have been assumed
in both cases. For equal vulnerability the specific weight of the inert gas radiator
varies from 8 to 5 times greater than that of metal vapor radiators as the number
of penetrations per year per kilowatt increases from 10^{-3} to 10^{-1}.

The number of probable penetrations to be allowed for in a given design will
be dependent upon the application and life expectancy of the particular space

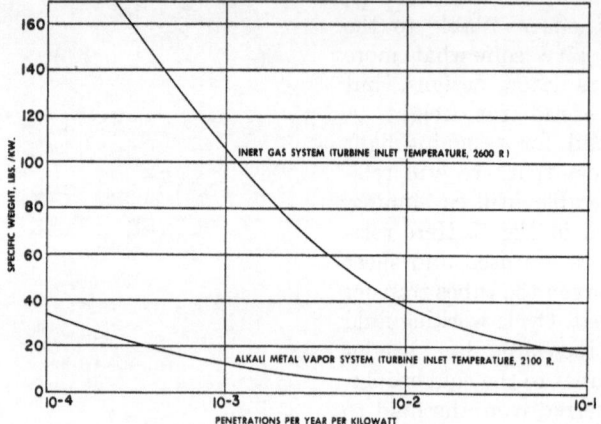

Fig. 9. Comparison of radiator vulnerability, inert gas and metal-vapor systems. Tube material, steel;
fin material, copper with steel cladding

vehicle. It is anticipated, however, that values not exceeding 10^{-2} will be the
maximum tolerable for most applications. Consequently, it can be readily seen
that for equal vulnerability the radiator weight for the alkali metal vapor system
is substantially less than that for the inert gas system. Moreover, since the radiator
may be a major portion of the total vehicle weight, these large differences will
also be required in launching system weights. As a result and in spite of added
complexity of the alkali metal system, the latter is considered to be the most
desirable for near future applications. As additional data become available on the
space environment, it may be desirable to reevaluate the systems. In the discussion
of design considerations that follows, however, only the alkali metal system will
be discussed.

B. Selection of Operating Temperatures, Pressures and Working Fluid

There are a large number of parameters to be considered in choosing the operating levels of pressure and temperature and working fluid for a two-phase system.
The most important, however, to the turbomachinery are the vapor characteristics
of the fluid. Shown in Fig. 10 is the relationship between vapor pressure and
temperature for a number of possible working fluids. The area of interest for any
design under consideration can be established by setting realistic upper and lower
limits on system pressures and temperatures. The maximum temperature in the
cycle is set by the maximum allowable turbine inlet temperature and for this
study will be assumed to be 2100° R. The minimum, or sink, temperature in the
cycle should be as high as possible in order to minimize radiator size and weight.
The upper pressure limit for practical systems appears to be in the order of 400 psia.

Higher pressures are feasible, but result in somewhat heavier systems. Lower pressure limits appear to be in the order of about 1 psia. Lower pressures in the system result in very large ducting and large diameter rotating machinery.

Establishing temperature and pressure limits results in identifying the most appropriate working fluids. From the figures, it is apparent that for the design limits chosen the alkali metals appear most suitable. The design conditions are beyond the regions where water might look attractive, where mercury is limited, and below those for lithium, bismuth or lead.

Among the possible fluids, rubidium and cesium are not very well known, are expensive, heavy, and require a great deal of technical development. Much more technology is available for sodium and potassium. Sodium has been used more extensively than potassium in the past because it is a better reactor cooling fluid and is somewhat cheaper. Use of sodium, however, restricts the maximum working pressure for any maximum available cycle temperature. Consequently, sodium systems have larger ducts, larger radiator tubes and larger diameter turbines. It appears, therefore, that at least for the conversion loop potassium is the best compromise among the possible working fluids.

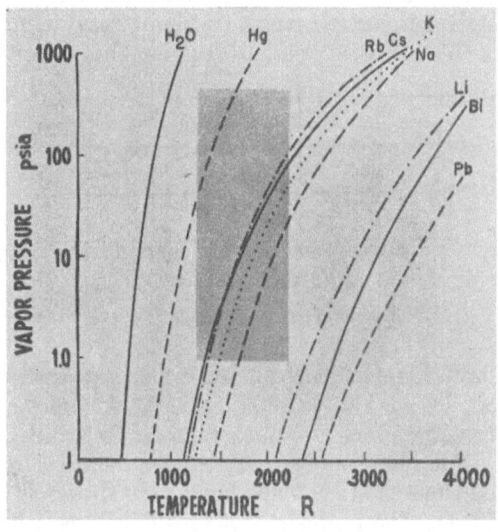

Fig. 10. Vapor-pressure-temperature characteristics of several working fluids

C. Turbogenerator

The power turbine is the most highly stressed, as well as nearly the highest temperature component of the space power plant. Because of the long life requirements of the power plant, creep of the turbine buckets becomes the limiting factor and the turbine design must be biased toward as low stress levels as possible while still giving high effi-

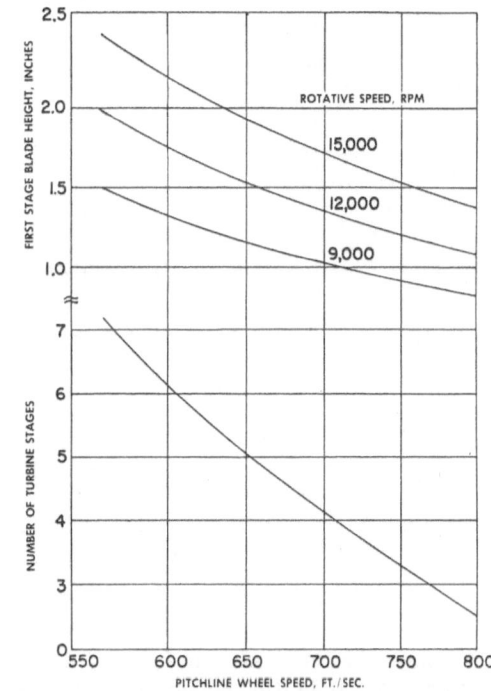

Fig. 11. Variation of required number of turbine stages with pitchline wheel speed

ciency. In addition, the turbine speed must be such as to favor low weight in the alternator. Thus, the turbine for the space power supply must be compromised between the conflicting requirements of high efficiency, low stress levels, low erosion rates, and minimum generator weight.

A family of multi-stage turbines has been designed which demonstrates the interrelationship among turbine stress, number of stages and alternator weight. In this series of related designs, the velocity diagrams for the first stages are identical on a non-dimensional basis. In the latter stages of these turbines, however, the level of through-flow velocity was adjusted so as to maintain a constant value of the ratio of annulus areas between adjacent stages. In each design each stage was designed at peak efficiency for its wheel speed and

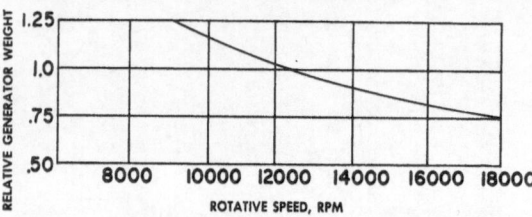

Fig. 12. Variation of generator weight with rotative speed

nozzle angle and the pitchline velocity was held constant. The family of turbines was designed to produce 1.25 MW for inlet conditions of 2100° R and 46 pounds per square inch absolute temperature and pressure respectively.

The number of turbine stages required to produce the needed turbine output and the first stage blade height are shown as functions of pitchline wheel speed in Fig. 11. The number of stages decreases with increasing wheel speed, as would be expected. Blading height is shown to vary inversely with wheel speed in feet per second and directly with rotational speed in RPM. Although minimum wheel heights are not large (approximately 1 inch), they are not small enough to cause any particular concern in this power size. The weight of the generator, as a function of rotational speed, is shown in Fig. 12 and is essentially inversely proportional to RPM. Consequently, the turbine designer would favor a compact light-weight, high rotational speed and high pitchline speed machine. There are certain limits imposed upon the design, however, by the last stage of the turbine. Because of the large increase of specific volume of the potassium vapor in passing through the

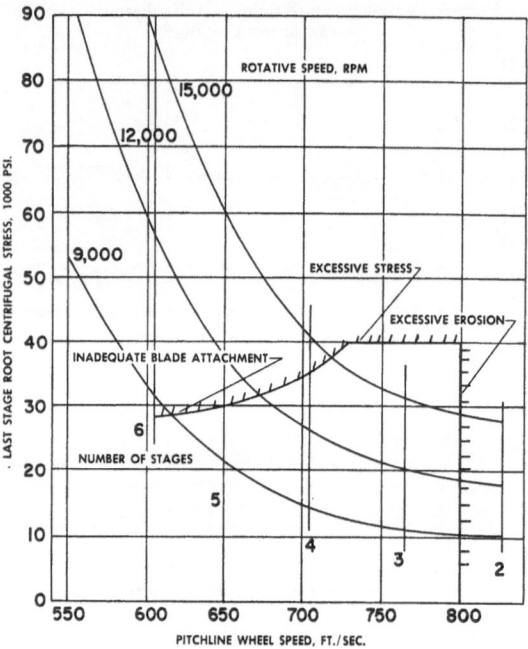

Fig. 13. Limitations on turbine design

turbine, the required annulus area for the last stage becomes very large. This produces two limiting effects. First, the hub to tip diameter ratio becomes so large that the hub may be too small for adequate blade attachment. Second, the blade root centrifugal stress becomes very large and may exceed allowable limits. Shown in Fig. 13 is the variation of root centrifugal stress with wheel speed for the last stage of the turbine. Also shown in the figure are the limiting conditions that must be considered before selecting a design. The upper horizontal line represents the allowable stress limit imposed by the temperature of the working medium and the selected material. To the left in Fig. 13 is shown the region in which the hub-tip ratio becomes so low that adequate blade attachment is not possible. The vertical line to the right represents the limit of pitchline speed above which erosion would be excessive. For the one megawatt power system a four-stage turbine operating at 12,000 RPM represents a conservative compromise.

VI. Description of 1 MW System

Fig. 14 shows the over-all arrangement of an alkali metal twophase 1 MW power supply. The over-all dimensions of the system are 12 ft. maximum diameter and 70 ft. in length. The largest component of the system is the conical radiator having 2000 sq. ft. effective area including a sub-cooler for generator cooling. The radiator is made up of a number of 2-inch diameter tubes spaced 3 inches on center, and interconnected by 0.11 inch fins. Depending on mission requirements and vehicle configuration, the power supply may be allowed to rotate or may be required to remain stationary relative to the vehicle. A rotating

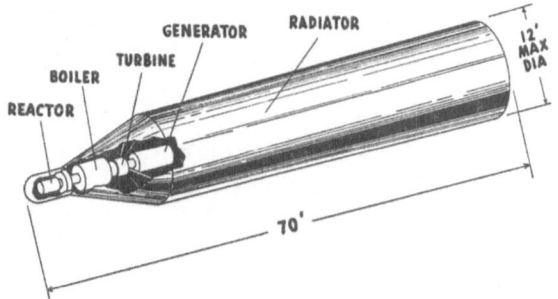

Fig. 14. Overall arrangement, two phase power supply. Power output, 1 MW

power supply simplifies the problems of liquid metal and vapor separation in the boiler and condenser. Many missions, however, will require a fixed non-rotating system. In such cases, viscous forces or internal centrifugal flow separators will be required to accept liquid vapor separation. The heat source and turbogenerator assembly are quite small relative to the over-all power supply dimensions and are located at one extreme end of the power pack. If the heat supply is a nuclear reactor, adequate space is thus provided for shadow shielding and a manned or instrument compartment may be attached at the opposite end of the power supply.

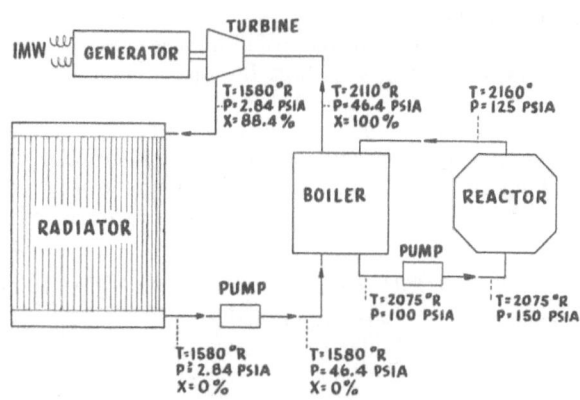

Fig. 15. Schematic diagram, two phase power supply. Working fluid, potassium

A schematic design, including working pressures and temperatures throughout the system, is shown in Fig. 15. The system contains two main loops. In the first, high temperature, high pressure, liquid metal transports heat from the heat source to the boiler heat exchanger. In this boiler the fluid flowing in the second loop is completely vaporized. The vapor then expands through a turbine which drives the electric generator and the pumps in the system. The vapor is then condensed in the radiator and is pumped back to the boiler. The significant temperatures in the system are a turbine inlet temperature of 2100° R and a radiator condensing temperature of 1500° R. Pressure at the turbine inlet is approximately 46 psia; pressure at the radiator outlet is approximately 2.8 psia. The weight flow of potassium is 300 lbs./sec. in the first loop and 6 lbs./sec. in the second loop.

The cycle shown is simplified in that it does not indicate the additional loops that are required for generator cooling and bearing lubrication, as well as an inert gas supply required for pressurizing the primary loop.

Fig. 16 shows the boiler, turbogenerator assembly designed for a rotating system. The boiler consists of a tube bundle containing the hot liquid metal coming

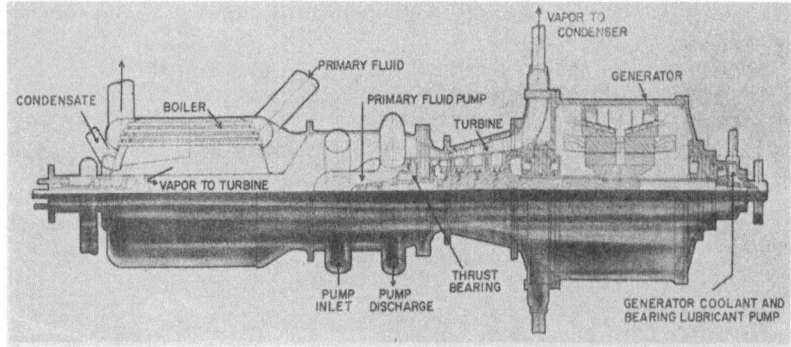

Fig. 16. Boiler, turbine, generator and pumps, two phase power supply

from the heat source. The tube bundle is surrounded by a cylindrical pressure vessel containing the working fluid of the second loop at a lower pressure. Because of the rotation of the entire system, the liquid metal remains at the periphery while the vapor collects at the center of the boiler. The vapor passes through a baffle system before it enters the turbine. These baffles serve the purpose of separating out any droplets of liquid that may not have been completely vaporized.

The turbine is a 4-stage impulse type built of molybdenum. It operates at 12,000 RPM. In as much as the turbine operates in the wet region, some condensation will occur within the turbine stages and special precautions must be undertaken in order to prevent erosion. For this reason, relative velocities are moderate and special provisions have been made to extract any condensed liquid after each turbine stage. The electric generator is of the brushless inductor type,

Table I. *Weight Estimate, Two Phase Power Supply. Power Output, 1 MW*

Reactor, unshielded	1000 lbs.
Boiler	300
Turbine	300
Pumps	150
Electric generator—12,000 RPM	4500
Controls	200
Ducting	250
Radiator, 2″ × 112″. SS—Cu, 10 punctures in 10,000 hrs.	5000
Structure	700
Liquid metal in system	600
Total	13000

generating 2000 cps AC at 3000 volts. The alternator-stator is cooled by liquid metal condensate in the secondary loop. The turbogenerator and pumps run on hydrodynamic liquid metal bearings.

Table I shows a weight breakdown for this two-phase system. The components that constitute major weight items in the system are the electric generator and the radiator. The total system weight, including these items and the heat source, but not including any shielding, is 13,000 lbs. These weights are based on a considerable number of design studies and are believed to represent conservative values.

Compressible Flat Plate Boundary Layer of an Electrically Conducting Fluid in the Presence of Magnetic Fields

By

Luigi G. Napolitano[1] and A. Pozzi[2]

(With 8 Figures)

(Received July 22, 1959)

Abstract — Zusammenfassung — Résumé

Compressible Flat Plate Boundary Layer of an Electrically Conducting Fluid in the Presence of Magnetic Fields. The characteristics of the boundary layer generated by an electrically conducting fluid over a flat plate are known only for incompressible fluids. The subject paper deals with the corresponding problem for compressible fluids.

The fundamental equations are written in their form valid for small values of the magnetic REYNOLDS number. The inviscid flow solution is determined first. Subsequently existence and limits of applicability of first integrals of the energy equations are investigated and discussed. The STEWARTSON-ILLINGWORTH transformation and concomitant hypotheses are then used to reduce the equations to "quasi-incompressible" forms. Series solutions in terms of a suitable "magnetic parameter" are found for both momentum and energy equations.

Results of the above outlined analysis are presented in terms of velocity profiles (determined up to third order terms in the "magnetic parameter"), enthalpy profiles (determined up to second order terms), skin friction coefficients, recovery factors and NUSSELTS number for the following situations of interest:

1. magnetic field fixed with respect to the plate
2. magnetic field fixed with respect to the free stream
3. electric conductivity as a function of the absolute temperature.

The characteristics of the subject magneto-fluid-dynamic boundary layer are finally compared with those pertinent to the conventional flat plate boundary layer of compressible, electrically conducting fluids. Conclusions and deductions therefrom are discussed.

Über die verdichtbare Grenzschicht einer elektrisch leitenden Flüssigkeit. Die Eigenschaften der Grenzschicht, die von einer elektrisch leitenden Flüssigkeit über einer flachen Platte erzeugt wird, waren nur für unverdichtbare Flüssigkeiten bekannt. Der Vortrag behandelt das entsprechende Problem für verdichtbare Flüssigkeiten.

Die fundamentalen Gleichungen sind in der für kleine Werte der magnetischen REYNOLDS-Zahl gültigen Form gegeben. Die Existenz und die Grenzen der ersten Integrale der Energiegleichungen werden untersucht und diskutiert. Die STEWARTSON-ILLINGWORTH-Transformationen und die damit verbundenen Hypothesen werden verwendet, um die Gleichungen zur „quasi-unverdichtbaren" Form zu reduzieren.

[1] Associate Professor of Aerodynamics, University of Naples, Department of Aeronautics, Naples, Italy.

[2] Research Associate, University of Naples, Naples, Italy.

Reihenlösungen für Bewegungsgröße und Energiegleichungen werden für einen geeigneten „magnetischen Parameter" gefunden.

Resultate der oben ausgeführten Analyse werden in Form von Geschwindigkeitsprofilen (bis zur dritten Ordnung in „magnetischen Parametern" bestimmt), Enthalpieprofilen (bis zur zweiten Ordnung bestimmt), Oberflächenreibungskoeffizienten, Wiedererlangungsfaktoren und Nusselt-Zahl für folgende interessante Situationen gegeben:

1. ein in Beziehung zur Platte festgesetztes magnetisches Feld,
2. ein in Beziehung zum freien Strom festgesetztes magnetisches Feld,
3. die elektrische Leitfähigkeit als eine Funktion der absoluten Temperatur.

Die Charakteristiken der magneto-hydro-dynamischen Grenzschicht werden schließlich mit denjenigen, die zu der konventionellen Flachplatten-Grenzschicht einer nichtleitenden Flüssigkeit gehören, verglichen und / oder mit denjenigen, die zu der Grenzschicht einer verdichtbaren, elektrisch leitenden Flüssigkeit passen. Schlußfolgerungen und Ableitungen daraus werden diskutiert.

La couche limite d'un fluide compressible électriquement conducteur. Les caractéristiques d'une couche limite de fluide électriquement conducteur, connues pour le cas incompressible, sont analysées pour le cas compressible.

Les équations fondamentales utilisées sont valides pour une faible valeur du nombre de Reynolds magnétique. La solution sans viscosité est d'abord établie. L'existence et les limites d'applicabilité d'intégrales premières des équations d'énergie sont ensuite examinées. La transformation de Stewartson-Illingworth et des hypothèses associées sont alors utilisées pour ramener les équations à la forme "quasi-compressible". Les équations de quantité de mouvement et d'énergie sont résolues en série dans un paramètre magnétique.

Les résultats sont présentés sous forme de profils de vitesse (jusqu'au troisième ordre dans le paramètre magnétique), profils d'enthalpie (jusqu'au second ordre), coefficients de frottement, facteurs de récupération et nombre de Nusselt dans les conditions suivantes:

1. champ magnétique fixe relativement à la plaque
2. champ magnétique en translation avec le courant amont
3. conductibilité électrique fonction de la température absolue.

Ces caractéristiques sont comparées avec celles de la couche limite conventionnelle d'un fluid compressible électriquement conducteur. Les déductions et conclusions de cette comparaison sont discutées.

Symbols and Definitions

e	specific internal energy	H	magnetic field
f, g	non-dimensional stream functions	J	current density vector
h	specific static enthalpy	L	reference length
m	magnetic parameter, eq. (18c)	M_0	Mach number
p	static pressure	P_m	magnetic pressure number, eq. (4d)
s_0	zeroth order enthalpy functions		
s_{1i}	first order enthalpy functions, eq. (32)	P_r	Prandtl number
		Q	microscopic heat flux vector
t	time	R	gas constant
u, v	velocity components	R_e	Reynolds number
v	velocity vector	R_m	Magnetic Reynolds number, eq. (4c)
x, y	space coordinates		
B	magnetic induction vector	S	stagnation enthalpy
B_0	imposed magnetic field	T	absolute temperature
C_f	skin friction coefficient	U, V	velocity components in the Stewartson plane
C_h	heat transfer coefficient		
E	electric field	T_0	initial temperature
G_i	velocity function, eq. (31)	T_{aw}	adiabatic wall temperature

V_0	reference velocity	ξ	non-dimensional parameter, eq. (4e)
X, Y	Stewartson-plane variables		
α, β	non-dimensional quantities defined in eqs. (18a, 18b)	ϱ	density
		σ	electrical conductivity
γ_0	reference specific heat ratio	τ	stress tensor
δ^+	displacement thickness in Stewartson-plane	φ, χ	non-dimensional parameters, eqs. (4a) and (4b)
η	similarity variable	ψ	stream function
λ	heat-conduction coefficient		$\dfrac{\partial}{\partial r} = \dfrac{\partial}{\partial x}\, \bar{\imath} + \dfrac{\partial}{\partial y}\, \bar{\jmath} + \dfrac{\partial}{\partial z}\, \bar{k}$
μ	viscosity coefficient		
μ_e	magnetic permeability		$\dfrac{D}{Dt} = \dfrac{\partial}{\partial t} + v \cdot \dfrac{\partial}{\partial r}.$
ν	kinematic viscosity		

Introduction

The study of mechanically dissipative regions of electrically conducting fluids in the vicinity of solid boundaries and in the presence of applied magnetic fields has received wide attention in the last couple of years and has already yielded some important results and conclusions. Solutions of the complex magneto-fluid-dynamic system of equations have been obtained, in general, by making one ore both of the following simplifying assumptions: a) the fluid-dynamic equations are uncoupled from the Maxwell's field equations; b) the fluid has constant properties.

Progresses in the understanding of the complicated patterns of interaction between fluid-dynamic fields and electromagnetic fields require the progressive abandonment of the above mentioned assumptions.

The main concern of the present paper is to remove the constant property assumption in the study of the flow of electrically conducting fluids along flat plates.

This study, besides adding to the knowledge of the characteristics of dissipative flows of electrically conducting fluids will furnish some new elements for the evaluation of the overall worthiness of fluid-magnetic interaction schemes as devices to reduce heat-transfer rates and to increase drags in connection with problems of hypersonic flights and re-entry.

In addition to this primary analysis, however, a number of side-investigations are carried out which aim to assess few fundamental questions arising in the study of magneto-fluid-dynamic dissipative flows.

These additional topics will be enumerated in what follows while briefly surveying the existing literature on the subject.

The flat-plate boundary layer for incompressible fluids has been treated by Rossow [1]. Compressibility effects in magneto-fluid-dynamic dissipative flows have been studied by Meyer [2] for stagnation flows; Bleviss [3] for Couette flow and by Napolitano [4] for the mixing of two semi-infinite streams. Some interesting properties of the pertinent equations were found in [3], and in [4]. It was deemed interesting, among other things, to see how many of these features derived from general properties of the basic equations rather than from the particular geometries considered.

One more interesting work in which compressibility was also accounted for should be mentioned. It is the paper by Lykoudis [5] which is wide in scope, by considering, as it does, several effects at the wall, such as: combustion, blowing, sublimation, dissociation, and so on. Some objections, however, have to be raised against the "similar" solutions whose existence was postulated in [5]. These objections are connected with the afore mentioned assumption (a). It implies some definite hypothesis on the order of magnitude of the non-dimensional param-eters entering the problem and, as a consequence, on the nature of the applied magnetic field. The bearing of these facts on the possibility of existence of similar flow fields is also discussed in the present work.

The basic magneto-fluid-dynamic equations are first subjected to order of magnitude analyses to assess the conditions under which the fluid dynamic equa-tions result uncoupled from the MAXWELL's equations and to establish the bound-ary-layer form of the equations.

Following a brief degression on the possibility of existence of similar solutions, the basic equations, in their uncoupled expressions, are specialized to the flat plate case, reduced to a quasi-incompressible form by means of the STEWARTSON transformation and solved, to a first approximation in the magnetic parameter for a unitary PRANDTL number and constant conductivity.

Results are presented in the form of first order universal functions for the velocity and total enthalpy profiles. A final section follows in which some of the results are proven to follow from general properties of the basic equations.

Finally the trends of the effects due to the magnetic fields are illustrated and discussed both in general and in relation to the present flow geometry.

The authors wish to thank GIOVANNI D'ELIA, Research Assistant at the Uni-versity of Naples, for his valid cooperation in obtaining the numerical solutions of the equations.

Basic Equations

In this section the boundary-layer approximation to the uncoupled fluid-dynamic equations will be derived and the conditions under which they hold will be discussed.

Under the assumptions that the fluid is perfect and in thermodynamic equilib-rium, that the reflection index of the medium is equal to one [6], that the motion is steady, and that relativistic effects are negligible the basic magneto-fluid-dynamic equations can be written as:

$$p = \varrho\, RT \tag{1a}$$

$$\frac{\partial}{\partial r} \cdot (\varrho v) = 0 \tag{1b}$$

$$\varrho\, \frac{Dv}{Dt} + \frac{\partial}{\partial r} \cdot [pU + \tau] = J \times B \tag{1c}$$

$$\varrho\, \frac{D}{Dt}\left(e + \frac{v \cdot v}{2}\right) + \frac{\partial}{\partial r} \cdot [Q + pv + v \cdot \tau] = J \cdot E \tag{1d}$$

$$\frac{\partial}{\partial r} \times E = \frac{\partial}{\partial r} \cdot E = 0 \tag{2a}$$

$$\frac{\partial}{\partial r} \times H = 4\pi J \tag{2b}$$

$$J = \sigma\, [E + v \times B]. \tag{2c}$$

Upon the further assumption that the cyclotronic frequency is much smaller than the frequency of collision, the electrical conductivity σ is a scalar [7] and the microscopic fluxes of momentum (τ) and heat (Q) are unaffected by the presence of electromagnetic fields and thus maintain their conventional expressions.

The fluid dynamic equations [eqs. (1)] and the electromagnetic equations [eqs. (2)] are strongly coupled and their simultaneous solution is a very difficult task even for the simplest geometries.

Two essential simplifications can be achieved upon an analysis of the order of magnitude of each term. The first one is based on a comparison of electromagnetic and fluid dynamic terms and will lead, under particular conditions, to the so-called "linearized" magneto-fluid-dynamic equations. The second one is based on the well known comparative analysis of the order of magnitude of microscopic and macroscopic fluxes and leads to the "boundary layer" form of the equations. To proceed in these analysis, we write the basic equations in a non-dimensional form as:

$$p^+ = \varrho^+ R^+ T^+ \tag{3a}$$

$$\frac{\partial}{\partial r^+} \cdot (\varrho^+ v^+) = 0 \tag{3b}$$

$$\varrho^+ \frac{Dv^+}{Dt^+} + \frac{\partial}{\partial r^+} \cdot \left[\frac{1}{\gamma_0 M_0{}^2} p^+ U + \frac{\tau^+}{R_e} \right] = \frac{P_m}{\chi} J^+ \times B^+ \tag{3c}$$

$$\varrho^+ \frac{D}{Dt^+} \left[e^+ + \gamma_0 (\gamma_0 - 1) M_0{}^2 \frac{v^+ \cdot v^+}{2} \right] + \frac{\partial}{\partial r^+} \cdot \left[\frac{\gamma_0}{P_r R_e} Q^+ + \right.$$

$$\left. + (\gamma_0 - 1) p^+ v^+ + \frac{\gamma_0 (\gamma_0 - 1)}{R_e} M_0{}^2 v^+ \cdot \tau^+ \right] = \xi J^+ \cdot E^+ \tag{3d}$$

$$\frac{\partial}{\partial r^+} \times E^+ = \frac{\partial}{\partial r^+} \cdot E^+ = 0 \tag{3e}$$

$$\frac{\partial}{\partial r^+} \times H^+ = \frac{4\pi J^+}{\chi} \tag{3f}$$

$$J^+ = \sigma^+ [\varphi E^+ + \chi R_m (v^+ \times B)] . \tag{3g}$$

The non-dimensional quantities, indicated with a superscript star, are defined in terms of the corresponding reference quantities (subscript zero). The only exceptions are the quantities r^+, τ^+, and Q^+ which are defined by:

$$r = L r^+ \qquad \tau = \frac{\mu_0 V_0 \tau^+}{L} \qquad Q = \frac{\lambda_0 T_0 Q^+}{L} .$$

The following new non-dimensional parameters appear in eqs. (3); besides those well known from conventional fluid-dynamics:

$$\varphi = \sigma_0 E_0 / J_0 \tag{4a}$$

$$\chi = H_0 / L J_0 \tag{4b}$$

$$R_m = \sigma_0 \mu_{e0} V_0 L \tag{4c}$$

$$P_m = \mu_{e0} H_0{}^2 / \varrho_0 V_0{}^2 \tag{4d}$$

$$\xi = \varphi P_m \gamma_0 (\gamma_0 - 1) M_0{}^2 / \chi^2 \quad R_m = J_0 E_0 L / \varrho_0 V_0 c_0 . \tag{4e}$$

These numbers are a measure of the relative importance of certain classes of phenomena in the over all balances expressed by eqs. (3). The numbers φ, χ and $R_m \chi$ measure the relative importance, with respect to the total current density, of the "ohmic" current, of the current due to the asymmetry of the magnetic field and of the current originated by magneto-fluid-dynamic interaction respectively. The number $P_m \chi^{-1}$ is a measure of the relative importance of electromagnetic body forces compared to inertia forces and, finally, the number ξ is a measure of the relative importance of the flux of electromagnetic energy compared to the flux of "mechanical" internal energy.

A comparative analysis of the order of magnitude of these numbers will make it readily possible to draw a number of conclusions on the general behavior of eqs. (3). Thus, for instance, the energy balance will involve only fluid-dynamic quantities not only when the imposed electric field is zero ($\varphi = 0$) [3, 4] or when $P_m \chi^{-1} \ll 1$[1] but also for low speed flows $(M_0{}^2 \ll 1)$[2].

In the subject case, however, the interest lies in the study of the possibility of uncoupling the fluid dynamic equations from the MAXWELL equations while still preserving the effects of the interactions between velocity and electromagnetic fields. A look at the basic equation shows that interaction effects are retained if the numbers $\dfrac{P_m}{\chi}$ and $R_m \chi$ are both of order of magnitude one. On the other hand, for the uncoupling to be possible, it is necessary and sufficient that the electromagnetic terms in the momentum and energy equations contain only the "imposed" electromagnetic fields[3].

In [9] it has been already shown that this, in turns, implies $R_m \ll 1$.

It proves convenient, however, to go over the analysis again in order to stress the very important point that the imposed field must be irrotational. Assume:

$$H^+ = H^{0+} + h^+$$

with $h^+ \ll 1$.

Letting, for simplicity, $E^+ = 0$, eqs. (3f) and (3g) yield:

$$\frac{\partial}{\partial r^+} \times H^{0+} + \frac{\partial}{\partial r^+} \times h^+ = 4\pi R_m \sigma^+ [v^+ \times B^{0+} + 0 \, (h^+)] \tag{5}$$

while the right hand side of eq. (3c) becomes:

$$J^+ \times B^+ = (v^+ \times B^{0+}) \times B^{9+} + 0 \, (h^+). \tag{6}$$

If the momentum equation has to be uncoupled from the MAXWELL's equations, it must be $R_m = 0 \, (h^+)$. Then H^{0+} is the imposed magnetic field and eq. (5) shows that $\dfrac{\partial}{\partial r^+} \times H^{0+} = 0$. As said, the imposed magnetic field must be "irrotational": *its components can only be harmonic functions of the coordinates.*

Then, by assuming, $R_m \ll 1$; $\chi = 0 \, (R_m)$; $P_m R_m \equiv 0 \, (1)$ one uncouples the fluid-dynamic equations from the MAXWELL's field equation, thus achieving a first essential simplification. We have now to investigate how the presence of the imposed magnetic field modifies the well known form of the boundary-layer approximation to the basic equations.

We consider the motion two-dimensional, (in the $x - y$ plane), assume the applied electric field to be zero, and apply the conventional order of magnitude analysis to eqs. (3). Since when the electric field is zero the energy equation has the same formal structure as in fluid-dynamics, one needs to investigate in detail only the momentum equation. It can be shown that, in the present case, its two components reduce to:

---

[1] No other case is possible since, by definition, $R_m \chi$ can be at most of order one.

[2] This is, of course, a trivial case since then e.m. terms disappear from the momentum equations also.

[3] It should be obvious that we are talking from the fluid-dynamics point of view. It is indeed possible to achieve a different type of "uncoupling" by requiring that the "coupling" term in the MAXWELL's equations contains only the "imposed" velocity field. This also calls for a "linearizing" process. This time, however, the linearization is made with respect to the velocity vector. The resulting equations can be used when one is interested in determining the changes in the magnetic field due to the presence of a velocity field. The order of magnitude of the parameters should be as follows:

$$\chi = 0 \, (1); \qquad R_m \chi = 0 \, (1); \qquad \frac{P_m}{\chi} \ll 1.$$

$$0 \, (1) + P_m R_m \, \sigma^+ \, \{u^+ \, [(B_2^{0+})^2 + (B_3^{0+})^2]\} = 0 \tag{7}$$

$$0 \, (\delta^+) + \frac{1}{\gamma_0 M_0^2} \, p_{y+}^+ - P_m R_m \, \sigma^+ \, u^+ \, B_1^{0+} \, B_2^{0+} = 0 \tag{8}$$

where the B_i^{0+} $(i = 1, 2, 3)$ are the components of the applied field and the symbols $0 \, (1)$ and $0 \, (\delta^+)$ include all the "mechanical" terms of the equations which are of order one and δ^+, respectively; δ^+ being the non-dimensional boundary layer thickness. Few important conclusions can be deduced from the above equations:

1) In general, it might no longer be true that $p_y = 0 \, (\delta^+)$ as in the conventional boundary layer[1].

2) In the present case, the imposed magnetic field cannot depend on z. Then upon the previous conclusions, the most general imposed field can only be given by[2]

$$H_1^{0+} = ay + d$$
$$H_2^{0+} = ax + b$$
$$H_3^{0+} = e \tag{10}$$

where a, b, d and e are constants. Thus p_y will certainly be of order δ^+ if $a = d = 0$ that is if the imposed field is constant. For either a or d different from zero the total change of pressure through the boundary layer will still be, at most, of order δ^+. However in this case the outer flow is no longer quasi-one-dimensional and some modification might be needed in the b. l. theory.

3) As it will be detailed later, the fact that the component H_2^{0+} of the imposed field can, at most, be a linear function of x strongly limits the possibilities of having similar flow fields.

4) To within the present approximations, the effects of applied magnetic fields of constant intensity are independent of the direction of the field itself (as long as it is normal to the free stream velocity, of course). Equal results are obtained, either with a magnetic field normal to the plate or with one parallel to it (span-wise).

To conclude: if $R_m \ll 1 \, [R_m \, P_m \equiv 0 \, (1)]$ of the basic magneto-fluid-dynamic system only the fluid dynamic equations need to be considered when one is interested in determining the velocity field. If, in addition, the imposed magnetic field is constant the usual boundary-layer form of the equations still hold with the only modification indicated by eq. (7).

The solution of these equations for the subject problem will be considered in the next section.

---

[1] The situation here is somewhat different from that treated in [3] (COUETTE flow). In terms of the present variables the two momentum equations for COUETTE flow [eqs. (10) and (11) of (3)] are:

$$\frac{\partial \tau_{xy}^+}{\partial x^+} = R_m P_m R_e \sigma^+ u^+ (B_2^+)^2$$

$$p_{y+}^+ = \gamma_0 M_0^2 R_m P_m \mu_e {}^+\sigma^+ u^+ B_2^+ h_1^+ . \tag{9}$$

Here B_2^+ is constant and coincides with the only component different from zero of the imposed field. The x-component of the magnetic field appearing in the second momentum equation is that due to the induced field. Thus when $R_m \ll 1$ it is again $p_y = 0$ since then also $h_1^+ \ll 1$. As it is pointed out in [3], however, in this particular case is it not necessary to assume $h_1^+ \ll 1$ (and thus $R_m \ll 1$) in order to uncouple the magneto-fluid-dynamic equations. The alternative is clearly shown in eqs. (10): it suffices to assume $(\gamma_0 M_0^2 R_m P_m) = R_m \mu_e H_0^2 \ll 1$.

[2] In deriving eqs. (9) it has been further imposed that $H_{2y}^0 = 0$.

Flat-Plate Boundary-Layer Equations

Imposed electric fields are absent and the imposed magnetic field has components only in the x, z-plane. The magnetic REYNOLDS number R_m is assumed

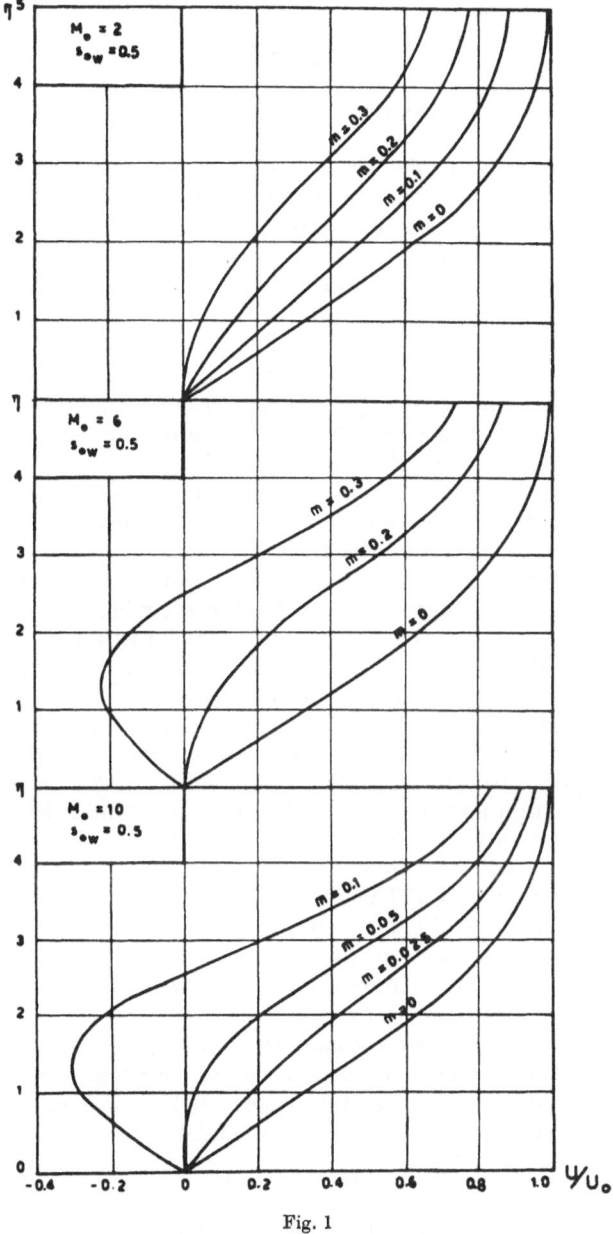

Fig. 1

to be much less than one and the magnetic pressure number P_m is assumed to be of order $1/R_m$. The motion is laminar, steady and, upon the results of the preceding section, isobaric.

The basic equations, then reduce to:

$$p = \varrho \, R \, T \tag{11a}$$

$$(\varrho \, u)_x \times (\varrho \, v)_y = 0 \tag{11b}$$

$$\varrho \, u \, u_x + \varrho \, v \, u_y + \sigma \, B_0^2 \, u = (\mu \, u_y)_y \tag{11c}$$

$$p_y = 0 \tag{11d}$$

$$\varrho \, u \, S_x + \varrho \, v \, S_y = \left\{ \frac{\mu}{Pr} \, [S_y + (Pr-1) \, u \, u_y] \right\}_y. \tag{11e}$$

Herein S is the total enthalpy $h + u^2/2$. The contribution to the magnetic induction B_0^2 can come, in general, from both the x and z components of the imposed magnetic field [see eqs. (7)]. When these components are required to be independent of y and z they can be, at most, linear functions of x [see eqs. (10)]. With regard to this point, it might be noteworthy to degrees a little in order to examine the possibility of having "similar" solutions for the subject problem.

It is well known that to seek "similar solutions" amounts to the determination of suitable velocity and length scales[1], so that differential equations, boundary and initial conditions can all be expressed in terms of a single similarity variable. It is also well known that when there exist additional characteristic lengths (besides the "mechanical" length related to the height of the dissipative region) a necessary (but, all the same, not always sufficient [8]) condition for the existence of similar solutions is that the new lengths be proportional to the "mechanical" one. In magneto-fluid-dynamics there is the additional length associated with the imposed magnetic field. Therefore, the existence of similar solutions will depend on the possibility of having an imposed magnetic field H_0 exhibiting such a functional dependence on x as to introduce a "magnetic characteristic length" proportional to the "mechanical" one. We have seen, however that H_0 can be at most a linear function of x: therefore the possibilities of having similar solutions are very limited. In particular, since it is possible to show that H_0 must be inversely proportional to the "mechanical" length, similar solutions can be obtained only in two cases [see eq. (10)]:1) when the "mechanical" length is constant (two-dimensional stagnation flow) $(a = 0)$; 2) when the "mechanical length" is proportional to x (this occurs when there is an axial pressure gradient proportional to x^5 (see [5]). In the latter case, however, the x-component of the magnetic field must be a linear function of y. This fact, besides being rather unrealistic from a practical point of view, modifies, in turn, the inviscid flow solution so that the entire boundary-layer analysis might have to be completely revised.

At any rate, it is certainly impossible to have similar solutions for isobaric fields.

Turning back to eqs. (11) we consider the case in which the imposed field is constant. In addition, we assume also a constant electrical conductivity σ. The modifications introduced by a variable σ are briefly outlined in Appendix A.

Eqs. (11) are subject to the following boundary conditions at the wall:

$$u \, (o) = v \, (o) = 0 \tag{12a}$$

$$C_1 \, S \, (o) + C_2 \, S_y(o) = C_3 \tag{12b}$$

where the C_i are constants. In particular we will be concerned with two sets of boundary conditions: one relative to constant wall temperature $C_1 = 1$; $C_2 = 0$

---

[1] For simplicity sake, we consider the constant fluid property case.

and another relative to adiabatic wall $(C_1 = C_3 = 0; C_2 = 1)$. The initial conditions are:

$$u(0, y) = u_0 = \text{const.} \tag{13a}$$

$$S(0, y) = S_0 = \text{const.} \tag{13b}$$

The remaining boundary conditions must be prescribed at infinity and should reflect the asymptotic joining of the dissipative region with the outer flow. The solution of this outer inviscid flow is thus required first.

In this region, by definition, there is no dissipation by "mechanical" causes, therefore, since $E = 0$, the flow will be homoenthalpic. Accordingly the solution of the energy equation is simply

$$S_e = c_p T_e + \frac{u_e^2}{2} = S_0 \tag{14}$$

where the subscript (e) indicates conditions in the outer inviscid flow and the subscript zero indicates conditions at $x = 0$. Since the motion is isobaric the density ratio can be obtained as function of the velocity ratio by combining eqs. (11a) and (14) to yield:

$$\frac{\varrho_0}{\varrho_e} = \frac{S_0}{h_0} - \frac{u_0^2}{2 h_0} \left(\frac{u_e}{u_0} \right)^2. \tag{15}$$

The momentum equation can then be written as:

$$\left(\frac{u_e}{u_0} \right)_x + \frac{\sigma B_0^2}{\varrho_0 u_0} \left\{ 1 + \frac{\gamma - 1}{2} M_0^2 \left(1 - \frac{u_e^2}{u_0^2} \right) \right\} = 0. \tag{16}$$

The solution of this equation, subject to the initial condition $u_e(0) = u_0$ is, for B_0 and σ constant,

$$\frac{u_e}{u_0} = \frac{(1 - a) + (1 + a) e^{-m\beta}}{\left(1 - \frac{1}{a} \right) + \left(1 + \frac{1}{a} \right) e^{-m\beta}} \tag{17}$$

where

$$a = \left[1 + \frac{2}{(\gamma - 1) M_0^2} \right]^{\frac{1}{2}} \tag{18a}$$

$$\beta = a (\gamma - 1) M_0^2 \tag{18b}$$

$$m = \frac{\sigma B_0^2}{\varrho_0 u_0^2} x. \tag{18c}$$

The parameter m is recognised to be the number $R_m P_m$ referred to initial free stream conditions and to the distance x from the leading edge of the plate. This parameter will be referred to as the magnetic parameter.

The expression for u_e/u_0 given by eq. (17) and valid for not too small values of u_e/u_0 is rather complicated. In view of the further developments of the analysis, it proves convenient to develop eq. (17) in power series in terms of $(m \beta)$. By retaining only the first order term one obtains simply:

$$\frac{u_e}{u_0} = 1 - m \tag{19}$$

much as in the case of incompressible flow. It is to be born in mind, however, that, while in the incompressible case eq. (19) is the exact solution for the outer flow, in the compressible case it is only a first approximation which neglects terms of order $(m \beta)^2$. This is an important point because it shows that the accuracy will become poorer and poorer not only as the magnetic parameter encreases but also as the free stream MACH number increases. This limitation will also be found in the viscous solution.

The additional boundary conditions can now be written as:

$$u\,(x,\,\infty) = u_e\,(x) \tag{12c}$$

$$S\,(x,\,\infty) = S_e = S_0 = \text{const.} \tag{12d}$$

To solve the system of eqs. (11) the STEWARTSON transformation [10] is first introduced to reduce it to a quasi-incompressible form. Series solutions in terms of the magnetic parameter m will be subsequently sought [1].

We introduce the following new independent and dependent variables:

$$X = x$$

$$Y = \int_0^y \frac{\varrho}{\varrho_0}\,dy$$

$$U = U_0\,\psi_y = u$$

$$V = -u_0\,\psi_x \tag{20}$$

where the stream function ψ is defined by:

$$\psi_y = \frac{\varrho u}{\varrho_0 u_0} \qquad \psi_x = -\frac{\varrho v}{\varrho_0 u_0} \tag{21}$$

we let, successively:

$$\eta = (u_0/v_0\,x)^{\frac{1}{2}}\,Y$$

$$\psi = [X\,v_0\,u_0]^{\frac{1}{2}}\,[f\,(\eta) + 2\,m\,g\,(\eta) + 0\,(m^2)]$$

$$\frac{S}{S_0} = s_0\,(\eta) + 2\,m s_1\,(\eta) + 0\,(m^2)\,. \tag{22}$$

Introducing these relations into eqs. (11), assuming $\varrho\,\mu = \text{const.}$, and $Pr = 1$ collecting and equating to zero the terms in the like powers of m results in:

$$2\,f''' + f f'' = 0 \tag{23a}$$

$$g''' = f'g' - f g'' - \frac{3}{2}\,g f'' + \frac{1}{2}\left\{\left[1 + \frac{\gamma-1}{2}\,M_0^2\right]s_0 f' - \frac{\gamma-1}{2}\,M_0^2\,f'^3\right\} \tag{23b}$$

$$s''_0 + \frac{Pr f}{2}\,s'_0 = \frac{(\gamma-1)\,M_0^2}{1 + \frac{\gamma-1}{2}\,M_0^2}\,(1 - Pr)\,[f'\,f''' + f''^2] \tag{24a}$$

$$\frac{s_1'''}{Pr} = f's_1 - f\frac{s_1'}{2} - \frac{3}{2}\,g s'_0 + \frac{(\gamma-1)\,M_0^2}{1 + \frac{\gamma-1}{2}\,M_0^2}\,\frac{1-Pr}{Pr}\,[f'g''' + g'f''' + 2\,f''g'']. \tag{24b}$$

The pertinent boundary conditions [eqs. (12)] become:

$$f\,(0) = f'\,(0) = 0 \qquad f'\,(+\infty) = 1 \tag{25a}$$

$$g\,(0) = g'\,(0) = 0 \qquad g'\,(+\infty) = -\frac{1}{2} \tag{25b}$$

$$C_1\,s_0\,(0) + C_2\,s_{0y}\,(0) = C_3 \qquad s_0\,(+\infty) = 1 \tag{26a}$$

$$c_1\,s_1\,(0) + c_2\,s_{1y}\,(0) = 0 \qquad s_1\,(+\infty) = 0 \tag{26b}$$

and the essential physical quantities result expressed as:

$$\frac{u}{u_0} = f' + 2\,m\,g' + 0\,(m^2)$$

$$-\frac{v}{u_0} = \frac{1}{2}\,(v_0/u_0\,x)^{\frac{1}{2}} \{f + 6\,mg - \eta\,(f' + 2\,mg') + 0\,(m^2)\}$$

$$\frac{\varrho_0}{\varrho} = \left(1 + \frac{\gamma-1}{2}\,M_0^2\right)[s_0 + 2\,m\,s_1 + 0\,(m^2)] - \frac{\gamma-1}{2}\,M_0^2\,[f'^2 + 4\,mf'g' +$$
$$+ 0\,(m^2)]\,. \tag{27}$$

The zeroth approximation for the momentum equation [eq. (23a)] is the well known BLASIUS function tabulated, for instance, in [11].

The solution of the system of eqs. (23b) and (24) is treated in the next section for $P_r = 1$.

Solution of the Equations $(P_r = 1)$

a) Constant Wall Temperature

For constant wall temperature the boundary conditions on the energy equations become:

$$s_0\,(0) = s_{0w} = \text{const.} \qquad s_1\,(0) = 0\,. \tag{28}$$

The zeroth approximation to the energy equation [eq. (24a)] for $P_r = 1$ becomes simply:

$$s_0'' + \frac{f}{2}\,s'_0 = 0 \tag{29}$$

subject to the boundary condition given by eqs. (28). Its solution is:

$$s_0 = 1 + (s_{0w} - 1)\,(1 - f')\,. \tag{30}$$

The first approximation to the momentum equation is a linear equation whose solutions can be determined by any standard numerical procedure and with the help of its asymptotic solutions [9]. Substitution of eq. (30) into eq. (23b) shows that the function g can be expressed as:

$$g = G_1 + \frac{1}{2}\left(1 + \frac{'\gamma-1}{2}\,M_0^2\right)[G_2 + (s_{0w} - 1)\,G_3]\,. \tag{31}$$

The functions G_i are tabulated, together with their first two derivatives, in Table I. They result from the solution of the following equations:

$$G_1''' = L\,(G_1) + \frac{1}{2}\,f'^3$$

$$G_2''' = L\,(G_2) + f' - f'^3$$

$$G_3''' = L\,(G_3) + f'\,(1 - f')$$

with

$$L\,(G_i) = f'\,G_i' - \frac{f\,G_i''}{2} - \frac{3}{2}\,G_i\,f''$$

and

$$G_i\,(0) = G_i'\,(0) = 0$$

$$G_1'\,(\infty) = -\frac{1}{2}$$

$$G_2'\,(\infty) = G_3'\,(\infty) = 0\,.$$

Table I. *Velocity Functions* G_i. — *Case* $P = 1$

η	G_1	G_2	G_3	G_1'	G_2'	G_3'	G_1''	G_2''	G_3''
0	0	0	0	0	0	0	−0.222	−0.449	−0.313
.2	−0.004	−0.009	−0.006	−0.044	−0.089	−0.062	−0.222	−0.442	−0.306
.4	−0.017	−0.035	−0.025	−0.089	−0.176	−0.122	−0.222	−0.423	−0.288
.6	−0.040	−0.079	−0.054	−0.133	−0.257	−0.177	−0.222	−0.390	−0.261
.8	−0.071	−0.138	−0.095	−0.178	−0.330	−0.225	−0.220	−0.347	−0.226
1.0	−0.111	−0.211	−0.144	−0.221	−0.396	−0.266	−0.218	−0.293	−0.184
1.2	−0.160	−0.295	−0.201	−0.265	−0.448	−0.299	−0.213	−0.232	−0.139
1.4	−0.217	−0.389	−0.263	−0.307	−0.488	−0.322	−0.207	−0.165	−0.092
1.6	−0.282	−0.490	−0.329	−0.347	−0.514	−0.336	−0.197	−0.095	−0.045
1.8	−0.355	−0.594	−0.397	−0.385	−0.526	−0.341	−0.184	−0.026	0
2.0	−0.436	−0.699	−0.465	−0.420	−0.525	−0.336	−0.167	+0.040	+0.041
2.2	−0.523	−0.803	−0.531	−0.452	−0.511	−0.324	−0.148	+0.099	+0.077
2.4	−0.616	−0.903	−0.594	−0.479	−0.487	−0.306	−0.126	+0.150	+0.106
2.6	−0.714	−0.998	−0.653	−0.502	−0.453	−0.282	−0.103	+0.189	+0.128
2.8	−0.816	−1.085	−0.707	−0.520	−0.412	−0.255	−0.079	+0.217	+0.142
3.0	−0.921	−1.163	−0.755	−0.534	−0.367	−0.226	−0.056	+0.232	+0.149

3.2	+0.150	+0.237	−0.034	−0.196	−0.320	−0.543	−0.797	−1.232	−1.029
3.4	+0.144	+0.231	−0.015	−0.166	−0.273	−0.547	−0.833	−1.291	−1.138
3.6	+0.134	+0.217	+0.000	−0.139	−0.229	−0.549	−0.874	−1.342	−1.247
3.8	+0.121	+0.197	+0.012	−0.113	−0.187	−0.548	−0.889	−1.384	−1.357
4.0	+0.106	+0.174	+0.021	−0.090	−0.150	−0.544	−0.909	−1.418	−1.466
4.2	+0.090	+0.149	+0.026	−0.071	−0.118	−0.539	−0.926	−1.445	−1.574
4.4	+0.075	+0.124	+0.028	−0.054	−0.091	−0.534	−0.938	−1.466	−1.681
4.6	+0.060	+0.100	+0.028	−0.041	−0.068	−0.528	−0.947	−1.482	−1.787
4.8	+0.047	+0.079	+0.026	−0.030	−0.050	−0.523	−0.954	−1.495	−1.892
5.0	+0.036	+0.061	+0.023	−0.022	−0.036	−0.518	−0.960	−1.504	−1.996
5.2	+0.027	+0.046	+0.019	−0.026	−0.026	−0.514	−0.963	−1.510	−2.099
5.4	+0.020	+0.033	+0.016	−0.018	−0.018	−0.510	−0.966	−1.515	−2.201
5.6	+0.014	+0.024	+0.013	−0.012	−0.012	−0.508	−0.968	−1.518	−2.303
5.8	+0.010	+0.017	+0.010	−0.008	−0.008	−0.505	−0.969	−1.521	−2.404
6.0	+0.007	+0.011	+0.007	−0.005	−0.005	−0.504	−0.970	−1.522	−2.505
6.2	+0.005	+0.008	+0.005	−0.003	−0.003	−0.503	−0.970	−1.524	−2.604
6.4	+0.003	+0.005	+0.004	−0.002	−0.002	−0.502	−0.970	−1.525	−2.706
6.6	+0.002	+0.003	+0.002	−0.001	−0.001	−0.501	−0.970	−1.525	−2.806
6.8	+0.001	+0.002	+0.002	−0.001	−0.001	−0.501	−0.970	−1.526	−2.906
7.0	+0.001	+0.001	+0.001	−0.000	−0.000	−0.500	−0.970	−1.526	−3.006

By substituting eq. (31) into eq. (24c) with the first approximation to the energy equation becomes

$$s_1{}'' = f' s_1 - f \frac{s_1{}'}{2} + \frac{3}{2} (s_{0w} - 1) g f''$$

subject to homogeneous boundary conditions. Its solution can be expressed as:

$$\frac{s_1}{s_{0w} - 1} = s_{11} + \frac{1}{2} \left(1 + \frac{\gamma - 1}{2} M_0{}^2 \right) [s_{12} + (s_{0w} - 1) s_{13}] . \qquad (32)$$

The functions s_{1i} and their derivatives are tabulated in Table II. They result from the solution of the following equations:

$$s_{1i}{}'' = M (s_{1i}) + \frac{3}{2} G_i f''$$

Table II. *Enthalpy Functions* s_{1i}.— *Case* $P_r = 1$

η	s_{11}	s_{12}	s_{13}	$s_{11}{}'$	$s_{12}{}'$	$s_{13}{}'$
0	0	0	0	0.099	0.153	0.102
.2	0.020	0.030	0.020	0.099	0.152	0.102
.4	0.039	0.061	0.041	0.098	0.151	0.109
.6	0.059	0.090	0.061	0.096	0.148	0.098
.8	0.078	0.120	0.080	0.093	0.141	0.093
1.0	0.096	0.135	0.098	0.089	0.130	0.086
1.2	0.113	0.171	0.114	0.082	0.115	0.076
1.4	0.129	0.193	0.128	0.072	0.096	0.063
1.6	0.142	0.210	0.139	0.060	0.073	0.047
1.8	0.153	0.222	0.146	0.046	0.047	0.029
2.0	0.160	0.228	0.150	0.031	0.019	0.011
2.2	0.165	0.229	0.151	0.014	—0.008	0.002
2.4	0.166	0.225	0.148	—0.004	—0.035	—0.025
2.6	0.163	0.216	0.141	—0.021	—0.058	—0.040
2.8	0.157	0.202	0.132	—0.037	—0.078	—0.053
3.0	0.149	0.185	0.120	—0.050	—0.093	—0.062
3.2	0.138	0.165	0.108	—0.061	—0.101	—0.067
3.4	0.124	0.122	0.094	—0.069	—0.105	—0.069
3.6	0.110	0.124	0.080	—0.073	—0.104	—0.069
3.8	0.096	0.104	0.067	—0.073	—0.098	—0.064
4.0	0.081	0.085	0.055	—0.071	—0.090	—0.058
4.2	0.066	0.068	0.044	—0.060	—0.079	—0.051
4.4	0.055	0.053	0.034	—0.053	—0.068	—0.044
4.6	0.042	0.041	0.026	—0.048	—0.056	—0.036
4.8	0.034	0.031	0.020	—0.044	—0.045	—0.029
5.0	0.024	0.022	0.014	—0.034	—0.035	—0.023
5.2	0.019	0.016	0.010	—0.029	—0.027	—0.017
5.4	0.011	0.011	0.007	—0.023	—0.019	—0.013
5.6	0.010	0.008	0.005	—0.018	—0.015	—0.009
5.8	0.004	0.005	0.000	—0.013	—0.010	—0.007
6.0	0.004	0.003	0.000	—0.009	—0.007	—0.005

with

$$M(s_{1i}) = f' s_{1i} - f \frac{s_{1i}'}{2}$$

and

$$s_{1i}(\infty) = s_{1i}(0) = 0 .$$

Obviously, in the corresponding incompressible case the solutions for g and s, are:

$$g = G_1 + \frac{1}{2} G_2$$

$$\frac{s_1}{s_{0w} - 1} = s_{11} + \frac{1}{2} s_{12} . \tag{33}$$

b) Adiabatic Wall

In this case, the boundary conditions at the wall are:

$$s_{0y}(0) = 0 \qquad s_{iy}(0) = 0 \tag{34}$$

for any i.

Then the zeroth order solution for the energy equation is simply:

$$s_0 = 1 . \tag{35}$$

All the higher order approximations s_i are identically zero since eq. (24c) is reduced to an homogeneous equation with homogeneous boundary conditions.

Thus, in this case as in the Couette flow [3] and in the incompressible case [1], the "adiabatic" wall temperature T_{aw} is constant, is not altered by the presence of the magnetic field and is given by:

$$\frac{T_{aw}}{T_0} = 1 + \frac{\gamma - 1}{2} M_0^2 . \tag{36}$$

As it will be shown in the next section this follows from a general property of the particular set of equations used.

The first approximation for the velocity is immediately obtained from eq. (31) by setting $s_{0w} = 1$ and results in:

$$g = G_1 + \frac{1}{2}\left(1 + \frac{\gamma - 1}{2} M_0^2\right) G_2 . \tag{37}$$

Results and Discussion

I. Inviscid Flow

In the subject case of no imposed electric field the motion is homoenthalpic since all dissipative effects are absent. The free stream velocity is a function of both the magnetic parameter m and the initial Mach number M_0. This function can be reduced to the linear dependence:

$$\frac{u_e}{u_0} = 1 - m \tag{19}$$

only for:

$$m(\gamma - 1) M_0^2 \left[1 + \frac{2}{(\gamma - 1)} M_0^2\right]^{\frac{1}{2}} \ll 1 . \tag{19a}$$

As a consequence, the values of m for which eq. (19) is valid get smaller and smaller as M_0 increases. In other words, the present first order solution becomes less accurate as M_0 increases.

II. Viscous Flow

a) Constant Wall Temperature

The dissipative flow field depends on the parameters m, M_0 and, in addition, on the ratios of wall enthalpy to stagnation enthalpy s_{0w}. The velocity profiles are given, to the first approximation in the magnetic parameter m, by:

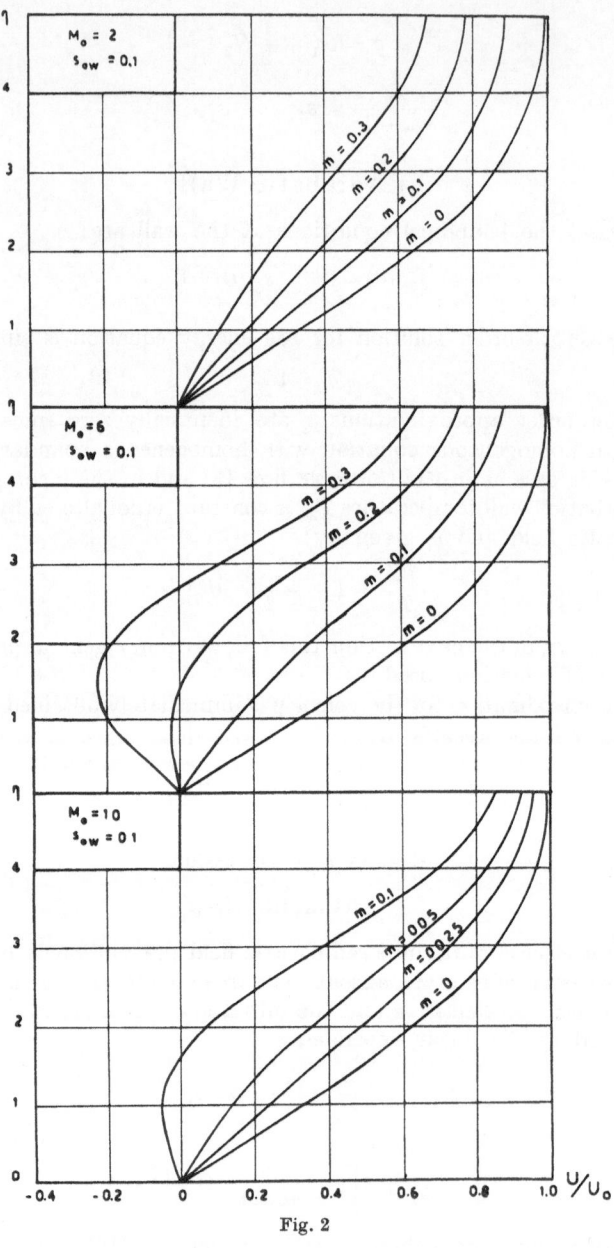

Fig. 2

$$\frac{u}{u_0} = f' + m\left\{2G_1 + \left(1 + \frac{\gamma-1}{2} M_0{}^2\right)[G_2 + (s_{0w} - 1)\,G_3]\right\}. \qquad (38\,a)$$

Velocity profiles are shown in Figs. 1 to 3 for several values of the magnetic parameter m and of M_0 and s_{0w}. It is seen that, for a given m, the effects of the applied
magnetic field increase as M_0 and s_{0w} increase. For given m and M_0 they are

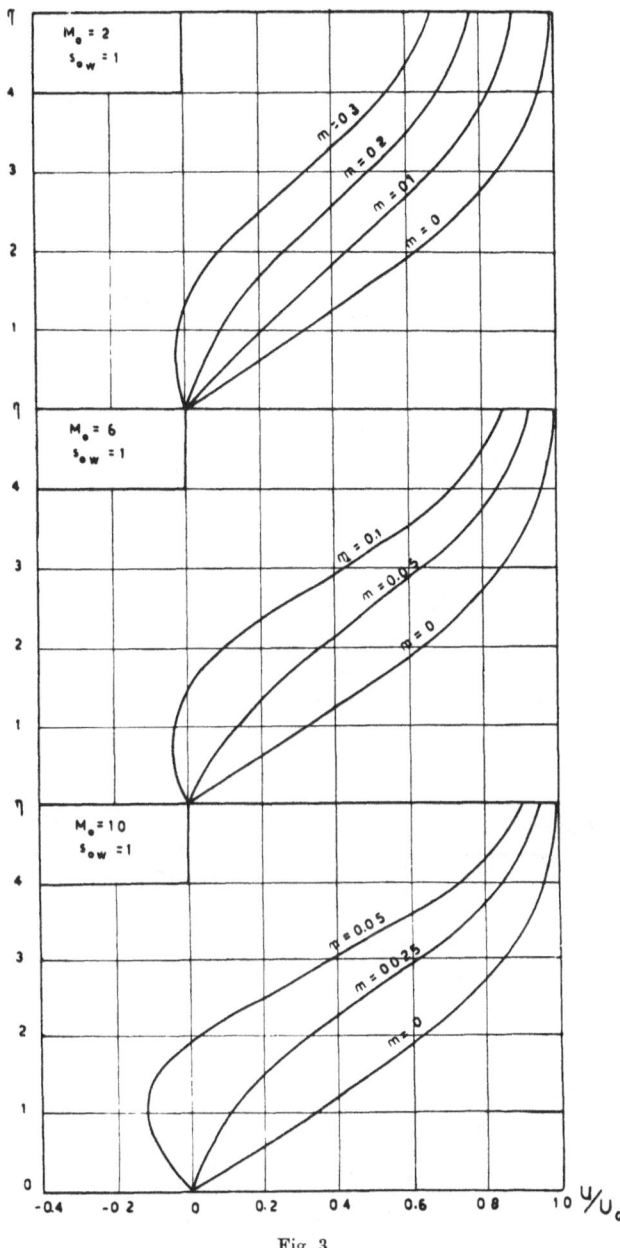

Fig. 3

minimum in the case of very cooled walls ($s_{0w} \to 0$). It is also apparent, from Figs. 1
to 3, how the results of this first order solution cease to be valid for values of m
which are the smaller the larger of M_0. This, of course, had to be expected since,
as said in a preceding section, the relative importance of the magnetic force is
measured by the local value of the number $P_m R_m$ which, for a given m (recall that

m is just the number $P_m R_m$ referred to the distance from the leading edge and to *initial* conditions), increases as the temperature increases (the process is isobaric).

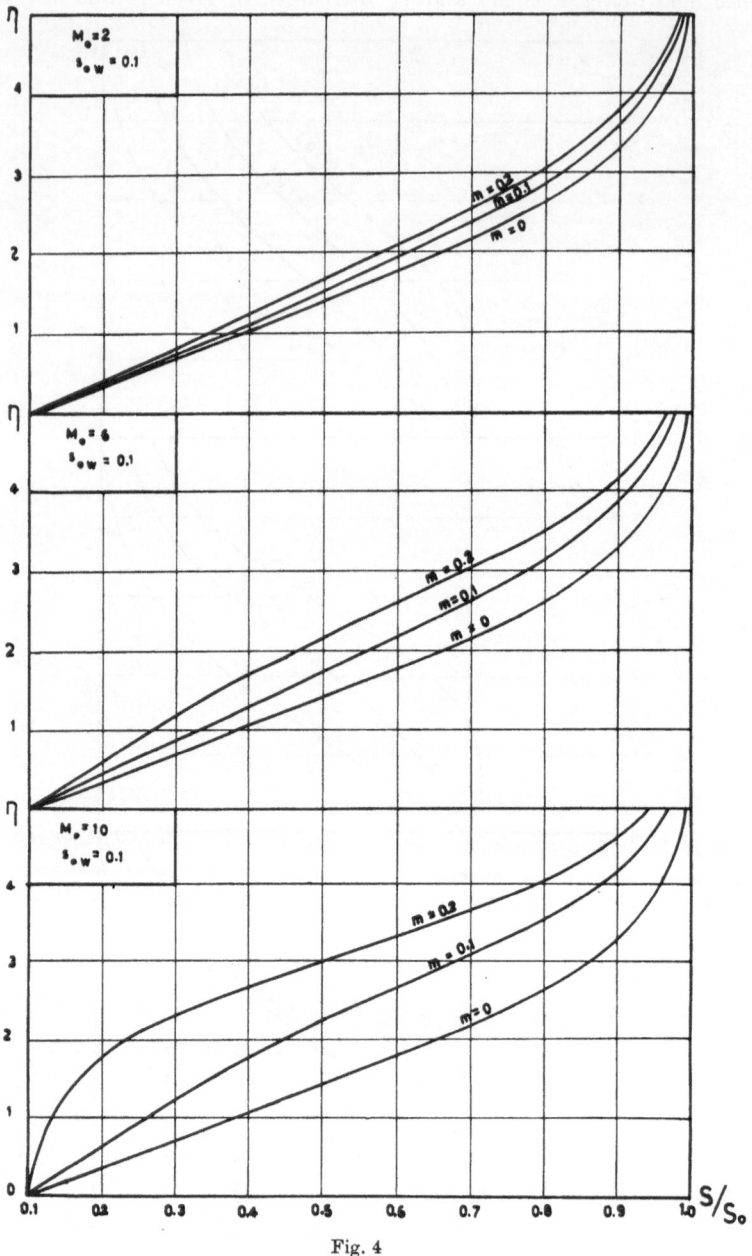

Fig. 4

Total enthalpy and temperature profiles are given, by:

$$\frac{S}{S_0} = 1 + (s_{0w} - 1)\left\{(1 - f') + m\left[2\,s_{11} + \left(1 + \frac{\gamma - 1}{2}M_0^2\right)(s_{11} + (s_{0w} - 1)\,s_{13})\right]\right\} \quad (38\,\text{b})$$

$$\frac{T}{T_0} = \left(1 + \frac{\gamma - 1}{2}M_0^2\right)\frac{S}{S_0} - \frac{\gamma - 1}{2}M_0^2[f'^2 + 4\,m\,f'\,g'] . \qquad (38\,\text{c})$$

Total enthalpy profiles are shown in Figs. 4 and 5 for two values of s_{0w} and several values of M_0. The effects of the magnetic field result in a decrease of s/s_0. One

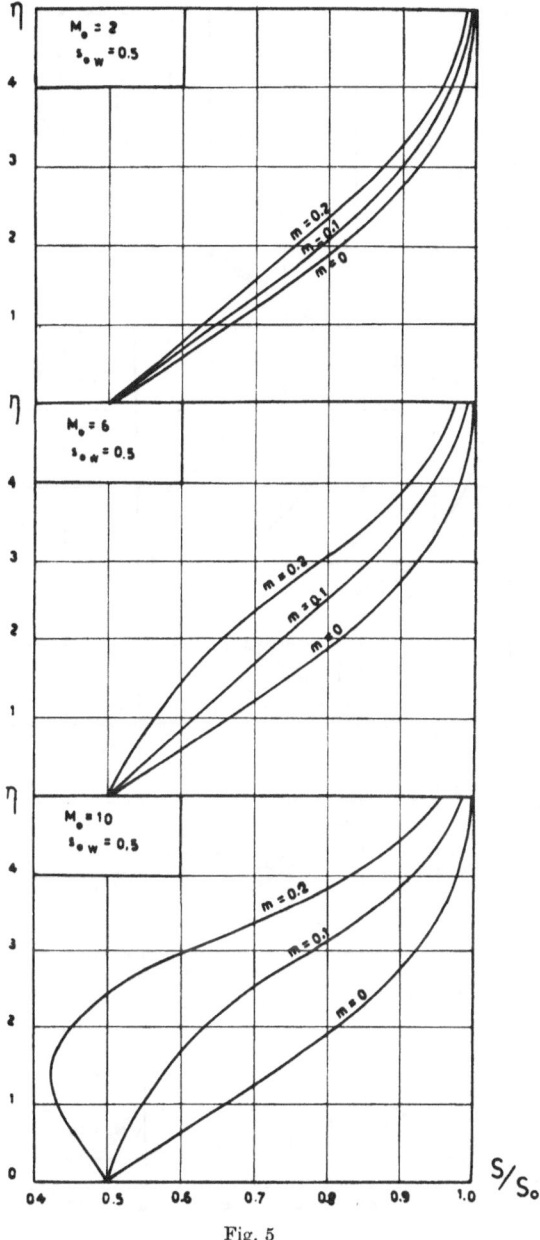

Fig. 5

finds again that these effects increase as M_0 and s_{0w} increase and that for increasing values of either the first order solution becomes increasingly poor in accuracy.

Temperature profiles are shown in Figs. 6 and 7. The temperatures are increased throughout except in the immediate vicinity of the wall. This is expectable since, except near the wall, the velocity reductions are larger than the total enthalpy

L. G. Napolitano and A. Pozzi:

reductions. The variations of these effects with M_0 and s_{0w} follow the same trends as described before.

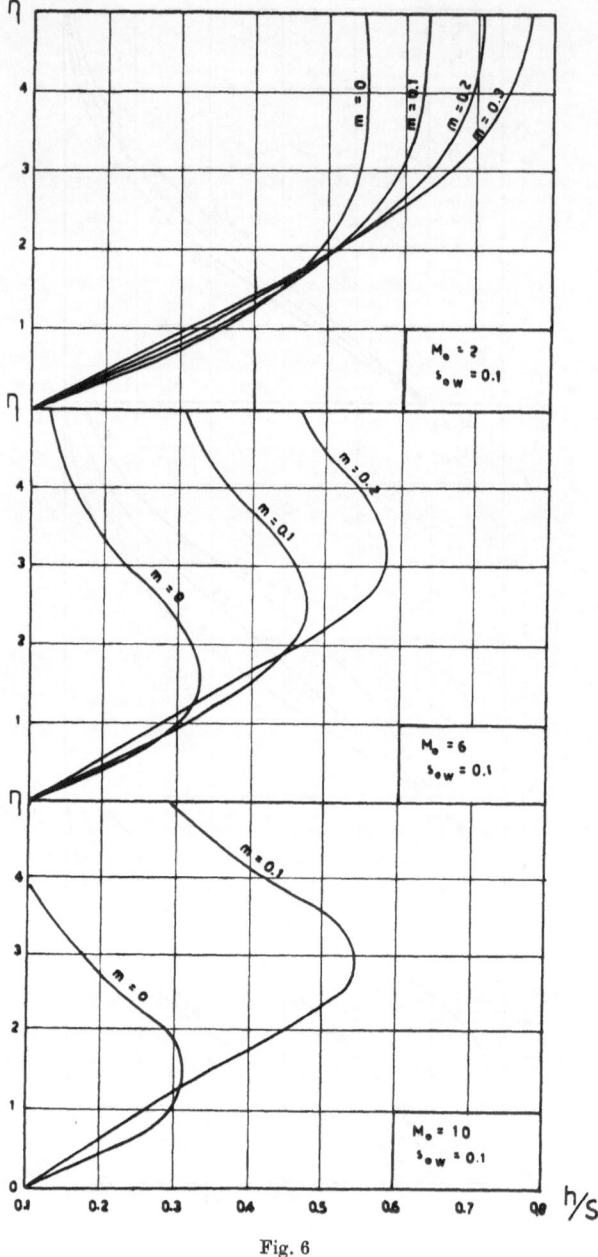

Fig. 6

Let the skin friction coefficient C_f and the heat-transfer coefficient C_h be defined in terms of initial flow properties as:

$$C_f = 2 \, (\mu \, u_y)_w / \varrho_0 \, u_0^2$$

$$C_h = - \, (\lambda \, T_y)_w / \varrho_0 \, u_0 \, (h_w - h_{aw}) \, . \qquad (39)$$

Performing the necessary substitutions and recalling that the adiabatic wall enthalpy is equal to the stagnation enthalpy [eq. (36)] one obtains:

$$C_f = \left(\frac{v_0}{u_0 x}\right)^{\frac{1}{2}} \left\{0.664 - m\left[0.888 + \left(1 + \frac{\gamma - 1}{2} M_0^2\right)(0.899 + 0.626\,(s_{0w} - 1))\right]\right\}$$

$$C_h = \left(\frac{v_0}{u_0 x}\right)^{\frac{1}{2}} \left\{0.332 - m\left[0,198 + \left(1 + \frac{\gamma - 1}{2} M_0^2\right)(0.153 + 0.152\,(s_{0w} - 1))\right]\right\}. \quad (40)$$

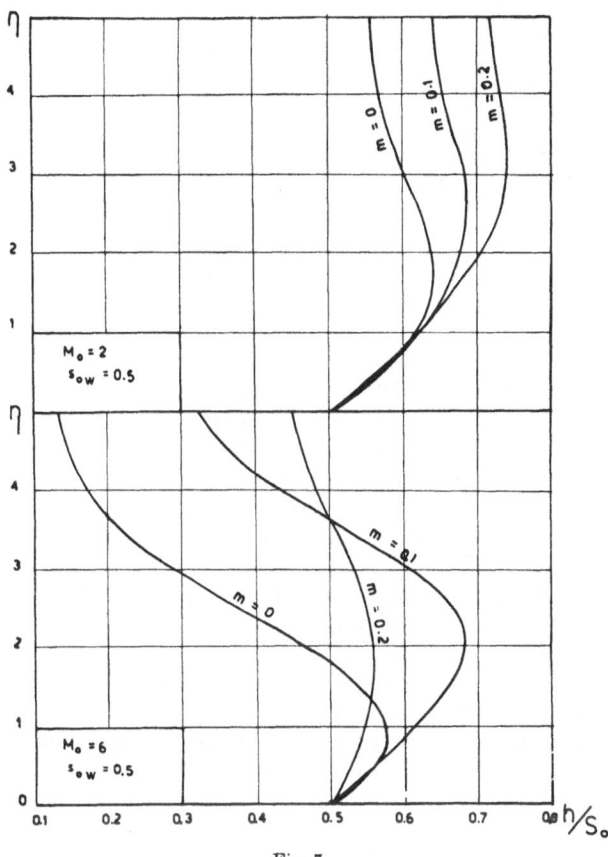

Fig. 7

Comparing these coefficients with those obtained for zero magnetic field yields:

$$\frac{C_f}{(C_f)_0} = 1 - m\left\{1.337 + \left(1 + \frac{\gamma - 1}{2} M_0^2\right)[1.354 + 0,943\,(s_{0w} - 1)]\right\}$$

$$\frac{C_h}{(C_h)_0} = 1 - m\left\{0.596 + \left(1 + \frac{\gamma - 1}{2} M_0^2\right)[0.461 + 0.307\,(s_{0w} - 1)]\right\} \quad (41)$$

where $(C_f)_0$ and $(C_h)_0$ are the friction and heat-transfer coefficients, for zero magnetic field, computed at the same REYNOLDS number $R_e = (u_0\,x/v_0/)$.

Both friction and heat transfer coefficients are decreased as a consequence of the interaction of the flow field with the magnetic field. The decrease of the heat transfer coefficient is less than that of the friction coefficient. They are, roughly, in the ratio 1 to 2. Both reductions follow the same pattern previously described: they are the larger the larger are M_0 and s_{0w}.

As pointed out in [1] and [3] is it no longer possible, in general, to establish a rigorous "Reynolds analogy" in magneto-fluid-dynamics[1].

The displacement thickness in the Stewartson plane, defined as:

$$\delta^+ = \int\limits_0^\infty \left(1 - \frac{u}{u_e}\right) dy$$

is given by:

$$\delta^+ = (\nu_0 x/u_0)^{\frac{1}{2}} \left\{1.73 + m\left[-0.988 + \left(1 + \frac{\gamma-1}{2} M_0^2\right) 1.526 + 0.970 \left(s_{0w} - 1\right)\right]\right\}.$$
(42)

It is, in general, larger than the one $(\delta^+)_0$ for $m = 0$ and computed at the some Reynolds number, although there can be some combinations of M_0 and s_{0w} for which $\delta^+ < (\delta^+)_0$ (for instance for $M_0 = 2$ and very cold wall.)

b) Adiabatic Wall

When the wall is insulated the stagnation enthalpy S is constant throughout the field and equal to its initial values S_0. The adiabatic wall temperature is then constant and it is not affected by the presence of the magnetic field. These features, as will be discussed later, are a direct consequence of the structure of the basic equations for $E = 0$ and do not depend on the assumption $R_m \ll 1$ nor on the geometry of the fields. Thus these results are also found in all other dissipative regions with no imposed electrical fields (see, for instance, [1] and [3]).

Velocity profiles, skin friction coefficient and all other boundary layer characteristics are immediately found for the adiabatic wall by setting $s_{0w} = 1$ in the relations previously derived. The effects of the magnetic field for given m depend now only on the initial Mach number M_0. As detailed before, they increase with M_0 and, for constant m and M_0, are larger than those relative to the cooled wall $(s_{0w} < 1)$. Velocity and temperature profiles are shown in Figs. 3 and 8 for several values of the Mach number. The same remarks on the validity of the first order approximation apply. In particular the maximum value of m for which results can be considered valid is, for given Mach number M_0, less than then one found for cooled wall.

Before concluding this section, it might be of interest to compare, as far as it is possible, the present results with those of other magneto-fluid-dynamic dissipative flows.

We first summarize the assumptions made in the present analysis as follows:

i) $R_m \ll 1$; $P_m R_m = 0$ (1). These assumptions are necessary and sufficient to uncouple the fluid dynamic equations for the Maxwell's field equation.

ii) Imposed electric fields are absent. This eliminates the term $J \cdot E$ from the energy equation.

iii) Flat plate boundary-layer approximation; constant imposed magnetic field, fixed with respect to the plate, constant electrical conductivity, unitary Prandtl number.

iv) Assumptions attendant to the power series solution, in terms of the magnetc parameter m, for both inviscid and viscous flow [cfr. eq. (19a)].

---

[1] This is well established a priori since, due to the fact that the energy balance does not involve e.m. terms, in no case it is possible to arrive at momentum and energy equations which are structurally similar.

Secondly, we notice, that the results found in the present analysis can be roughly classified in the following three groups:

1) results which are a direct consequence of the structure of the magneto-fluid-dynamic equations which have been used and, thus, are independent of the particular problem studied; 2) results which are directly predictable on the basis of assumptions made on the order of magnitude of the terms involved in the basic equations; and 3) results which are pertinent to the particular problem studied.

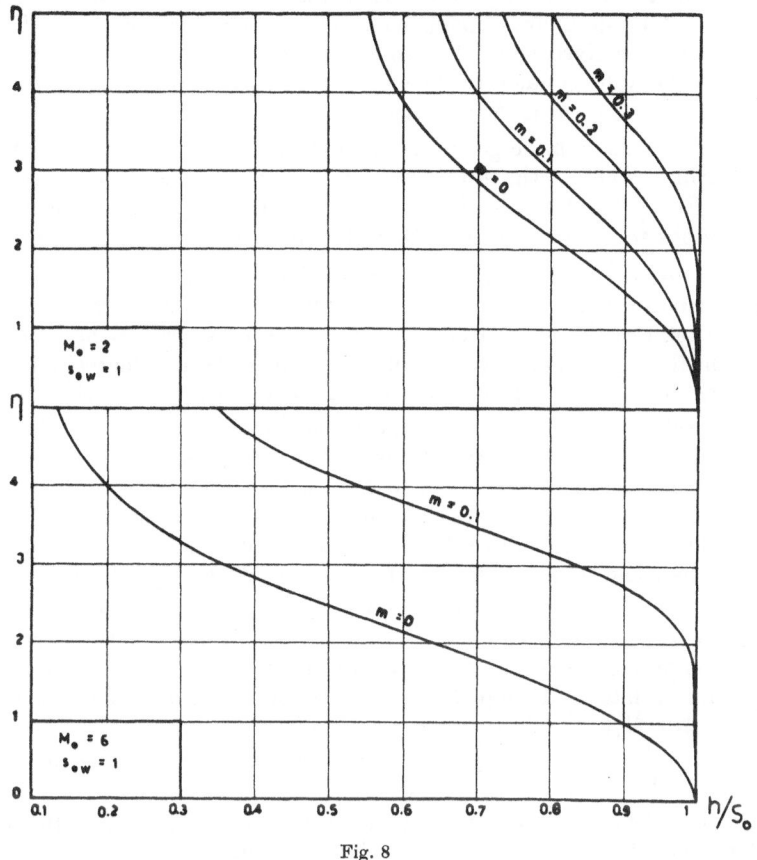

Fig. 8

To the first group belong the results about the constance of S_0, the independence of T from B_0, and the impossibility of the REYNOLDS analogy. Indeed from the entropy equation [12] one derives:

$$\frac{De}{Dt} + p\,\frac{D(1/\varrho)}{Dt} = \frac{1}{\varrho}\left[\frac{J^2}{\sigma} - \frac{\partial}{\partial r}\cdot Q - \tau:\frac{\partial v}{\partial r}\right]$$

and the momentum equation multiplied scalary by v yields (in the steady case with $E=0$):

$$\varrho\,\frac{Dv^2/2}{Dt} + \frac{Dp}{Dt} + v\cdot\frac{\partial}{\partial r}\cdot\tau = -\frac{J^2}{\sigma}\,.$$

Combining these equations results in:

$$\varrho\,\frac{DS}{Dt} = -\frac{\partial}{\partial r}\cdot[Q + v\cdot\tau]\,. \tag{43}$$

Thus the solution of the energy equation in the inviscid region is always $\varrho = $ const. along a streamline. Furthermore, in the cases in which one can let

$$v^2 \simeq u^2 ; \qquad Q \simeq -\lambda T_y \overline{\jmath} ; \qquad \tau \simeq -\mu u_y ; \qquad \frac{\partial}{\partial r} \simeq \frac{\partial}{\partial y} \overline{\jmath} ; \qquad P_r = 1 \quad (44)$$

eq. (43) admits always one and only one particular solution $S = $ const. $= S_0$.

It follows that: a) the solution is the solution valid for adiabatic wall; b) the adiabatic wall temperature is independent of the magnetic field; c) no REYNOLDS analogy can be derived since the proportionality between heat and momentum transfer at the wall is related to the existence of a solution of the energy equation of the form $s = a u + b$ (a, b constants).

This last result is more general. The first two, for boundary layer flows, are independent of whether the flow is isobaric or not, whereas they do not hold any longer if $Pr \neq / $ ([1]). It is also important to notice that we have never assumed the equations to be uncoupled so that the above conclusions hold for any R_m.

Concerning the results which fall within the second group, a basic conclusion of the dimensional analysis has to be recalled: the relative importance of the magnetic term in the momentum equation is proportional to the local value of the parameter $R_m P_m$. For a given values of the magnetic parameter m, it increases, in isobaric flows, as M_0 increases and as the temperature level increases.

The decelerating action of the magnetic field thus increases with M_0 and with the temperature level everything else being the same. Conversely, it can be stated that to achieve a determined effect with an imposed magnetic field, its intensity must be the smaller the larger: a) the MACH number M_0, the electrical conductivity σ of the fluid, the temperature level (which, in turns, increases σ thus having a double enhancing effect on $R_m P_m$ through σ and through the density ϱ).

The presence of the magnetic field definitely increases the temperature level. A detailed description of this increases, however, will depend on the wall conditions and on the problem considered. For adiabatic walls, for instance the increase in temperature will depend uniquely on the reduction of kinetic energy and thus will increase with M_0.

These conclusions naturally agree with the results of [3] since they are independent on the geometry considered.

Few remarks must now be made on the results which depend on the particular problem studied. Since it is now clear from what previously discussed that compressibility tends to increase all the magnetic effects, no further comparison with the incompressible counterpart of the present problem [1] need to be made. As to the comparison with the results obtained for COUETTE flow we will say that, to within the accuracy of the present linearized analysis, the trends predicted in [3] are verified. The presence of the magnetic field decreases the heat transfer at the wall: this decreases, however, might be higher than that estimated in [3]. The decrease in skin friction coefficient is more pronounced in the present case.

---

[1] The situation is slightly different for COUETTE flow. In that case all convective terms are identically zero and the expressions for Q and τ given by eqs. (44) are now exact. The solution of eq. (43) for zero heat transfer at the wall is $\lambda T_y + (u^2/2)_y = 0$ and it can be integrated to yield $h + P_r u^2/2 = $ const. upon the only assumption $P_r = $ const. In [3], where this result was first derived, it is also mentioned that it ceases to be valid if the flow is not isobaric. Upon the present demonstration, it appears that this happens only when h depends on both T and p.

No comparison can be established for the drag coefficient since in the present analysis the electrical conductivity has been assumed constant and this leads, with a magnetic field fixed to the plate, to an infinite drag.

Another important point to notice is that the results relative to the present geometry are markedly different from these relative to stagnation flow [2, 13]. The differences are condensed in the following points: 1) both skin friction and heat transfer coefficients are reduced; 2) these reductions are caused by the actions of imposed magnetic fields on both inviscid and viscous flow, these actions being of the same order of magnitude[1].

These results point to the facts that although a qualitative estimate of magnetic effects is rather easily derivated from the "a priori" analysis of the basic equations, the actual quantitative effects are strongly dependent on the geometry of the body analyzed and on the relative geometry of the imposed field. This must be born in mind when passing judgements upon the worthiness of fluid-magnetic interactions as means to reduce heat-transfer rates and to increase over-all drags.

Conclusion

In the present paper the compressible flat-plate boundary layer of an electrically conducting fluid in the presence of magnetic fields has been investigated under the assumptions of small magnetic REYNOLDS number, unitary PRANDTL number and constant electrical conductivity. During the main analysis it has been possible to establish also some properties of the basic magneto-fluid-dynamic equations which hold in more general cases.

The main results can be summarized as follows. In the absence of applied electric fields the boundary-layer form of the energy equation admits one and only one particular solution [total enthalpy constant throughout] if either the PRANDTL number is equal to one or the MACH number is much less than one. Since the outer inviscid flow is always homoenthalpic in the steady case, the above mentioned solution applies when the wall is adiabatic. As a consequence: a) the adiabatic wall temperature is not effected by the magnetic field; b) the REYNOLDS analogy never holds for magneto-fluid-dynamic dissipative flows.

The above conclusions are general and do not depend on the order of magnitude of the magnetic REYNOLDS number R_m.

When $R_m \ll 1$ the fluid dynamic equations are uncoupled from the MAXWELL's field equations. In the plane two-dimensional case the transverse component of the imposed field can be, at most, a linear function of x and thus similar flow fields can only be obtained for stagnation flows and for isobaric flows characterized by a pressure parameter $\beta = 3/2$.

When the imposed magnetic field is constant its effect on the velocity field do not depend on whether it acts in the y-direction or span-wise.

Turning to the flat-plate case, it has been found that:

i) the influence of the magnetic field on the inviscid stream depends on both the magnetic parameter $m = \dfrac{\sigma B_0^2 x}{\varrho_0 u_0}$ and on the initial MACH number M_0. The flow is homoenthalpic.

---

[1] Unless, of course, the electrical conductivity of the free stream is very low due to its low temperature level.

ii) the influence of the magnetic field on the viscous region depends, in addition on the ratio of wall enthalpy to total enthalpy (s_{0w}).

iii) Both skin friction and heat transfer coefficients appear reduced when compared to the corresponding ones for zero magnetic field and for the same Reynolds number.

iv) These reductions are caused by the actions of the imposed magnetic field on both inviscid and viscous flow, and these actions are of the same order of magnitude.

v) The above effects, for given m, and to within terms of order m, increase as M_0 and s_{0w} increase. Results, based on solution of the equations up to terms of order m become thus less and less accurate at high Mach numbers and large s_{0w}.

Appendix A

Case of Variable Conductivity

The electrical conductivity σ is a scalar if the cyclotron's frequency is much less than the collision frequency [7].

Its dependence on the macroscopic properties of the fluid can be established by means of kinetic theory [7].

In the present case of isobaric motion, in order to estimate the effects of its variation on the velocity and temperature field the following simple relation can be assumed:

$$\sigma = \sigma_0 \, (T/T_0)^\omega \tag{A1}$$

where the constant σ_0 and ω can be determined to fit the more exact curve $\sigma = \sigma\,(T)$ in the temperature range of interest.

By substituting eq. (A1) into the electromagnetic term of the momentum equation one gets, for isobaric motion

$$\frac{\beta_0^2 \sigma}{\varrho_0 u_0} \frac{u}{u_0} \frac{\varrho_0}{\varrho} = m_1 \left(\frac{\varrho_0}{\varrho} \right)^{1+\omega} \frac{u}{u_0} = \frac{m}{x} \left[\frac{S_0}{h_0} - \frac{u_0^2}{2 h_0} f'^2 \right]^{1+\omega} f' + 0 \, (m^2) \tag{A2}$$

where:

$$m_1 = \frac{\sigma_0 \beta_0^2}{\varrho_0 u_0} . \tag{A3}$$

Thus the first order velocity function g satisfies now the equation:

$$g''' = f'g' - fg''/2 - \frac{3}{2} gf'' + \frac{1}{2}\left[\left(1+\frac{\gamma-1}{2} M_0^2\right) s_0 - \frac{\gamma-1}{2} M_0^2 f'^2\right]^{1+\omega} f' \tag{A4}$$

subject to the boundary conditions given by eqs. (25b). Due to the fact that ω is not an integer, the general solution for g can no longer be given in terms of universal functions. The non-homogeneous part of eq. (A4) must be solved separately for each value of M_0 and s_{0w}. The homogeneous solution of eqs. (A4) is naturally the same as the solution of eq. (23b) and it has been tabulated, for convenience, in Table III.

Table III. *Velocity Functions. Homogeneous Solution*

η	γ_0	γ_0'	γ_0''	η	γ_0	γ_0'	γ_0''
0	0	0	0.100				
.2	0.002	0.020	0.100	3.2	0.524	0.348	0.152
.4	0.008	0.040	0.100	3.4	0.597	0.380	0.169
.6	0.018	0.060	0.100	3.6	0.677	0.415	0.188
.8	0.032	0.080	0.100	3.8	0.764	0.455	0.210
1.0	0.050	0.100	0.100	4.0	0.859	0.500	0.234
1.2	0.072	0.120	0.100	4.2	0.964	0.549	0.260
1.4	0.098	0.140	0.100	4.4	1.079	0.604	0.287
1.6	0.128	0.160	0.101	4.6	1.206	0.664	0.315
1.8	0.162	0.180	0.102	4.8	1.345	0.730	0.344
2.0	0.200	0.201	0.104	5.0	1.498	0.802	0.374
2.2	0.243	0.222	0.108	5.2	1.667	0.880	0.403
2.4	0.289	0.244	0.112	5.4	1.851	0.963	0.432
2.6	0.341	0.267	0.119	5.6	2.052	1.052	0.460
2.8	0.397	0.292	0.127	5.8	2.272	1.147	0.488
3.0	0.458	0.319	0.139	6.0	3.511	1.247	0.515

References

1. V. I. ROSSOW, On Flow of Electrically Conducting Fluids Over a Flat Plate in the Presence of a Transverse Magnetic Field. NACA, TN 3971 (1957).
2. R. C. MEYER, On Reducing Aerodynamic Heat Transfer Rates by Magneto-hydro-dynamic Techniques. Paper presented at the 26th I.A.S. Meeting in New York, Preprint No. 816.
3. Z. O. BLEVISS, Magneto-gas-dynamics of Hypersonic COUETTE Flow. J. Aero/Space Sci. 25, No. 10 (1958).
4. L. G. NAPOLITANO, Magneto-fluid-dynamic Mixing of two Compressible Streams. Paper presented at the 1959 Symposium on Rockets and Astronautics, Tokyo, May 1959.
5. P. S. LYKOUDIS, On a Class of Compressible Laminar Boundary Layers with Pressure Gradient for an Electrically Conducting Fluid in the Presence of a Magnetic Field. Proceedings of the IXth International Astronautical Congress, Amsterdam 1958, p. 168. Wien: Springer, 1959.
6. B. T. CHU, Thermodynamics of Electrically Conducting Fluids and its Application to Magneto-Hydromechanics. Brown University, Providence, R. I.
7. L. SPITZER, JR., Physics of Fully Ionized Gases. New York: Interscience Publ. Inc., 1956.
8. L. G. NAPOLITANO and A. POZZI, Magnetofluidodinamica dei getti. (In Italian.) Missili 1, No. 3, 17—19 (1959).
9. L. G. NAPOLITANO, Magneto-Fluid-Dynamics of Two Interacting Streams. Proceedings of the IXth International Astronautical Congress, Amsterdam 1958, p. 570. Wien: Springer, 1959.
10. C. B. COHEN and E. RESHOTKO, Similar Solutions for the Compressible Laminar Boundary Layer with Heat Transfer and Pressure Gradient. NACA Report 1293, 1956.
11. H. SCHLICHTING, Boundary Layer Theory. London: Pergamon Press, 1955.
12. L. G. NAPOLITANO, Contributo alla Magnetofluidodinamica. (In Italian.) Missili 1, No. 1, 15—34 (1959).
13. N. H. KEMP, On Hypersonic Stagnation-Point Flow with a Magnetic Field. Readers' Forum. J. Aero/Space Sci. 25, No. 6, 405—407 (1958).

Unsteady Compressible Magnetic Laminar Boundary Layers in Hypersonic Flow

By

Paul S. Lykoudis[1, 2] and John P. Schmitt[1]

(With 8 Figures)

(Received June 22, 1959)

Abstract — Zusammenfassung — Résumé

Unsteady Compressible Magnetic Laminar Boundary Layers in Hypersonic Flow.
The unsteady hypersonic flow of a compressible, viscous, thermally and electrically
conducting fluid is considered in the presence of a magnetic field. The class of non-
steady flows is studied for which the velocity at the edge of the boundary layer varies
hyperbolically with time.

It is shown that under reasonable restrictions the equations of conservation of
total mass, energy, and momentum may be brought into similarity form. An approximate transformation is found which reduces their solution to that of the known non-
magnetic steady case. The basic equations are integrated by means of an electronic
analogue computer and the results are compared with the approximate solution.
Velocity and enthalpy profiles, wall shear, and heat transfer rates are presented for
different decelerations and magnetic field intensities. The effects of the magnetic
field and the deceleration rate upon the bow shock wave stand-off distance are also
evaluated. It is demonstrated that for reasonable decelerations the steady-state solutions are good approximations.

Unstetige verdichtbare magnetische Grenzschicht in hypersonischer Strömung.
Unstetige hypersonische Strömungen einer verdichtbaren, zähen, thermisch und elektrisch leitfähigen Strömung in Gegenwart eines magnetischen Feldes werden hier
untersucht, und zwar insbesondere solche Strömungen, bei denen sich die Geschwindigkeit am Rande der Grenzschicht hyperbolisch mit der Zeit ändert.

Es wird gezeigt, daß unter vernünftigen Bedingungen die Gleichungen der Erhaltung der Gesamtmasse, der Energie und der Bewegungsgröße in eine Ähnlichkeits-
form gebracht werden können. Eine Annäherungs-Transformation wurde gefunden,
welche ihre Lösung auf die des bekannten unmagnetischen Stetigkeitsfalles reduziert.
Die Grundgleichungen werden mit Hilfe einer elektronischen Analogie-Rechen-
maschine integriert; die Resultate werden mit den Näherungslösungen verglichen.
Geschwindigkeits- und Enthalpieprofile, Wandreibungsschub und Wärmetransport
werden für verschiedene Verzögerungen und magnetische Feldstärken angegeben. Für
mäßige Verzögerungen erweisen sich die Näherungslösungen des stetigen Zustandes
als ausreichend.

**Couches limites laminaires en écoulement hypersonique non stationnaire en présence
d'un champ magnétique.** L'article étudie la classe d'écoulements non stationnaires

---

[1] Allison Division, General Motors Corporation, Indianapolis, Indiana, U.S.A.

[2] Also Associate Professor of Aeronautical Engineering, Purdue University,
Lafayette, Indiana, U.S.A.

pour lesquels la vitesse à la limite de la couche a une variation hyperbolique avec le temps.

Sous des hypothèses restrictives raisonnables, les équations de conservation peuvent être mises sous une forme à similitude. Une transformation approchée réduit leur solution à celle du cas connu: permanent et sans champ magnétique. Les équations de base sont intégrées sur machine électronique analogue et les résultats sont comparés avec la solution approchée. Pour diverses valeurs de la décélération et du champ, des résultats sont donnés pour les profils de vitesse, de tensions tangentielles d'enthalpie et de taux de transfert de chaleur. Pour des valeurs raisonnables de la décélération les solutions permanentes constituent une bonne approximation.

Nomenclature

a	Acoustic velocity	X	Transformed longitudinal coordinate:
$\vec{B}$	Magnetic induction vector		
B_x, B_y	x and y components of $\vec{B}$		$$X = \int\limits_0^x \frac{a_e\,p_e}{a_o\,p_o}\,dx$$
C_f	Friction factor defined by Eq. (24)		
C_0	Stagnation point velocity gradient at zero time	x	Longitudinal coordinate measured from stagnation point
C_t	Stagnation point velocity gradient at general time	Y	Transformed normal coordinate:
c	Radius of curvature of bow shock wave		$$Y = \frac{a_e}{a_o} \int\limits_0^y \frac{p_e}{p_o}\,dy$$
c_1	Proportionality factor defined by Eq. (A-1)	y	Normal coordinate
c_2, c_3	Constants appearing in Eq. (A-2)	Z	Boundary value of f' defined by Eq. (15)
f	Velocity function defined through Eq. (6)	α	Non-steadiness parameter defined by Eq. (4)
f_o	Value of f corresponding to non-magnetic case	β	Pressure gradient parameter
h	Enthalpy	$\varDelta$	Bow shock wave stand-off distance parameter defined by Eq. (29)
k_1	Proportionality factor defined by Eq. (A-4)	ζ	Magnetic parameter defined by Eq. (10)
k_2	Proportionality factor defined by Eq. (A-3)	η	Similarity parameter defined by Eq. (6)
k_0, k_t	Density functions defined in Eqs. (31) and (32)	$\overline{\eta}$	Similarity parameter defined by Eq. (B-3)
m	Power appearing in Eq. (21)	η_o	Value of η according to Eq. (A-7)
Nu	Nusselt number defined by Eq. (27)	λ	Transformation variable defined by Eq. (B-2)
Pr	Prandtl number	μ	Dynamic viscosity
p	Static pressure	ν	Kinematic viscosity
R	Radius of curvature of blunt vehicle nose	ϱ	Density
Re	Reynolds number defined by Eq. (25)	σ	Electrical conductivity
		τ	Shear stress
S	Stagnation enthalpy function defined by Eq. (6)	*Subscripts* (except those defined above)	
T	Temperature	e	Condition at the edge of the boundary layer
t	Time		
U	Velocity	o	Free stream stagnation value behind bow shock wave
U^*	Velocity corresponding to absence of magnetic field	s	Stagnation value inside the boundary layer
u	Longitudinal velocity component		
$\vec{u}$	General velocity vector	w	Condition at the wall
v	Normal velocity component	∞	Condition ahead of the bow shock wave

Other Notation

Primes denote differentiation of a function with respect to its argument.

An underlined quantity such as $f''(0)$ or $S'(0)$ is evaluated for the steady, compressible, non-magnetic case of [3] for zero wall temperature and with the pressure gradient parameter, β, given by Eq. (B-8).

Introduction

In a number of recent papers [2, 5, 7, 8, 9, 10, 12] it has been found that application of a magnetic field fixed relative to a blunt body moving at hypersonic speed alters the flow behind the bow shock wave. The magnetic lines offer resistance to the ionized and electrically conducting medium, thereby decelerating the flow outside the boundary layer and reducing both the wall shear stress and the heat transfer rate at the surface. Furthermore, it results that the bow shock wave is detached at a greater distance from the blunt nose than in the non-magnetic case, a fact that has been observed experimentally [12].

The solutions referred to above have been obtained for the steady state. In an actual case, however, if a magnetic field is applied at some time during the steady motion of a vehicle, the flow will be decelerated and becomes time dependent. The non-steady nature of the problem is also immediately apparent when one considers re-entry calculations.

Since the theory of non-steady boundary layers is rather complex, solutions seem to be possible only under a number of restrictive assumptions. In [11] it is shown that for the case of incompressible, two-dimensional, non-magnetic stagnation flow, a similarity solution may be obtained if the velocity at the edge of the boundary layer is assumed to vary hyperbolically with time. These results are extended in the present paper to the compressible case including the additional effects of a magnetic field acting normal to the motion of the conducting fluid; the wall temperature is considered to be much lower than the free stream temperature behind the shock wave.

It is shown that under reasonable restrictions the equations of conservation of momentum, energy, and total mass may be brought into similarity form. An approximate transformation is found which reduces their solution to that of the known steady, non-magnetic case. The basic equations are integrated by means of an electronic analogue computer and the results compared with the approximate solution. Velocity and enthalpy profiles, wall shear stress, and heat transfer rates are presented for different decelerations and magnetic field intensities. The effects of the magnetic field and the deceleration rate upon the bow shock wave stand-off distance are also evaluated.

In general, it is found that the effects of both deceleration and the magnetic field are those that would be expected from an adverse pressure gradient. The magnetic field changes essentially the inviscid flow, the variation being reflected proportionately throughout the boundary layer profile. The effect upon the heat transfer, as well as upon the relative growths of the momentum and thermal boundary layers, is quite similar to what one would expect from a lowered PRANDTL number. It is also apparent that the structure of the boundary layer and the trend of its behavior with increasing deceleration rate are markedly different from those of the incompressible case. The steady-state solutions are shown to be good approximations for practical decelerations.

The Equations

The assumptions under which the conservation equations are written are the same as those of [7], apart from discussion of the non-steady terms; for convenience, they are restated here.

The equations are written for an ideal gas mixture. The electron and ion concentrations are assumed to be low enough that their contributions to the energy transfer may be neglected, yet of sufficient magnitude to permit appreciable interaction between the fluid and the imposed magnetic field. This condition is met at hypersonic flight.

Let $\vec{B}$ be the intensity of the magnetic field imposed normal to the flow direction, x. It is assumed that the electric field in a system of reference fixed relative to the moving body is zero. This assumption is reasonable because no external electric field is applied and the effect of polarization of the ionized fluid may be expected to be small if the conditions in the ionized layer are two-dimensional. With these assumptions, the retarding magnetic force per unit volume acting on each fluid element is equal to $\sigma\,(\vec{u} \times \vec{B})$.

Under the usual PRANDTL boundary layer assumptions, the equations of conservation take the following form:

Conservation of Momentum

$$\varrho\left(\frac{\partial u}{\partial t} + u\,\frac{\partial u}{\partial x} + v\,\frac{\partial u}{\partial y}\right) = -\frac{\partial p}{\partial x} + \frac{\partial}{\partial y}\left(\mu\,\frac{\partial u}{\partial y}\right) - \sigma\,B_y{}^2\,u + \sigma\,v\,B_x\,B_y\,, \qquad (1)$$

Conservation of Energy

$$\varrho\left(\frac{\partial h_s}{\partial t} + u\,\frac{\partial h_s}{\partial x} + v\,\frac{\partial h_s}{\partial y}\right) = -\frac{\partial p}{\partial t} + \mu\left(\frac{\partial u}{\partial y}\right)^2 + u\,\frac{\partial}{\partial y}\left(\mu\,\frac{\partial u}{\partial y}\right) +$$
$$+ \frac{\partial}{\partial y}\left(\frac{\mu}{Pr}\,\frac{\partial h}{\partial y}\right). \qquad (2)$$

Conservation of Mixture-Mass

$$\frac{\partial \varrho}{\partial t} + \frac{\partial (\varrho\,u)}{\partial x} + \frac{\partial (\varrho\,v)}{\partial y} = 0\,. \qquad (3)$$

At this point, the further assumption is made that the magnetic REYNOLDS number is small so that the induced currents flowing in the medium do not appreciably distort the imposed magnetic field; this condition seems to be valid for actual hypersonic flight. Therefore, the last term in Eq. (1) may be neglected in comparison with the previous one since the magnetic field component B_z is much smaller than the component B_y and the normal velocity component v is much less than u.

Since the magnetic field vector, $\vec{B}$, has been assumed everywhere normal to the x-direction and the electric field has been assumed zero, solution of MAXWELL's equations becomes redundant.

Considering next the equation of mass conservation, it is assumed that density does not change with time at any point in the coordinate system fixed relative to the moving body; the first term of Eq. (3) is therefore zero. This assumption is a reasonable idealization which later permits the use of a stream function in defining transformations. Compressibility aspects are retained through the more significant spatial variation of density.

Simplifications made in the equation of energy conservation are more easily discussed in terms of the transformed variables; therefore, this point will be deferred.

It is now possible, by using a modified Stewartson transformation [3] and by assuming a particular velocity variation at the edge of the boundary layer, to determine under what conditions the set of conservation equations may be brought to similarity form.

It is assumed that the inviscid velocity variation in the neighborhood of the stagnation point is given by the following relation:

$$U_e^* = \frac{C_0 x}{1 + a C_0 t} . \tag{4}$$

The quantity C_0 represents physically the velocity gradient at the stagnation point at zero time, the velocity gradient at any other time being given by

$$C_t = \frac{C_0}{1 + a C_0 t} . \tag{5}$$

It is apparent from Eq. (4) that the nondimensional factor a is a measure of the non-steadiness of the flow. For values $a > 0$ of practical interest, as time increases, the velocity U_e^* decreases. A physical interpretation of this velocity will follow from the next discussion.

The pressure distribution in the inviscid region may be assumed to be that calculated in the absence of a magnetic field. The validity of this assumption has been discussed in [8], where the results of [2] have been analyzed in this respect. It was found that the approximation is a reasonable one when compared to the assumption that the magnetic field remains constant in the direction perpendicular to the surface. Physically, a magnetic field originating from a coil within the nose of the vehicle diminishes with a certain inverse power away from the surface. (For an appreciable distance away from the ends of a magnetic dipole, this power is equal to three.) At the wall, where the magnetic field is strongest, the velocity is zero so that the electromagnetic body force is also zero; on the other hand, beyond the edge of the boundary layer and behind the shock wave, where both the fluid velocity and electrical conductivity are highest, the magnetic field is relatively weak. Numerical calculations in [8] have shown that the effect of the diminishing of the magnetic field is of greater influence than the altering of the pressure distribution due to the presence of the magnetic field. A very satisfactory compromise may be made by considering B_y to be an average of the magnetic field intensity taken between the stagnation point and the apex of the bow shock wave while assuming the pressure distribution in the inviscid region to be that of the non-magnetic case[1].

It follows from the above arguments that U_e^* of Eq. (4) represents the velocity that would have prevailed at the edge of the boundary layer in the absence of a magnetic field. It will be shown later that this velocity is very simply related to U_e, the actual velocity at the boundary layer edge.

Transformation of the conversation equations from the (x, y, t) to the (x, η, t) coordinate system may now be effected in terms of the following definitions:

$$f'(\eta) = \frac{U}{U_e^*}, \qquad S(\eta) = \frac{h_s}{h_o}, \qquad \eta = Y \sqrt{\frac{U_e^*}{\nu_o X}} \tag{6}$$

and the approximation

$$\frac{\varrho_e}{\varrho} \approx \frac{h_s}{h_o} = S(\eta) \tag{7}$$

---

[1] Actually, the variation of the pressure distribution may be taken into account merely by basing the calculation of the inviscid velocity upon the shock curvature rather than upon the curvature of the blunt nose.

which is justifiable at hypersonic speeds. The momentum and energy equations then become, respectively,

$$f''' + f f'' = (f')^2 - S - a\left(\frac{\eta}{2} f'' + f' - S\right) + \zeta S f' \tag{8}$$

and

$$S'' + Pr\left(f + \frac{a\eta}{2}\right) S' = 0, \tag{9}$$

where

$$\zeta = \frac{\sigma B_y^2 (1 + a C_0 t)}{\varrho_o C_0}. \tag{10}$$

In order that the above equations admit of a similarity solution, it is obvious that the nondimensional parameter ζ (the ratio of ponderomotive force to inertial force) must be a constant independent of time. One manner of holding ζ constant throughout all time of interest would be to cause the magnetic field intensity B_y to vary as the inverse square power of $(1 + a C_0 t)$. In this case, however, according to FARADAY's Law, there would be an induced electric field in contradiction to an assumption previously stated. Such an artifice does not seem necessary; over an actual re-entry trajectory, the electrical conductivity of the medium behind the bow shock wave may be expected to decrease with time as the strength of the shock decreases. Thus, on a qualitative basis, one may assume that

$$\sigma \sim \frac{1}{1 + a C_0 t}, \tag{11}$$

so that the magnetic parameter ζ is effectively a constant.

It is now appropriate to return to a discussion of the equation of energy conservation and simplifications that have been made in obtaining Eq. (9). The term $Pr \frac{a\eta}{2} S'$ appearing in this equation gives the contribution of the time-variation of the stagnation enthalpy arising from the spatial coordinate transformation, y to η[1]. Apart from this term, one must consider the contributions of the time variation of stagnation enthalpy and pressure appearing in Eq. (2) as the first and fourth terms, respectively. In their transformed form, they become the first and second terms in the following expression:

$$Pr S \frac{U_e^{*2}}{h_o} a + Pr S \frac{U_e^{*2}}{h_o} a (a - 1) = Pr S \frac{U_e^{*2}}{h_o} a^2. \tag{12}$$

However, since $h_o \gg U_e^{*2}$ and $a^2 < a,^2$ this expression is obviously negligible when compared with the terms retained in Eq. (9).

It has, therefore, been shown that the basic conservation equations may be written in their similarity form as Eqs. (8) and (9) and that, under the assumed inviscid velocity variation of Eq. (4), these equations admit of a similarity solution.

The Boundary Conditions

From the definitions of the non-dimensional velocity f' and stagnation enthalpy S, and for the case of no slip or mass transfer at the wall, the following boundary conditions are appropriate:

$$f(0) = f'(0) = 0,$$
$$S(0) = S_w, \tag{13}$$
$$S(\infty) = 1.$$

---

[1] From the definition of the similarity parameter η in Eq. (6), one may note that η contains the time, t, through the velocity U_e^*.

[2] It will be shown later that $a < 1$ always.

At the edge of the boundary layer, as the similarity parameter η tends to infinity, the function $f'(\eta)$ assumes a constant value and the second and third derivatives of $f(\eta)$ correspondingly approach zero. As a result, Eq. (8) yields

$$f'^2(\infty) + (\zeta - a) f'(\infty) - (1 - a) = 0. \tag{14}$$

The positive solution of this quadratic gives the final boundary condition:

$$Z = f'(\infty) = \frac{a - \zeta}{2} + \sqrt{\left(\frac{a - \zeta}{2}\right)^2 + (1 - a)}. \tag{15}$$

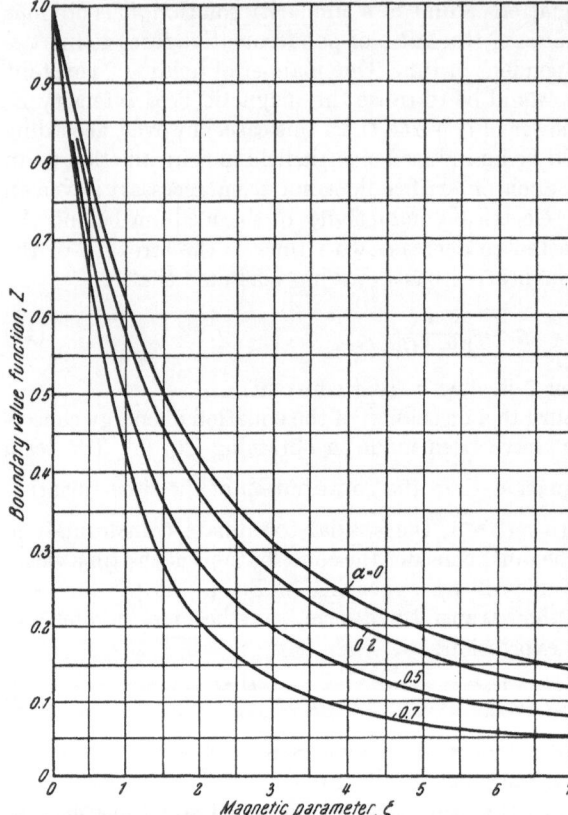

Fig. 1. Boundary value function Z for different values of non-steadiness parameter a and magnetic parameter ζ

One immediately concludes from Eq. (15) and the definition of the function $f'(\eta)$ that the velocity U_e actually prevailing at the edge of the boundary layer in the unsteady magnetic flow is given by the simple relation:

$$U_e = Z U_e{}^*. \tag{16}$$

As will be demonstrated later, the parameter a may take only positive values less than unity. Therefore, the discriminant of Eq. (15) is always positive so that the boundary condition for $f'(\eta)$ at infinity always admits of a positive root[1]. For zero value of the magnetic parameter ζ, the function $Z(a, \zeta)$ is always unity regardless of the value of a. This function is shown in Fig. 1. It may be seen that, for the same value of ζ, values of Z are lowered with increasing values on the non-steadiness parameter a, as one would expect from physical argument.

Solution of the Equations

The system of conservation Eqs. (8) and (9), may now be solved in terms of the boundary conditions given by Eqs. (13) and (15). Before taking up the results of the analogue computations, it is worth while to investigate the nature of the expected solutions by an examination of the differential equations themselves.

An approximation to the final results is always desirable. In the case of an analogue computer which recognizes the similarity parameter η as physical time, integration starts at "time" zero and proceeds from the boundary conditions at

---

[1] Inspection of Eq. (14) shows that the other root is always negative and hence of no physical significance.

the wall toward satisfaction of the boundary conditions at infinity. Therefore, it becomes necessary to guess two additional boundary conditions at the wall for solution of the fifth degree coupled equations. The two additional quantities introduced are $f''(0)$ and $S'(0)$; thus it becomes necessary to guess the quantities which are precisely of greatest interest, representing the wall shear stress and heat transfer rate, respectively.

The behaviour of the function $f'(\eta)$ is first examined in the neighborhood of the wall. Here, the momentum equation yields

$$f''' = -S_w(1-a). \tag{17}$$

Since S_w is always positive, f''' is always negative for decelerated flow. Therefore, the curve of f' vs. η is concave downward, and neither the magnetic parameter ζ nor the non-steadiness parameter α influence the concavity. This means that, although the magnetic fields acts like an adverse pressure gradient, the flow may be decelerated but not separated. From these arguments, one would expect to observe a similarity between the profiles $f'(\eta)$ for $\zeta=0$ and $f'(\eta)$ for $\zeta \neq 0$ for the same value of α. The magnetic field would be expected to strongly influence the inviscid flow, the change being reflected proportionally throughout all values of η, in exactly the same manner as in the magnetic steady case discussed in [7].

As a first approximation, one may state that

$$f'(\eta) = f_o'(\eta) Z, \tag{18}$$

where $f_o'(\eta)$ is the velocity corresponding to the non-magnetic case for which $f_o'(\infty)=1$. Considering this approximation, the momentum equation may be written in terms of $f_o'(\eta)$ as follows:

$$f_o''' + f_o f_o'' = (f_o')^2 - S - \frac{a\eta}{2} f_o'' ,$$

or

$$f_o''' + \left(f_o + \frac{a\eta}{2}\right) f_o'' = (f_o')^2 - S . \tag{19}$$

According to Appendix A, one may set, approximately,

$$f_o = \frac{\eta}{3} ,$$

so that Eq. (19) becomes

$$f_o''' + (1 + 1.5a) f_o f_o'' = (f_o')^2 - S .$$

It is shown in Appendix B that the solution of this equation for $f_o''(0)$ is given by

$$f_o''(0) = \sqrt{1 + 1.5a}\ f''(0) ,$$

where $f''(0)$ is the wall shear stress function calculated in [3] for steady, compressible, non-magnetic flow with PRANDTL number unity, wall temperature zero, and the pressure gradient parameter given by $\beta = \dfrac{1}{1 + 1.5a}$.

Hence, from Eq. (18), one finds as a first approximation that

$$f''(0) = \sqrt{1 + 1.5a}\ Z\ f''(0) . \tag{20}$$

Use of these results in the energy equation yields

$$S'' + Pr(1 + 1.5a) Z f_o S' = 0 .$$

It is apparent that the effect of the non-steady term and the magnetic parameter are combined in the product $(1+1.5\alpha)\,Z$ and have the same influence as the PRANDTL number. It is now a relatively simple matter to establish that

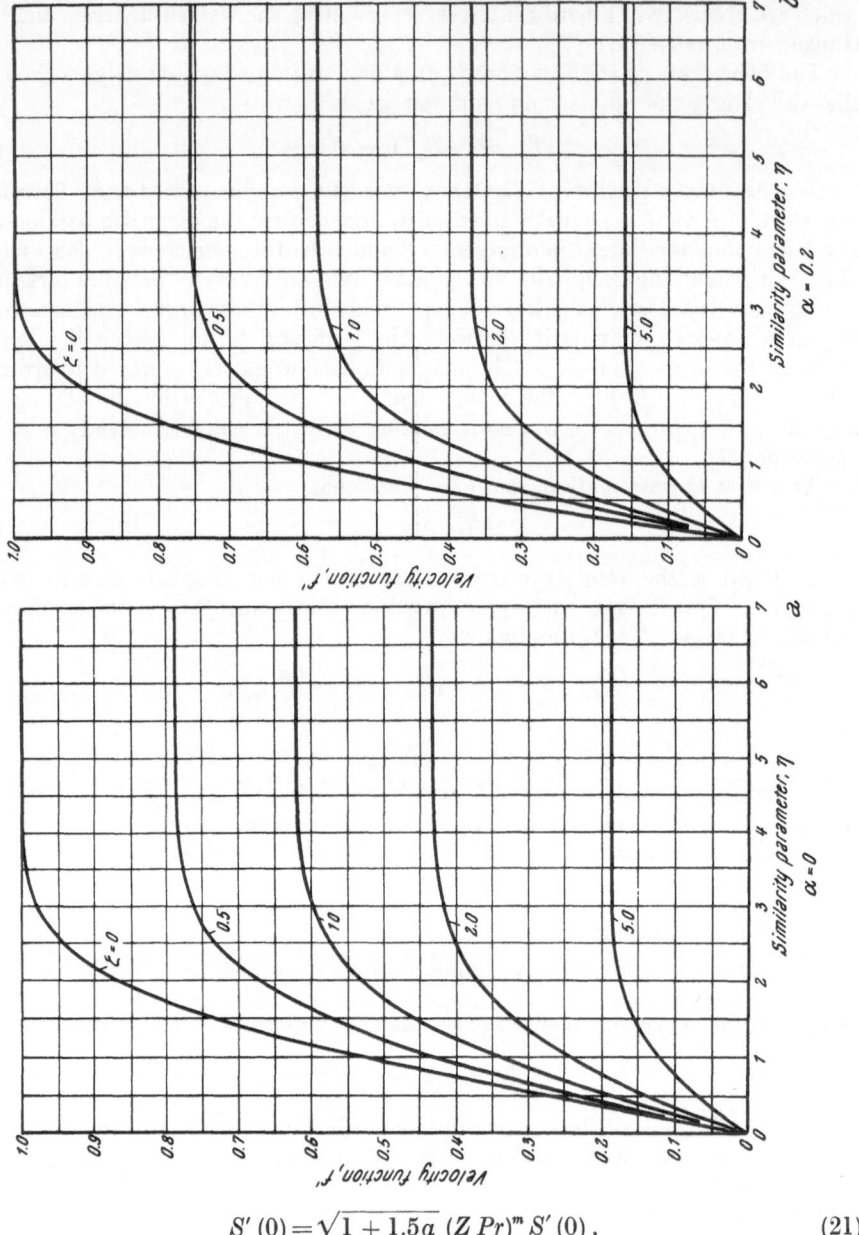

$$S'(0)=\sqrt{1+1.5\alpha}\;(Z\,Pr)^{m}\,\underline{S}'(0)\,,\tag{21}$$

where m is a power dependent upon the pressure gradient.

The approximations obtained for $f''(0)$ and $S'(0)$ are shown, respectively, in Figs. 4 and 5 for different values of the parameters ζ and α.

It is well known that the power m depends upon the effective pressure gradient, a measure of which is the parameter β. For $\alpha=0$, corresponding to steady flow,

β is unity and it has been established that m is approximately 0.4 for this case. On the other hand, in the extreme case of $\alpha = 1$, β becomes 0.4 and one may set m roughly equal to $1/3$. This means that for stronger decelerations and higher

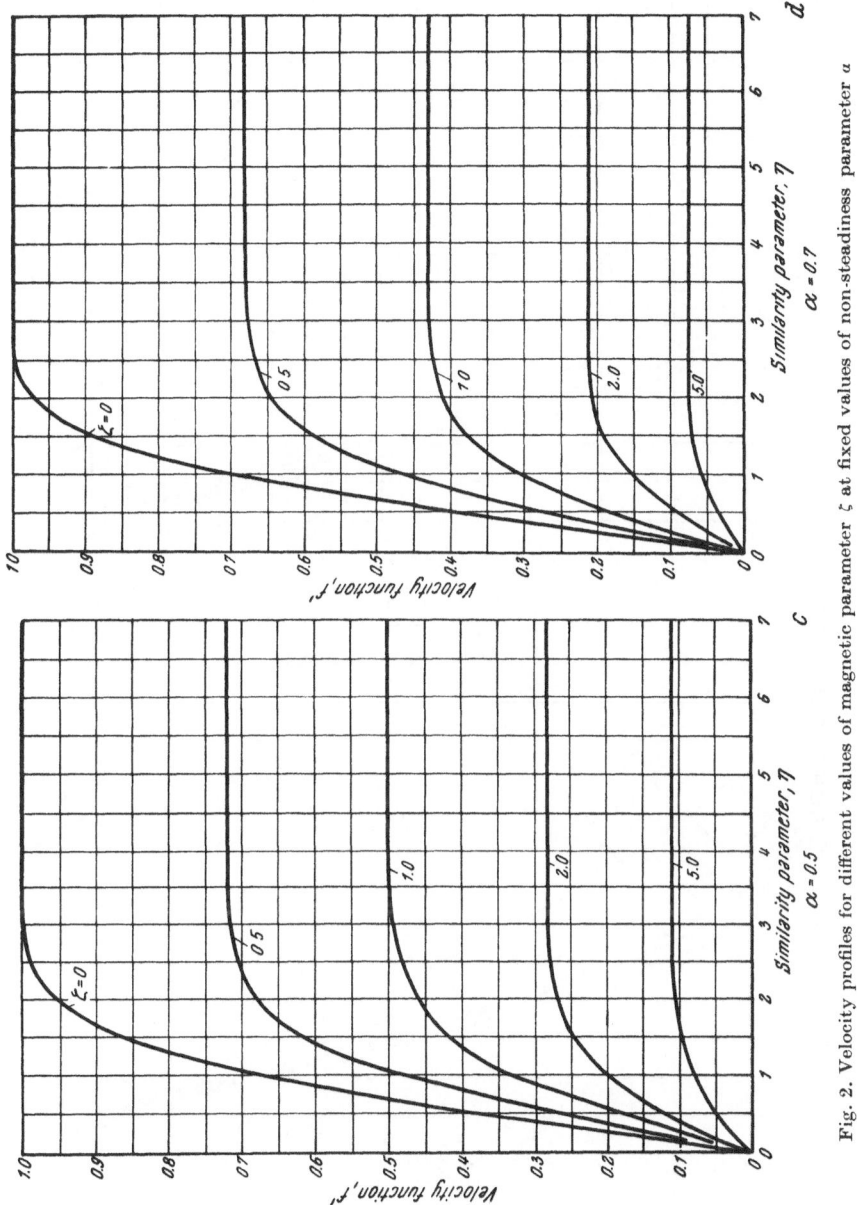

Fig. 2. Velocity profiles for different values of magnetic parameter ζ at fixed values of non-steadiness parameter α

values of the magnetic field, the effective pressure gradient becomes less favorable, exactly as one would expect on physical grounds. Assuming a parabolic variations of m with α over the range $0 \leqslant \alpha \leqslant 1$, one calculates

$$m = 0.4 - \frac{a(2-a)}{15}. \tag{22}$$

From the above discussion, it is seen that the solution of the magnetic non-steady problem may be reduced approximately to the known solution of the non-magnetic steady case.

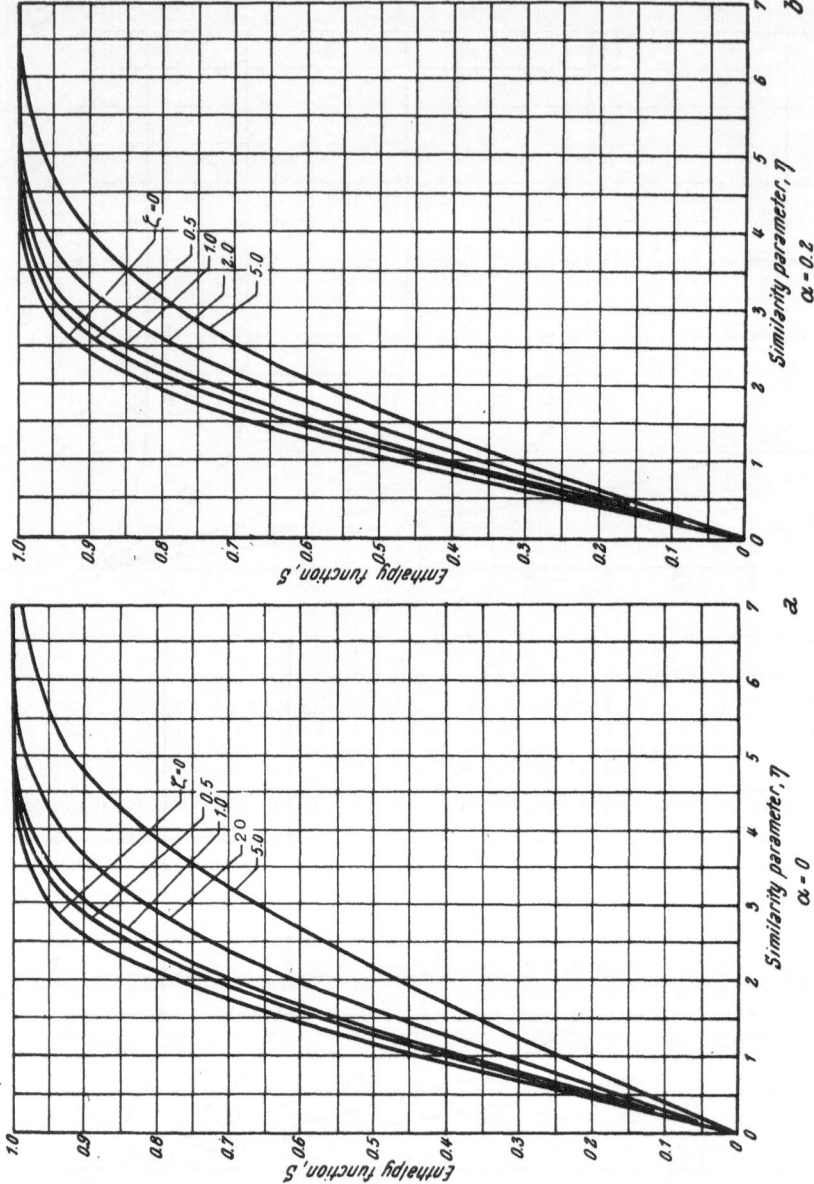

The basic conservation equations (8) and (9) were programed for solution on Allison's Berkeley EASE Electronic Analogue Computer. Sample velocity and stagnation enthalpy profiles are shown in Figs. 2 and 3, respectively[1]. The shear and enthalpy gradient functions, shown in Figs. 6 and 7, may be compared

---

[1] In all calculations the PRANDTL number was taken as 0.7 and S_w was set equal to zero.

with their theoretical predictions in Figs. 4 and 5. Agreement is seen to be good for low values of a and all values of ζ as well as for low values of ζ and all values of a.

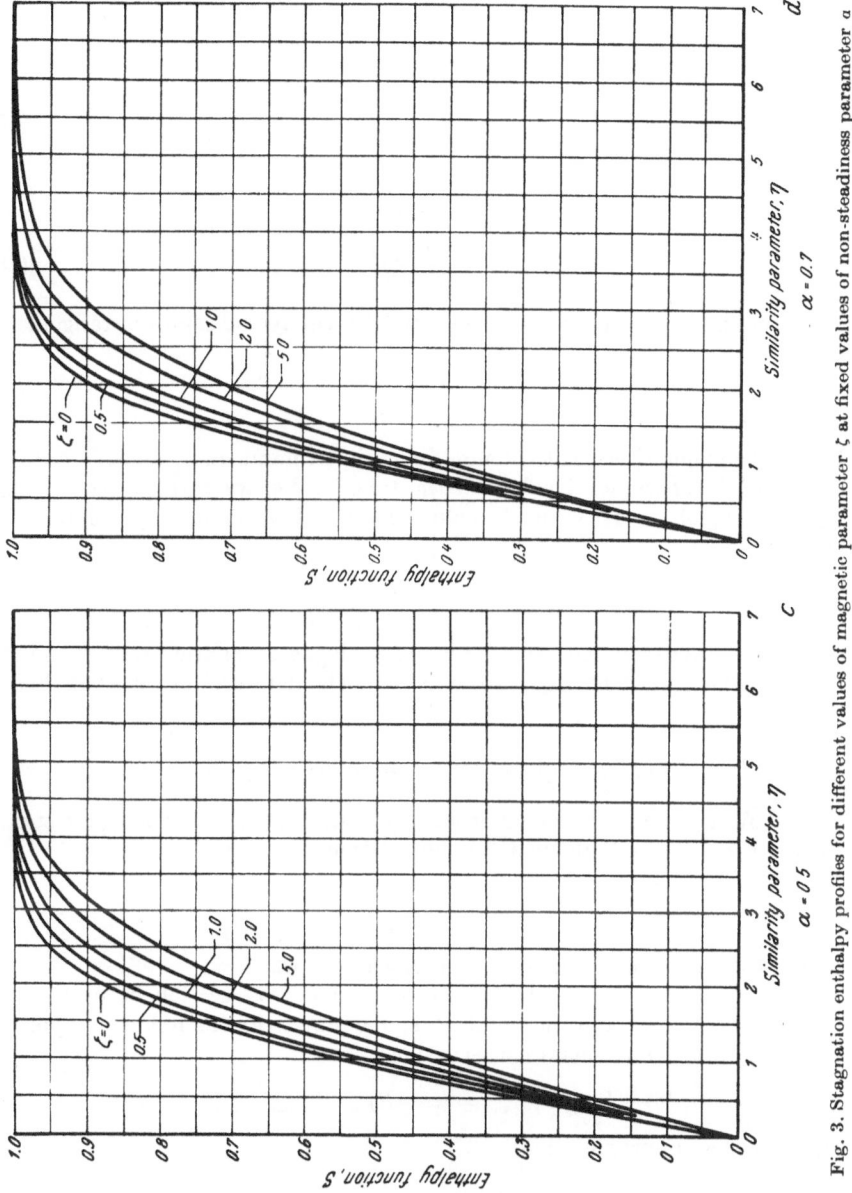

Fig. 3. Stagnation enthalpy profiles for different values of magnetic parameter ζ at fixed values of non-steadiness parameter a

Calculation of Drag and Heat Transfer

The actual shear stress at the wall may be calculated from

$$\tau_w = \mu \left[\frac{\partial u}{\partial y} \right]_w = \mu_o \sqrt{\frac{U_e^*}{\nu_o X}} \; U_e^* f''(0). \tag{23}$$

The friction coefficient is, therefore, given by

$$C_f \equiv \frac{\tau_w}{\frac{1}{2}\varrho_0 U_e^{*2}} = 2\,Re^{-\frac{1}{2}} f''(0)\,, \tag{24}$$

where the Reynolds number has been defined as

$$Re = \frac{U_e^* X}{\nu_0}\,. \tag{25}$$

The Nusselt number is then formed as

$$Nu = \frac{X\left[\frac{\partial T}{\partial y}\right]_w}{T_0 - T_w}\,, \tag{26}$$

which becomes, after calculation of the partial derivative and some rearrangement,

$$Nu = \frac{S'(0)\,Re^{\frac{1}{2}}}{1 - S_w}\,. \tag{27}$$

It is to be emphasized that the non-dimensional numbers C_f and Nu are calculated in terms of the local values of the time- and space-variable parameters.

The validity of the Reynolds analogy may be investigated by defining the parameter

$$\frac{C_f\,Re}{Nu} \equiv 2\,(1 - S_w)\,\frac{f''(0)}{S'(0)}\,. \tag{28}$$

This quantity was calculated using both the analogue calculation results and the theoretical values given by Eqs. (20) and (21). The data are presented in Fig. 8 as a function of α and ζ.

The Bow Shock Wave Stand-Off Distance

The following is an order of magnitude calculation of the bow shock wave stand-off distance parameter $\varDelta$ defined by the relation

$$\varDelta \equiv \frac{c - R}{R}\,, \tag{29}$$

where c is the radius of curvature of the bow shock wave.

For an incompressible fluid, and considering a cross-section of unit depth between the bow shock and the blunt nose, one may write the following approximate equation of overall mass conservation:

$$(c - R)_{mag}\,U_{avg,\,mag} = (c - R)_{non-mag}\,U_{avg,\,non-mag}\,.$$

Then, since as an approximation

$$\frac{U_{avg,\,mag}}{U_{avg,\,non-mag}} = Z\,,$$

one finds that

$$\frac{\varDelta_{mag}}{\varDelta_{non-mag}} \approx \frac{1}{Z}\,. \tag{30}$$

The inverse of this function is given as Fig. 1.

For high magnetic field intensities and for increasing rates of deceleration, the function $Z(\alpha, \zeta)$ decreases. Therefore, the above stand-off distance ratio increases as one would expect and has been observed experimentally.

The above approximation is valid only for small values of the magnetic parameter ζ.

Discussion of the Results

Before discussing the results of the calculations, it is appropriate to examine the permissible range of values of the non-steadiness parameter α.

From [6] it may be estimated that

$$C_t = \frac{U^*_{\infty,t}}{R} \sqrt{\frac{2\varrho_{\infty,t}}{\varrho_{0,t}}} = \frac{U^*_{\infty,t}}{R} k_t \qquad (31)$$

and

$$C_0 = \frac{U^*_{\infty,0}}{R} \sqrt{\frac{2\varrho_{\infty,0}}{\varrho_{0,0}}} = \frac{U^*_{\infty,0}}{R} k_0, \qquad (32)$$

where the second subscript and that on k refers to the time at which the velocity gradient is calculated; R is considered to be the radius of curvature of the blunt

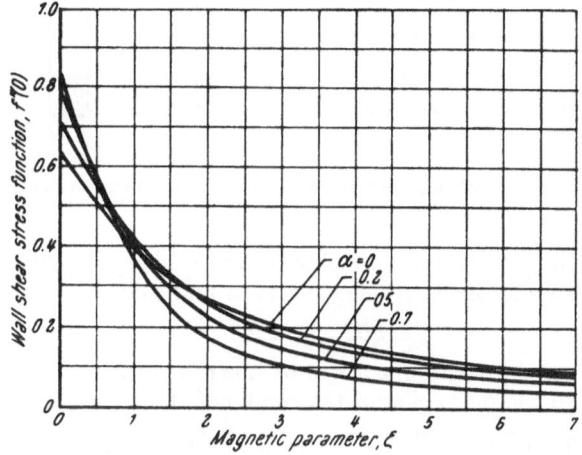

Fig. 4. Predicted values: wall shear stress function $f''(0)$ for different values of non-steadiness parameter α and magnetic parameter ζ

nose rather than that of the shock wave. If one solves Eq. (5) for α and substitutes the above expressions, the result is

$$\alpha = \frac{R}{t} \left[\frac{1}{U^*_{\infty,t} k_t} - \frac{1}{U^*_{\infty,0} k_0} \right].$$

Therefore, α will be greatest when the initial velocity $U^*_{\infty,0}$ is large in comparison with the instantaneous velocity $U^*_{\infty,t}$ as would be the case for high rates of deceleration. The maximum possible value of α may be estimated, then, by setting

$$\alpha_{max} = \frac{R}{t\, U^*_{\infty,t}\, k_t}.$$

Taking k_t to be approximately 0.25[1] and assuming that α_{max} might be as high as 1.0, it follows that

$$t\, U^*_{\infty,t} \approx 4R.$$

---

[1] The factors k_0 and k_t differ very little in comparison with the variation in the velocity. For example, k_0 corresponding to a velocity of 26,000 ft/sec at an altitude of 200,000 ft may be calculated from [4] to be 0.32; k_t corresponding to 7,000 ft/sec is approximately equal to 0.20 for all altitudes.

Physically, this condition implies that α will be of the order of unity when the distance traveled with the terminal velocity $U_{\infty,\,t}^{*}$, during the total deceleration

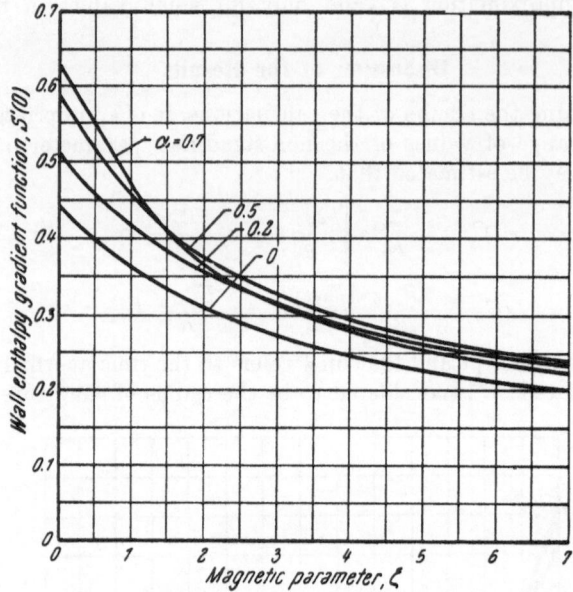

Fig. 5. Predicted values: wall enthalpy gradient function $S'(0)$ for different values of non-steadiness parameter α and magnetic parameter ζ

time t is approximately twice the diameter of the vehicle. Obviously, this would be an extreme case and, instead, one expects to find that

$$t\,U_{\infty,\,t}^{*} \gg 4\,R$$

so that for practical decelerations, α will be much less than unity.

Turning now to the discussion of the results of the present paper, it may be

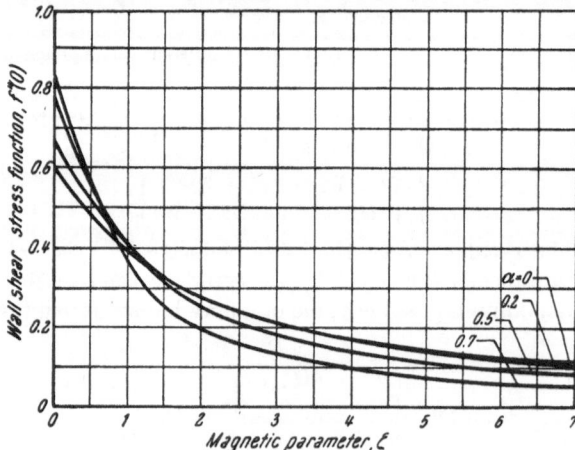

Fig. 6. Computed values: wall shear stress function $f''(0)$ for different values of non-steadiness parameter α and magnetic parameter ζ

noted from Figs. 4 and 6 that for zero magnetic parameter, ζ, the wall shear stress function $f''(0)$ takes higher values for increasing values of α. This trend is exactly the reverse of that reported in [11] for the incompressible case. In other words, the compressibility in unsteady flow not only changes the numerical value of

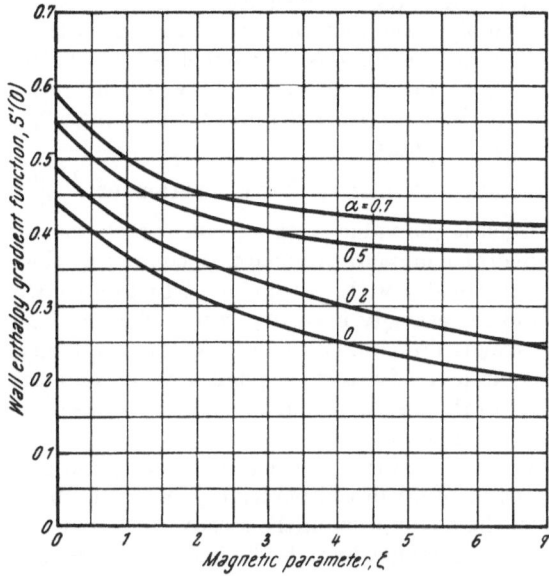

Fig. 7. Computed values: wall enthalpy gradient function $S'(0)$ for different values of non-steadiness parameter a and magnetic parameter ζ

$f''(0)$ but also reverses the trend of its dependence upon the non-steadiness parameter a. Of course it may be seen from Eq. (23) that the actual shear stress at the

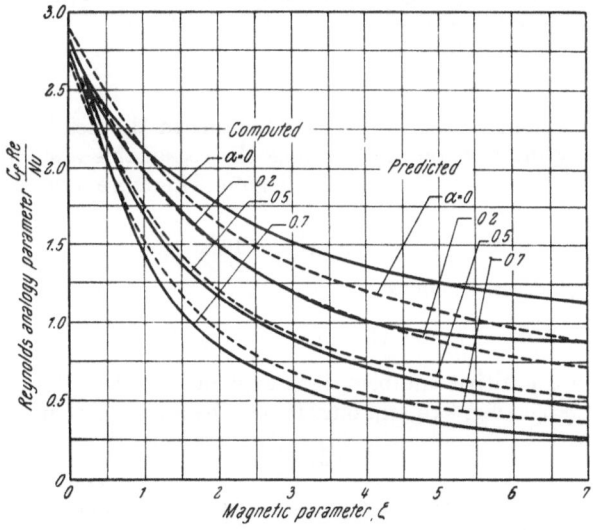

Fig. 8. Predicted and computed results: REYNOLDS analogy parameter $\dfrac{C_f\,Re}{Nu}$ for different values of non-steadiness parameter a and magnetic parameter ζ

wall must decrease due to the decrease of the local inviscid velocity. The stagnation enthalpy gradient at the wall is seen from Figs. 5 and 7 also to increase with a for zero magnetic field.

For constant values of the deceleration parameter, the above figures show that stronger magnetic fields act to decrease both the wall shear stress and the heat transfer rate. The predicted rates of decrease for high ζ (above 2) and high a (above 0.2) are appreciably lower than those resulting from the complete analogue solution of the conservation equations. Nevertheless, the theoretical predictions exhibit the correct trend. Better agreement between theoretical and analogue results is shown for the Reynolds analogy parameter than for the shear stress and enthalpy gradient functions.

From Fig. 8, one may also observe that the parameter $\dfrac{C_f Re}{Nu}$ assumes values smaller than 2.8, which is the value for $a = \zeta = 0$, exactly as would be expected to happen for less favorable pressure gradients. In comparing this reduction of the Reynolds analogy parameter with the similar reduction shown in [3] for zero wall temperature and increasingly negative values of β, one must bear in mind that, although the trends are similar there is a significant distinction between them. In the non-magnetic steady case, the point of flow separation will eventually be reached; however, in the magnetic non-steady case as discussed in the present paper, this point is only approached asymptotically.

Finally, it may be concluded that, for reasonable decelerations similar to those reported in [1], the value of a will be appreciably less than unity. Therefore, calculations based upon the assumption of the steady state, but using values of the flow parameters which prevail locally in both time and space, may be expected to yield reasonably satisfactory results.

Acknowledgement

The authors wish to gratefully acknowledge the help of Mr. D. R. Zimmerman of the Allison Division Applied Physics Section who accomplished the analogue solution of the basic conservation equations.

Appendix A

Since the behaviour of the unknown functions in the neighbourhood of the wall is of primary interest, one may utilize the reasonable assumption for cool walls that S varies linearly with η in this region. Therefore, one may write from Eq. (8) that

$$f''' = -c_1 (1-a)\,\eta,\qquad\qquad\text{(A-1)}$$

which, upon integration, becomes

$$f = -(1-a)\,c_2\,\eta^4 + c_3\,\eta^2.\qquad\qquad\text{(A-2)}$$

Since c_2 and c_3 are positive numbers, it is evident from the above that f varies with a power of η between two and one. It may be assumed that

$$f_o = k_2\,\eta^2,\qquad\qquad\text{(A-3)}$$

so that if one wishes to express an equivalent linear variation as

$$f_o = k_1\,\eta,\qquad\qquad\text{(A-4)}$$

the coefficient of proportionally k_1 is to be evaluated from the following condition:

$$\frac{1}{\eta_o}\int_0^{\eta_o} k_1\,\eta\,d\eta = \frac{1}{\eta_o}\int_0^{\eta_o} k_2\,\eta^2\,d\eta,\qquad\qquad\text{(A-5)}$$

where η_o is a finite value denoting the edge of the boundary layer. This condition yields, approximately,

$$k_1 = \frac{2}{3} k_2 \eta_o , \tag{A-6}$$

where the value of η_o must be calculated according to its definition from

$$f_o'(\eta_o) = 1 = 2 k_2 \eta_o . \tag{A-7}$$

Therefore, one concludes that

$$k_1 \approx \frac{2}{3} \times \frac{1}{2} = \frac{1}{3} . \tag{A-8}$$

Appendix B

Consider the differential equation

$$f_o''' + (1 + 1.5 a) f_o f_o'' = (f_o')^2 - S . \tag{B-1}$$

Define the transformation

$$f_o \equiv \lambda f(\bar{\eta}) , \tag{B-2}$$

where

$$\bar{\eta} = \frac{\eta}{\lambda} \tag{B-3}$$

and λ is a constant independent of η. It follows that

$$f_o' = f' , \quad f_o'' = \frac{1}{\lambda} f'' , \quad f_o''' = \frac{1}{\lambda^2} f''' . \tag{B-4}$$

Upon substitution, Eq. (B-1) becomes

$$\frac{1}{\lambda^2} f''' + (1 + 1.5 a) f f'' = (f')^2 - S . \tag{B-5}$$

If one now sets

$$\lambda = \frac{1}{\sqrt{1 + 1.5 a}} , \tag{B-6}$$

Eq. (B-5) becomes

$$f''' + f f'' = \frac{1}{1 + 1.5 a} [(f')^2 - S] . \tag{B-7}$$

From Eqs. (B-2), (B-3) and (B-4) it is obvious that the boundary conditions for the solution of Eq. (B-7) are precisely those used for solving the compressible, non-magnetic, steady problem discussed in [3]. In fact, this equation corresponds exactly to the case in which the pressure gradient parameter β of this reference is expressed as

$$\beta = \frac{1}{1 + 1.5 a} . \tag{B-8}$$

For the present study, this parameter assumes values between 1.0 for $a = 0$, and 0.4 for $a = 1$.

One notes that by increasing values of the deceleration parameter a, the effective pressure gradient becomes less favorable. Nevertheless, for cool walls (for which $S_w=0$) the changes in $f''(0)$ and $S'(0)$ over the range of β are small. The shear function at the wall is then given by

$$f_0''(0) = \sqrt{1 + 1.5a}\, f''(0),   \tag{B-9}$$

in which $f''(0)$ is taken from [3] for zero wall temperature and the pressure gradient parameter given by Eq. (B-8).

Table I. *Boundary Value Function Z for Different Values of Non-Steadiness Parameter a and Magnetic Parameter ζ*

a	ζ	Z	a	ζ	Z
0	0	1.000	0.5	0	1.000
	0.5	0.781		0.5	0.707
	1.0	0.618		1.0	0.500
	2.0	0.414		2.0	0.281
	5.0	0.193		5.0	0.111
	7.0	0.140		7.0	0.077
0.2	0	1.000	0.7	0	1.000
	0.5	0.757		0.5	0.657
	1.0	0.580		1.0	0.418
	2.0	0.369		2.0	0.200
	5.0	0.161		5.0	0.070
	7.0	0.116		7.0	0.048

Table II. *Predicted and Computed Values: Wall Shear Stress Function $f''(0)$ for Different Values of Non-Steadiness Parameter a and Magnetic Parameter ζ*

a	ζ	$f''(0)$ Eq. (20)	$f''(0)$ Computed	a	ζ	$f''(0)$ Eq. (20)	$f''(0)$ Computed
0	0	0.637	0.608	0.5	0	0.801	0.776
	0.5	0.497	0.475		0.5	0.566	0.551
	1.0	0.394	0.382		1.0	0.401	0.401
	2.0	0.264	0.277		2.0	0.225	0.247
	5.0	0.123	0.147		5.0	0.0889	0.115
	7.0	0.0892	0.113		7.0	0.0617	0.0864
0.2	0	0.713	0.678	0.7	0	0.849	0.837
	0.5	0.540	0.513		0.5	0.558	0.556
	1.0	0.414	0.402		1.0	0.355	0.365
	2.0	0.267	0.275		2.0	0.170	0.193
	5.0	0.115	0.142		5.0	0.0595	0.0785
	7.0	0.0827	0.108		7.0	0.0408	0.0581

Analogue computer values were obtained from readings of a four-place digital voltmeter.

Table III. *Predicted and Computed Values: Wall Enthalpy Gradient Function $S'(0)$ for Different Values of Non-Steadiness Parameter a and Magnetic Parameter ζ*

a	ζ	$S'(0)$ Eq. (21)	$S'(0)$ Computed	a	ζ	$S'(0)$ Eq. (21)	$S'(0)$ Computed
0	0	0.443	0.438	0.5	0	0.584	0.546
	0.5	0.402	0.398		0.5	0.517	0.500
	1.0	0.366	0.366		1.0	0.458	0.468
	2.0	0.312	0.313		2.0	0.374	0.424
	5.0	0.229	0.234		5.0	0.270	0.381
	7.0	0.202	0.200		7.0	0.238	0.372
0.2	0	0.510	0.486	0.7	0	0.630	0.588
	0.5	0.460	0.444		0.5	0.546	0.540
	1.0	0.418	0.408		1.0	0.469	0.500
	2.0	0.355	0.363		2.0	0.365	0.454
	5.0	0.263	0.282		5.0	0.255	0.418
	7.0	0.233	0.243		7.0	0.225	0.410

Analogue computer values were obtained from readings of a four-place digital voltmeter.

Table IV. *Predicted and Computed Results: Reynolds Analogy Parameter $\dfrac{C_f Re}{Nu}$ for Different Values of Non-Steadiness Parameter a and Magnetic Parameter ζ*

a	ζ	$\dfrac{C_f Re}{Nu}$ Predicted	$\dfrac{C_f Re}{Nu}$ Computed	a	ζ	$\dfrac{C_f Re}{Nu}$ Predicted	$\dfrac{C_f Re}{Nu}$ Computed
0	0	2.874	2.776	0.5	0	2.742	2.842
	0.5	2.472	2.386		0.5	2.188	2.204
	1.0	2.152	2.086		1.0	1.750	1.712
	2.0	1.629	1.768		2.0	1.202	1.164
	5.0	1.074	1.256		5.0	0.658	0.602
	7.0	0.880	1.130		7.0	0.522	0.465
0.2	0	2.796	2.790	0.7	0	2.694	2.846
	0.5	2.346	2.310		0.5	2.042	2.058
	1.0	1.980	1.970		1.0	1.512	1.460
	2.0	1.504	1.514		2.0	0.930	0.850
	5.0	0.874	1.006		5.0	0.462	0.374
	7.0	0.712	0.888		7.0	0.364	0.282

Analogue computer values were obtained from readings of a four-place digital voltmeter.

References

1. H. J. ALLEN and A. J. EGGERS, A Study of the Motion and Aerodynamic Heating of Missiles Entering the Earth's Atmosphere at High Supersonic Speeds. NACA Technical Note 4047, October 1957.
2. W. B. BUSH, Magnetohydrodynamic-Hypersonic Flow Past a Blunt Body. J. Aero/Space Sci. **25**, 685 (1958).

3. C. B. Cohen and E. Reshotko, Similar Solutions for the Compressible Laminar Boundary Layer with Heat Transfer and Pressure Gradient. NACA Report 1293, 1956.
4. S. Feldman, Hypersonic Gas Dynamic Charts for Equilibrium Air. AVCO Research Laboratory, January 1957.
5. N. H. Kemp, On Hypersonic Stagnation-Point Flow with a Magnetic Field. J. Aeronaut. Sci. (Readers' Forum) 25, 405 (1958).
6. T. Y. Li and R. E. Geiger, Stagnation Point of a Blunt Body in Hypersonic Flow. J. Aeronaut. Sci. 24, 25 (1957).
7. P. S. Lykoudis, On a Class of Compressible Laminar Boundary Layers with Pressure Gradient for an Electrically Conducting Fluid in the Presence of a Magnetic Field. Proceedings of the IXth International Astronautical Congress, Amsterdam 1958, p. 168. Wien: Springer, 1959.
8. P. S. Lykoudis, The Matching of the Viscid and Inviscid Regions for the Stagnation Magnetic Flow. J. Aero/Space Sci. (Readers' Forum) 26, 315 (1959).
9. J. L. Neuringer and W. McIlroy, Incompressible Two-Dimensional Stagnation-Point Flow of an Electrically Conducting Viscous Fluid in the Presence of a Magnetic Field. J. Aeronaut. Sci. 25, 194 (1958).
10. V. J. Rossow, Magnetohydrodynamic Analysis of Heat Transfer near a Stagnation Point. J. Aeronaut. Sci. (Readers' Forum) 25, 334 (1958).
11. K. T. Yang, Unsteady Laminar Boundary Layers in an Incompressible Stagnation Flow. American Society of Mechanical Engineers paper no. 58-A-3, presented at ASME Annual Meeting, New York, December 1958.
12. R. W. Ziemer, Experimental Magneto-Aerodynamics. American Rocket Society paper no. 707-58, presented at ARS 13th Annual Meeting, New York, November 1958.

Etude d'un échangeur de chaleur par convection et rayonnement pour un autopropulseur fissiothermique

Par

P. Perrier[1]

(Avec 10 Figures)

(Reçu le 8 juin 1959)

Résumé — Zusammenfassung — Abstract

Etude d'un échangeur de chaleur par convection et rayonnement pour un autopropulseur fissiothermique. Les échanges de chaleur entre un réacteur nucléaire et le propulsif et, par conséquent, les performances de la fusée correspondante pourraient être augmentés en opacifiant celui-ci et utilisant ainsi l'énergie émise sous forme de rayonnement par les parois portées à haute température. On montre qu'il est essentiel de tenir compte dans ces échanges radiatifs de la répartition spectrale de l'énergie et des coefficients d'absorption. On évalue numériquement sur un exemple concret le gain possible pour un autopropulseur après avoir choisi les paramètres de façon optimale. Le propulsif utilisé comprend des particules de carbone en suspension dans de l'ammoniac, on montre que l'énergie totale transmise par rayonnement $\sum W\,(\theta_c, N, r)$ à une particule qui dépend de la température de celle-ci θ_c, de la concentration N, et du rayon r est toujours très loin du maximum correspondant au corps noir. L'intégration numérique des équations du problème montre que les gains sont obtenus pour une très faible concentration en particules.

Untersuchung eines durch Konvektion und Strahlung arbeitenden Wärmeaustauschers für ein mittels Fissionsreaktors beheiztes Triebwerk. Der Wärmeaustausch zwischen einem kernchemischen Reaktor und dem Treibstoff, infolgedessen auch die Leistung der Rakete, könnten erhöht werden, indem letzterer künstlich undurchsichtig gemacht wird und die von den hocherhitzten Behälterwänden ausgestrahlte Energie daher ausgenutzt wird. Man zeigt, daß es bei diesem Wärmeaustausch durch Strahlung wichtig ist, die spektrale Energieverteilung und die Absorptionskoeffizienten zu berücksichtigen. Es wird am Zahlenbeispiel der mögliche Gewinn berechnet, der bei optimaler Parameterwahl für ein Triebwerk möglich ist. Der benutzte Treibstoff enthält Kohlenstaubsuspension in Ammoniak. Es wird gezeigt, daß die durch Strahlung an ein Teilchen übertragene Gesamtenergie $\sum W\,(\theta_c, N, r)$, die von der Partikeltemperatur θ_c, von seiner Konzentration N und von seinem Radius r abhängt, noch immer weit von dem einem schwarzen Körper entsprechenden Maximum liegt. Die numerische Auswertung der Gleichungen des Problems erweist, daß schon bei sehr schwacher Partikelkonzentration Gewinne erzielbar sind.

Study of a Convection and Radiation Heat Exchanger for a Fissiothermic Propulsion Unit. The exchange of heat between a nuclear reactor and the propellant can be augmented by opacifying the latter and utilizing thus the emitted energy in the form of radiation through the walls at high temperature. It is shown that it is essential to take into account in radiative exchanges the spectral division of the energy and the absorp-

---

[1] Ingénieur de Recherches au Laboratoire d'Aérothermique du Centre National de la Recherche Scientifique, 4ter route des Gardes, Meudon (S.-et-O.), France.

tion coefficients. A numerical evaluation of an example shows the possible gain for a propulsion system after having chosen the optimum parameters. The propellant used consists of particles of carbon in suspension in NH_3. It is shown that the total energy transmitted by radiation $\sum W (\theta_e, N, r)$ to a particle which depends on the temperature of the latter θ_e, on the concentration N, and on the radius r is always far from the maximum corresponding to the black body. The numerical integration of the equations of the problems shows that gains are obtained for a very small concentration in particles.

I. Position du problème

L'énergie P fournie par le réacteur nucléaire d'une fusée au fluide propulsif, de débit μ et de vitesse d'éjection w, s'écrit:

$$P = \frac{1}{2}\mu w^2 = \frac{1}{2}Fw = \frac{1}{2}M_0\Gamma_0 w$$

si $F = \mu w$ est la poussée du propulseur, M_0 la masse de la fusée au départ et Γ_0 l'accélération correspondante. L'utilisation de la propulsion nucléaire dès le départ de la Terre exige que:

$$\frac{P}{M} > \frac{1}{2}gw \qquad g, \text{ accélération de la pesanteur.}$$

A fortiori P/M_p, rapport de la puissance développée par le réacteur à la masse de celui-ci, devra être plus grand que $\frac{1}{2}gw$, soit 22500 kW/tonne pour de l'ammoniac éjecté à 4500 m/s.

Corrélativement la transmission au fluide, par convection, de fortes quantités de chaleur devient difficile sans augmenter les surfaces d'échange et le poids du propulseur. Il y a donc un gain à espérer de l'utilisation de l'énergie rayonnée

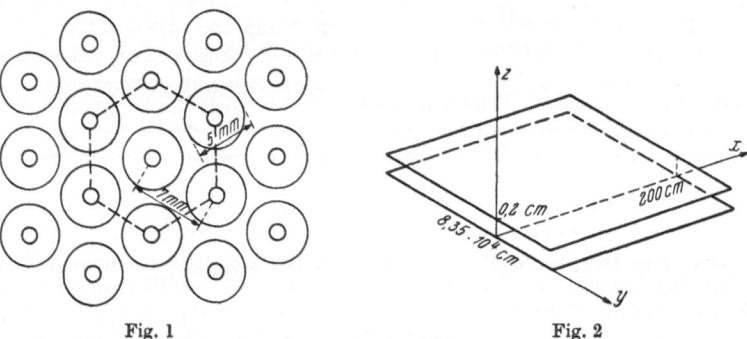

Fig. 1   Fig. 2

par les parois du réacteur au moyen de l'opacification du propulsif par de petites particules de carbone. Ce processus a été préconisé par M. l'Ingénieur en chef Barré au Congrès International d'Astronautique de 1958 [1]. On a repris ici les caractéristiques du réacteur nucléaire proposé pour chiffrer les gains, les calculs relatifs à l'échange radiatif de chaleur sont cependant de portée générale.

Le réacteur utilise de l'Uranium 235 enrichi et a la forme d'un cylindre de 1 mètre de rayon et de 2 mètres de hauteur; il est constitué d'un faisceau tubulaire de 74 000 éléments de 5 millimètres de diamètre extérieur; chaque élément possède une âme de 1 mm de diamètre en carbure d'uranium enrobée de graphite. La distance des axes de deux éléments voisins est de 7 mm, ce qui laisse un intervalle d'au moins 2 mm entre les tubes, (Fig. 1). Le fluide utilisé est de l'ammoniac.

La complexité d'un tel ensemble a conduit à considérer le tout comme équivalent à deux plans parallèles, à une distance de 2 mm l'un de l'autre, dont la surface totale serait égale à celle des éléments.

Les caractéristiques de ce réacteur schématique sont les suivantes:

— entreplan, $h = 0,2$ mm,
— longueur, $L = 200$ cm
— largeur, $8,35 \cdot 10^4$ cm, section d'entrée, $1,67 \cdot 10^4$ cm²,
— débit d'ammoniac à l'entrée: $q = 371$ kg/s,
— température du gaz ammoniac à l'entrée: 300° K,
— pression à l'entrée $p_0 = 50$ kg/cm²,
— vitesse d'entrée $u_0 = 645$ cm/s.

Les parois ont une température T qui dépend du flux local de chaleur entre propulsif et combustible nucléaire. Cependant, nous admettons tout d'abord que l'on a réglé la réaction pour maintenir la température des éléments à leur maximum possible. Le graphite se sublimant à 3825° K, il a semblé préférable de se limiter à 3000° K, quoique la valeur de 3500° K soit plus intéressante.

L'origine O_x des axes est sur l'une des plaques planes dans le plan d'entrée du fluide (Fig. 2); l'axe ox a la direction et le sens de l'écoulement. On voit qu'il sera toujours possible de considérer les deux plans comme illimités pour le calcul des échanges.

L'état du propulsif sera caractérisé à l'abscisse x par:

u_f vitesse du fluide
θ_f température maximale dans le plan de symétrie
ϱ_f masse spécifique correspondante,
p_f pression pour le fluide.
L'état des particules de carbone est caractérisé, à l'abscisse x, par:
u_c vitesse
θ_c température.

Nous désignerons par r le rayon des particules de carbone et par N leur nombre par cm³ à l'abscisse x. Nous introduirons les deux paramètres sous la forme suivante:

$m = \dfrac{r}{r_o}$, rapport du rayon de la particule à la valeur $r_o = 0,125$ microns de ce rayon qui réduirait au maximum les pertes dans la tuyère du propulseur.

$N_o = N \cdot \dfrac{u_f}{u_o}$, la concentration initiale en particules.

Choisir le réacteur utilisant au mieux les échanges radiatifs consiste à choisir de façon optimale les deux paramètres m, N_o.

Les six inconnues u_f, θ_f, ϱ_f, p_f, u_c, θ_c sont reliées aux grandeurs m, N_o et x par les relations suivantes:

— conservation de la masse:

$$\varrho_f\, u_f = c^{te} = 22,2 \text{ C. G. S.} \tag{1}$$

— échange de quantité de mouvement entre le fluide et la paroi, d'une part, le fluide et les particules d'autre part:

$$\frac{d\,p}{d\,x} = -\frac{1}{2}\,\frac{\varrho_f u_f{}^2}{n} \times 0,079\; Re_h{}^{-1/4} \tag{2}$$

$$u_c\,\frac{du_c}{d\,x} = \frac{1}{2}\,\frac{\varrho_f}{\varrho_c} \times \frac{3}{8\,r}\,(u_f - u_c)^2\,\frac{24}{Re_{2r}}\;. \tag{3}$$

— équation de l'énergie pour le fluide ou pour une particule:

$$\frac{d\theta_f}{dx} = f\left(T - \theta_f, \theta_c - \theta_f, N_o\right) \tag{4}$$

$$\frac{d\theta_c}{dx} = f\left(\theta_c, r, N_o, u_f\right) \tag{5}$$

— équation d'état du fluide

$$f\left(p_f, \varrho_f, \theta_f\right) = 0. \tag{6}$$

Dans ces équations, Re désigne le nombre de Reynolds rapporté soit au double de l'entreplan Re_h soit au diamètre d'une particule Re_{2r}.

L'éq. (3) donnée par la mécanique des suspensions [2], peut se mettre sous la forme:

$$\left(u_f - u_c\right)\frac{d\left(u_f - u_c\right)}{dx} - \frac{d\left(u_f - u_c\right)}{dx}u_f - \left(u_f - u_c\right)\left[\frac{du_f}{dx} - \frac{18\,\mu}{\varrho\,c\,d^2}\right] = 0.$$

On verrait facilement que $\dfrac{du_f}{dx}$ est petit devant $\dfrac{18\,\mu}{\varrho\,c\,d^2}$ et, supposant u_f grand devant $u_f - u_c$, il vient:

$$-u_f\frac{d\left(u_f - u_c\right)}{dx} = -\frac{18\,\mu}{\varrho\,c\,d^2}\left(u_f - u_c\right).$$

Le retard maximum en vitesse ne dépasse pas alors $u_f - u_c \simeq 10^{-2}$ cm/s. soit $Re_{2r} \simeq 10^{-5}$; on le négligera ailleurs que dans la tuyère.

D'autre part, on peut voir facilement que $\dfrac{dp}{dx}$ est assez faible pour pouvoir considérer p_f comme constant à quelques pour cent près.

Nous allons donc évaluer les équations exprimant les échanges d'énergie par convection et rayonnement dans ces hypothèses.

II. L'échange convectif entre parois et fluide

Le calcul de l'échange convectif exige la connaissance des propriétés physiques du fluide, que nous ne supposerons pas modifiées par la surcharge en particules solides.

Considérons tout d'abord le gaz ammoniac à l'entrée de l'échangeur ($u_c = 645$ cm/s, $\theta_f = 300°$ K). Le nombre de Reynolds relatif à l'écoulement entre deux plans distants de h intervient comme pour un tuyau de diamètre double. Nous caractériserons l'écoulement par

$$Re_{2h} = \frac{\varrho_f\,u_f \cdot 2\,h}{\mu} = \frac{8{,}88}{\mu}.$$

Quand θ_f passe de 300 à 3000° K, le nombre de Reynolds reste compris entre 10^4 et 10^5. Compte tenu de la turbulence préalable du propulsif, due à son passage dans le circuit de refroidissement et à travers les pompes, on pourra toujours considérer l'écoulement comme turbulent. L'analogie de Reynolds fournit alors le nombre de Nusselt local suivant:

$$Nu = 0{,}023\ Re^{0,8}\ P^{0,4}.$$

Remarquons que la longueur de référence est toujours $2\,h = 0{,}4$ cm. Il faut calculer λ, μ et c_p pour le fluide à $p = 50$ kg/cm² et pour θ_f variant de 300 à 3000° K; la courbe peut ainsi être tracée à priori sans faire intervenir l'abscisse x, puisque l'on a considéré comme négligeable la variation de p avec x.

Le gaz ammoniac se dissocie en molécules N_2 et H_2 et atomes N et H. Pratiquement, la dissociation en molécules est complète à 1200° K et commence dès

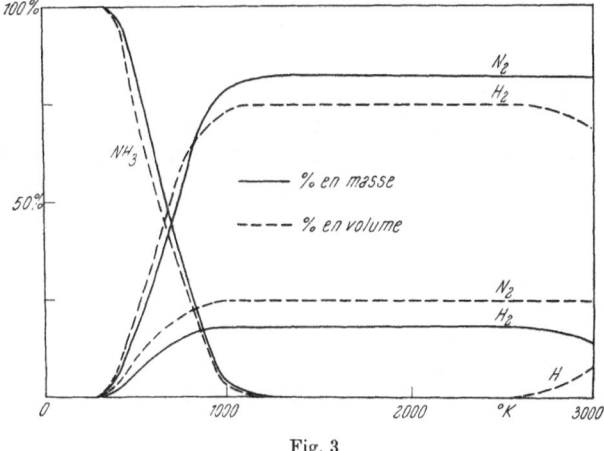

Fig. 3

400° K; seul H_2 se dissocie de façon sensible en dessous de 3000° K. On a les valeurs suivantes (pourcentage en masse, Fig. 3).

	NH_3	N_2	H_2	N	H
300°	97 %	2,5 %	0,5 %	0	0
600°	60 %	31 %	9 %	0	0
1000°	1,5 %	80,5 %	18,0 %	0	0
2000°	0 %	82 %	18 %	0	0
3000°	0 %	82 %	17 %	0	1 %

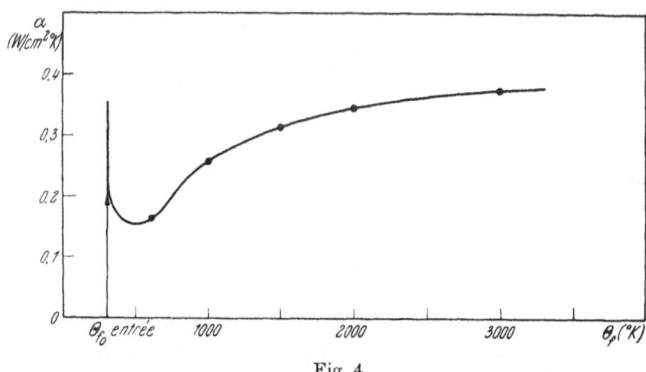

Fig. 4

Le diagramme entropie-enthalpie du mélange gazeux fournit la chaleur spécifique à 50 atmosphères du gaz. On indique en annexe I, la méthode de calcul de μ et λ pour le mélange dissocié. Finalement, on obtient le tableau suivant pour le coefficient de convection (Fig. 4):

θ_f ° K	300	600	1000	2000	3000
$\alpha \cdot$ watt/cm². ° K	0,189	0,164	0,258	0,345	0,375

III. Echange radiatif

Nous allons calculer l'énergie totale apportée à une particule de carbone par rayonnement. Soit ΣW cette énergie qui se décompose en:

1. énergie rayonnée directement par les parois sur la particule W_p,
2. énergie rayonnée par les particules environnantes sur la particule,
3. énergie perdue par rayonnement par la particule elle-même.

L'ensemble de ces énergies fait intervenir le coefficient d'absorption du fluide opacifié par les particules de carbone. Considérons un faisceau d'intensité I traversant dans la direction $\vec{dr}$, un volume $dA\,dr$; l'énergie est diminué à sa traversée de $dI = -k\,dr$ où:

avec
$$k = k_A + k_R + k_D$$

k_A coefficient d'absorption
k_R coefficient de réfléxion
k_D coefficient de diffusion.

Roessler montre [7] que l'on peut toujours négliger k_R et k_D pour les particules qui nous occupent et dont les dimensions sont de l'ordre de grandeur de la longueur d'onde de la lumière. Il en résulte que les coefficients d'absorption et d'émission sont égaux, et que l'énergie absorbée par une particule dans le schéma précédent est, si l'on désigne par N le nombre de particules par unité de volume,

$$\frac{dI}{dN} = -\frac{k\,dr}{N\,dA\,dr} = -\frac{k}{N\,dA}.$$

Etudions maintenant l'effet de la turbulence de l'écoulement sur la température θ_c d'un corpuscule. L'énergie absorbée n'étant pas la même pour une particule près de la paroi ou au milieu de l'intervalle entre plans, il en résulterait, dans un écoulement laminaire, un échauffement inégal des particules; θ_c dépendrait donc de x et de z. Au contraire, si nous admettons un taux de turbulence $\Delta v/v = 2\%$ pour une vitesse d'écoulement de 10 m/s, il suffit de 5 cm pour qu'une particule passe du centre à l'un des plans. On pourra alors admettre que chaque particule reçoit, pendant un temps Δt correspondant à $\Delta x = 5$ cm, la moyenne des énergies $\Sigma W\,(z)\,\Delta t$ transmises, et il est légitime, à la limite, d'écrire que la particule reçoit $\varepsilon \cdot W\,(o)$, où $W\,(o)$ est l'énergie à la paroi et

$$\varepsilon = \frac{\int_0^h W\,(z)\,dz}{W\,(o) \cdot h}.$$

Dès lors pour évaluer l'énergie totale transmise par les parois, il suffira de faire la somme des produits $\varepsilon \cdot W\,(o)$ pour l'ensemble des longueurs d'onde. Il est en effet facile de voir que, l'absorption variant beaucoup avec la longueur d'onde, on ne peut considérer les radiations de toutes les longueurs d'ondes comme absorbées également. Il n'en serait ainsi que pour une opacification par très grosses particules dont la masse serait prohibitive.

Appelons b_λ la brillance spectrale dans l'intervalle de longueur d'onde λ, $\lambda + d\lambda \cdot b_\lambda^{\theta_c}$ sera la brillance du corps noir à la température θ_c, $b_{\lambda_m}^{\theta_c}$ la valeur maxima de cette brillance.

Les deux derniers termes de ΣW peuvent être évalués simultanément en remarquant que l'énergie totale apportée par rayonnement à une particule est nulle pour $\theta_c = T = 3000°$ K. L'énergie perdue par rayonnement par la particule de surface S est:

$$-S\,\pi \int_0^\infty k_A\,b_\lambda^{\theta_c}\,d\lambda.$$

De même l'énergie reçue des particules environnantes peut s'exprimer de la façon suivante:

$$S' \pi \int_0^\infty k_A \, b_\lambda^{\theta_c} \, d\lambda \, .$$

Le coefficient S' sera donné comme nous l'avons remarqué par la relation:

$$\Sigma W = W_p + S' \pi \int_0^\infty k \, b_\lambda^{T} \, d\lambda - S \pi \int_0^\infty k_A \, b_\lambda^{T} \, d\lambda = 0 \, .$$

Soit:

$$S - S' = \frac{W_p}{\pi \int_0^\infty k_A \, b_\lambda^{T} \, d\lambda} \, .$$

Nous pourrons alors mettre ΣW sous la forme:

$$\Sigma W = W_p \left[1 - \frac{\int_0^\infty k_A \, b_\lambda^{\theta_c} \, d\lambda}{\int_0^\infty k_A \, b_\lambda^{T} \, d\lambda} \right] .$$

L'expression de W_p calculée en annexe II conduit ainsi à l'expression numérique suivante pour $T = 3000°$ K:

$$\Sigma W_{\text{Watt}} = 3,89 \cdot 10^{-8} \, \text{m}^3 \cdot \left[1 - \left(\frac{\theta_c}{T} \right)^5 \frac{\int_0^\infty \frac{b_\lambda^{\theta_c}}{b_{\lambda_m}^{\theta_c}} \cdot \frac{k_A}{N \pi r^2} \cdot d\lambda}{\int_0^\infty \frac{b_\lambda^{3000}}{b_{\lambda_m}^{3000}} \cdot \frac{k_A}{N \pi r^2} \cdot d\lambda} \right] \cdot \int_0^\infty \frac{b_\lambda^{3000}}{b_{\lambda_m}^{3000}} \frac{k_A \, \varepsilon}{N \pi r^2} \frac{du}{u^2}$$

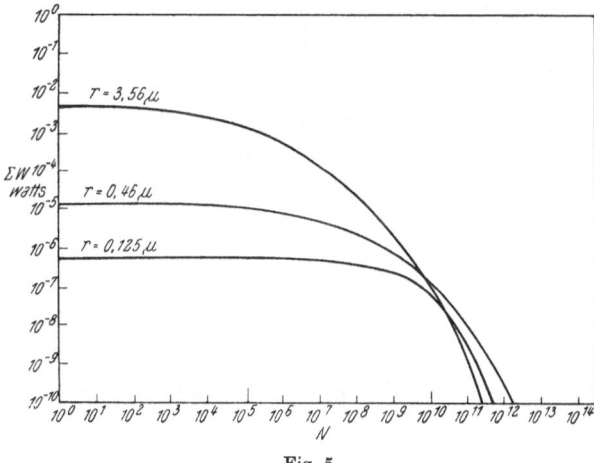

Fig. 5

où $\frac{b_\lambda^0}{b_{\lambda_m}}$ représente le rapport de la brillance du corps noir à la longueur d'onde λ à la brillance maximum correspondant à la température θ. On sait que $b_\lambda / b_{\lambda_m}$ est une fonction de λ / λ_m seul, λ_m étant la longueur d'onde du maximum de brillance.

On donne Fig. 5 et en annexe III les valeurs de ΣW en fonction de N pour divers θ_c.

IV. Les échanges entre particules et fluide

On a vu plus haut que les nombres de Reynolds rapporté au diamètre d'une particule et à la vitesse relative par rapport au fluide ne dépassaient pas 10^{-5}. Il en résulte que les échanges se font uniquement par conduction; le calcul peut

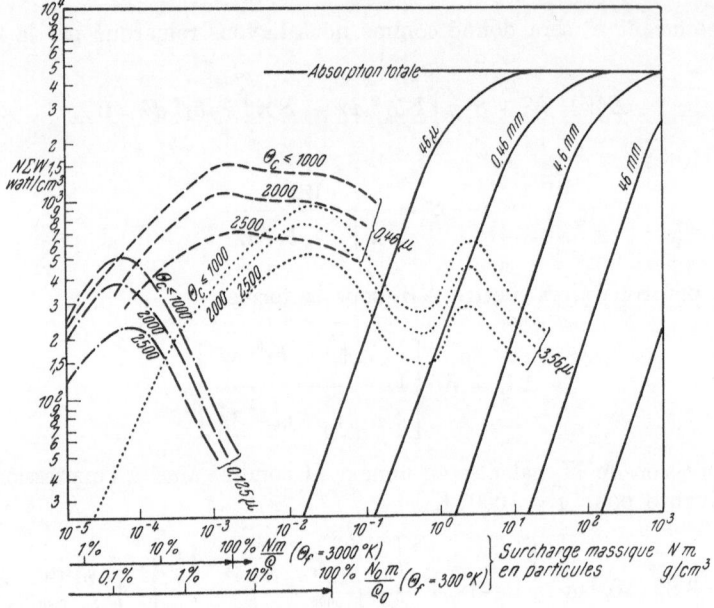

Fig. 6

être fait à l'aide de la théorie cinétique des gaz. Nous admettrons pour valeur du coefficient d'accomodation thermique lors d'un choc moléculaire: $\sigma = 0,1$. Dans ces conditions l'énergie cédée pendant dt est :

$$d\varphi = 0,1 \, e \, v \, r^2 \left(\frac{8 \pi M}{R} \right)^{1/2} \frac{P}{\sqrt{\theta_f}} \, (\theta_c - \theta_f) \, dt \, .$$

La masse moléculaire moyenne des composantes chimiques du fluide varie de 17 (ammoniac non dissocié) à 4,25 si l'on suppose N_2 et H_2 non dissociés. $\frac{d\varphi}{dt}$ variera donc de:

$$\frac{d\varphi}{dt} = - \, 151 \cdot 10^{-8} \, m^2 \, \frac{\theta_f - \theta_c}{\sqrt{\theta_f}} \quad \text{à}$$

$$\frac{d\varphi}{dt} = - \, 76 \cdot 10^{-8} \, m^2 \, \frac{\theta_f - \theta_c}{\sqrt{\theta_f}} = k \, (M) \, \frac{\theta_f - \theta_c}{\sqrt{\theta_f}} \, .$$

V. Bilan thermique

Ecrivons les bilans thermiques pour l'unité de masse du fluide et pour une particule de carbone. Il vient:

$$\begin{cases} c_p \dfrac{d\theta_f}{dx} = \dfrac{a \, (T - \theta_f)}{h \, \varrho_f + u_f} + \dfrac{k \, (M) \, N}{\varrho_f \, u_f} \cdot \dfrac{\theta_c - \theta_f}{\sqrt{\theta_f}} \\[2ex] m_c \dfrac{d\theta_c}{dx} = \dfrac{1}{u_f} \sum W - \dfrac{k \, (M)}{u_f} \cdot \dfrac{\theta_c - \theta_f}{\sqrt{\theta_f}} \end{cases}$$

ou encore:

$$\frac{d\theta_l}{dx} = 0.225 \frac{a}{c_p} (3000 - \theta_l) + \frac{1,25 \cdot 10^{-3}}{c_p u_l} \frac{\theta_c - \theta_l}{\sqrt{\theta_l}} \cdot N_0 m^2$$

$$\frac{d\theta_c}{dx} = 0,295 \cdot 10^{10} \cdot \frac{N \Sigma W}{m^3 N_0} - \frac{0,369 \cdot 10^{-3}}{u_l} \frac{\theta_c - \theta_l}{\sqrt{\theta_l}} \times \frac{1}{m} .$$

Il est évident que $\frac{d\theta_c}{dx}$ doit être petit devant les termes figurant au second membre. Supposons qu'on puisse le négliger il vient l'expression suivante de $\varkappa$ après intégration:

$$x = \int_{300}^{\theta_l} - \frac{c_p \, d\theta_l}{0,225 \, a \, (3000 - \theta_l) + \frac{N \Sigma W}{u_l}} .$$

En réalité cette expression devient imprécise quand θ_l tend vers $3000°$ car θ_c donc $N \Sigma W$ n'est pas défini avec précision et il faut calculer exactement la solution du système différentiel précédent. On peut cependant faire quelques remarques importantes:

Il y a intérêt à choisir $N \Sigma W$ maximum. On a porté Fig. 7, $N \Sigma W$ en fonction de la surcharge en particules par cm³. Il est intéressant de porter ainsi qu'on l'a

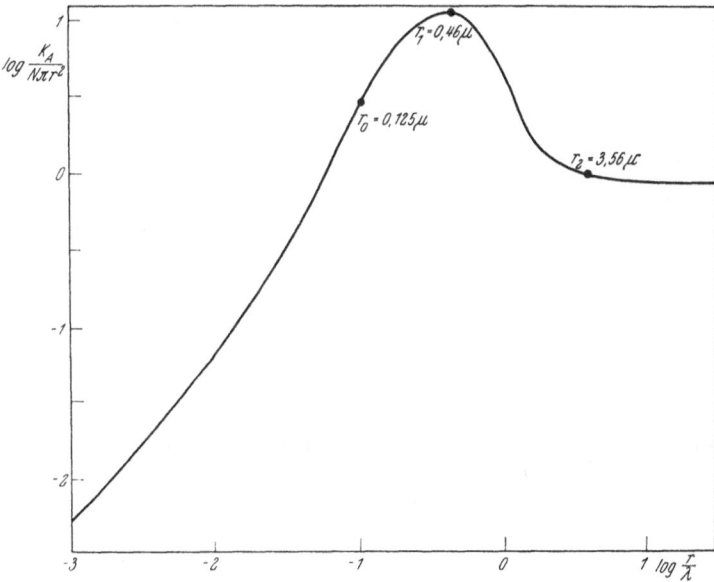

Fig. 7. Courbe de ROESSLER-PEPPERHOFF

fait le pourcentage en masse par rapport au gaz à 300 et $3000°$ K. On voit ainsi de façon très nette qu'il y a intérêt à se placer au maximum de la courbe de ROESSLER $(r = 0,46\,\mu)$, un rayon trop petit conduisant à un effondrement de la puissance ainsi qu'un rayon trop grand. Remarquons cependant que, dans le cas où l'on n'envisage pas une détente, une surcharge en grosses particules n'est pas gênante, et il est possible d'avoir une plus grande latitude de granulométrie. Réciproquement l'existence d'un maximum situé vers les faibles concentrations indique que l'effet de très faibles surcharges (0,1 % en masse) n'est pas négligeable avec de petites particules.

Nous avons indiqué plus haut qu'il était intéressant de régler la réaction nucléaire pour que la température reste presque partout égale à 3000° K. Ceci est évident sur l'expression ci-dessus de x. Supposer un dégagement constant d'énergie revient à écrire:

$$0,225\,\alpha\,(T-\theta_f)+\frac{N\Sigma W}{\varrho_f\,u_f}=c^{te}=A\ .$$

Soit

$$T=\theta_f+\frac{A}{0,225\,\alpha}-\frac{N\Sigma W}{0,225\,\alpha\cdot 22,2}\ .$$

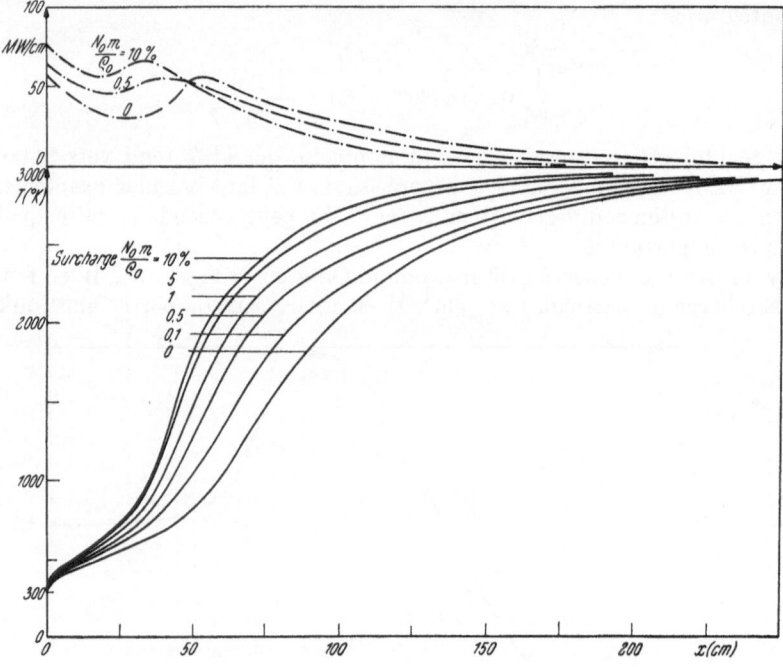

Fig. 8 (en haut). Flux d'énergie à la paroi. $r=0,46\ \mu$
Fig. 9 (en bas). Température du propulsif dans l'échangeur

Pour augmenter A, il faut augmenter T, car $N\Sigma W$ décroit très vite (en T^4) avec T l'intérêt de la surcharge disparait. Il semble donc intéressant d'adapter le réacteur pour avoir T maximum, cependant, fixant cette valeur de T à 3000°, on pourrait admettre que vers la sortie de l'échangeur une température de 3500° puisse être atteinte, multipliant $N\Sigma W$ par 1,85, en augmentant la poussée d'au moins 20%.

VI. Application à un engin

On a porté, Fig. 9, l'ensemble des courbes de températures du propulsif dans l'échangeur pour diverses valeurs de la surcharge en particules, de 0,1 à 10%. Pour choisir la concentration optima, on a étudié l'amélioration apportée par la chauffe radiative à un engin en regardant l'action sur le gain de vitesse:

$$\Delta V=w\log\cdot\frac{M}{m}$$

où w est la vitesse d'éjection
   M la masse au départ
   m la masse en fin de propulsion.

La fusée de départ a le devis de poids:

$m = 19$ t. (1 t. masse utile, 5 t. blindage, 8 t. propulseur, 3 t. réservoir, 2 t. auxiliaires);

$M = 93$ t. (74 t. d'ammoniac).

On a tenu compte de l'augmentation de m et M du à la surcharge et de la perte de poussée provoquée par la présence des particules. La Fig. 10 indique que

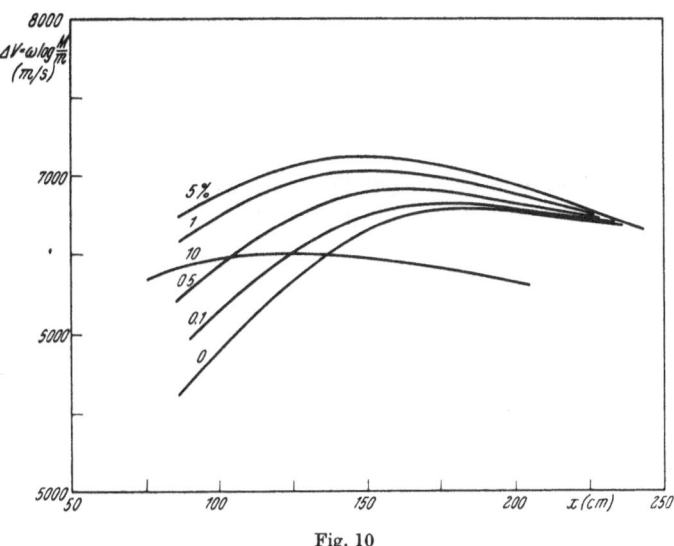

Fig. 10

le gain absolu n'est pas très grand mais qu'il correspond à une diminution des dimensions du réacteur ce qui est toujours intéressant lors du départ.

Remarquons cependant que l'on est loin d'employer au mieux la chauffe radiative, une température de 3500° au lieu de 3000° K améliorant de 80% son efficacité, le processus est donc celui des très hautes températures d'autant plus que l'optimum se déplace alors vers les très faibles pourcentages avec rendement moyen ou en conservant les mêmes dimensions de particules vers les rendements plus intéressants des grandes valeurs de r/λ. Corrélativement, les dimensions du réacteur et son poids peuvent diminuer beaucoup: $\varkappa = 120$, gain 30% à $T = 3500°$ K, ce qui augmente l'accélération initiale possible Γ_0 de 30%.

Annexe I

On a extrapolé les valeurs de la viscosité et de la conductibilité thermique du gaz ammoniac en utilisant pour le mélange complètement dissocié, les valeurs théoriques fournies par la théorie moléculaire.

Si σ_{ij}, ε_{ij}/k, T_{ij} sont les caractéristiques des forces moléculaires des gaz seuls ou en mélange; Ω_{ij}^*, A_{ij}^* les coefficients de Lennard-Jones, x_1, et x_2 les concentrations des deux gaz, on a:

$$\mu_{\text{mélange}} = \frac{1 + {}^3/_5 A_{12}\left[x_1^2 \dfrac{M_2}{M_1} + 2 x_1 x_2 \left\{\dfrac{(M_1 + M_2)^2}{4 M_1 M_2}\left(\dfrac{\eta_{12}}{\eta_1} + \dfrac{\eta_{12}}{\eta_2}\right) - 1\right\} + x_2^2 \dfrac{M_1}{M_2}\right]}{\dfrac{x_1^2}{\eta_{11}} + \dfrac{2 x_1 x_2}{\eta_{12}} + \dfrac{x_2^2}{\eta_2} + \dfrac{3}{5} A_{12}\left\{\dfrac{x_1^2}{\eta_2}\dfrac{M_1}{M_2} + \dfrac{2 x_1 x_2}{\eta_{12}}\dfrac{(M_1 + M_2)^2}{4 M_1 M_2}\dfrac{\eta_{12}^2}{\eta_1 \eta_2} + \dfrac{x_2^2}{\eta_2}\dfrac{M_2}{M_1}\right\}}$$

avec:

$$\eta_{ij} = 266{,}93 \cdot 10^{-7} \frac{\sqrt{M_{ij}\, T}}{\sigma_{ij}^2\, \Omega_{ij}\,(T^*)} \, .$$

Annexe II

Supposons que les parois planes s'étendent à l'infini. Leur rayonnement à T° K est sensiblement analogue à celui du corps noir à T° K. La variation dI d'intensité due à l'absorption des particules situées entre les plans z et $z+dz$ est, pour un faisceau lumineux faisant l'angle θ avec la normale aux plaques:

$$d I = I_0 \cos \theta \cdot k_A \, e^{-k_A \frac{z}{\cos \theta}} .$$

I_0 est l'intensité suivant la normale. Or,

$$I = \frac{d \Phi}{d s \cdot 2 \pi \sin \theta \, d \theta}$$

où $d\Phi$ est le flux lumineux émis par l'élément de surface ds dans la direction faisant l'angle θ avec la normale au plan émissif. La variation $d^2\Phi$ de ce flux entre les plans z et $z+dz$ est:

$$\frac{d^2 \Phi}{d s} = - I_0 \cos \theta \, k_A \, e^{-k_A \frac{z}{\cos \theta}} \cdot 2 \pi \cdot \sin \theta$$

correspondant à une absorption par particule:

$$d W = \frac{d^2 \Phi}{d s \cdot d v \, N} = \frac{I_0 \cos^4 \theta \, k_A \cdot e^{-k_A \frac{z}{\cos \theta}}}{N z^2} .$$

Puisque K_A et I_0 dépendent de la longueur d'onde, il est préférable d'introduire la brillance spectrale b_λ et d'écrire:

$$d W = \frac{b_\lambda \cos^4 \theta \, k_A \, e^{-k_A \frac{z}{\cos \theta}}}{N z^2} \cdot d s .$$

Intégrons pour tout le plan; il vient:

$$W_\lambda = \frac{2 \pi b_\lambda k_A}{N} \int_0^{\pi/2} e^{-\frac{k_A z}{\cos \theta}} \sin \theta \cos \theta \, d \theta .$$

Soit, en posant $\dfrac{1}{\cos \theta} = u$:

$$W_\lambda(z) = \frac{2 \pi b_\lambda k_A}{N} \int_1^\infty \frac{e^{-k_A z u}}{u^3} \, d u .$$

Pour avoir l'énergie moyenne $W_{p\lambda}$ reçue par une particule de l'ensemble des deux plans il suffit de remarquer que:

$$\int_0^h W_\lambda(z) \, d z = \int_0^h W_\lambda(h-z) \, d z \frac{h}{2} \, W_{p\lambda}$$

et par conséquent:

$$W_{p\lambda} = \frac{\int_0^h W_\lambda(z) \, d z + \int_h^0 W(h-z) \, d z}{h} = \frac{2}{h} \, \frac{2 \pi b_\lambda k_A}{N} \int_0^h \int_1^\infty \frac{e^{-k_A z u}}{u^3} \, d u \, d z$$

$$W_{p\lambda} = \frac{2 \pi b_\lambda k_A}{N} \cdot \varepsilon(k_A h) .$$

On obtient facilement les valeurs numériques de $\varepsilon\,(k_A\,h)$ en intégrant par parties l'expression de $W_\lambda\,(z)$ et introduisant l'intégrale exponentielle

$$\int_{k_A z}^{\infty} \frac{e^{-v}}{v}\,dv = E\,(k_A\,z)$$

$$W_\lambda\,(z) = \frac{\pi\,b_\lambda\,k_A}{N}\,e^{-k_A z}\,[1 - k_A\,z + (k_A\,z)^2 \times e^{k_A z} \times E\,(k_A\,z)].$$

Il ne reste plus qu'à intégrer $W_{p\lambda}$ pour les diverses longueurs d'onde du spectre :

$$W_p = 2\pi \int_0^\infty b_\lambda\,\frac{k_A}{N}\,\varepsilon\,(k_A\,h)\,d\lambda = 2\pi\,r_0^2 \left(\frac{r}{r_0}\right)^2 \pi\,b_{\lambda m} \int_0^\infty \frac{b_\lambda}{b_{\lambda m}}\,\frac{k_A}{N\pi\,r^2}\,\varepsilon\,d\lambda$$

en introduisant le diamètre de référence $2\,r_0 = 0{,}25\,\mu$ et la brillante maxima à T^0. Numériquement pour $T = 3000°$ il vient en watts :

$$W_p = 3{,}89 \cdot 10^{-8} \left(\frac{r}{r_0}\right)^3 \int_0^\infty \frac{b_\lambda}{b_{\lambda m}} \cdot \frac{k_A}{N\pi\,r^2} \cdot \varepsilon \cdot \frac{du}{u^2}$$

où $u = r/\lambda$, et $k_A/N\pi\,r^2$ est donné en fonction de u par la courbe de F. Roessler (Fig. 4).

Remarque: Pour $z = 0$. $W_\lambda\,(0) = \dfrac{\pi\,b_\lambda}{N}\,k_A$ correspond à la particule isolée. et

$$\left(\frac{dW}{dz}\right)_{z=0} = -\frac{2\pi\,b_\lambda}{N}\,k_A\,e^{-k_A z}$$

ce qui indique que l'énergie décroit deux fois plus vite qu'en supposant le rayonnement normal.

Le quotient

$$\frac{\displaystyle\int_0^\infty b_\lambda\,\frac{k_A}{N\pi\,r^2}\,\varepsilon\,d\lambda}{\displaystyle\int_0^\infty b_\lambda\,d\lambda}$$

représente l'équivalent d'un coefficient d'absorption, il représente un rendement du processus d'opacification, les résultats numériques montrent qu'il est toujours très petit pour les dimensions choisies des particules.

Annexe III

Valeurs de ΣW pour :

		$N = 1$	10^9	5.10^9	10^{10}	5.10^{10}
	$1\,000°$	5.10^{-7}	$2,45.10^{-7}$	$8,4.10^{-8}$	$4,90.10^{-8}$	$4,5.10^{-9}$
$r = 0,125\mu$	$2\,000°$	$3,9.10^{-7}$	$1,94.10^{-7}$	$6,5.10^{-8}$	$3,40.10^{-8}$	$3,3.10^{-9}$
	$2\,500$	$2,4.10^{-7}$	$1,17.10^{-7}$	$4,05.10^{-8}$	$2,37.10^{-8}$	$2,2.10^{-9}$
	$1\,000°$	2.10^{-6}	$1,05.10^{-6}$	$2,85.10^{-7}$	$1,32.10^{-7}$	$2,7.10^{-8}$
$r = 0,46\mu$	$2\,000$	$1,5.10^{-6}$	$7,8.10^{-7}$	$2,17.10^{-7}$	$1,0.10^{-7}$	$2,0.10^{-8}$
	$2\,500$	$9,4.10^{-6}$	$5,1.10^{-7}$	$1,43.10^{-7}$	$6,2.10^{-8}$	$1,6.10^{-8}$
	$1\,000°$	$3,5.10^{-3}$	$3,85.10^{-7}$	$6,0.10^{-8}$	$6,0.10^{-8}$	$6,3.10^{-9}$
$r = 3,65\mu$	$2\,000$	$2,4.10^{-3}$	$2,8.10^{-7}$	$4,4.10^{-8}$	$4,4.10^{-8}$	$4,6.10^{-9}$
	$2\,500$	$1,6.10^{-3}$	$1,83.10^{-7}$	$2,9.10^{-8}$	$2,9.10^{-8}$	$3,05.10^{-9}$

Bibliographie

1. J. J. Barré, Autopropulseurs fissiothermiques. Comptes rendus du IXᵉ Congrès International d'Astronautique, Amsterdam 1958, p. 333. Wien: Springer, 1959.
2. E. Brun et F. Vasseur, Mécanique des suspensions. G.R.A. Rapport n° 42/149 (1949).
3. Tables of Thermal Properties of Gases. National Bureau of Standards (1956).
4. Tables des propriétés thermodynamiques de l'ammoniac. CEPA RA n° 9 856 (1956).
5. J. Knudsen et D. Katz, Fluid Dynamics and Heat Transfer. New York: McGraw-Hill, 1958.
6. J. O. Hirschfelder, R. B. Bird et C. F. Curtiss, Molecular Theory of Gases and Liquids. New York: J. Wiley, 1954.
7. F. Roessler, Propriétés optiques du carbone. Rapport LRSh 4/52 (1952).
8. P. Dillon et L. E. Line, Jet Propulsion 26, 1091 (1956).

Personnel Selection and Training for Space Flight[1]

By

Don Flickinger[2]

(Received August 27, 1959)

Preliminary Considerations

In preparing for the first manned orbital flight, the factor of crew reliability and safety is one of major importance. From every standpoint—humanitarian, economic, scientific, and that of national prestige—we are faced with the prime requirement of insuring the complete attainment of mission objectives, including the safe return of the astronaut—and on the first try.

When one pauses to reflect on the time and energies required to progress from the X-1 to the X-15 research aircraft (1949—1959), comparing the unknowns involved and new materials and techniques required for those exploratory flights, to those involved in the first manned orbital flight, the magnitude and scope of the challenge assumes more realistic proportions.

In every man-machine system operating in a potentially hazardous environment the possibility of mission failure or accident always exists. The accident potential of each operating system is a dynamic entity with the major governing factors being the integrity of the machine, the effectiveness of the human component and the spectrum of surrounding environmental forces. No one would deny that we have still much to learn and apply toward the improvement of flight safety in conventional aircraft. Yet it would be equally facetious to argue against man's attempt to personally explore space until all hazards were eliminated from conventional flight. Progress is simply not possible under such a philosophy.

The basic problem for discussion, therefore, can be stated rather simply. How can we select and train pilots for the first orbital missions who will be capable of accepting the psycho-physiologic stresses of the total mission and at the same time perform effectively those functions required of him under both routine and emergency conditions? If we are to apply ourselves to this problem with reasonable intelligence and logic we must first define in objective terms the various personnel requirements for crew survival and performance from which we can intelligently extract the selection and training criteria to be used in the program.

Basic Requirements for Space Flight

1. That the pilot be free of any intrinsic medical defect potentially capable of causing serious disability or discomfort during the period of the mission.

---

[1] Summary and Conclusions, see p. 648.

[2] Brigadier General, USAF (MC), Assistant for Bioastronautics and Surgeon, Headquarters, Air Research and Development Command, U.S. Air Force, Andrews Air Force Base, Washington 25, D.C., U.S.A.

2. That the pilot be capable of accepting the predicted psycho-physiologic stresses of the total mission without significant performance decrement or irreversible injury.

3. That the pilot be capable of performing all assigned tasks which are critical to the success of the mission and his own safety.

For the purposes of discussion, it may be useful to further simplify these requirements by referring to them as medical fitness, psycho-physiologic stamina and performance capability. Let us consider them now in that order.

Medical Fitness

Once having decided upon those segments of the population from which we wish to solicit our volunteer applicants, our first and most basic step in the selection process is that of the medical or clinical evaluation. The procedures used in the selection of the National Aeronautics and Space Administration astronauts and in our Air Force test pilots are ones jointly developed by the Air Force School of Aviation Medicine, San Antonio, Texas, under Dr. LAWRENCE LAMB, and the Lovelace Clinic, Albuquerque, New Mexico, under Dr. W. R. LOVELACE, II. Basically, it represents an extension of the diagnostic and examining techniques developed for flying, executive, and special project personnel. The examinations carried out during this medical evaluation are listed as follows:

A. History and Physical Examination
    1. Proctoscopic
    2. Genito-urinary
    3. Orthopedic
    4. Dental

B. Medical Specialty Examinations
    1. Ophthalmology
    2. Otolaryngology
    3. Cardiology
    4. Neurology

C. Laboratory Tests
    1. Hematology
    2. Biochemistry
    3. Bacteriology

D. Special Physiologic Examinations
    1. Physical competence test
    2. Pulmonary function test
    3. Total body radiation count
    4. Specific gravity of whole body
    5. Blood volume and total circulating hemoglobin
    6. Detection of minute congenital openings between the chambers of the heart
    7. Total body water determination.

As performed at the Lovelace Clinic for use in the National Aeronautics and Space Administration astronaut selection program, a number of innovations were introduced into the medical evelation which merit special mention, to wit:

1. All data obtained on each candidate were placed upon color-coded machine record cards (International Business Machines Corporation type) using a "mark-sense" electrographic pencil. This reduced greatly the administrative burden imposed upon the physician, insured accuracy and comparability of records, and provided ready accessibility of data for further studies. Both the cards and techniques were developed by Dr. A. W. SCHWICHTENBERG of the Lovelace Clinic.

2. A method was developed to detect an artrial septal defect or patent foramen ovale without resorting to cardiac catheterization. This entails the continuous recording of arterial oxygen saturation (using a very accurate ear oximeter) which shows significant reduction after the patient performs the Valsalva maneuver in the presence of a right to left shunt of venous blood.

3. Through the use of supersensitive intensifying screens, shielding and ultrafast film, radiation exposures during the X-ray examinations were reduced to less than one-half of previous levels. Bullae on the lungs were particularly looked for because of their etiological relation to spontaneous pneumothorax and an X-ray moving picture of the heart was studied in a search for early coronary sclerosis.

4. The physical competence test as developed by Dr. ULRICH C. LUFT of the Lovelace Clinic was performed as follows: Graded work is performed on by bicycle ergometer ranging from 300 mkg/minute to 1200 mkg/minute; the rate of increase being 75 mkg/minute[1]. Measurements of heart rate, blood pressure, and respiratory volume, and gas exchange are made at frequent intervals until a heart rate of 180 beats/minute is reached or other signs of acute overload become evident. Oxygen consumption recorded at point of greatest work load is used as the criterion of aerobic work capacity. Each individual is rated in comparison with norms established for age, height, weight, and whole body specific gravity.

One week is devoted to this phase of the evaluation and this is not at all excessive particularly in the event several re-checks have to be made. The entire clinical team is composed of skilled specialists in their respective fields and at the end of each day the findings in each case are reviewed and discussed by the chiefs of each department. Upon completion of the total procedure, a final evaluation is made by the head of the examining team in conjunction with the psychiatrist, the flight surgeon in charge of the examinees, the physiologist responsible for the cardio-vascular pulmonary function tests, and the chief of the stress testing unit from the Aerospace Medical Laboratory. In the event a disqualifying and uncorrectible defect has been found the candidate is notified of such and apprised that he will be dropped from further consideration.

Psycho-physiological Stress Testing Procedures

The first requirement in setting up this phase of the program is to determine:

a) The quantitative and qualitative parameters of the various stresses, both physical and mental, to which the individual may be exposed during the total space mission under both routine and emergency conditions, and,

b) How can we simulate most accurately both the single and combined stress patterns?

Fortunately, in the case of the orbital flight using a vehicle of known configuration and characteristics, the scientists and engineers are able to give us very precise data on not only the type of energy force which will be imposed upon the occupant but also the time intensity profile to be expected for both routine and emergency phases of the space flight. From these data, then, we can set up patterns of exposure to acceleration forces on the centrifuge and to thermal flux in the heat chamber which will duplicate quite accurately the actual conditions of exposure during the space mission.

Insofar as the factor of reduced ambient pressure is concerned we must approach it from two aspects. First, is the question of the normal capsular gaseous

---

[1] Meter-kilogram per minute.

environment which for engineering reasons is maintained at 5 psi or one-third atmosphere, using 100 percent oxygen to provide an equivalent sea level partial pressure of oxygen; all of which can be easily duplicated in a low pressure chamber. The second is the question of tolerance to full pressure suit wear in the event the cabin life support system fails or in the extremely unlikely case of the capsule being ruptured by a meteorite. We know from considerable past experience that wearing of the inflated pressure suit is of itself a stressful situation to the individual with some, apparently quite healthy and fit, showing either unfavorable cardio-vascular response leading to circulatory collapse or emotional unacceptance of the situation.

In the area of weightlessness we are faced with the impossible task of simulating the true condition for the time it would supervene in one orbital pass (90 minutes). The Air Force has been studying the psycho-physiologic effects of this unique aspect of space flight using the F-100f (45—60 seconds) and the C-131 (15—20 seconds) over the past 2-year period and though many interesting findings have been reported, we have not felt justified in using any of the existing techniques in our battery of crew selection tests.

The problem of simulating the psychological stresses of space flight is almost as formidable as in the case of the weightless phenomenon. There is literally no possible means of reproducing the total spectrum of psychological factors intrinsic to first orbital flight—the excitement, the ever-present potential danger, the "aloneness" in orbit—short of the actual experience itself. However, on the basis of considerable work done in both university and Armed Services life science laboratories there has been included in our test battery a period of exposure in the isolation chamber.

Once having determined the extent to which we can duplicate the individual and collective psycho-physiologic stress factors which are predicted will exist in space flight we, then, turn to the problem of measuring and evaluating the candidate's total reaction pattern to these exposures. The following is a résumé of the procedures which we have used thus far in this work and which, we feel, have stood the "test of time"; at least until we can begin studying individuals *after* they have returned from a space flight:

Schedule of Stress Testing Procedures at the USAF Aerospace Medical Laboratory

A. Measures of Motivation and Personality
1. Psychiatric Interview
2. RORSCHACH Test
3. Thematic Apperception Test
4. Draw-a-Person Test
5. Sentence-Completed Test
6. Minnesota Multiphasic Personality Inventory
7. Who Am I
8. Peer Ratings.

B. Measures of Intellectual Functions
1. WECHSLER Adult Intelligence Scale.

C. Response to Isolation
1. Each subject is confined to a locked, dark, sound-proof room for two hours. This exposure serves to identify subjects who have a low tolerance to reduced sensory inputs, enclosure in small spaces, and enforced inactivity.

D. Anthropometric Measurements

1. Recording of 50 selected direct body measurements.

2. Stereophotographs from which accurate body diameters, circumferences, surface areas, cross sections, and volume data can be derived.

3. Somatotype photograph for body type analysis.

4. Skin fold thickness for determination of body density using the multiple regression equation.

5. Photograph of subject—seated, side view with right arm horizontal; for determination of work space dimensions.

E. Exposure to Low Ambient Pressure

1. Harvard Step Test

2. Treadmill Maximum Workload Test (BALKE's Method)

3. Cold Pressor Test (McGUIRE's Modification)

4. Tilt Table Test

5. Flack Test

6. MC-1 Partial Pressure Suit Test—Subject in MC-1 partial pressure suit denitrogenates for two hours on 100 percent oxygen during which time pulse, blood pressure, and electrocardiogram are recorded. After control readings are stabilized, subject is taken to 65,000 feet equivalent altitude and remains there for one hour. Premature termination of test (and disqualification) occurs under any of the following circumstances:

a) Subject requests termination.

b) Medical monitor advises termination.

c) Dysbarism or unconsciousness.

d) Blood pressure drops below 80 mm Hg systolic or 60 mm Hg diastolic.

e) Pulse rate exceeds 160 beats per minute.

f) Significant electrocardiogram abnormality develops.

F. Exposure to Accelerations

1. Determine tolerance to longitudinal (head to foot) positive G force and correlate pulse and blood pressure with blackout threshold. Maximum exposure 7 G's.

2. Determine tolerance to transverse G force using "contoured-couch" and end points of dyspnea, chest pain or black-out with correlation of pulse and blood pressure. Maximum exposure 12 G's.

G. Exposure to Heat: 130 degrees Fahrenheit; Relative Humidity 20 percent; Duration two hours.

1. Measurements taken include: total weight loss; evaporative loss; skin and rectal temperatures; heart rate; blood pressure and electrocardiograph.

2. From data recorded during the exposure, an evaluation of the psycho-physiologic strain endured by the individual is made using a modification of the CRAIN index of strain formula in which the terminal heart rate/100, total weight loss/hour, and the total rectal temperature rise/hour in Centigrade degrees are added together. The resultant figure ranges between one and six and the higher the number, the greater is the strain evidenced by the subject in maintaining thermal balance. Comparison with the average (age, body mass, and density) can then be made, with the individual's relative efficiency in handling increased heat loads expressed as either a score or percentile.

H. Exposure to Noise and Vibration

1. Subject is placed in an equilibrium chair which is programmed to pitch and roll on its horizontal axis at the same time vertical vibrations at frequencies of 3, 5, 7, 9, 12, and 15 cycles per second are introduced into the rig. Subject, after a short period of training on the device, is blind-folded and told to maintain

a "straight and level" position through proper manipulation of the control stick, during runs with and without being subjected to the vibratory movements. A score is evolved which indicates the degree to which any given vibratory frequency affects his ability to maintain a horizontal attitude. Subjective sensations such as vertigo, chest pain, respiratory difficulty, etc. are recorded as well as heart, blood pressure respirations, electrocardiogram during the test. Blood and urine samples are taken before and after the test.

2. The subject is exposed to a high intensity noise field of 130 decibels with sufficient dampening by muffs to prevent ear pain. During the exposure the subject is required to perform a psycho-motor task and the ratio of this score to his own control score recorded under no-noise conditions is used for comparison with the established normal range.

General Comments on Stress Testing Program

In practice at the Aerospace Medical Laboratory, Wright Air Development Center, Wright-Patterson Air Force Base, Ohio, we have found that a 1-week period is required for the stress tolerances testing procedures and this, though tight, does allow for some re-testing on those cases where an unusual or questionable test response has been recorded.

For those interested in obtaining more data and in greater detail regarding the test procedures and the experiences of our team under the direction of Lieutenant Colonel WILLIAM H. TURNER, USAF, MC, I can state that a full technical report is now being compiled by the Aerospace Medical Laboratory, Wright Air Development Center, Dayton, Ohio, with expected publication date near the end of this year, 1959.

A few personal and general observations on the operation of this phase in a Space Crew Selection Program as reflected in our own experiences to date may have some limited usefulness to those contemplating similar activities.

Initially and of great importance is the requirement for skilled, well-trained, and highly-motivated clinicians, scientists, and technicians to make up the testing and evaluation team. Willingness to work overtime, flexibility in thought and action, plus the ability to work congenially and understandingly with others are basic attributes vital to the success of the untertaking.

In a sense one can say that to be successful in identifying best qualified individuals to accept and respond favorably to simulated space flight stresses you must have on your examining team individuals who possess comparable psycho-physiologic qualifications.

Next, I would point out that although some cross-correlations between various test results appear valid with the limited sample dealt with, yet we have by no means developed any type of mathematical formula which can substitute for painstaking, detailed clinical observation of the individual as he undergoes the full test battery with subsequent careful analysis and evaluation of the results.

Finally, we must ask the question: "What have we produced from our efforts and how reliable is the product?" An honest answer to this question must include and admit the following points:

1. Candidates showing abnormal patterns of personality, behavior, and physiologic response to predicted and simulated space flight stresses are disqualified and excluded from further participation. The criteria for establishing normality of response patterns are based upon fairly traditional concepts of emotional stability and physical stamina.

2. Candidates found qualified are those who have demonstrated a capacity to accept the imposition of simulated space stresses, adapt themselves realistically and maturely to their existence, and compensate for their effects productively and economically in respect to essential body functions.

3. Since we are incapable of specifying and simulating the total spectrum of performance requirements and stress burdens contained in space flight, it is obviously impossible for us to practise positive selection procedures which would require the establishment of criteria which have been proven to be valid in predicting effective performance during a space mission. We, therefore, have practised basically a technique of "negative" selection, ruling out the patently unsuitable and qualifying those successful as being, in our opinions, individuals best fitted to meet both the expected and unexpected psycho-physiologic hazards of space flight.

The Training and Indoctrination of Space Crews

Once having qualified the candidates in respect to their physical and mental fitness and their demonstrated stamina and durability under simulated stresses of space flight, our next problem is that of education and indoctrination. Our objective here is to equip them with the wide range of technical knowledge and skills which will be required of them in operating and maintaining the space vehicle during the mission

Initially, and as we have stated before, in the problem of setting up the physical and mental standards for qualification, the fact must be accepted that we have no factual knowledge of space vehicle performance requirements upon which to base our training objectives and standards either qualitatively or quantitatively. This lack of precise knowledge on performance requirements for space flight was the basis for the decision to recruit volunteers for first manned orbital flight from the ranks of the military experimental test pilots; and one can certainly argue convincingly the point that the only true process of selection, in the strict sense of the word, occurred at the time this decision was made. In actual practice, it may well turn out that candidates with completely different occupational backgrounds may possess attributes and capabilities which fit them equally well for space crew duty. However, if one attempts to list, in general terms, those capabilities which will be most useful, or likely to be drawn upon, during man's pioneer space flight, it becomes logically apparent that experimental test pilots are most likely to possess, by nature and training, the majority of these capabilities as a single group.

Let us for the sake of completeness, list briefly those performance capabilities which we estimate will be required.

1. Capability to analyze information inputs and respond effectively.

2. Capability to make and record accurate observations.

3. Capability to exercise good judgment and make intelligent decisions when confronted with a completely unique situation having no counterpart in past experience.

Adding to these the obvious requirement for an individual who has demonstrated repeatedly in the past a high degree of courage, self-confidence, and flexibility, the choice of the test pilot becomes fairly inescapable. Regardless, however, of the relative merits of past experiences and occupationally developed skills in preparing the candidate to meet the demands of the projected space mission, there still remains the requirement to provide him with such additional knowledge and skills as are considered specific and pertinent to the optimal success of the mission.

What then are these specific requirements which we must meet in the indoctrination program?

1. To thoroughly understand the design and engineering principles of the space vehicle, its components and all ground based supporting equipment to a degree that will insure its effective operation and maintenance.

2. To be capable of making all required observations and control adjustments necessary during all phases of the flight.

3. To maintain himself in a top condition of mental and physical fitness and stamina, thoroughly familiar with all factors and stresses intrinsic to his effective performance and safe return.

4. To have sufficient knowledge of both the terrestrial and spatial environments as will enable him to observe accurately, reason logically, and act constructively in all predicted situations.

Once having established these broad categories of training and indoctrination requirements we can then proceed to specify the content and activities of the program in greater detail; and which on first approximation could include the following general configurations:

A. Mission Task Simulation
  1. Display and Control
     a) Vehicle
     b) Environment
     c) Emergency
  2. Observations
     a) Engineering
     b) Astrophysical
     c) Biomedical

B. Engineering and Test Participation
  1. Satellite System
     a) Booster
     b) Orbital Vehicle
     c) Life Support
     d) Instrumentation
     e) Communications
  2. Ground Complex
     a) Pre-launch
     b) Tracking
     c) Recovery

C. Physical Fitness
  1. Stress Familiarization
     a) Human Centrifuge—G-Forces
     b) Heat Chamber—Thermal Flux
     c) Altitude Chamber—Pressure Suit Wear
     d) Parabolic Flights—Weightlessness
     e) Disorientation Simulators
  2. Orbital Hygiene
     a) Eating, Drinking, and Waste Disposal
     b) Preventive Medicine and First Aid
  3. Physical Conditioning
     a) Hobbies and Sports
     b) Skin Diving
     c) Health Hygiene
  4. Survival Techniques
     a) Practical Training in Land, Sea, Jungle, and Arctic Survival

D. Academic Education
   1. Orbital Dynamics
   2. Astrophysics
   3. Space Biomedicine
   4. Meteorology
   5. Geography

A considerable body of knowledge on training methods and devices has been developed over the past 15-20 years in response to the many and varied requirements emanating from the Armed Services. Computor technology has advanced rapidly since World War II and every major military and commercial aircraft has its counterpart ground simulator in which crews are trained to high degrees of proficiency with no hazard and minimal cost. New simulator techniques were required for the X-15 program to train pilots on the use of ballistic rather than aerodynamic controls to maintain proper attitude of the vehicle. The large United States Navy centrifuge at Johnsville, Pennsylvania, was modified and integrated with a computor to provide for the first time a dynamic closed loop system in which the X-15 pilot could actually fly a launch and re-entry pattern.

With the inauguration of the National Aeronautics and Space Administration's astronaut program, the Space Task Group at Langley Field, Virginia, has developed an excellent, well-rounded training program which meets admirably every requirement stated above; and even more. Considerable imagination and ingenuity are evidenced in the new devices and techniques developed by this group to simulate the display and control problems contained in the orbital flight and re-entry modes of the space mission. Procedure trainers are being developed simultaneously with prototype sub-system such as the life support system which will provide realistic training for the astronaut in the full range of routine and emergency operational situations.

On the biomedical side, we have mentioned the progress being made in adapting existing centrifuge, low-altitude and environmental facilities with the Armed Forces research and development complexes to provide realistic and accurate familiarization of the trainee-astronaut in the physiologic stresses of orbital flight. Although these are creditable steps in the right direction, many new ideas and advances in techniques are needed if we are to closely simulate dynamic environmental energy and force fields and record accurately psychobiologic responses thereto. There are, of course, many phenomena of the space environment with potential influence and hazard to biologic and behavioral function which must await finite definition before terrestrial simulation for purposes of both biologic-effects studies and familiarization can be accomplished. Even as fundamental a question as "what useful functions can man perform in space" must await significant advances in our simulation and analytical techniques.

On the more applied side of physical fitness, Lieutenant Colonel WILLIAM K. DOUGLAS, USAF, MC, the Astronaut Flight Surgeon, has reported unexpected side benefits from training the astronauts in skin diving techniques. Besides the obvious improvement in general muscular fitness as a result of swimming appreciable distances under water, the men have developed more efficient breathing techniques, which are equally applicable in other closed or semi-closed respiratory gas systems. Transporting themselves in a 3-dimensional body-supporting medium has taught them greater awareness of disorientation phenomena and the need to check sensory impressions of position and direction against indicators of proven accuracy and reliability.

As the trainee progresses through the program of stress familiarization, physiologic training, and physical conditioning, data will be recorded on his vital functions, health status, physical competence, or work-efficiency, psychomotor performance, and mental attitude toward his job. These records, carefully validated and evaluated, will provide the attendant flight surgeon with a dynamic profile indicating the normal range of psycho-physiologic function in each individual trainee. These data will serve a number of important purposes; first, to correlate with similar vital data recorded during initial medical evaluation and stress tolerance testing; second, to compare with post-space flight data to determine any immediate or delayed psycho-biologic effects of the space mission exposure; and lastly as comparative data with that being telemetered from the space vehicle during flight to the monitoring ground stations. At these stations, trained medical observers will determine by comparison of these two sets of data whether or not the spaceman is exceeding his normal range of 'vital functions and whether an emergency re-entry is required.

The question of whether or not during the medical aspects of the indoctrination program the attendant space crew surgeon can determine the readiness of an individual for a specific time scheduled mission or the relative readiness of one individual over other in the group is an interesting one. As has been mentioned before in discussing the application of formulae to decide whether a volunteer qualifies or fails in the stress tolerance tests, it is doubtful that any similar approach to final mission fitness of the trainee would prove fruitful. Rather, one would expect the final decision to be based upon the best clinical judgment of the space surgeon, the considered estimate of operational fitness by the Space Crew Commander, and finally the personal estimate of his own readiness by the astronaut concerned.

Summary and Conclusions

The problems of organizing and implementing a program for the selection and indoctrination have been briefly discussed. An outline of procedures has been presented which reflects the United States Air Force and to a limited degree the National Aeronautics and Space Administration experiences in this field to date; being considered under the three general headings of medical evaluation, stress tolerance testing, and indoctrination. It was pointed out that the sole selection process contained in the entire procedure was that one which specified that all candidate-volunteers be recruited from those Service pilots who had graduated from experimental test pilot schools. Medical and psycho-physiologic evaluations were considered more as qualifying procedures than selective.

The determination of an individual's psycho-physiologic fitness for a specific task involving a vehicle operating in a hostile and potentially hazardous environment comprises a continuing task of major proportions to the behaviorist, the physiologist, and the flight surgeon. Single, spectacular man-in-space events are unlikely to provide valid answers to the many unknown which now exist. Once man's survival and safe recovery from space been assured, the use of the space environment to augment and complement terrestrial laboratories in the study of living processes and human behavior will doubtless prove a most valuable adjunct. The potential benefits to be derived for the health and well-being of future generations are likely to prove commensurate with the time, effort, and money expended.

References

1. Mercury Astronaut Selection Fact Sheet. NASA Release No. 59—113, National Aeronautics and Space Administration, Washington, D.C., April 9, 1959.
2. Astronaut Program Outlined. NASA Release No. 132, National Aeronautics and Space Administration, Washington, D.C., May 12, 1959.
3. E. L. Brown, Aero Medical Laboratory, Wright Air Development Center, Wright-Patterson Air Force Base, Ohio, Research on Human Performance During Zero G.
4. W. H. Turner, Aero Medical Laboratory, Wright Air Development Center, Wright-Patterson Air Force Base, Ohio, Astronaut Stress Testing Program. (Unpublished report.)
5. L. E. Lamb, Chief, Department of Internal Medicine, School of Aviation Medicine, USAF, Brooks Air Force Base, Texas, Space Crew Selection Program. (Personal communication.)
6. W. Randolph Lovelace, A. H. Schwichtenberg, and U. C. Luft, Lovelace Clinic and Foundation, Albuquerque, New Mexico, Selection Program for Astronauts for the National Aeronautics and Space Administration.
7. D. K. Trites, Personnel Laboratory, Wright Air Development Center, USAF, and W. O'Connor, US Naval School of Aviation Medicine, Comments on Objective Psychological Testing.
8. J. L. Brown, Biodynamics of Launch and Re-Entry. Naval Research Reviews, May 1959, Office of Naval Research, Washington, D.C.
9. T. F. McGuire, Aero Medical Laboratory, Wright Air Development Center, Wright-Patterson Air Force Base, Ohio, Stress Tolerance Studies. (In publication.)
10. G. E. Ruff, Medical Criteria in Space Crew Selection. Presented at the American Medical Association Meeting, Atlantic City, New Jersey, June 9, 1959.
11. D. Flickinger, Biomedical Aspects of Space Flight. Delivered before the Eighth General Assembly, AGARD, NATO, Copenhagen, Denmark, October 23, 1958, and published in three installments in Aviat. Week, December 22, 1958, December 29, 1958, and January 5, 1959.

Impulsive Midcourse Correction
of an Interplanetary Transfer

By

R. J. Gunkel[1], D. N. Lascody[1] and D. S. Merrilees[1]

(With 33 Figures)

(Received June 25, 1959)

Abstract — Zusammenfassung — Résumé

Impulsive Midcourse Correction of an Interplanetary Transfer. A survey of possible ballistic trajectories from Earth to another planet, with particular attention to sensitivity to initial condition errors, indicates that initial condition tolerances can be maximized by careful choice of the trajectory to be used. The possibility of midcourse correction to compensate for initial errors is considered and the relationship between correction impulse requirements and sensitivity to initial conditions is established. A method of utilizing available correction impulse to obtain the desired amount of correction is determined.

Inferential methods based on a two-dimensional model are used for the basic survey with the effect of third dimension estimated a posteriori. A more detailed analysis of some "maximum tolerance" trajectories is obtained by using a three-dimensional open integration simulation of the solar system. A comparison of results shows the suitability of approximate methods for determining basic trends, but indicates the necessity for using more accurate solutions for design purposes. This extends to these interplanetary trajectories a conclusion reached in a previous paper concerning a lunar shot.

Unterwegs erfolgende Impulskorrektion eines interplanetaren Schusses. Eine Prüfung der möglichen ballistischen Übergangsbahnen von der Erde zu einem anderen Planeten, unter besonderer Berücksichtigung der Empfindlichkeit gegen durch die Anfangsbedingungen verursachte Fehler, zeigt, daß die Toleranzen der Anfangsbedingungen durch sorgfältige Auswahl der zu benutzenden Bahn auf ein Maximum gebracht werden können; die Beziehung zwischen den Erfordernissen eines Korrektionsimpulses und der Empfindlichkeit gegen die Ausgangsbedingungen wird ermittelt und eine Methode zur Verwendung einer verfügbaren Impulskorrektion entwickelt, um den gewünschten Korrekturbetrag zu erhalten.

Es wurden Inferenzmethoden, die sich auf ein zweidimensionales Modell gründen, für eine Voruntersuchung benutzt, wonach der Einfluß der dritten Dimension a posteriori abgeschätzt wurde. Eine eingehendere Analyse einiger Bahnen mit „maximaler Toleranz" wird erhalten, wenn eine dreidimensionale offene Integrationssimulierung des Sonnensystems verwendet wird. Ein Vergleich von Ergebnissen zeigt die Eignung von Näherungsmethoden zur Bestimmung der wesentlichen Züge, weist aber auf die Notwendigkeit hin, für Konstruktionszwecke genauere Lösungen zu benutzen. Dies bestätigt für diese interplanetarischen Bahnen eine ähnliche Schlußfolgerung, die in einer früheren Arbeit über einen „Schuß zum Mond" erhalten wurde.

---

[1] Missiles and Space Systems Engineering, Douglas Aircraft Company, Inc., Santa Monica Division, Santa Monica, California, U.S.A., Engineering Paper No. 804.

Impulsion correctrice à mi-course pour transfert interplanétaire. Une étude des possibilités de trajectoires balistiques de la Terre à une autre planète montre que les tolérances sur les conditions initiales peuvent être rendues maximum par un choix judicieux de la trajectoire utilisée. La compensation des erreurs initiales par une correction à mi-course est envisagée et les corrections d'impulsion mises en relation avec la sensibilité aux erreurs initiales. Une méthode pour l'utilisation de l'impulsion disponible en vue d'obtenir la correction désirée est établie.

Une étude préliminaire est basée sur un modèle à deux dimensions, les effets de la troisième dimension étant estimés a posteriori. Une analyse plus détaillée de quelques trajectoires à tolérance maximum est obtenue par une intégration pas à pas du problème tridimensionnel du système solaire. La comparaison des résultats montre que les méthodes approchées rendent compte de façon satisfaisante des tendances générales mais que des solutions plus précises sont nécessaires pour les besoins d'une application spécifique. Ce résultat étend aux trajectoires interplanétaires une conclusion déjà formulée dans un article précédent pour une trajectoire lunaire.

Introduction

As a result of recent developments of rocket propulsion, extra-terrestrial flight of man-made objects has passed within the last two years from the state of scientific speculation to reality. Many scientific satellites have been launched into orbits around the Earth; several lunar probes have been launched; and both Russia and the United States have successfully launched payloads which after close passage of the Moon have escaped the Earth's gravitational field and are now in orbit around the Sun. The next objective in this new era of space flight will be to send a vehicle to another planet such as Venus or Mars. The next opportunity for near minimum energy transfer to Mars is late 1960 and to Venus early 1961.

The first vehicles which will attempt interplanetary flight will undoubtedly be scientifically instrumented probes projected from Earth to either Mars or Venus. Their purpose will be to obtain scientific environmental data, as well as develop the equipment and techniques employed in later manned interplanetary expeditions. Such space exploration at this time, and probably for several years ahead, is characterized by its dependence on vehicle design. Accomplishments in interplanetary flight will be limited largely by the degree of excellence of the solution to the transportation problem.

It is recognized that advanced capabilities come from the appropriate integration of one specialized field with another, with the balancing of one technology with another. Flight mechanics is an indispensable science, as is the study of advanced propulsion techniques. But flight mechanics, properly meshed with the intimate details of advanced propulsion, is both an art and science, and can lead to sensible and practical vehicular capabilities which neither specialty could produce alone. The same may be said for the impact of guidance limitations and capabilities upon the field of flight mechanics. One of the purposes of this paper is to illustrate some of the effects of flight mechanics on propulsion and guidance through the proper selection of the launch conditions for interplanetary transfer. This will be accomplished by exhibiting the possible transfer orbits as a function of the vehicular velocity capabilities, by determining the sensitivity of these transfer orbits to slight deviations of initial conditions, and by demonstrating the possible improvements in these sensitivities by use of midcourse correction.

Discussion

The capability of a particular vehicle for accomplishing interplanetary flight can be described most easily in terms of the vehicle's performance, i.e., the vehicle's velocity capability, and relating this to the interplanetary missions which can be achieved. It is necessary not only to establish the spectrum of transfer orbits possible with this velocity, but also to examine the perturbations which may occur. Propulsion and guidance cut-off errors will result in deviations in the magnitude and direction of the launch velocity. Transfer trajectories should be selected which minimize the consequence of these errors on mission objectives. Other sources of errors in the heliocentric departure conditions exist. These include an uncertainty in Earth's escape velocity of ± one foot per second, resulting from the uncertainty in the knowledge of the combined mass of the Earth-Moon system, and an uncertainty in Earth's heliocentric velocity of ±33 feet per second, resulting from the uncertainty in the Astronomical Unit when converted to engineering units [1, 2].

General investigations of unpowered ballistic trajectories in two- and three-dimensional Earth-Sun-Planet systems were conducted by the Missiles and Space Systems Engineering Department of the Douglas Aircraft Company, Inc. These studies established the required launch parameters and permissible errors in these parameters to accomplish successful interplanetary transfer, as well as the improvement in the sensitivity to errors in these initial launch conditions resulting from the use of a burst of propulsive impulse applied during the mid-course part of travel. A summary of the more important results is given in this paper.

Two-dimensional Analysis

The solution of an n-body problem may often be inferred by successive applications of analytical solutions of the classical two-body problem. Application of such inferential techniques has given approximate solutions to the inter-

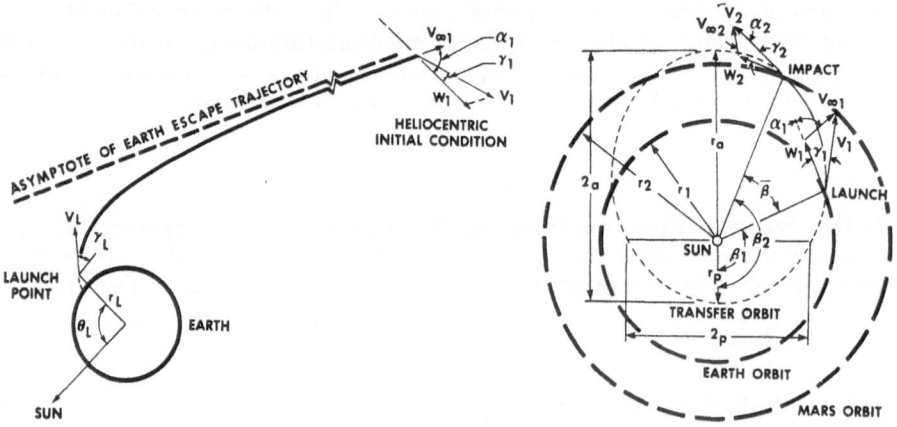

Fig. 1. Earth escape trajectory. Two-dimensional. Earth — Mars transfer

Fig. 2. Orbital diagram

planetary transfer problem from which much insight into the character of the exact solution has been obtained. The details of the method of analysis are presented in Appendix A.

The variable quantities entering the geometry of this simplified system are illustrated in Figs. 1 and 2. The four geocentric initial conditions are defined as:

radial distance from the center of Earth, r_L, where Earth's hyperbolic escape flight begins; the flight path angle elevation, γ_L, measured from the local horizontal; the launch velocity, V_L, and the local hour angle at launch, θ_L. All calculations were performed for an initial radial distance from the center of the Earth, r_L, of 3742 nautical miles; since this represents an altitude 300 nautical miles above the Earth's surface, is above the sensible atmosphere, and most likely is beyond burnout of the launching vehicle. The four heliocentric initial conditions are defined as: radial distance from the center of the Sun. r_1. where the interplanetary transfer begins; the initial velocity. V_1; the angle. β_1. measured from perihelion of the transfer orbit; and the time of launch. t_1.

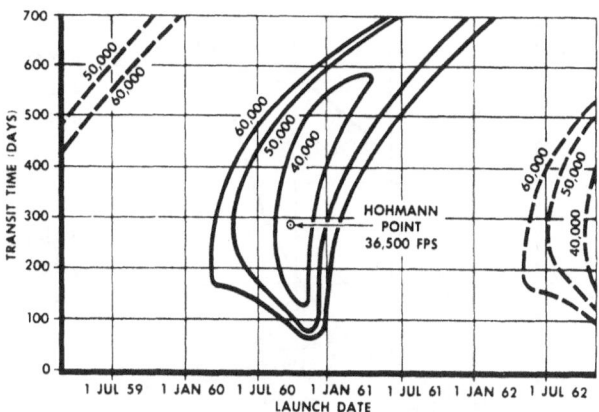

Fig. 3. Effect of launch velocity on Earth—Mars transfer orbits. Two-dimensional analysis

Simple calculations of approximate Earth-Mars trajectories show it would require 290 days to reach Mars on the HOHMANN transfer ellipse with the minimum velocity (36,500 feet per second) necessary, and that a projectile employing such a transfer can be launched on only one day during each synodic period of these planets. The influence of increased launch velocity is illustrated in Fig. 3, which is an all-inclusive plot of transfer time versus launch date. At near minimum launch velocities, there is only a small region of possible

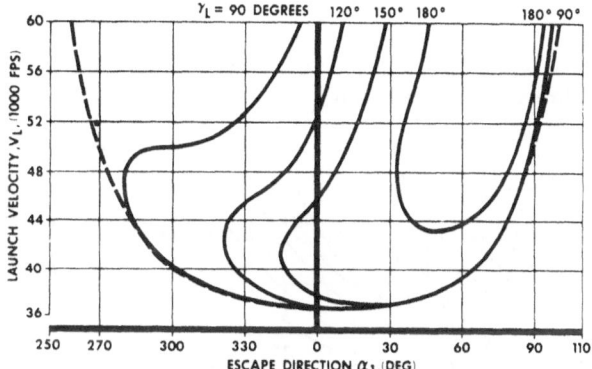

Fig. 4. Initial launch conditions for velocity cancellation. Earth—Mars transfer

launch dates and transfer times, but these regions increase rapidly with increasing launch velocity. Before interplanetary flight can occur on any date regardless of the astronomical configuration, a velocity of 50,000 feet per second is required of the launch vehicle.

Additional information on the characteristics of these transfer orbits is presented in Appendix B.

The permissible errors in both the magnitude and direction of the launch velocity vector for Earth-Mars transfer trajectories were calculated. The results suggest an advantage of certain families of trajectories which possess "velocity cancellation"; i.e., moderate deviations in the flight path and flight time of the transfer orbit from the intended, due to slight errors in the magnitude of the launch velocity, are neatly canceled by the orbital motion of Mars. The launch conditions corresponding to these transfer orbits are shown in Fig. 4. The region of maximum tolerance varies for different launch conditions and thus allows considerable latitude in the selection of a transfer orbit with fairly large velocity tolerance if the escape direction is properly chosen. In the case of sensitivity to the direction of the launch velocity vector the situation is different. For near minimal energy transfer orbits a similar convergence occurs since the projectile will remain in the vicinity of aphelion for an appreciable length of time, thus allowing Mars to capture it more easily.

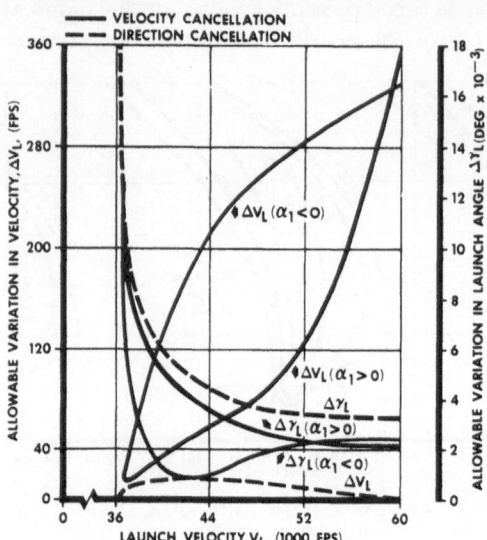

Fig 5. Maximum allowable errors in velocity and flight path angle. Earth—Mars transfer. $\gamma_L = 90$ deg.

A family of trajectories for which aphelion of the transfer orbit is near Mars' orbit, possess this minimum sensitivity to deviations in the direction of the launch velocity vector. The permissible error in the direction of the launch velocity is maximum for the Hohmann transfer, of the order of a tenth of a degree; however, the permissible error in the magnitude of the launch velocity is extremely small, only of the order of a few tenths of a foot per second. As the launch velocity is increased for these "direction cancellation" trajectories, the allowable direction errors decrease rapidly, and are of the order of a thousandth of a degree, whereas the magnitude errors increase considerably, but are still an order of magnitude less than those of the "velocity cancellation" trajectories. This is illustrated in Fig. 5 for the special cases of "velocity" and "direction cancellation" transfer orbits which are vertically launched from the Earth. The permissible errors in both the magnitude and direction of the launch velocity vector corresponding to impact on Mars are

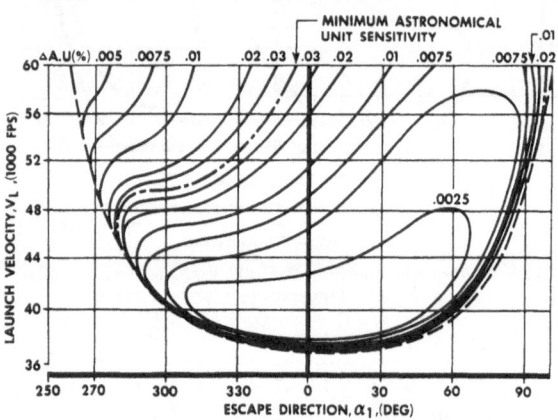

Fig. 6. Maximum permissible errors in astronomical unit cancellation for Martian impact. Earth—Mars transfer

shown as a function of launch velocity for both families of transfer orbits. These errors were determined by finite difference calculations rather than using the first derivative method. The desirability of the "velocity cancellation" trajectories is apparent from this comparison.

The uncertainty of conversion of astronomical and engineering units may be expressed as the uncertainty of the mean distance from the Earth to the Sun in engineering units. Present uncertainty is of the order of .03% as seen in [1, 2]. This uncertainty in engineering units can be replaced by an uncertainty in the residual velocity of the projectile in astronomical units if all calculations are done in astronomical units, since both represent only a basic uncertainty in the conversion of the distance units between the two systems. This procedure is valid only for cases where the launch flight path angle is vertical to the Earth. The allowable uncertainty in the parameter has been evaluated and is presented in Fig. 6. As in the case of permissible errors in the magnitude of launch velocity the values in the region of maximum tolerance have been determined using finite differences methods.

On the basis of this two-dimensional analysis, a comparison of the propulsive and guidance requirements is made for two interplanetary transfer trajectories. One is the minimum energy HOHMANN ellipse, and the other is the author's "preferred trajectory"; so called preferred because the characteristics of this transfer orbit include:

a) "velocity cancellation".

b) maximum tolerance to uncertainty in the astronomical unit,

c) near minimal energy requirements.

These two trajectories are illustrated in Fig. 7 and a comparison of the important features is presented in Table I.

Table I. *Summary of Launch Conditions and Permissible Deviation for Martian Impact*

Item	HOHMANN	Preferred
Launch Date	1 Oct. 1960	10 Aug. 1960
Transfer Central Angle, $\bar{\beta}$, deg.	180°	185°
Nominal Launch Conditions		
Velocity, V_L, fps.	36,500	38,650
Flight Path Angle, γ_L, deg.	90	90
Permissible Errors		
Velocity, ΔV_L, fps.	$\pm$ .19	$\pm$ 95
Flight Path Angle, $\Delta \gamma_L$, deg.	$\pm$ .03	$\pm$ .002
Astronomical Unit, $\frac{\Delta AU}{AU}$, %	$\pm$ .0011	$\pm$ .147
Transfer Time, t, days	290	236

It can be seen that choosing a transfer orbit with a modest increase in the velocity of the launching vehicle over that required for minimum energy transfer considerably improves the tolerances to errors in the launch conditions, as well as those due to uncertainty in astronomical measurements. Thus, vehicle propulsive requirements of the launching vehicle are thereby made more stringent, but the velocity cut-off requirements are relaxed.

An analysis was conducted to determine the effect resulting from the use of an impulse correction applied during the midcourse portion of the transfer trajectory. It is convenient to consider the planetary miss in terms of the central angle at the time when both the projectile and Mars are equidistant from the Sun. This allows determination of the required burst of propulsive impulse to be treated as a unique parameter in the two-dimensional analysis. The necessary condition for Martian impact of a trajectory with midcourse correction is

$$\overline{\beta} = \overline{\beta}_a + \overline{\beta}_b = \delta_1 + (t_a + t_b)\,\omega_2,$$

where "a" indicates the segment of the transfer ellipse from launch to point of application of correction, and "b" the remaining segment as shown in Fig. 8. For small magnitudes of correction velocity, the miss which may be corrected for a given point of application is approximately proportional to the magnitude

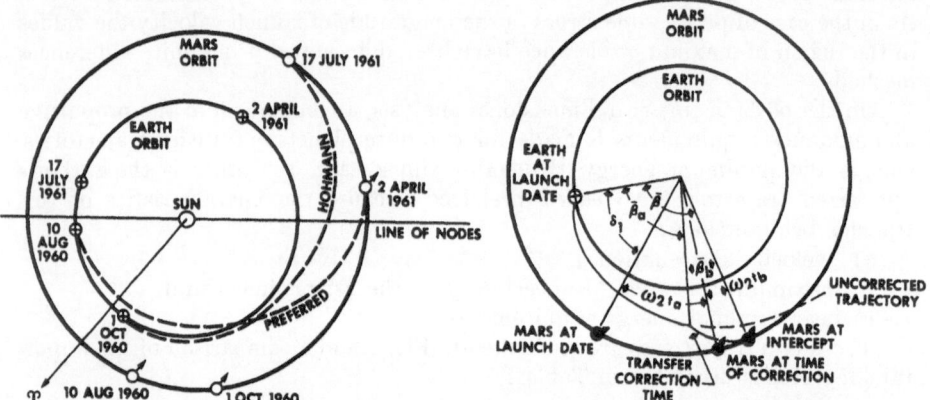

Fig. 7. Preferred and Hohmann trajectories

Fig. 8. Midcourse correction impact requirement. Earth—Mars transfer

of the correction. Thus, both the change in orbit geometry after correction, which results in a displacement of the point of intersection of the transfer and target planet orbits, and the change in remaining transfer time are properly taken into consideration.

The specific correction, i.e., the distance corrected per unit velocity applied, varies with the direction in which it is added in a sinusoidal manner as shown in Fig. 9. For application of this correction at a specific time, there is a direction of application for which the available correction is maximized. Thus, if it is predetermined that correction is to be made at a given time, the amount of correction obtained can be determined by selection of the direction of velocity applied. The amount of correction velocity capability required for injection at a given time can be determined by evaluating the probable errors in the system, and therefore the probable miss, and relating the necessary correction to the maximum specific correction.

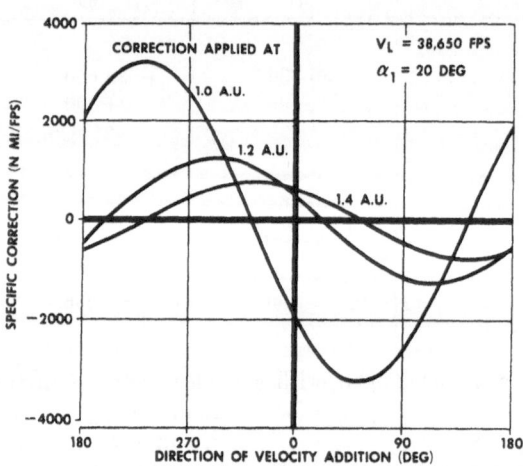

Fig. 9. Effect of corrective direction. Earth—Mars transfer

On any particular transfer orbit the maximum specific correction is a function of the point of application, and for transfers from Earth to Mars is greatest at the initiation of the trajectory. This is illustrated in Fig. 10. For those trajectories which pass through perihelion (indicated by a negative escape direction angle) the maximum specific correction decreases rapidly during the early part of the trans-fer. Velocity capability requirement can be determined as a function of point of application. The velocity required to compensate for a given amount of error will be least for correction early in the trajectory. Based on this factor alone, it seems desirable to plan correction early in flight. Other factors, such as possible improvement of information during flight, guidance system simplicity, etc., however, may also impose certain requirements on the time of correction and the best choice must be determined for various cases in the light of all of these factors.

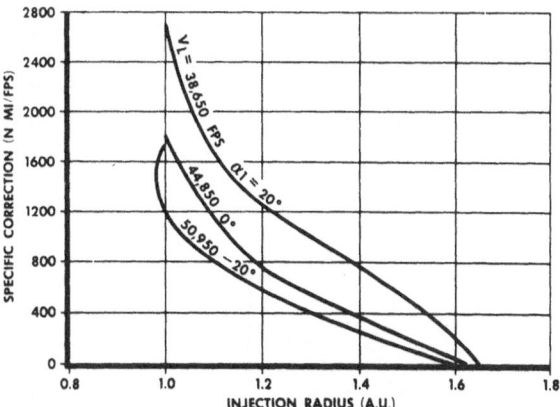

Fig. 10. Effect of correction position. Earth—Mars transfer

Available correction is a function only of initial velocity, escape direction, and point of application, and decreases, for a given escape direction, with increasing initial velocity, as shown in Fig. 11. The amount of correction velocity required to compensate for unit errors in launch velocity and flight path angle is shown in Figs. 12 and 13. It is apparent that the most economical correction for errors in these quantities does not occur for the same trajectory. If economy of correction is to be a criterion for choice of launch conditions, the expected errors must be evaluated and a choice made to minimize the sum of the correction velocities required. Comparison of the Hohmann ellipse and

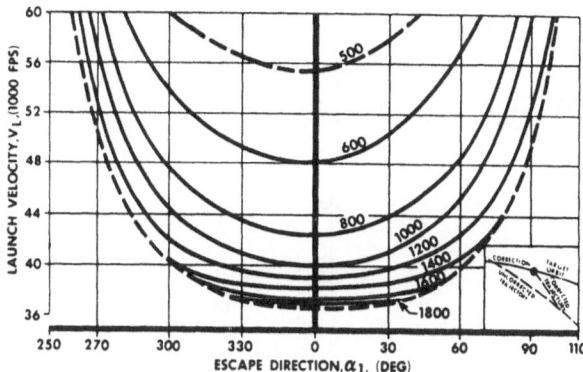

Fig. 11. Available correction along target orbit $\left(\dfrac{\text{N MI}}{\text{FPS}}\right)$. Correction applied at 1.2 AU's. Earth—Mars transfer

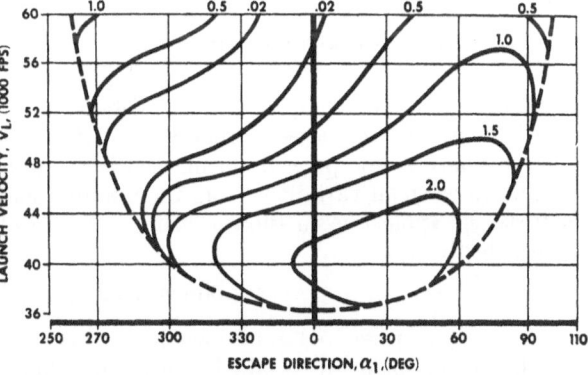

Fig. 12. Ratio of required correction velocity to launch velocity error. Correction applied at 1.2 AU's. Earth—Mars transfer

the author's "preferred trajectory" indicates that the preference previously established is not altered. For the "preferred trajectory" approximately twice as much correction velocity is required to compensate for initial direction errors as for the HOHMANN ellipse, but no correction will be required to compensate for initial velocity error if such errors can be held to less than 95 feet per second.

An adaptation of classical aircraft interceptor techniques will prove useful in formulating a self-contained guidance system which may be used to actuate a correction device. One method for successful interception indicates that the direction of the line of sight from interceptor to target remains fixed and may be determined from knowledge of the velocities involved and the angle between the paths of the two vehicles. It can be shown that this is not affected by the presence of uniform potential fields. In order to utilize an interplanetary guidance system based on this principle, it will be necessary to limit the use of the system to a region where the existing gravitational fields have equal, or nearly equal, effects on both the target planet and the projectile. This will exclude the region near the planet itself to eliminate the planetary gravitational effect on the vehicle, thus prescribing that the resulting system is truly midcourse, and not terminal, correction. The effect of heliocentric gravity is dependent on distance from the Sun.

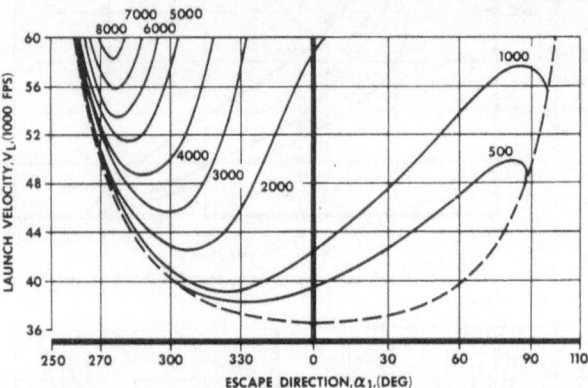

Fig. 13. Ratio of required correction velocity to escape direction error. Correction applied at 1.2 AU's (FPS/deg.). Earth—Mars transfer

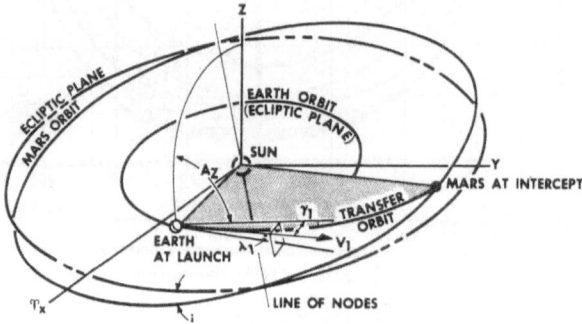

Fig. 14. Three-dimensional intercept conditions

For Earth to Mars trajectories, the distance of target planet and projectile vehicle from the Sun is such that the gravitational acceleration on each will differ by only a few percent during the last few hundred hours of the flight. Despite these small variations from the uniform field condition, the basic theory may still be applied. A method utilizing a system of this type for Earth flights is described in [3].

Three-dimensional Analysis

The analyses presented so far have been restricted to the case of two-dimensional, co-planar planetary orbits. However, in reality, the planet orbits are non-co-planar ellipses. As a result of the inclination between the plane of Mars orbit and that of Earth, the transfer orbit will not be contained in the ecliptic plane, but as shown in Fig. 14. The inclination, λ, of the transfer orbit plane to the ecliptic was determined and is presented in Fig. 15 as a function of the initial launch conditions. It is interesting to note for those cases where the position of the Earth at launch and Mars at impact are near opposition, i.e., near 180°, the inclination angle approaches 90°, as shown in Fig. 16.

The velocity increment over the two-dimensional launch velocity required to compensate for the inclination of the transfer orbit is greatest where the incli-

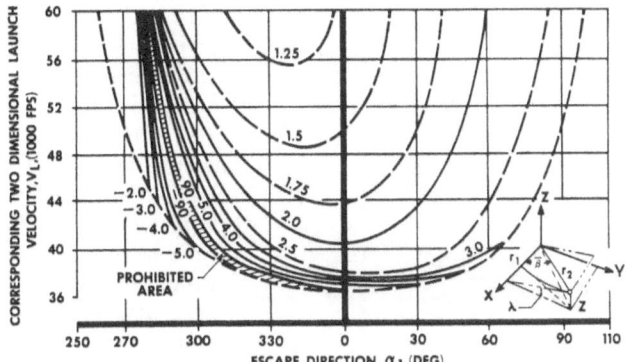

Fig. 15. Effect of initial conditions on transfer orbit inclination. Three-dimensional analysis. Earth—Mars transfer λ (deg.)

nation is greatest. Thus, the minimum energy transfer of the two-dimensional analysis (HOHMANN case) is that which has a high inclination and therefore a large velocity requirement in the three-dimensional case. The trajectories to which further attention will be directed are those with low velocity increment required and low inclination between transfer and ecliptic planes. This is illustrated in Fig. 17, which was determined inferentially using a three-dimensional model with elliptic orbits.

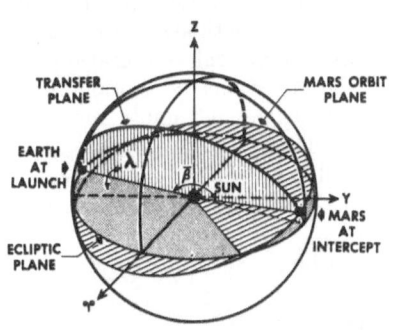

Fig. 16. Orbit inclination schematic

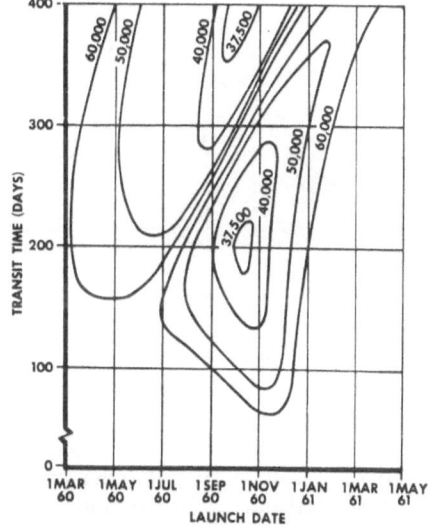

Fig. 17. Effect of launch velocity on Earth—Mars transfer orbits. Three-dimensional analysis

It can be seen that a realistic minimum energy transfer occurs at approximately the same date as the HOHMANN transfer. This minimum velocity of approximately 37,000 feet per second is only slightly greater than the HOHMANN value of about 36,500 feet per second. However, the transit time is decreased from 290 days to 200 days if a minimum energy transfer is desired. For moderate increases in launch velocity, the transfer time can be decreased considerably. Similar minimum energy transfers recur with a period of approximately 780 days; the launch velocity and transfer times are a function of the astronomical configuration.

The consequence of the inclination of the transfer orbit plane to the plane of Mars' orbit is to reduce the permissible errors in the magnitude of the launch velocity. Larger tolerances are otherwise achievable by choice of "velocity cancellation" trajectories for which deviations in the flight path and transfer time from those intended due to these errors are canceled by Mars' orbit motion. In any actual case, only the component of the deviation projected into Mars' orbital plane is compensated as in the two-dimensional case; however, there also exists a component of miss perpendicular to this plane which is not compensated. The faster projectile will pass directly over Mars, the slower projectile directly below Mars, as shown in Fig. 18. Desirable trajectories are those for which the inclination of the transfer plane to the ecliptic plane or Martian orbit plane are low and which are in the region of minimum two-dimensional sensitivity.

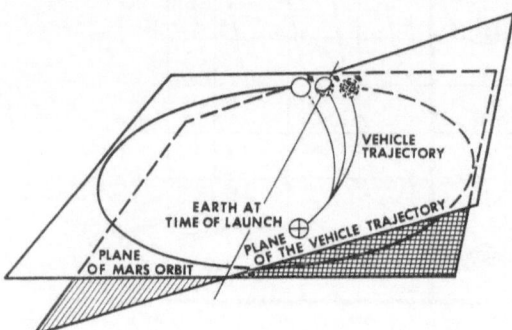

Fig. 18. Typical interplanetary trajectories

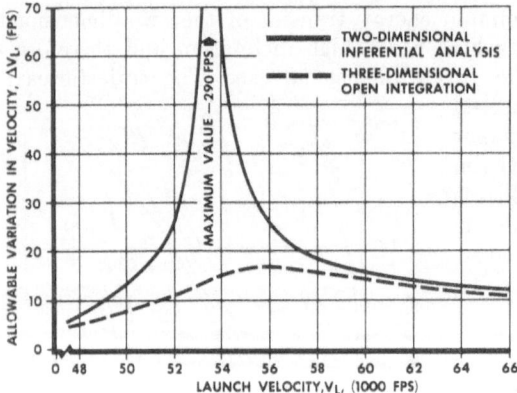

Fig. 19. Allowable variation in velocity

More precise results than could be realized by the inferential methods were obtained by numerical integration of the equations of motion for a vehicle of negligible mass on an IBM 704 Electronic Data Processing Machine. These calculations, designed to allow for the three-dimensional nature of the solar system, properly considered the following important factors which were neglected in the previous analyses:

a) the inclination of Mars' orbit to the Earth's orbit;
b) the eccentricity of Earth's orbit;
c) the latitude and hour angle of the launch site;
d) the gravitational fields of the other planets in addition to Earth-Sun-Mars.
The details of this model are given in [4].

A three dimensional analysis was conducted to determine the effect of transfer trajectory inclination to Mars' orbit, and the respective location, movement, and gravitational fields of the planets and Sun in accordance with the latest ephemerides [2]. Detailed calculations were made to determine the sensitivities of launch velocity, flight path angle, and azimuths for Martian impact for a family of transfer orbits. Launch velocities between 48,000 and 66,000 feet per second, a flight path angle of 90 degrees, an azimuth angle of 90 degrees, and a longitude of launch location of 80.565 degrees west were chosen as launch conditions. The latitude of launch location was variable with launch velocity. This family of trajectories corresponds to a two-dimensional family launched with the same magnitude of launch velocity and an escape direction of 335 degrees. The permissible errors in both the magnitude and direction of the launch velocity vector corresponding to impact on Mars are shown in Figs. 19 and 20, and are compared to the corresponding two-dimensional data. This family was selected as representative of typical three-dimensional transfer orbits, one of which fulfills the following conditions:

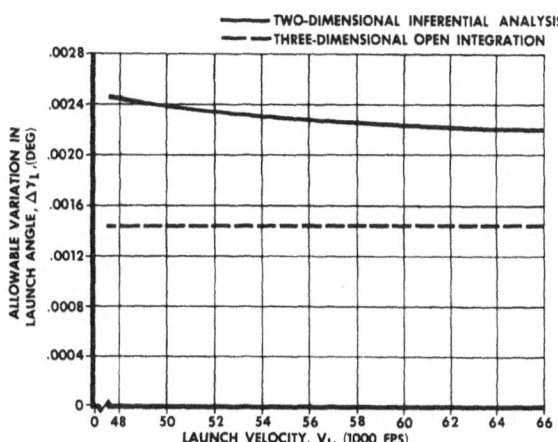

Fig. 20. Allowable variation in escape direction. Earth – Mars transfer. $\gamma_L = 90°$, $a_1 = 335°$

a) low inclination of the plane of the transfer orbit with the planes of the ecliptic and the orbit of Mars;

b) maximum tolerance to uncertainty in the astronomical unit;

c) "velocity cancellation."

A launch velocity of 54,819 feet per second was selected as the nominal transfer orbit. Although this launch velocity is considerably above that of minimum energy, it is in the region of near minimum energy for a trajectory satisfying the above conditions. A comparison of results of the two-dimensional inferential technique with the more precise three-dimensional open integration method is presented in Table II.

Table II.

Summary of Launch Conditions and Permissible Deviation for Martian Impact

Item	Two-dimensional	Three-dimensional
Launch Date	21 September 1960	21 September 1960
Launch Time	12^h	13^h 16^m 20^{s}05
Transfer Central Angle, β, deg.	94	91.5
Nominal Launch Conditions		
Velocity, V_L, fps.	54,819	54,819.15
Flight Path Angle, γ_L, deg.	90	89.8
Azimuth Angle, Az_L, deg.	---	89.95
Inclination Angle, λ_1, deg.	—	1.30
Permissible Errors		
Velocity, ΔV_L, fps.	310	16.25
Flight Path Angle, $\Delta\gamma_L$, deg. ..	0.00230	0.00146
Azimuth Angle, ΔAz_L, deg.	—	0.33
Astronomical Unit, $\frac{\Delta AU}{AU}$, % ..	0.41	0.0245
Transfer Time, t, days	88.8	90.5

The results of these calculations indicate that deviations of the launch velocity in magnitude and errors in azimuth predominantly change the "latitude", but those in flight path angle predominantly change the "longitude" of the impact on Mars. Although the maximum permissible error in magnitude of the launch velocity for the three-dimensional case has been considerably reduced from the two-dimensional, it appears to be adequate. When expressed in geocentric residual velocity or Astronomical Unit length this error corresponds to ± 0.024 per cent. This is only slightly less than the currently expected error of 0.030 per cent in Astronomical Unit length.

Detailed calculations were made to determine the permissible errors in the magnitude, azimuth, and elevation of the intended launch velocity vector for other regions of the domain near the two-dimensional "velocity cancellation". The results indicated maximum permissible errors in the launch velocity magnitude of less than one foot per second for launch velocities of 40,000 feet per second or less. In view of the difficulty of propulsion systems to generate velocities in excess of 40,000 feet per second, the difficulty of guidance system to control the burn-out velocity vector to within one foot per second, and the uncertainty in astronomical data, the importance of midcourse correction is emphasized.

In order to confirm the conclusions concerning mid-course correction, a more detailed analysis of a trajectory with a launch velocity of 39,123 feet per second, an initial flight path angle of 24.40 degrees, and an azimuth of 29.71 degrees was made using a three-dimensional, open-integration model. This trajectory corresponds to a two-dimensional trajectory with 38,650 feet per second launch velocity and an escape direction angle of 0 degrees. Table III presents the comparison of the results of two- and three-dimensional analyses of the trajectory prior to any midcourse correction.

The miss-distances which occur as a result of deviations from the standard initial launch velocity magnitude and direction and launch time are illustrated

Table III.

Summary of Launch Conditions and Permissible Deviations for Martian Impact Nominal Trajectory Used in Midcourse Correction Analysis

Item	Two-dimensional	Three-dimensional
Launch Date	22 September 1960	29 September 1960
Launch Time	12^{h}	2^{h} 13^{m} 7^{s}
Transfer Central Angle, $\bar{\beta}$, deg.	124	126.5
Nominal Launch Conditions		
Velocity, V_L, fps.	38,650	39,123
Flight Path Angle, γ_L, deg.	24.40	24.40
Azimuth Angle, Az_L, deg.	—	29.71
Inclination Angle, γ_1, deg.	—	2.0
Permissible Errors		
Velocity, ΔV_L, fps.	0.42	0.25
Flight Path Angle, $\Delta \gamma_L$, deg.	0.0049	0.00245
Azimuth Angle, ΔAz_L, deg.	—	0.00245
Astronomical Unit, $\dfrac{\Delta AU}{AU}$, %	0.000465	0.00028
Transfer Time, t, days	163.3	157.0

in Fig. 21. The amount of correction velocity required to compensate for miss-distances was determined for each of several perturbed trajectories as a function of time of correction and time of impact. A minimum required correction velocity was determined for each of the perturbed trajectories as shown in Fig. 22. The specific correction was obtained from these minimum velocities. A band containing the resulting values is presented in Fig. 23 with the comparable results of the two-dimensional inferential analysis. The three-dimensional results appear as a band rather than a single line because of the varying effects of corrections out of the transfer plane. These results, however, verify the conclusions of the two-dimensional analysis regarding approximate magnitude of the correction velocity required.

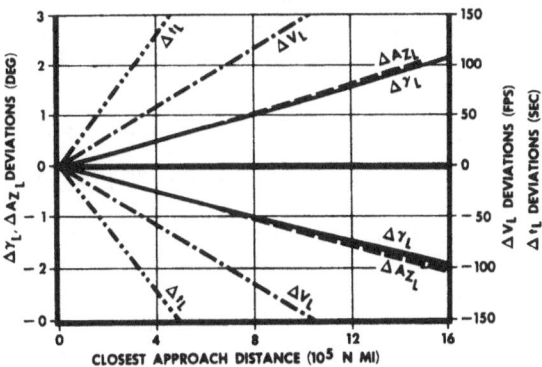

Fig. 21. Closest approach distance for deviations in nominal launch conditions. Three-dimensional. Earth—Mars transfer

A small rocket engine with a velocity capability of 1,000 feet per second, if utilized at a time halfway through the flight, will allow considerable improvement in the allowable errors in initial conditions. By this method the tolerance on initial conditions can be increased to approximately 125 feet per second in launch velocity and 1.15 degrees in initial flight path angle and azimuth.

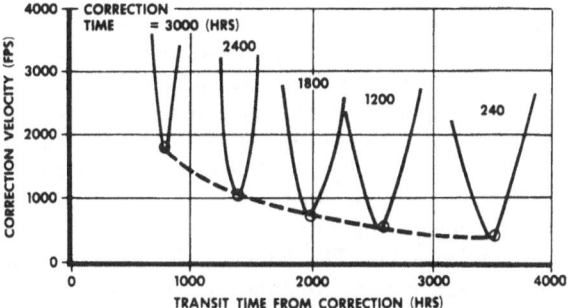

Fig. 22. Effect of transit time on correction velocity and time. Trajectory perturbed by a one degree error in γ_L. Earth—Mars transfer. ○ refers to intercept time for unperturbed nominal trajectory

Acknowledgement

The authors wish to acknowledge the assistance of R. D. SHINKLE, G. K. TOMLIN and J. T. WINTLER in the preparation of this report.

Conclusions

The analysis and data presented show that the general trends and approximate magnitudes of orbital transfer parameters and permissible deviations in the launch conditions from the intended can be established by two-dimensional infer-ential techniques with the effect of the third dimension estimated a posteriori.

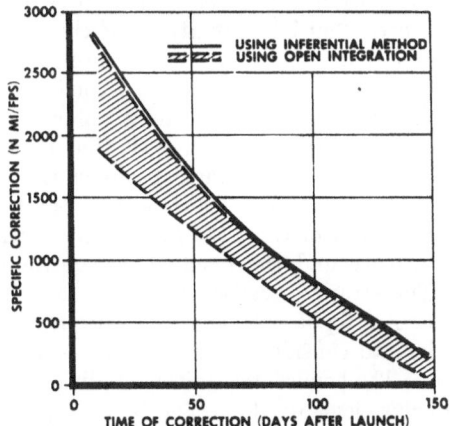

Fig. 23. Effect of specific correction on correction time. Earth—Mars transfer

The feasibility of impulsive midcourse correction to compensate for errors in the launch conditions, in particular, was evaluated by the inferential technique. Furthermore, this simple technique was shown to be adequate by comparison of results obtained with those of a more precise method. The use of a midcourse correction system as a means of relaxing the initial guidance requirements was demonstrated.

Appendix A

Inferential Method of Analysis

The solution of an n-body problem may often be inferred by successive applications of analytical solutions of the classical two-body problem. Application of such inferential techniques has given approximate solutions to the interplanetary transfer problem from which much insight into the character of the exact solution has been obtained. The equations of motion of the two-body central force problem may be found in many dynamics or mechanics text books (e.g. [5]).

For purpose of illustration, transfer from Earth to Mars during late 1960 and early 1961 is considered. The basic astronomical information used for the analysis presented herein are taken from [2] unless stated otherwise. It was assumed that the orbit of Earth was circular (one Astronomical Unit from the Sun) and that the orbit of Mars was eccentric but coplanar with that of Earth.

The capabilities of a particular vehicle for accomplishing interplanetary missions can be described most easily in terms of the vehicle's performance, i.e., the vehicle's velocity capability, and by relating this launch velocity to the interplanetary missions which can be achieved with various velocity magnitudes. The launch velocity of a hypothetical Earth-Mars vehicle is treated as the basic independent variable. The launch velocity magnitude, V_L, is that which would be measured at a radial distance, r_L, of 3,742 nautical miles from and relative to the center of the Earth, since this represents an altitude of 300 nautical miles above the Earth's surface. The other launch conditions required are: the flight path angle elevation, γ_L, measured from the local horizontal; and the local hour angle, θ_L, of the launch position with respect to the instantaneous position of the Sun, as shown in Fig. 1. These geocentric launch conditions are converted to the heliocentric "initial" conditions of the interplanetary transfer orbit by the classical equations of the two-body problem. These initial conditions as shown in Fig. 2 are: the initial heliocentric velocity, V_1; the initial distance, r_1, from the center of the Sun; and heliocentric angle, β_1, measured from perihelion of the transfer orbit. It can be seen (Fig. 1) that there is a unique correspondance of the escape direction, α_1, the angle between the asymptote of the Earth-escape trajectory and the normal with respect to the instantaneous position of the Sun, and the launch velocity to the heliocentric initial velocity, as well as between the launch velocity and the residual geocentric velocity, $V_{\infty 1}$, after escape from Earth. The escape direction, however, may be obtained from various combinations of launch flight path angle elevations and positions angles, and it is quite inconvenient to consider these parameters as independent variables.

The heliocentric initial conditions obtained completely determine the transfer orbit elements. For each "admissible" set of initial conditions there is an astronomical configuration, and hence, a launch time which will allow impact on the target planet. The criterion for "admissible" is, of course, that the aphelion of the transfer orbit is at least as far from the Sun as is the target planet.

Heliocentric conditions upon arrival at the target planet can be easily obtained and terminal conditions, e.g., with respect to the center of the planet of concern, also can be determined in a manner similar to that in which the initial conditions were determined.

Successful flight to other planets is dependent upon the capability of hitting a small moving target at a great distance. Mars, for example, subtends a maximum arc of only one-thirtieth of a milliradian. For near minimum energy flights which will be the earliest attempts, the subtended arc will be about one-fourth of this maximum value. Impact on Mars will be the equivalent of hitting a moving target of less than one inch diameter at a range of one thousand yards. It is necessary to determine the initial condition accuracy required for impact in order to evaluate the chance for success of such a mission.

Although initial conditions may vary slightly from nominal and the resulting transfer trajectory miss the intended intercept point by a wide margin, it is not always true that the target planet will also be missed by this same margin. The fact that the target is moving in space will sometimes cancel all or part of this

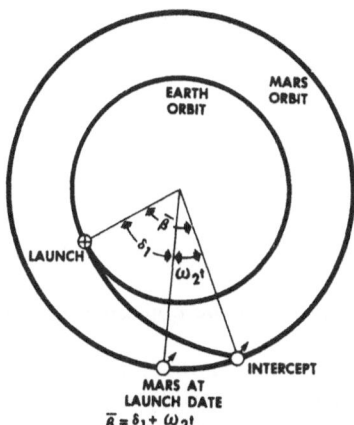

Fig. 24. Impact requirement. Earth–Mars transfer

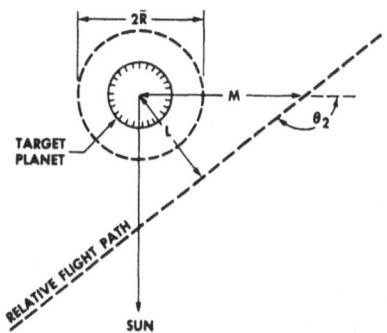

Fig. 25. Terminal conditions. Earth—Mars transfer

error and the miss distance must be measured in a moving reference system. The condition for perfect hit is that the total displacement angle of the transfer, $\bar{\beta}$, exactly equal the sum of the initial constellation angle, δ_1, and the central angle through which the target moves during the transfer time, $\omega_2 t$. This is illustrated in Fig. 24 and may be expressed by:

$$\bar{\beta} = \delta_1 + \omega_2 t .$$

For fixed launch time (fixed initial constellation) the distance from vehicle to target planet center, M, ignoring the gravitational attraction of the target, at the time when the vehicle crosses the target planet orbit, may be expressed by:

$$M = r_2 \left(\frac{\partial \bar{\beta}}{\partial x} - \omega_2 \frac{\partial t}{\partial x} \right) \Delta x .$$

Note that this orbital miss distance, M, is the closest approach to the target planet only if the angle, θ_2, between the vehicle relative flight path and target planet orbit is 90 degrees. Otherwise the closest approach is $L = M \sin \theta_2$ as shown in Fig. 25.

The two-body orbit calculations, neglecting the effects of the planet's gravitation, establish the relative velocity of approach, $V_{\infty 2}$, and position at crossing of the planet's orbit, M. The influence of the planet's gravitation attraction was approximated by considering the target to be a "virtual planet" having an effective gravitational cross sectional radius, R_1

$$\bar{R} = \sqrt{R^2 + \frac{2\mu}{V_\infty{}^2}} \ .$$

This effective radius defines an imaginary sphere about the target planet center such that any vehicle approaching the target planet with a velocity $V_{\infty 2}$,

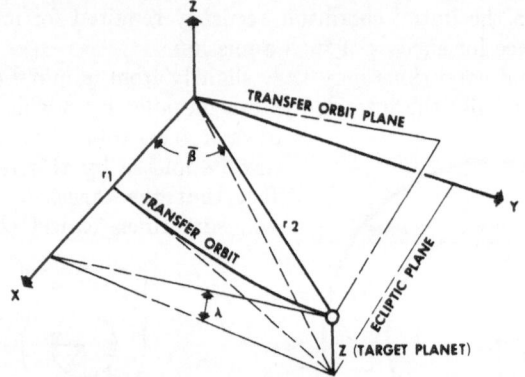

Fig. 26. Transfer orbit inclination

will impact the planet if, and only if, the asymptote to the approach hyperbola intersects this imaginary sphere.

A tolerance value on the arbitrary variable, x, may now be defined as the amount of deviation which exactly corresponds to a closest approach distance of one effective radius. Thus, the tolerance value, Δx, is expressed by

$$\Delta x = \frac{\bar{R}}{r_2 \left(\dfrac{\partial \bar{\beta}}{x} - \omega_2 \dfrac{\partial t}{\partial x} \right) \sin \theta_2} \ .$$

For transfer between non-coplanar orbits, the full benefit of the Earth's velocity cannot be realized, and the launch velocity must be increased to com-

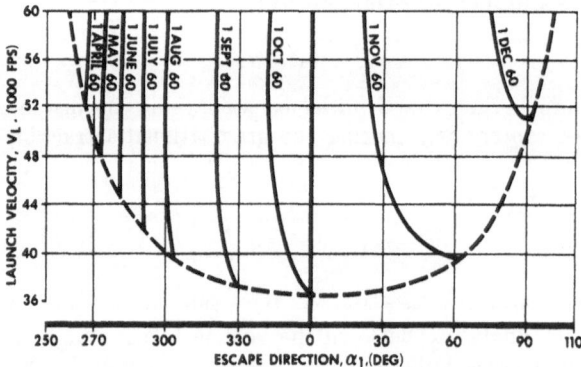

Fig. 27. Effect of initial conditions on launch date. Two-dimensional analysis. Earth—Mars transfer

pensate for this loss of "free" energy. In order to estimate the amount of additional energy (velocity) required it is necessary to define a correspondence between two- and three-dimensional transfers. In this correspondence the heliocentric transfer orbit elements (eccentricity, semi-major axis, longitude of perihelion and epoch) are the same in the transfer plane of each case. The inclination of the transfer orbital plane to the ecliptic can then be determined by

$$\lambda = \sin^{-1}\left(\frac{Z\,\text{Target Planet}}{r_2 \sin \overline{\beta}}\right),$$

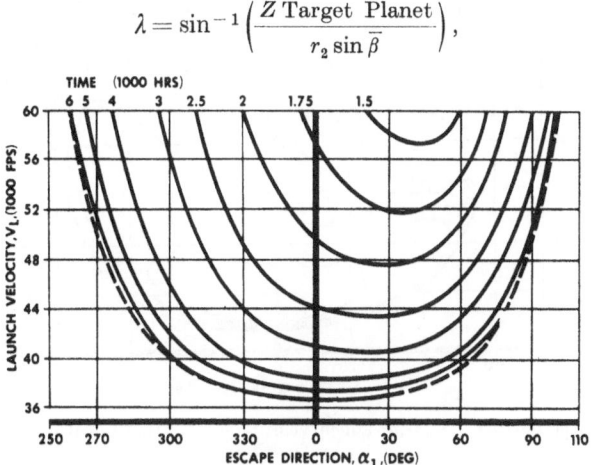

Fig. 28. Effect of initial conditions on transit time. Earth—Mars transfer

and is shown in Fig. 26. The launch velocity required can be determined by subtracting vectorially the velocity of the Earth from the heliocentric initial velocity of the projectile.

Appendix B

Transfer Orbits and Their Sensitivities

Additional information on the characteristics of the transfer orbits are presented in this appendix for reference purposes. As discussed earlier, the independ-

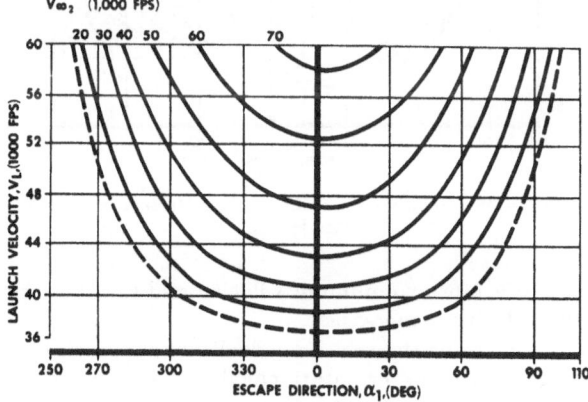

Fig. 29. Effect of initial conditions on approach velocity. Two-dimensional analysis. Earth—Mars transfer

ent variables chosen for this analysis are the launch velocity and the escape direction. The data contained in Fig. 3 are also presented in Fig. 27, where the launch date is shown as a function of the launch velocity and escape direction.

The limiting curve illustrating the band of admissible escape directions as a function of launch velocity is included. The maximum heliocentric initial velocity for a given launch velocity may be obtained when the escape direction is parallel to the heliocentric velocity vector of the Earth; however, as shown in Fig. 28, this

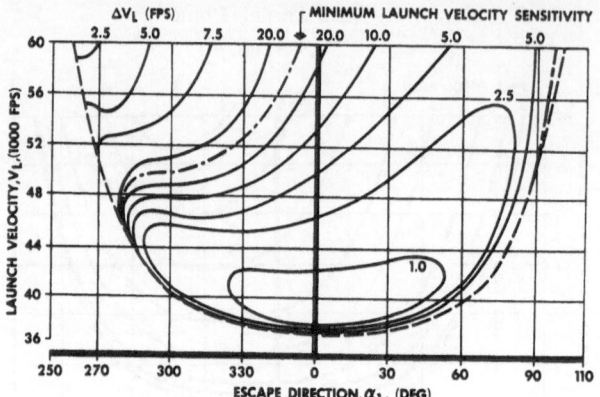

Fig. 30. Maximum permissible errors in launch velocity for Martian impact. Earth—Mars transfer. $\gamma_L = 90$ deg.

does not always correspond to minimum transfer time. The escape direction for minimum time of flight at a given launch velocity is parallel to the Earth's

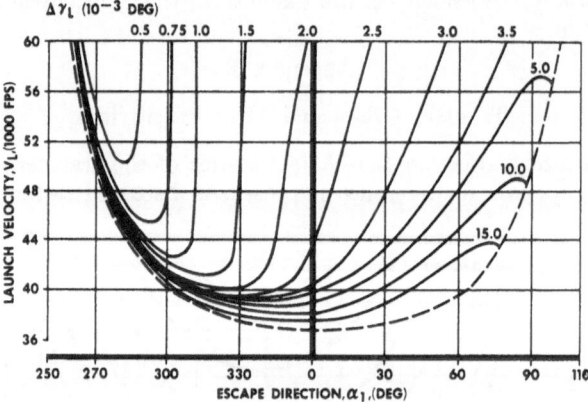

Fig. 31. Maximum permissible errors in escape direction for Martian impact. Earth—Mars transfer. $\gamma_L = 90$ deg.

heliocentric velocity vector at minimum velocity and tends to approach the perpendicular as the velocity is increased.

The terminal velocity, or the relative velocity of approach to Mars and before potential energy due to Mars' gravitational field has been properly taken into account, is presented in Fig. 29. The terminal velocity increases quite rapidly with modest increase in launch velocity.

The permissible errors in both the magnitude and the direction of the launch velocity vector have been evaluated for Martian impact trajectories and are presented in Figs. 30 and 31. On the basis of this first order approximation, certain families of transfer trajectories were found where the maximum permissible error in launch velocity approaches infinity. Although this "infinite" tolerance is, of course, unattainable, it is reasonable to believe that these trajectories are least sensitive to errors in the launch velocity. These trajectories are, therefore, of interest, and finite difference calculations have been employed to evaluate their permissible errors.

The permissible errors in launch velocity are, for most trajectories, well above one foot per second and range up to several hundred feet per second in the region

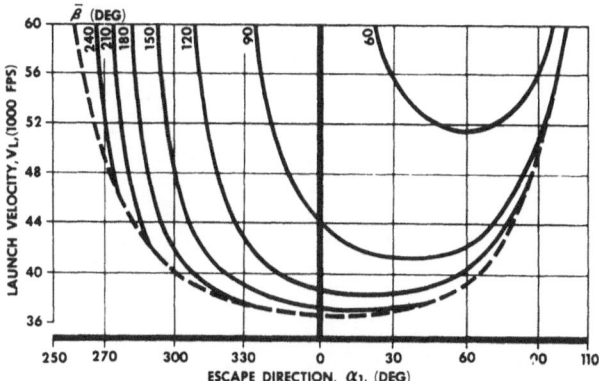

Fig. 32. Effect of initial conditions on transit angle. Earth—Mars transfer

of maximum tolerance. These data presented are for vertical ascent from the Earth, but similar values have been calculated for other initial angles. In the case

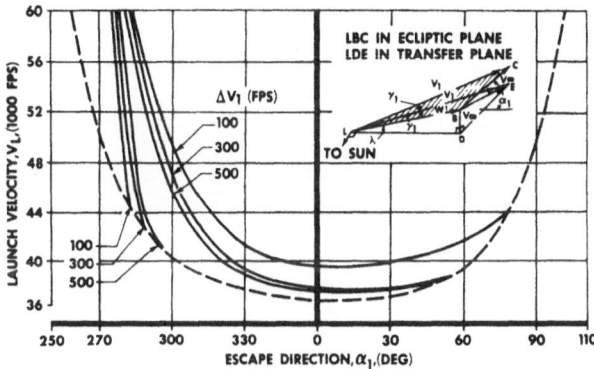

Fig. 33. Velocity increment required to compensate for transfer orbit inclination. Three-dimensional analysis. Earth—Mars transfer

of sensitivity to the direction of the launch velocity vector, the permissible errors are of the order of one tenth to one thousandth of one-degree throughout most of the domain of launch conditions.

The heliocentric central angle corresponding to these transfer orbits is presented in Fig. 32. These data were used to calculate inclination between the plane of the transfer orbit and the ecliptic. The velocity increment over the two dimensional launch velocity required to compensate for this inclination of the transfer plane is greatest where the inclination is greatest and some transfer trajectories are prohibited completely as shown in Fig. 33.

References

1. C. W. ALLEN, Astrophysical Quantities. University of London: Athlone Press, 1955.
2. The American Ephemeris and Nautical Almanac for the Year 1960. Washington, D.C.: Nautical Almanac Office, United States Government Printing Office.
3. M. W. HUNTER, W. B. KLEMPERER and R. J. GUNKEL, Impulsive Mid-Course Correction of a Lunar Shot. Douglas Aircraft Company, Inc., Engineering Paper 674, August 1958. Proceedings of the IXth International Astronautical Congress, Amsterdam 1958, p. 626. Wien: Springer, 1959.
4. J. C. WALKER, Interplanetary Trajectory Simulation. Douglas Aircraft Company, Inc., Report SM-27742, 10 September 1958.
5. H. GOLDSTEIN, Classical Mechanics. Reading, Mass.: Addison-Wesley Publ. Comp. Inc., 1957.

Multi-Directional G-Protection During Experimental Sled Runs

By

Harald J. von Beckh[1], SAI, DGRR, ARS

(With 14 Figures)

(Received August 27, 1959)

Abstract — Zusammenfassung — Résumé

Multi-Directional G-Protection During Experimental Sled Runs. The Author described at the VIIIth I.A.F. Congress (Barcelona 1957) [5] a device termed "Anti-G Capsule" which would protect a human operator or a test animal against high G loads during space flight.

For testing this new principle of G protection an "Anti-G platform" was designed and mounted on catapult and rocket-driven sleds. The accelerations and decelerations which acted on the test animals were electronically detected and continuously recorded.

It was shown that this device satisfactorily transforms, by its positioning qualities, longitudinal G's into better tolerated transverse G's.

Mehrseitiger Beschleunigungsschutz in Raketenschlittenversuchen. Der Autor beschrieb beim VIII. IAF Kongreß (Barcelona 1957) [5] eine sogenannte „Anti-G-Kapsel", die den Insassen oder ein Versuchstier in einem Raumfahrzeug gegen hohe Beschleunigungen schützen würde.

Um dieses neuartige Prinzip des Beschleunigungsschutzes zu erproben, wurde „ad hoc" eine „Anti-G-Plattform" konstruiert und an Versuchsschlitten, die durch Katapulte oder Raketenmotoren angetrieben waren, angebracht. Die Beschleunigungen und Verzögerungen, die auf die Versuchstiere einwirkten, wurden elektronisch gemessen und kontinuierlich registriert.

Es zeigte sich, daß diese Vorrichtung durch ihre positionsändernden Eigenschaften longitudinale Beschleunigungen in besser zu ertragende Querbeschleunigungen umwandelt.

Protection multidirectionnelle anti-G au cours d'expériences effectuées au moyen de traineaux. L'auteur a décrit au Huitième Congrès International Astronautique (Barcelona 1958) [5], un dispositif, appelé "Capsule anti-G", destiné à protéger les occupants des engins spatiaux contre les fortes accélérations.

Pour essayer ce nouveau principe de protection contre les accélérations, une "plate-forme anti-G" fut réalisée et montée sur des traineaux expérimentaux propulsés par catapulte ou moteurs fusées. L'action des accélérations et décélérations sur des animaux utilisés pour l'expérience fut mesurée électroniquement et continuellement enregistrée.

Il a été montré que ce dispositif — qui présente l'avantage de pouvoir changer la position du sujet, — transforme les accélérations longitudinales en accélérations transversales, mieux tolérées.

---

[1] Space Biology Branch, USAF Aeromedical Field Laboratory, Holloman Air Force Base, New Mexico, U.S.A.

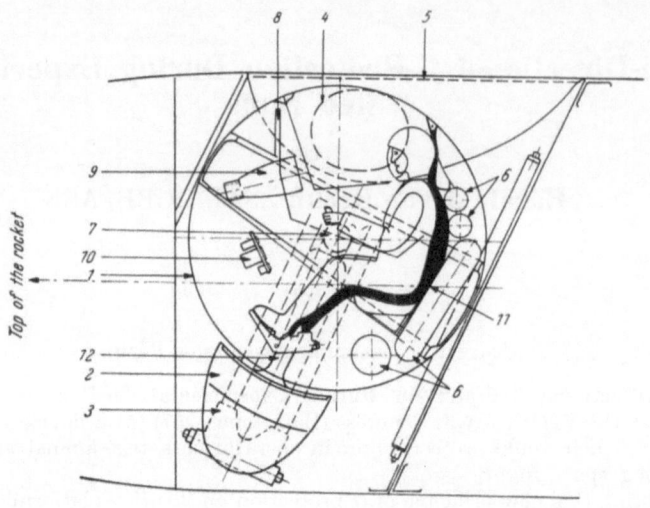

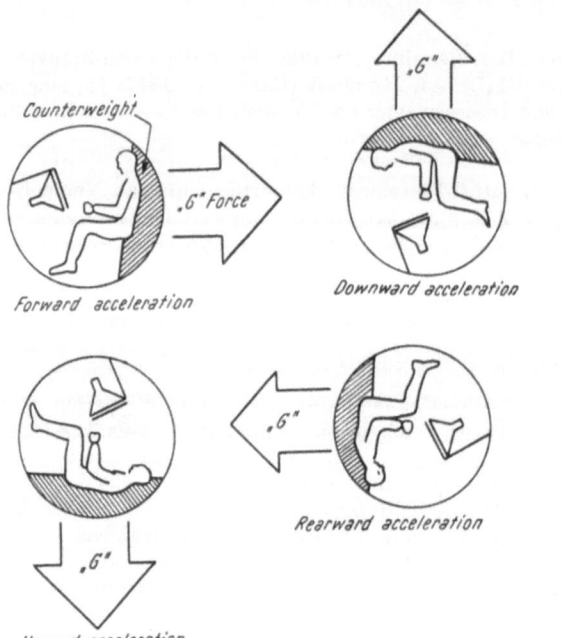

Fig. 3. Daisy Track. At the far end the water-inertia brake is shown. On the sled in the foreground a human subject is ready for a backward facing deceleration experiment

Fig. 4. The High Speed Sled "Sonic Wind II" with Astrodyne "Megaboom" booster developing a thrust of 10,000 pounds for ten seconds

such that the resultant of all acting accelerations is perpendicular to the heart-head line of the subject.

The ejection and stabilization mechanism of this capsule also affords an analogous G protection during and after emergency separation of the life compartment from a disabled space craft.

Twenty nine sled runs with an ad hoc designed experimental device, termed "Anti-G platform" served to show the advantage of this positioning mechanism.

II. Methods

1. Test Vehicles

For these experiments a catapult sled track was used, which was developed for the USAF Aeromedical Field Laboratory and designated "Daisy Track" (Fig. 3).

Fig. 5. At the end of the 35,000 foot Captive Missile Track the sled has just entered the braking area. It is decelerated by scooping water which is dammed between the rails of the braking area

Two rails, 120 feet in length, spaced 5 feet apart at rail centers, allow the rapid displacement of a chromium-molybdenum steel sled, weighing 2000 pounds.

The sled is propelled by one or two M1A1 ejection-seat catapults. At the end of the track is a water-inertia brake mounted on a rigid welded-steel base. The brake contains many apertures through which the water is displaced to the exterior. They may be adjusted according to the deceleration pattern desired. The sled stopping distance may be varied from $\frac{1}{2}$ inch to 4 feet. Peak-G impact forces varying from 5 G to 200 G can be obtained with rates of onset of 100 G/sec to 12,000 G/sec.

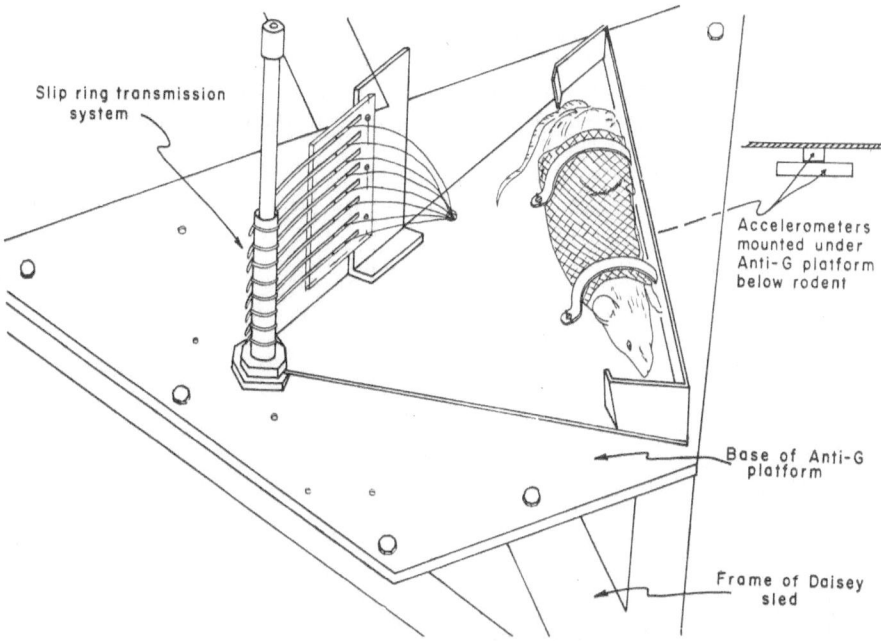

Fig. 6. Sketch of an experimental Anti-*G* device (Anti-*G* platform). The test animal (Albino rat) is located on a turning triangular platform. Accelerometers record the acting *G* loads in animal's longitudinal and transverse axis

Fig. 7. The Anti-*G* platform (without test animal) is mounted on the forward section of the catapult-driven Daisy-sled. On the vertical axis are shown the slip-rings. They transmit the data from the accelerometers and from a potentiometer. The latter records the rotatory movements of the platform

For the final run (No. 29) a high speed deceleration vehicle was used which was designed for the Aeromedical Field Laboratory by Northrop Inc. and was designated "Sonic Wind II". It was powered by a Astrodyne rocket motor, type 10 KS 100,000, designated "Megaboom", and developed a thrust of 1,000,000 pounds in 10 seconds (Figs. 4 and 5).

The experiment took place on the new 35,000 foot Captive Missile Track of the Air Force Missile Development Center. This facility provided a Mach 2 run

Fig. 8. Side view of the Daisy sled. In the seat is a dummy, subject of another non-interference experiment. Behind and to the left of the Anti-G platform is shown a container with the control animal, which is unprotected and receives the G load in its spinal axis. Both animals are encased in wire netting

with acceleration and deceleration patterns of long duration. The technical details of this unique test facility are described in a Technical Report entitled "Holloman Track Capabilities" [1].

2. Anti-G Platform

The Anti-G platform consisted of an Aluminium sheet, in form of an isoceles triangle. Its base measured 8 in. and its height 9.5 in. The base of the triangle was adapted to tie down a rat measuring up to 7.5 in. The platform was pivoted on its apex by a vertical axis which was fixed on the structure of the sled and allowed free rotation through 360 degrees (Figs. 6 and 7).

A simple frictional damping device was located at the shaft under the platform. Two accelerometers were fixed on the under side of the platform in positions which corresponded exactly to the animals spinal and transverse-to-spine direction. These axis intersected at a point under the animals center of gravity and at the same distance from the shaft.

The shaft carried eleven slip-rings for transmission of acceleration- and other data. Later a potentiometer was added, which recorded the rotary movement of the device.

3. Instrumentation

a) On the Track

Velocity measurements were obtained from make-circuit devices along the track, connected through a thyratron tube to Berkeley counters.

Fig. 9. In the rear part of the sled "Sonic Wind II" are located the Anti-*G* platform and above the nonmovable container with the control animal. Both animals are anesthetized immediately before the run

b) On the Sled

Several Statham accelerometers were mounted on the sled structure to measure accelerations in the longitudinal, vertical and lateral axis of motion.

c) On the Anti-G Platform

Acceleration data was obtained by the mentioned accelerometers. For the parallel-to-spine *G* load a Statham accelerometer type Fa-25-120 was used with a range of plus minus 25 *G*. The transverse-to-spine *G* load was obtained with a Statham accelerometer Fa-50-120 with a range of plus minus 50 *G*.

The potentiometer used was a Fairchild type 764 with a resistance of 60 K Ohm plus minus 5% and an available rotation of 360 degrees. Fiber gears formed the connection between potentiometer and platform shaft.

All sled-borne data was fed through a trailing cable and over a calibration box into a 50 channel Consolidated C.E.C. Oscillograph, model 5-119 P 4—50 and recorded on photosensitive Lino-Writ paper.

d) Cinematographic Observation

The runs were covered by two Fastex Cameras (1600 frames/sec) placed so as to record the side and top views of the experiment. A conventional movie camera (64 frames/sec) recorded the head on view.

4. Description of Test Runs

a) Preparation

The rails are cleaned and then oiled with a lightweight aircraft oil. The brake cylinder is sealed with a polystyrene disc and filled with water. The rear of the sled is attached to the M1A1 catapults. Calibrations are recorded for each accelerometer.

Fig. 10. The Daisy sled during the run

The test animal (Albino rat, measuring approximately 7 in.) is tied down by wire netting parallel to the base of the platform. The control animal is tied down in a nonmovable wire netting cage in an identical head forward position. Immediately before the run both animals are anesthetized (Figs. 8 and 9).

b) Procedure of the Run

The catapults are fired. The sled accelerates, then travels in a free run with slightly decreasing velocity. The sled is abruptly decelerated when the braking piston ruptures the polystyrene disc and meets the resistance of the water in the water brake (Figs. 10 and 11).

At the onset of the acceleration the Anti-G platform swings backwards and makes—corresponding to the degree of damping—one or more revolutions.

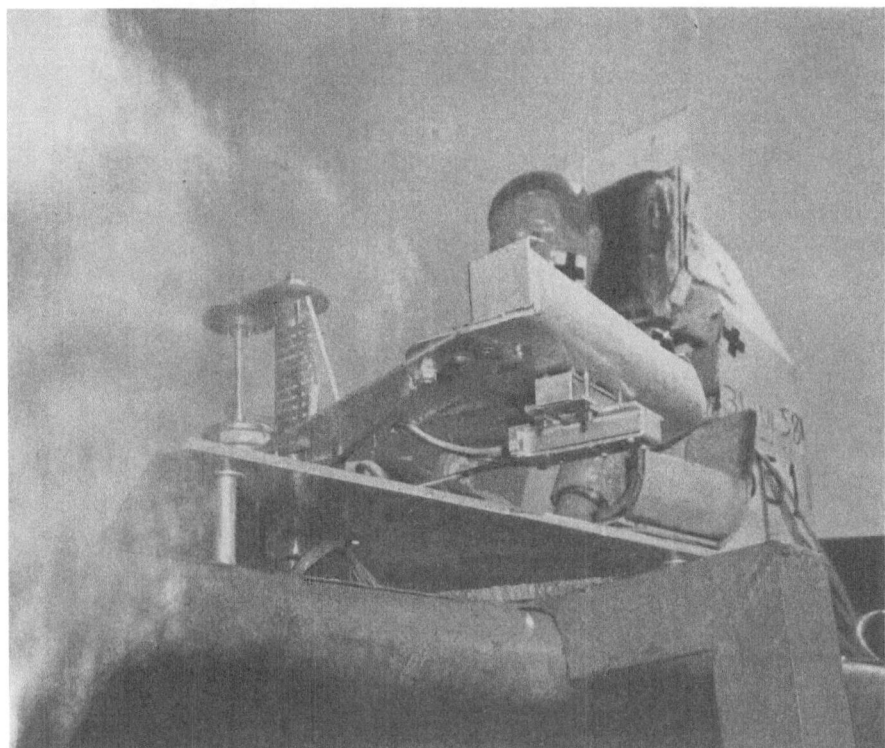

Fig. 11. The Daisy sled entering in the braking area. On the underside of the platform are shown the two accelerometers, which record the *G* loads in animal's transverse and longitudinal axis

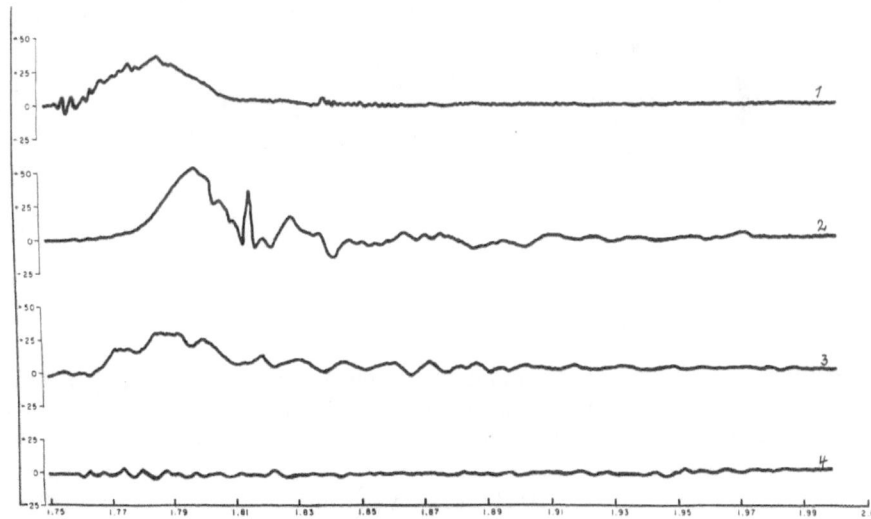

Fig. 12. Results of "Daisy" sled run No. 357 (Experiment No. 26). Deceleration versus time:
1 Deceleration recorded on the seat of the dummy. Peak *G* load: 36 *G*.
2 Deceleration recorded on the dummy. Peak *G* load: 54 *G*.
3 Deceleration recorded on "Anti-*G* platform" in animal's transverse axis. Peak *G* load: 31 *G*.
4 Deceleration recorded on "Anti-*G* platform" in animal's longitudinal axis. Peak *G* load: 3 *G*.
It was shown that the "Anti-*G* platform" satisfactorily transforms longitudinal *G* loads in better tolerable transverse *G* loads

An analogous movement of the Anti-G platform takes place in the decelerative phase (impact) which will be described later.

III. Results

1. Accelerative Phase

The data of all runs showed that the test animal received practically no G load in the vulnerable parallel-to-spine axis. The G load in the better tolerable transverse-to-spine axis was generally somewhat lower than the linear sled accelera-

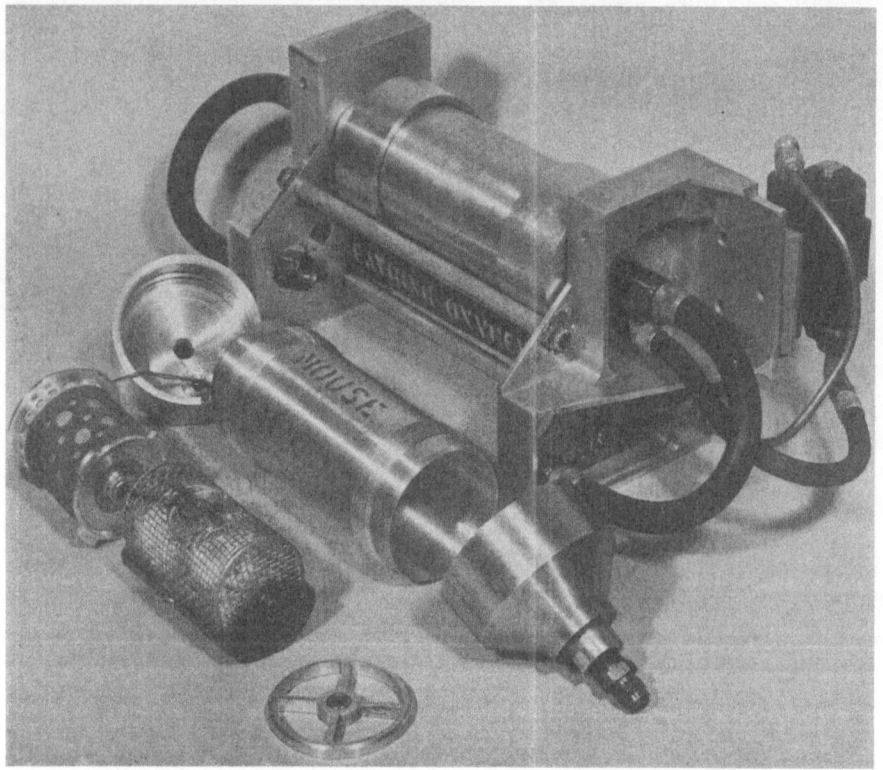

Fig. 13. Mouse package as used in the M.I.A. ("Mouse in Able") project. The assembly was developed by "Space Technology Laboratories". The mouse cradle turns about its longer axis, when direction of G load changes. (Illustration by courtesy of Space Technology Laboratories)

tion. Only when little or no damping was applied, the transverse-to-spine G load acted with equal or somewhat higher intensity, due to the fact that by the rotation of the Anti-G Capsule an additional centrifugal G load was build up.

2. Decelerative Phase

During deceleration only insignificant parallel-to-spine G loads were recorded (Fig. 12).

Also the transverse-to-spine G loads were generally smaller than the sled deceleration. Its value depends upon the degree of damping and upon the position of the Anti-G platform in the moment of the impact:

When the angle between the platform's apex bisector with the line of motion is 90 or 270 degrees, the momentum of rotation is greatest. This results in the build up of an additional centrifugal *G* load. However, by adequate damping this additional load can be minimized.

When the above mentioned angle is 0 or 180 degrees, the transverse-to-spine *G* value equals the sled deceleration.

The control animal, as tied down in a non movable cage, is exposed in both phases to the same *G* load as the sled structure. It receives the accelerative load in head-tail and the decelerative load in tail-head direction.

IV. Discussion

It is evident that the platform's freedom of swinging forward and backward avoids the action of significant parallel-to-spine *G* loads.

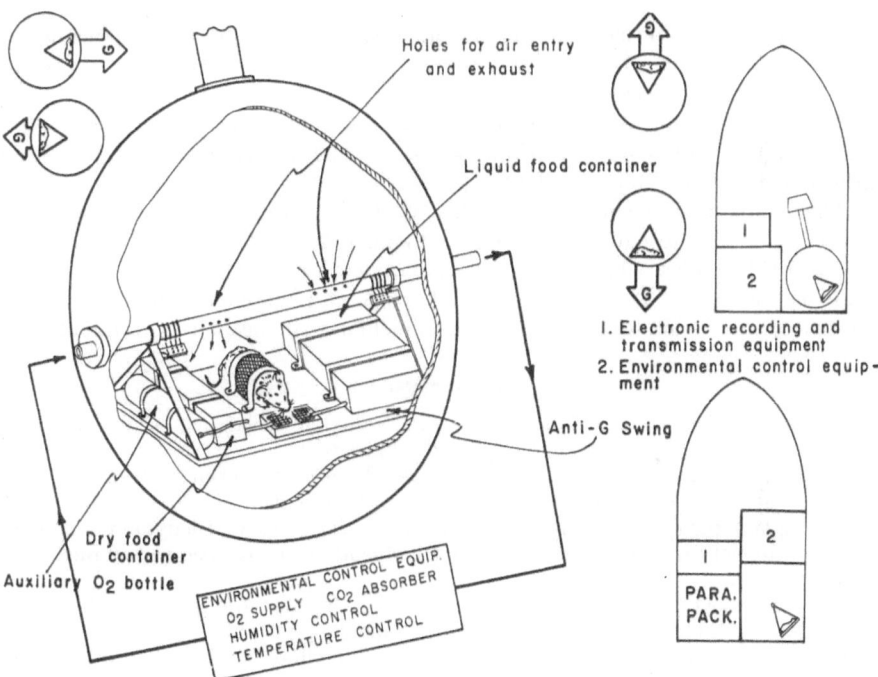

Fig. 14. Artist's conception of a mouse carrying Anti-*G* Capsule

The fact that the transverse-to-spine *G* load can under certain circumstances be somewhat higher than the original sled deceleration seems not a disadvantage, because it is doubtlessly favorable to "trade in" dangerous parallel-to-spine loads in even slightly higher, but better tolerable transverse-to-spine loads. In addition, the build up of this additional centrifugal loads can be avoided by adequate damping of the rotation.

Multi-directional *G* protection was already used—although in a simplified modification—in the M.I.A. ("Mouse in Able") bio-ballistic flights.

This was a non interference experiment in conjunction with the Project Able re-entry test program. The mouse container and its accessories were developed by Space Technology Laboratories [4].

The USAF Aeromedical Field Laboratory provided for this project assistance and physiological background information.

The mouse was located with its longer axis parallel to the axis of a turnable cradle. Fig. 13 shows the partially assembled M.I.A. package. At the extreme left is to be seen the mouse container, consisting of a cylindrical wire cage turnable around its longitudinal axis.

This arrangement does not fulfill a "multi-directional" G protection in its complete sense, because accelerations which act in the direction of the container's axis would not be converted to transverse-to-spine accelerations and therefore act in the animal's vulnerable axis.

Only when the animal is located perpendicularly to the turning axis (Fig. 14), it is safe from all parallel-to-spine G loads. Space restrictions, however, required this compromise in the M.I.A. project.

V. Conclusions

During re-entry, impact landing and emergency separations of the life compartment high G loads can be expected. By malfunctioning of guidance systems they even could become excessive.

An important contribution to G protection would be made if the G loads— what ever their intensity, duration and rate of onset—by no means act on the subject in its vulnerable parallel-to-spine axis.

The tested device showed that this kind of protection would be possible. However, further testing with more sophisticated and larger Anti-G platforms and Anti-G Capsules, which allow experiments with primates and later with men, would be necessary. The incidence of labyrinthine phenomena due to the short, but fast rotatory movements should be elucidated in human experiments.

Acknowledgement

The author gratefully acknowledges the valuable assistance of the following personnel of the USAF Aeromedical Field Laboratory:

Dr. J. E. COOK, Captain USAF (VC), Chief, Veterinary Services Branch, for the procurement, care, and examinations of the test animals.

A. V. ZABOROWSKI, Lieutenant USAF, Chief, Test Section, Biodynamics Branch, for the technical direction of the tests and his generous assistance in technical areas of this program.

The personnel of the Aeromedical Group of Land Air Inc. for gathering all the electronic data and for improving the test equipment.

Last but not least Mr. DONALD S. BRISTOW, who willingly fabricated in the workshops of the Aeromedical Field Laboratory several prototype Anti-G devices.

References

1. T. G. ASHMORE, Holloman Track Capabilities. Technical Report AFMDC-TR-59-4, (April 1959).
2. E. L. BEEDING, JR., Daisy Track and Supporting Systems. Technical Note AFMDC-TN-57-8 (June 1957).
3. J. P. STAPP and W. C. BLOUNT, Effects of Mechanical Force on Living Tissue. J. Aviat. Med. 28, 281 (1957).
4. F. L. VAN DER WAL and F. L. YOUNG, A Preliminary Experiment with Recoverable Biological Payloads in Ballistic Rockets. Project M.I.A. Space Technology Laboratories, Los Angeles, Sept. 1958; Also published: J. Amer. Rocket Soc. 29, 716 (1959).
5. H. J. VON BECKH, Multi-Directional G-Protection in Space Vehicles. VIIIth International Astronautical Congress, Barcelona 1957, Proceedings, p. 37. Wien: Springer, 1958; J. Aviat. Med. 29, 335 (1958).

Laboratory Experimental Studies in Re-entry Aerothermodynamics

By

Walter R. Warren[1]

(With 22 Figures)

(Received August 27, 1959)

Abstract — Zusammenfassung — Résumé

Laboratory Experimental Studies in Re-entry Aerothermodynamics. The integrated aerodynamic and thermodynamic problems of atmospheric re-entry are discussed in terms of their relationship to current and future re-entry vehicles. The overall problem area is divided into separate areas that are amenable to experimental studies on a laboratory scale. The capabilities in these problem areas of existing test facilities— wind tunnels, shock tunnels, continuous arc heaters, etc.—are discussed.

Detailed performance characteristics of two high enthalpy test devices—the shock tunnel and the air arc facility—are presented and it is shown that they are complementary to each other in allowing study of many of the re-entry problems of present interest. Test facilities of these types have been in development and operation for several years at the Missile and Space Vehicle Department of the General Electric Company. To illustrate their usefulness, examples of experimental data obtained in the 6 Inch Shock Tunnel and in several arc heated facilities, including a 2500 KW (test gas power) unit, are presented. These include the results of shock tunnel studies on flow field visualization, surface pressure distributions, surface heat transfer distributions, and aerodynamic force measurements and the results of arc studies on materials suitable for use in ablation heat protection systems.

In conclusion, a critique is made of the status of present laboratory test facilities in terms of the re-entry problems that will require future investigation. It is postulated that the arc heated wind tunnel, because of its many desirable characteristics, has a higher potential than other laboratory facilities in the study of re-entry aero-thermodynamic problems. To support this viewpoint, the performance of a relatively large arc wind tunnel, the design of which appears feasible on the basis of current state of the art considerations, is presented and discussed briefly.

Laboratoriumsuntersuchungen der Wiedereintritts-Aerothermodynamik. Die Gesamtprobleme der Aero- und Thermodynamik des Wiedereintritts in die Atmosphäre werden mit Beziehung auf ihre Bedeutung für gegenwärtige und zukünftige Raumfahrzeuge besprochen. Das Gesamtproblemgebiet wird in Teilgebiete zerlegt, die der experimentellen Erfassung im Laboratorium zugänglich sind. Die Anwendbarkeit vorhandener Versuchsanlagen — wie Windkanäle, Schocktunnel, Lichtbogenmaschinen — wird erörtert.

Zwei Arten von Versuchsgeräten mit hohem Wärmeumsatz — nämlich der Schocktunnel und die Lichtbogen-Luftstrahlanlage und ihre Arbeitsweise — werden ausführlich beschrieben. Es zeigt sich, daß sie einander gegenseitig ergänzen, da sie

---

[1] Aerosciences Laboratory, Missile and Space Vehicle Department, General Electric Comp., Philadelphia 4, Pennsylvania, U.S.A.

verschiedene Seiten der Wiedereintrittsprobleme zu studieren gestatten. Derartige Versuchsanlagen sind seit mehreren Jahren bei der Geschoß- und Raumfahrzeugabteilung der General Electric Co. in Entwicklung und in Betrieb. Ihre Anwendung wird an Hand von Beispielen experimenteller Ergebnisse des 15-cm-Schocktunnels und mehrerer Lichtbogenstrahlanlagen (eine davon mit 2500 kW Stärke) erläutert. Dabei sind Studien über die Sichtbarmachung des Strömungsfeldes, Druckverteilungen, Wärmedurchtrittsverteilungen und Luftkraftmessungen inbegriffen sowie auch Messungen an Materialien, die als Wärmeschutz durch Ablation geeignet sind.

Zum Schluß wird eine Kritik des gegenwärtigen Standes der existierenden Versuchseinrichtungen für das Studium von Wiedereintrittsvorgängen, die noch der Untersuchung harren, gegeben. Es steht zu erwarten, daß der lichtbogenbeheizte Windkanal sich auf diesem Gebiete allen anderen Apparaturen wegen seiner vielen Vorteile als überlegen erweisen wird. Um diese Meinung zu erhärten, wird ein Projekt einer besonders großen solchen Anlage, die durchaus im Bereich des Möglichen zu liegen scheint, vorgeführt.

Etudes expérimentales en laboratoire sur les problèmes aérothermodynamiques de la ré-entrée. Les problèmes aérodynamiques et thermodynamiques en interférence dans une ré-entrée dans l'atmosphère sont discutés en relation avec les véhicules actuels et futurs. Le problème est scindé en groupes sur lesquels des études peuvent être menées à l'échelle du laboratoire. Les possibilités des installations expérimentales existantes — souffleries, tubes à choc, réchauffeurs continus à arc, etc. — sont analysées.

Les détails de performances de deux dispositifs d'essai à enthalpie élevée — le tube à choc et l'arc électrique — sont présentés. Ils ont un caractère complémentaire dans l'étude des problèmes d'intérêt actuel. Des installations de ce genre ont été développées et opérées depuis plusieurs années au département des engins et véhicules spatiaux de la General Electric Company. Leur utilité est mise en évidence par des exemples de données expérimentales obtenues dans le tube à choc de 6 pouces et plusieurs installations à arc électrique, dont une unité de 2500 KW. Ces exemples comportent des études de visualisation d'écoulements, des distributions de pression en surface, des mesures de forces aérodynamiques, des distributions de transfert de chaleur et des résultats d'essais sur matériaux utilisables pour une protection thermique par ablation.

En conclusion on fait une analyse critique des facilités expérimentales actuelles dans leurs applications aux problèmes de ré-entrée qui requièrent des études plus poussées. Il semble que le tunnel à arc électrique ait un potentiel de recherche supérieur grâce à plusieurs caractéristiques adéquates. Ce point de vue est appuyé par la présentation et discussion des performances d'une soufflerie à arc relativement grande, dont la réalisation apparaît possible dans l'état actuel de nos connaissances techniques.

Nomenclature

d	diameter	H	altitude
h	enthalpy	L	lift
$\dot{m}$	mass rate flow of air	M	MACH number
p, P	pressure	ΔP	pressure change
$\dot{q}, q$	heat transfer rate	Q	total heat transferred to one ft.² surface area
r	nose radius		
t	time	R	universal gas constant
u	velocity	R	base radius
x	axial distance	Re	REYNOLDS number
A	cross-sectional area	T	temperature
C_D	drag coefficient	W	weight
D	drag	μ	micro
E	power added to test gas		

Introduction

The re-entry of vehicles at near orbital velocities has led to the current interest in high stagnation enthalpy or high temperature experimental aerodynamics. It has fostered, therefore, the development of several new and advanced laboratory devices which aid both in the study of basic aerothermodynamic problems and in establishing the feasibility and increasing the efficiency of engineering design approaches.

Consider the re-entry missions of interest at present and in the predictable future. Fig. 1 illustrates some of these. First, there is the long range ballistic missile which re-enters the atmosphere at a steep, but far from vertical, angle.

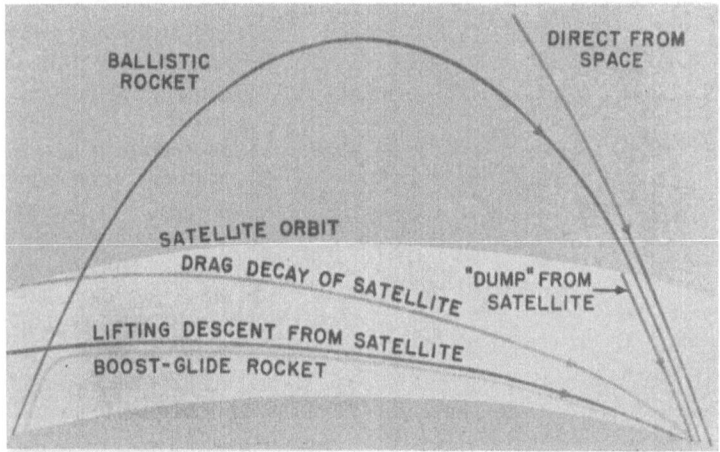

Fig. 1. Various types of atmospheric entry

This type of vehicle is characterized by a high weight-to-drag ratio and a high drag deceleration, generally exceeding that which a man can endure. A second interesting vehicle is the re-entering satellite. This body is usually non-lifting and, therefore, follows a ballistic trajectory through the atmosphere. It has a small re-entry angle—the order of 1°—and a relatively low maximum drag deceleration during re-entry and thus is compatible with the manned re-entry concepts of current interest. A third vehicle of importance is the boost-glide missile. This body spends a large part of its flight time in the atmosphere at high altitudes and derives its lift from a combination of centrifugal and aerodynamic forces. From the point of view of re-entry aerothermodynamic problems, each of the other vehicles illustrated in Fig. 1 can be considered similar to one of these three.

One problem associated with re-entry that has received a great amount of publicity is that of aerodynamic heating. The character of the heating problems

associated with the vehicles just discussed can be illustrated with the aid of Fig. 2.
Shown are typical stagnation point heating rate curves for these bodies. Zero
time is taken arbitrarily at the 250,000 ft. altitude point for each vehicle. It is
seen that the peak heating rate for the ballistic missile is 40 to 50 times greater
than that for either the satellite or the glide vehicle. However, the ballistic missile
and the satellite take only approximately 1 % and 10 %, respectively, of the time
required by the glide vehicle to impact from this altitude. These considerations
lead to the total heating ratios shown; that is, total stagnation point heating per
square foot for a glide body is approximately 20 times that of a satellite and 5
times that of a ballistic missile. It can be seen that wide differences are already
indicated in the types of problems that will be encountered by these vehicles.

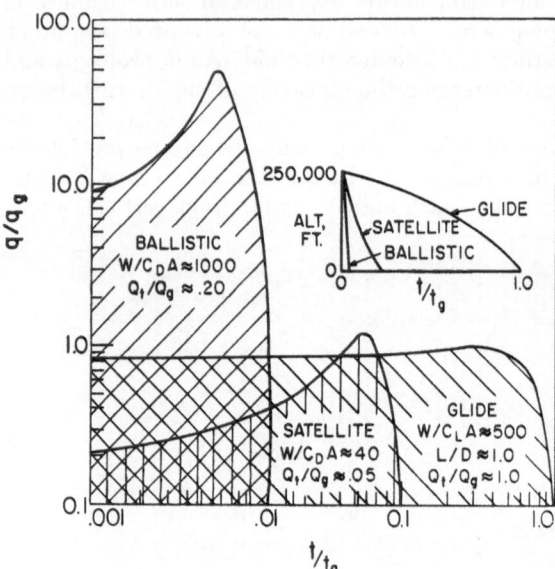

Fig. 2. Typical re-entry heating cycles

Although heating is certainly a problem of major importance, there are many others, the solutions of which are also critical to the success of re-entering vehicles. For example, the shape must be aerodynamically stable over the complete re-entry trajectory and may require the use of lift and control; also, the re-entry problem is complicated by the need for compatibility between the heat protection and aerodynamic systems. It is seen that the situation is indeed complex when one sonsiders that a typical body during its mission: (1) experiences flight conditions varying from the low density aerodynamics regimes at MACH numbers of
greater than 20 to transonic and subsonic velocities at altitudes of from 0 to 50,000 ft;
(2) is often cooled by mass transfer from the surface to the boundary layer; (3)
is for a major part of its critical flight in a region in which the gas surrounding it
is reacting and perhaps not in equilibrium; (4) is covered by turbulent boundary
layers for a large portion of the flight time; (5) is, in general, oscillating through an
appreciable angle of attack range and rolling; and has many other complications
resulting from its interaction with the continuously changing flow environment.

Fortunately, it has been shown through the last several years that it is possible
to use combined analytical and experimental approaches to attack successfully
this problem area. The one general technique is to divide the problem area into
smaller problems that can be approached both on an analytical basis and with
the experimental tools that are available.

A discussion of the analytical approaches to re-entry aerothermodynamic
problems is beyond the proposed scope of this paper. It suffices to say that the
many excellent contributions that have been made in the past few years—such
as the work by LEES [1], FAY and RIDDELL [2], and SCALA [3] on chemically
reacting boundary layers and by SCALA and SUTTON [4] on stagnation point

ablation—have done much to organize and to specify the requirements of experimental research in the re-entry aerodynamics field.

Experience has shown that the general problem area may be divided into

Test Duration	Type of Test or Data	Rel- ative Value[1]	Rel- ative Utility
5 μsec ..	Aerodynamic flow patterns—pictures—no instrument	10	10
1 ms	Aerodynamic Studies, Pressure and heat flux distribution	10	20
1 sec ...	Aerodynamic Studies, Force Balance Data	10	30
1 min ..	Aerodynamic Studies, Programmed Tests	5	35
1 min ..	Basic Materials Studies	10	45
1 min ..	Material—Shape Tests	10	55
1 min ..	Heat Protection Systems	15	70
1 min ..	Aerothermoelastic Tests	15	85
1 min ..	Programmed Trajectories	10	95
1 hr	Fatigue and Life Tests	5	100

[1] Test duration and relative value are order of magnitude only and are not intended for design purposes.

Fig. 3. Re-entry aerothermodynamic problems

smaller, but still broad, problem areas as shown in Fig. 3 (Table 2 of [5]). In this table is shown also the order of magnitude requirement on test duration for studies in these areas. It can be seen that, as the available test time increases, more problems can be studied.

Actually, seven parameters can be considered to be of basic importance in this type of experimental work. These are the gas total enthalpy, the gas pressure, the flow MACH number, the flow chemistry, the flow and model geometry, the test dimensions, and the test duration. Other useful parameters, such as REYNOLDS number, follow from these properties. Of course, no laboratory facility can provide correct conditions in all respects simultaneously;

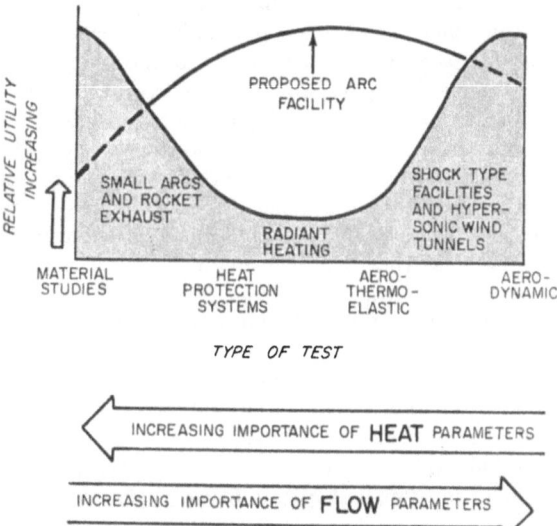

Fig. 4. Ground facilities for hypervelocity simulation

however, greater degrees of flight simulation are continuously being obtained as improved facilities are developed. It is believed that the properties of test section size as well as flow duration are currently the areas in which the most test facility improvement is required.

An attempt has been made to define the relative utility of various types of test facilities in terms of the important problem areas; this is shown in Fig. 4.

The list of problems shown in Fig. 3 has been reduced to four major groups in this figure. Also indicated here are the relative importances of flow parameters and heat parameters in the consideration of facility usefulness. Only the basic types of facilities are included. The supersonic or hypersonic wind tunnel in which the air is heated mechanically or by a heat exchanger—pebble bed heaters, strip heaters, etc.—is shown to have a reasonable utility in the aerodynamics areas. However, the real gas effects cannot be obtained with this device. Short duration, high enthalpy, blow-down wind tunnels, such as shock tunnels and "hotshot" tunnels, although operating on different principles, are competitive in the types of problems they can be used to study. Both have the capability of reproducing the flow MACH numbers, stagnation enthalpies, and in some cases, pressures of re-entry flight vehicles, but because of their short flow duration they cannot be used to study the responses of surfaces or structures to re-entry environments. Long duration, high enthalpy facilities, typified by arc heated devices, have great potential over the complete range of problems as indicated by the "proposed arc facility" line on Fig. 4. However, they have not yet been widely exploited except in the materials study area where rocket motors have also been useful tools. Even in what appears to be the next generation of arc devices—the large hypersonic air arc wind tunnel—the utility will probably fall somewhat below the short duration, high temperature facilities in the aerodynamics problem area because of practical problems, such as continuous vacuum pumping capabilities, which will restrict their MACH number range somewhat. Thus, it is believed that short duration and long duration high temperature facilities will both be required laboratory tools, at least for the next several years.

Notably missing from consideration in the above discussion is the aerodynamics or ballistic range. This is certainly a useful device that is complementary to the others mentioned, particularly because of its capabilities in the dynamic stability area and in the study of problems that require a non-expanded free stream gas. However, and again to keep the scope of this paper within reason, the ballistics range will not be further considered.

The purposes of this discussion have been to outline briefly the aerothermodynamics problem areas associated with the re-entry concepts of current interest and to introduce the types of test facilities being used in the laboratory today to study these problems. The object of the remainder of the paper is to describe the progress that has been made in this experimental field with particular reference to the work being done at the General Electric Missile and Space Vehicle Department, where both shock tunnels and arc heated facilities have been under development and in operation for several years. Discussed in the following sections are: first, some of the detailed capabilities of shock tunnels and air arc heated devices; second, typical results of experimental studies in both types of facilities; and third, what are believed to be the developmental requirements of future experimental tools in view of the expected problem requirements and anticipated facility development possibilities.

Facility Performance

Shock Tunnel

The shock tunnel is actually a combination of a shock tube and a wind tunnel in which the shock tube provides high temperature, high pressure gases that are expanded to the desired test section conditions. Because of the short duration of steady conditions that is inherent in shock tube operation, the resulting test flow is generally only of the order of a few milliseconds. (In the General Electric

6 Inch Tunnel, shown in Fig. 5, quasi-steady flow times vary between 1 and 5 milliseconds depending upon the test conditions.) Although this short testing time is usually considered to be a disadvantage, note that two major simplifications result from the short time operation. First, although the gases enclosed in the tunnel are often in excess of 5000° K, there is not sufficient time for the structure to heat to temperatures at which any consideration must be given to cooling;

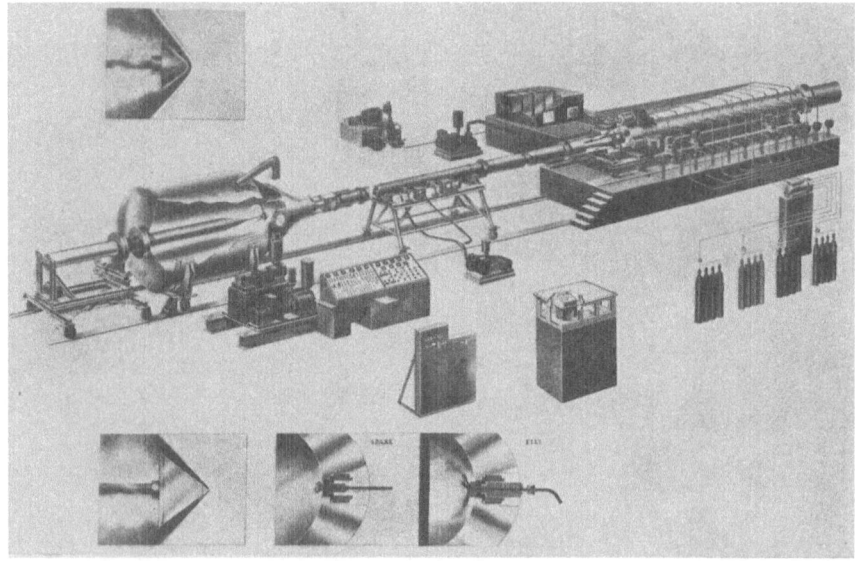

Fig. 5. 6 Inch Shock Tunnel operation

and second, the vacuum system required to support test section MACH numbers of high values—greater than 20—and correspondingly very low test section pressure values—a few microns—is relatively simple since its required capacity is not large.

Fig. 5 is a schematic representation of the General Electric 6 Inch Shock Tunnel. To initiate operation of the tunnel, air in the 6 inch diameter driven tube is adjusted to the proper pressure value—usually between 10 mm of mercury and several atmospheres. A combustible mixture of 70 % helium, 20 % hydrogen, and 10 % oxygen (mole percentages) is injected into the driver and ignited through the simultaneous discharge of 20 spark plugs. The resultant high pressure breaks a metal diaphragm between the driver and driven tubes and generates a shock wave that moves through the air in the driven tube. This wave compresses, heats, and moves the air towards the test section end of the tube where it is used in the experiment.

In this facility, there are three general methods that may be used for quasi-steady aerodynamic studies of flows over models. In the first, a model is located in the straight portion of the driven tube and tests are conducted after the incident shock wave passes the model and until subsequent waves change the flow conditions. This flow is generally characterized by a low free stream MACH number—about 2—and relatively high REYNOLDS numbers. In the second test configuration —the reflected nozzle shock tunnel—the incident wave "reflects" from the closed end of the tube, thus further heating and compressing the test gas through a reflected shock wave. The stagnation gas then flows through a small port (nozzle throat) in the reflected region to the test section through a large area ratio conical nozzle. Conditions at the test section may be varied over a wide range by varying the initial pressure in the driven tube, the incident shock MACH number, and the throat diameter of the conical nozzle. The third test configuration is the non-reflected nozzle or straight-through nozzle shock tunnel. In this system, the incident wave is allowed to pass through an expanding nozzle to the test section. When the flow starting processes in the nozzle are completed, the experiment is

Fig. 6. 6 Inch Shock Tunnel

conducted in the expanded gas that was initially compressed by the incident wave. For this facility, the reflected nozzle technique provides high MACH numbers (6 to greater than 20) at relatively low REYNOLDS numbers while the non-reflected technique gives moderate MACH numbers—approximately 5—at moderate to high REYNOLDS number values. Each of the three test configurations can provide extremely high stagnation enthalpies in its test section, since this property depends primarily upon the velocity of the incident shock wave.

It should be noted here that another short time test facility, the "hotshot" tunnel [6] operates in a range of test flow properties which is similar to that of the shock tunnel although it functions on a completely different principle. The main difference between the two facilities is that the shock tunnel uses a shock wave to compress and heat the test gas while the hotshot facility heats the working gas directly through the discharge of a short duration electrical arc. In general, the hotshot tunnel has a somewhat longer run time than the shock tunnel.

Fig. 6 is a photograph of the downstream end of the 6 Inch Shock Tunnel.

The test section diameter is 30 inches and there is a two axis optical system available for flow observation. A Fastax camera is used to monitor the flow processes on the vertical axis and sub-microsecond schlieren or shadow photographs are taken on the horizontal axis.

The approximate range of operation of the shock tunnel facility is shown in Fig. 7. It is apparent that wide ranges of test conditions are available. If a shock tunnel is run at relatively low stagnation enthalpies (shock MACH numbers of the order of 2 to 3) but at high pressure levels, high REYNOLDS number—MACH number test flows may be obtained. However, if it is desired to operate a facility at high stagnation temperatures, one may choose to employ high shock MACH numbers in the shock tube at the expense of test flow MACH number and REYNOLDS number. At this time, the 6 Inch Tunnel functions over a range of conditions between flow MACH numbers of approximately 5 and 22 and between test section REYNOLDS numbers of between 10 and 10^6 per inch. Actual run conditions can be extended to higher MACH numbers and higher stagnation temperatures than those shown in Fig. 7. However, instrumentation sensitivity

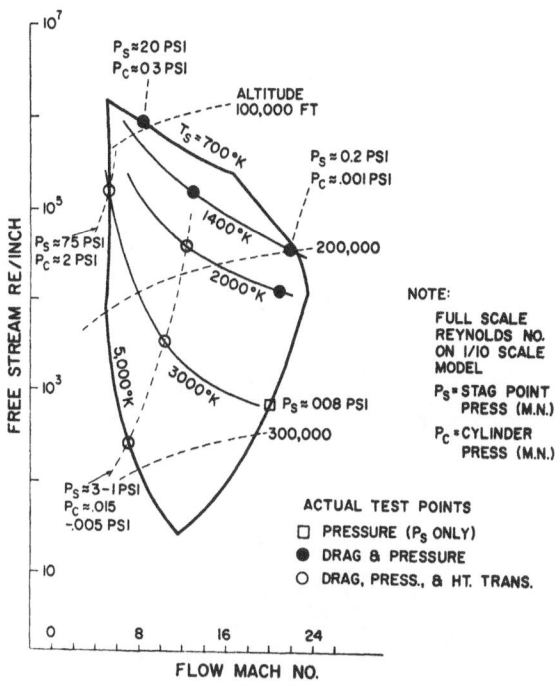

Fig. 7. Range of test flow conditions for General Electric 6 Inch Shock Tunnel

limitations have reduced the desirability of doing this to date. The test condition ranges shown include only the reflected and non-reflected nozzle test configurations.

The development and operational characteristics of the shock tunnel are discussed more fully in [7, 8, 9].

Air Arc Facilities

The major disadvantage of the shock tunnel or hotshot tunnel type of facility is the short testing time available. Although it is possible to observe aerodynamic, or external, flow properties in such a facility, it is virtually impossible to study the interaction of the surface or structure with the external flow. Because it is important to understand this interaction in many of the aerothermodynamic problem areas, a facility must be found that provides not only the flow properties encountered during re-entry but also does this for flow times of the order of minutes. The facility that currently comes closest to meeting this general requirement is the continuous air arc heated facility. The main property of the arc facility that suits it for re-entry studies is that it can produce, on a long time basis, test gas enthalpies corresponding to flight at near-orbiting velocity at widely varying pressure levels.

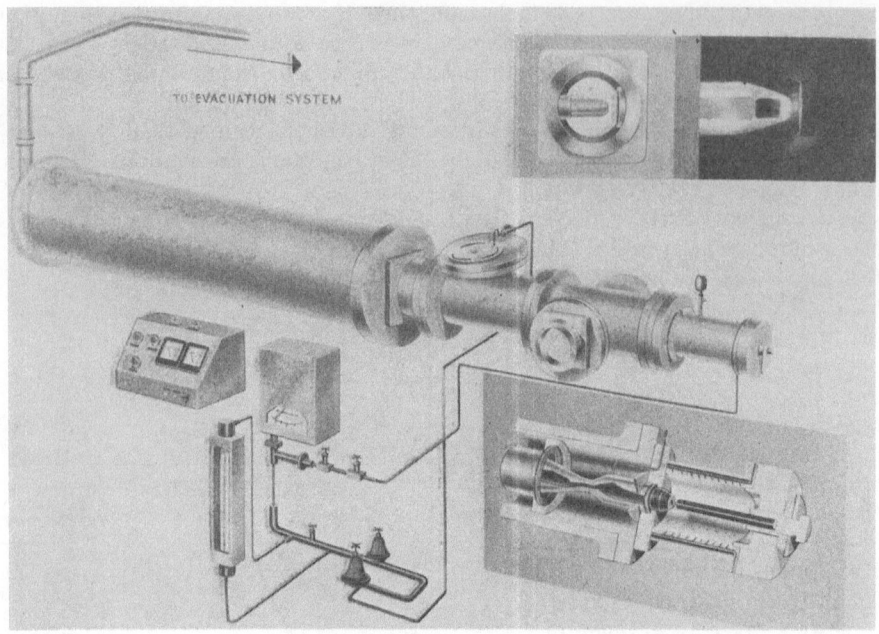

Fig. 8. Electric arc heated supersonic wind tunnel

Fig. 9. Arc wind tunnel facility

The evolution of the air arc test facility was to a large extent initiated by the studies of MAECKER [10] and WEISS [11] in Germany several years ago. Since their original work on water stabilized arcs, much work has been done in this country on the adaptation of the constricted arc principle to many aerodynamics related studies. At General Electric, work was begun in 1955 on the development of arc facilities and on their application to the study of materials suitable for re-entry vehicles [12]. Initial work was along lines similar to those of the original German work on water arcs; however, test techniques soon were developed that employed air stabilized arcs because of the desire to closely simulate the environment of a re-entering missile.

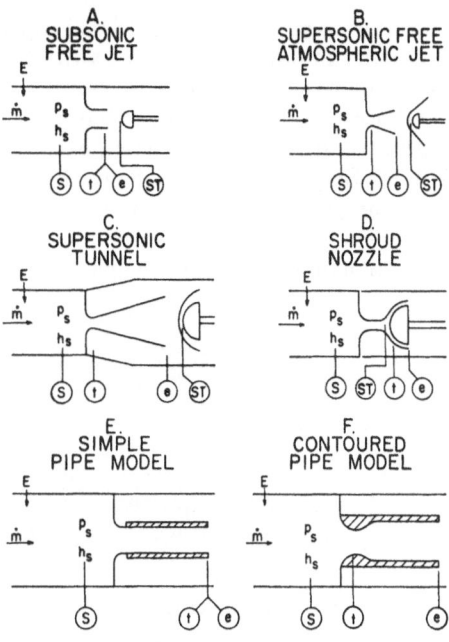

Fig. 10. Test configurations

Fig. 8 illustrates an early air arc heated supersonic wind tunnel. The insert shows a sectional view of the arc chamber, electrodes, plenum chamber, primary throat, and conical nozzle. During operation, air is injected tangentially into the arc chamber, the arc is struck between the two electrodes, and the gas passes through the annular electrode into the plenum chamber and expands through the nozzle to the test section. After interacting with the test model, the gas

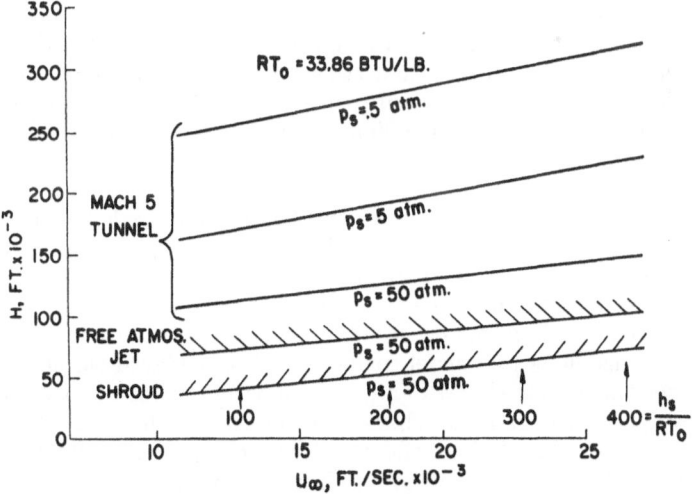

Fig. 11. Stagnation point property simulation of flight conditions in arc heated facility

is exhausted through a vacuum system. In this unit, all sections except the electrodes are cooled by a forced convection water system. The electrodes are consumed at a slow rate during operation; much of the particulate material is deposited along the walls of the plenum chamber but some remains in the heated air, thus contaminating it somewhat. The facility shown in Fig. 8 operates at 90 KW DC electrode power and has a 2.6 inch diameter test section. Nominal test conditions for this tunnel are a free stream Mach number of 4.5 and a stagnation enthalpy corresponding to 18,000 ft./sec. flight velocity. The stagnation point pressure simulates flight at an altitude of 230,000 ft. at this velocity. A photograph of this test facility is shown in Fig. 9.

Since the arc unit is essentially a gas heater, it may be used in conjunction with several types of test flows, each of which has certain advantages in re-entry studies. Six of these test configurations are shown in Fig. 10. In these, it is assumed that the power, E, has been added to the gas flow, m, upstream of the plenum chamber through a continuous arc discharge. Fig. 10 A shows a sketch of a configuration in which the heated gas is expanded only to subsonic velocities. Because

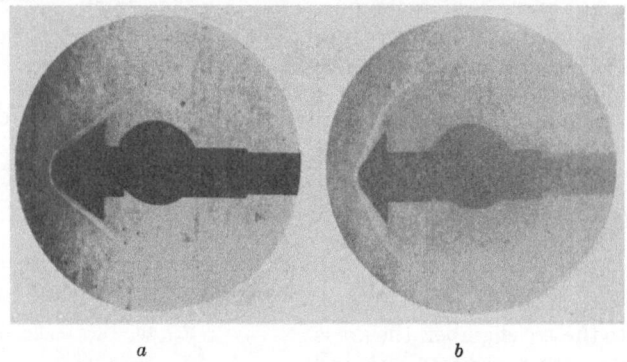

a   b

Fig. 12. Shock tunnel test flows. a Sphere-cone ($^1/_2$ angle $= 35°$). $M_m = 11.4$, $T_s = 2470°$ K; b Sphere-cone ($^1/_2$ angle $= 55°$). $M_m = 11.6$, $T_s = 2320°$ K

of the poor flow simulation in this case, such a device has been used only as a screening tool for materials studies. In Fig. 10 B it is assumed that the stagnation chamber pressure, p_s, is greater than twice the ambient pressure. The test flow is expanded to ambient pressure and reaches a moderate supersonic free stream test flow. Thus, some degree of free flight simulation is achieved about the model. Fig. 10 C is a representation of the supersonic tunnel previously discussed. Here, relatively large models may be tested for a given arc heater in a relatively good simulation of atmospheric flight conditions. The shroud nozzle flow configuration shown in Fig. 10 D is an adaptation of the test technique suggested by Ferry and Libby [13]. It has several advantages in the materials testing area: it allows a large percentage of the heated gas to interact with the model and, therefore, allows the testing of a large model; there is an appreciable latitude in the selection of flow conditions along the model surface which is achieved through the design of the model and nozzle shapes; and, also, it is not necessary to experience a total pressure loss through a bow shock wave in front of the model. The final two test configurations, the straight and the contoured pipe models, Figs. 10 E and 10 F are included to illustrate how an arc heated facility may be modified for the purpose of studying special problems: in this case, turbulent flow.

In the discussion of possible arc facility test configurations—Fig. 10—it has been implied that each has a particular value in re-entry studies. This is indicated in Fig. 11 in which the stagnation property simulation of free atmospheric flight is shown for the free jet, the shroud, and the supersonic tunnel test configurations. It is seen that the shroud can be considered most useful for low altitude-high velocity testing whereas the supersonic tunnel derives its primary applications in the high altitude—satellite, glide—types of re-entry. The Mach 5 test flow has been used here only for illustrative purposes.

Experimental Techniques and Results

Shock Tunnel

The development of the shock tunnel for use in the study of re-entry aerothermodynamic problems has involved a considerable amount of investigation. In general, development problems may be divided into two areas: those that are

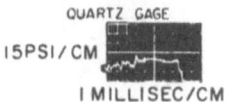

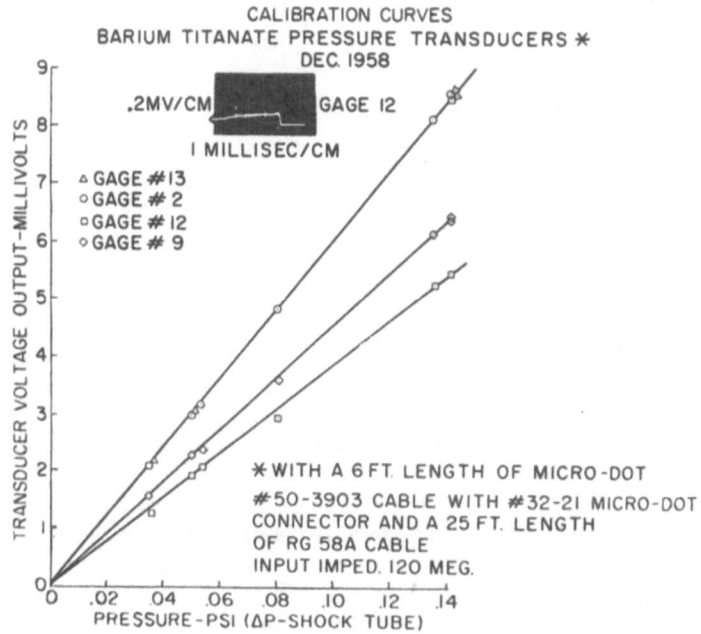

Fig. 13. Pressure gage response and calibration curves

directly associated with the operation and monitoring of the facility and those that are related to the obtaining of aerodynamic data on models located in the test flow. Items in the first category—driver ranges of operation and performance, incident shock wave attenuation, shock velocity measuring techniques, flow property measuring techniques, etc.—will be considered as beyond the scope of this paper. Some of the work done in this area of shock tunnel development at General Electric at other laboratories is presented in [7, 8, 9, 14].

Four types of model data measuring methods have been investigated and developed for use in the 6 Inch Shock Tunnel. These are schlieren and shadow optical techniques, surface pressure measurement techniques, surface heat transfer measurement techniques, and aerodynamic force and moment measurement techniques. The first three of these instrumentation methods have been well established, although possible improvements are constantly being studied. Current development work in the model instrumentation area is directed towards the measurement of model normal forces and pitching moments. Several techniques already have been developed for the measurement of axial forces over wide ranges of tunnel conditions.

Schlieren photographs of two blunted cones in the reflected nozzle shock tunnel are shown in Fig. 12. Data of this type have been useful in explaining the characteristics of the flow field generated about such bodies when used in combination

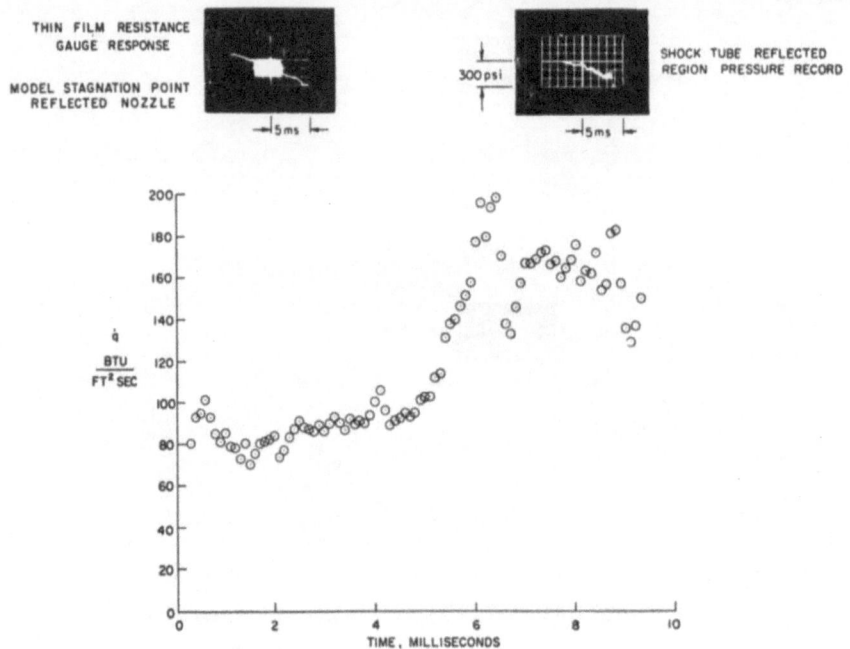

Fig. 14. Representative model heat transfer data

with surface pressure distributions [9]. The present optical system has been designed to be extremely fast and has recently been used to obtain photographs of the flow about bodies at MACH numbers in air in excess of 20.

A considerable effort has been expended in the development and evaluation of pressure measuring transducers suitable for use in the shock tunnel. The short time of steady flow during a test has necessitated the use of fast response time instrumentation. Piezoelectric crystal sensors have been investigated over the wide ranges of tunnel conditions and have been found to be satisfactory [14]. Fig. 13 shows the responses of two types of gages that have been employed, a quartz crystal gage and a barium titanate crystal gage. In general, the quartz crystals have been used to measure high pressure values—1 psi to 10,000 psi (the higher values are driver tube pressures). The barium titanate crystals, although more difficult to use than quartz because of their poor stability, have been found to be satisfactory low pressure gages. Calibration curves for several barium titanate instruments, shown in Fig. 13, demonstrate their linearity at low pressure levels. Because of the low noise to signal ratio experienced in the experimental responses shown in the figure, it appears that barium titanate crystals will be useful to values as low as .001 psi and calibration systems are now being checked out for the purpose of evaluating these gages in that pressure region. The accurate measurement of pressures of this low value will allow experimental studies to be conducted in the shock tunnel at MACH numbers and simulated altitudes of approximately 20 and 300,000 ft., respectively.

Heat transfer measurements are obtained through the use of sub-micron thick platinum films which are sputtered on a quartz substrate and used as resistance thermometers. These provide the surface temperature history of the quartz during the experiment and, from this, the heat transfer history can be calculated [15]. An IBM 704 computer is used for this calculation. The range of operation of this technique is approximately 1 to 1000 BTU/ft.²-sec. At high heat transfer rates, the calorimeter or thick film gage technique is employed [16]. The experi-

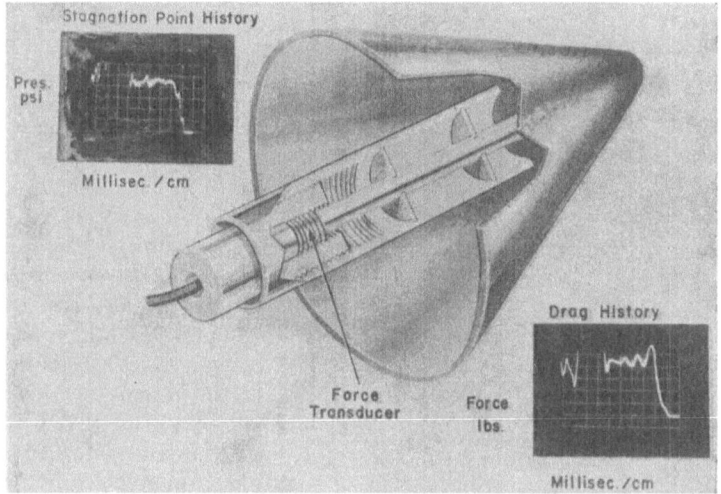

Fig. 15. Aerodynamic drag balance and response

mental response of one of the thin film gages is shown in Fig. 14. Also shown are the reflected region pressure and the reduced heat transfer rate histories. A quasi-steady state is seen to exist for approximately 3.5 milliseconds. The slight rise of heat transfer to the body during this period is consistent with the reflected region pressure variation.

As mentioned previously, force and moment measuring techniques in shock tunnel applications are now in their development phases. Three techniques appear to be worthwhile for investigation. In the first, a model is allowed to move freely after the start of the tunnel flow[1]. This is accomplished by supporting the model with fine threads which are destroyed during the passage of the incident waves. The desired data is then obtained by observing the motion of the model through high speed photography. Such a system requires the use of extremely light models and relatively high test pressure levels and is subject to inaccuracies in the translation of displacement information into force data. However, it does appear to have a range of usefulness, particularly in the study of the aerodynamic stability of ballistic missiles in the low MACH number and high REYNOLDS number flow

---

[1] This technique was first attempted in a shock tunnel at the U.S. Naval Ordnance Laboratory [17].

regimes. The second method consists of a model fastened rigidly to a cantilevered beam; the deflection of the beam at various stations is measured by strain gages and this information provides the desired force and moment data. This system is attractive because of its high sensitivity; thus, high altitude studies may be contemplated. However, the reduction of the response time of a satisfactory beam-model combination to values that are small compared to the flow times has proven to be difficult and development studies are still continuing. The third method, suggested by VIDAL [14], incorporates a model that is softly supported on a beam. The forces tending to move the model during the test flow are recorded by internally mounted accelerometers. In principle, this technique is quite sensitive and should allow the study of high altitude aerodynamics problems.

Techniques have been developed which allow the measurement of axial forces on models located in shock tunnel flows. Fig. 15 is a schematic representation of a piezoelectric drag balance with which many experiments have been conducted. In this device, the force is transmitted to the surface of the gage through a beam which is mounted to the main sting on two flexible membranes. The sensitivity of the balance can be varied by changing the stiffness of the membranes.

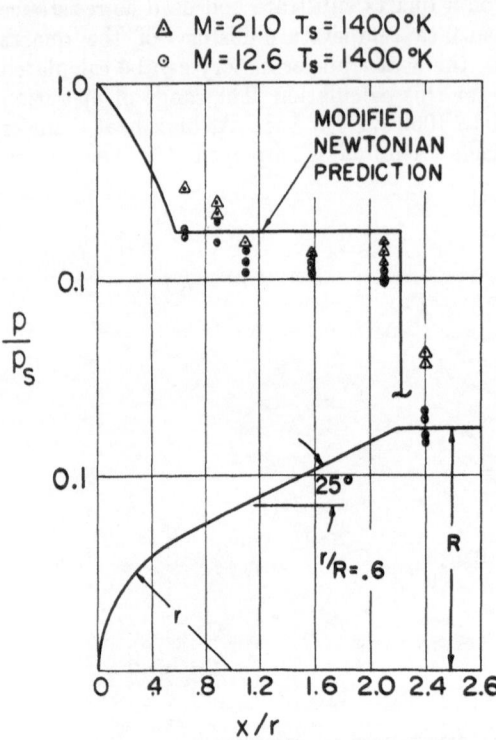

Fig. 16. Surface pressure distribution on hemisphere-cone-cylinder

Fig. 16 shows pressure distributions measured on a blunted cone in the 6 Inch Shock Tunnel at two relatively high flow MACH numbers. These data have been included to illustrate the types of experimental results that are obtainable with this facility.

Air Arc Facilities

Several types of test configurations that may be incorporated into an air arc facility were described in an earlier section of the paper. In this section, a brief description of several existing experimental devices and their application to ablation studies will be made.

Fig. 17 shows a small shroud nozzle air arc in operation. The electrode power is approximately 90 KW DC and the model being tested is a 2 inch diameter hemisphere cylinder. The main purpose of test in this unit is to study the stagnation point ablation characteristics of various materials. For this particular experiment, the shroud nozzle contour was chosen to correspond to a streamline that would exist for an incompressible potential flow about a sphere at the stagnation conditions of the gas stream. As the model ablates during a test, it is moved into the nozzle by the test operator so that test conditions are maintained at constant values. Surface pressure measurements near the stagnation point were shown to agree with the predicted potential values within a small percentage scatter. The stagnation point non-ablating, or calorimeter, heat transfer rates were also measured and. when used in conjunction with the measured pressure gradients and the theoretical results of SCALA [3], checked the facility enthalpy level to within 10 %.

Spectroscopic measurement of the test gas temperature without a model present in the flow checked the experimentally determined enthalpy level to the same accuracy. Other experimental values that are measured during a test are

Fig. 17. Flow in small shrouded air arc

the plenum chamber pressure, the air mass flow through the facility, and the electrode material mass lost during a test. This device is useful in that it provides a well known experimental flow over a model of large size considering the facility power level; it is a high heat transfer rate facility, the stagnation heat transfer rates generally being greater than 1000 BTU/ft.²-sec.

An ablation test conducted in a shroud nozzle facility will generally produce a steady state ablation process because of the high heat transfer rates. For many applications in which ablation can be considered for use as a heat protection system, low heat transfer rates will be experienced so that the ablation process will be non-steady; that is, for heat transfer rates below approximately 100 BTU/ft.²-sec., the damage caused by heat transferred into the ablating material will proceed at a faster rate than does the removal of the surface. This process is quite complicated and must be studied experimentally under the conditions of interest. For such a study, a low density wind tunnel of the type shown in Fig. 8 is an ideal tool. This facility provides flows at high stagnation

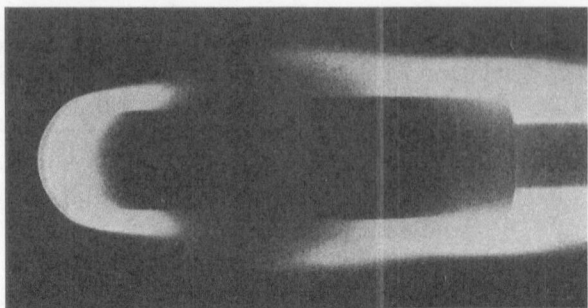

Fig. 18. Flow in air arc tunnel

enthalpies but at reduced heat transfer rates because of the low pressure values. Fig. 18 is a photograph of the flow in the arc tunnel about a 1 inch diameter hemi-

Fig. 19. Large A.C. air arc test facility

sphere-cylinder model. Measurements that have been made in this facility are the non-ablating heat transfer rate to the model stagnation point and the stagnation pressure.

To obtain large scale tests which are desirable for many of the problems outlined earlier, it is necessary to have large scale arc facilities. At General Electric, work was started some time ago on the development of high power arc facilities. The first step in this program was the study of a 2500 KW air arc [18]. The power level referred to is that added to the test gas; that is, a 2500 KW facility will require approximately 6000 KW of electrode power. A unit of this size is shown in Fig. 19. Actually this facility has been run a few times at electrode power levels approaching 20 megawatts. It operates on 3 phase AC power and in its first development stage employed carbon electrodes. The electrodes are located in the lower portion of the cylindrical chamber: the upper portion acts as a plenum chamber and the flow exhausts through a nozzle at the top of the unit. Fig. 20 shows this facility in operation.

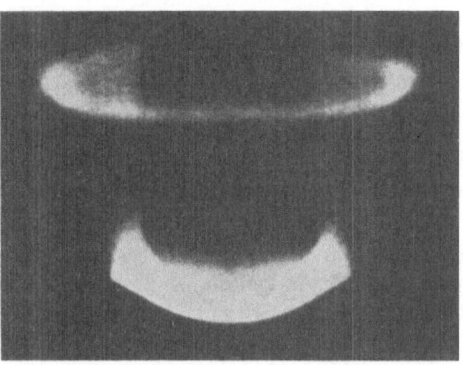

Fig. 20. Flow in large air arc test facility $h_s/RT_o = 200$, $M_t = 2.2$, $d_m = 1.0$ inch

Experiments have been conducted in both the free jet and the shrouded nozzle configuration. Current development work is centered about the replacement of

a   b

Fig. 21. Comparison of steady and non-steady ablation processes. Melamine-glass, $h_s \approx 6200$ BTU/lb.
a) $\dot{q}_s = 75$ BTU/ft.²-sec., $t = 60$ sec.,
b) $\dot{q}_s = 1050$ BTU/ft.²-sec., $t = 10$ sec.

the carbon electrodes by cooled metallic electrodes which are essentially non-consumable. It appears that this will be an achievable goal, at least over certain ranges of power levels and test conditions.

To illustrate the types of experimental work that have been done in air arc facilities, Fig. 21 has been included. Shown are models of identical material tested at the same enthalpy level but at two widely different stagnation point heat transfer rates. In the high heat transfer rate test conducted in the shroud nozzle facility, there exists a very thin char layer followed by a somewhat thicker layer of pyrolized plastic. This cross-sectional view is the same at any time shortly after the start of ablation. For the low heat transfer rate tests, it is seen that the heat has "soaked" deeply into the material. Also, the char layer is thick compared to that in the high heat transfer rate specimen. For a glass reinforced plastic of the type shown here, it has been found that this glass layer forms after about 30 seconds of exposure to the test environment. Thus, it is seen that not only does the ablation process occur in a non-steady fashion, but the surface mechanism of heat blockage actually changes drastically during the process.

Future Developments

This has been a brief coverage of a very broad field. It is clear that, although much progress has been made in the investigation of re-entry aerothermodynamic problems, much work remains to be done. The central region of the problem spectrum shown in Fig. 4—the general areas of aerothermoelastic and heat protection system problems which require relatively long test times, large test sections, and good aerodynamic simulation—has been relatively untouched and only a part of the needed work has been done in the materials and aerodynamics areas. Each experimental device that now exists is consistently being extended to its operational limits in the investigation of the more complicated effects of re-entry, and therefore, improvements in their capabilities are being sought constantly.

The shock tunnel or short duration type of facility can be improved in several respects. Better and new instrumentation techniques and high pressure operation can extend the capabilities of a shock tunnel considerably. High pressure operation—i.e., driver pressures between 50,000 and 100,000 psi—will permit the attainment of high MACH number—high REYNOLDS number flows never before available to the experimenter; improved instrumentation sensitivity will allow the study of very high altitude flows—in the slip, transition, and rarified gas flow regimes—at the correct flight MACH numbers; and new instrumentation methods, such as direct density measuring and spectroscopic techniques, for measuring the flow properties about models of complicated shapes will extend the knowledge of the aerodynamic processes that determine the model-environment interaction.

Long duration, high enthalpy facilities such as the arc heated wind tunnel are still in their infancy and as discussed above, it is in their application area that a large number of studies are required. It is possible, and perhaps probable, that they will become the most nearly universal types of test facilities available; however, to reach this goal, many improvements will be required. Three major practical limitations restrict the capabilities of an arc heated wind tunnel. These are the need for an extremely high performance vacuum system, the high pressure limitation on continuous arcing (it is believed that this is at present the order of 1000 psi in units of appreciable size), and the difficult cooling problem at the throat of the test section nozzle at high plenum pressures. Fortunately, if the test requirement of MACH number is relaxed, but model stagnation flow conditions are still produced that simulate re-entry flight, the three limitations mentioned above are alleviated. That is, a reduction in required test flow MACH number from, say, 20 to 8 reduces the arcing pressure and throat pressure values by approximately two orders of magnitude and increases the test section pressure by approximately one order of magnitude. Progress has been made recently in several of the other problems of arc facilities: the reduction of contamination to negligible amounts with air as the test gas appears now to be an achievable goal; plenum chambers have been employed to reduce flucutations imparted to the test gas by the arcing process; and heat exchangers have been used to improve the efficiency of vacuum systems.

On the basis of considerations described only briefly here, it appears that the next type of useful arc heated facility will be an arc wind tunnel of an appreciable size with a moderate hypersonic MACH number test flow range. Fig. 22 shows an artist's conception of a preliminary design of this facility. It is believed that this device will have about a one square foot test section of either axially symmetric

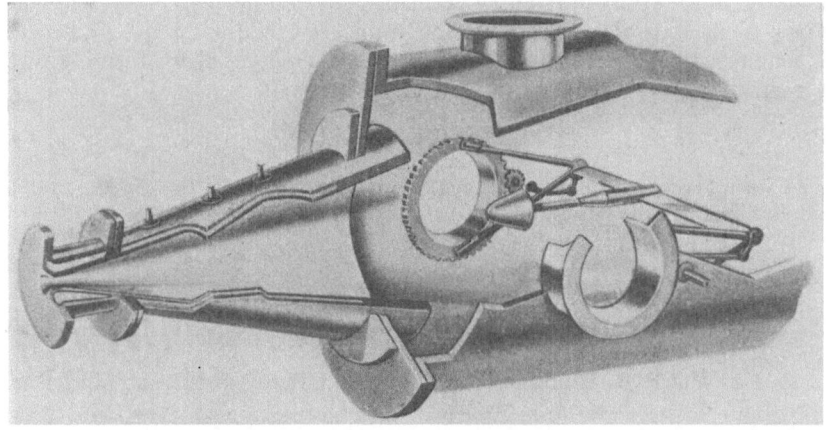

a

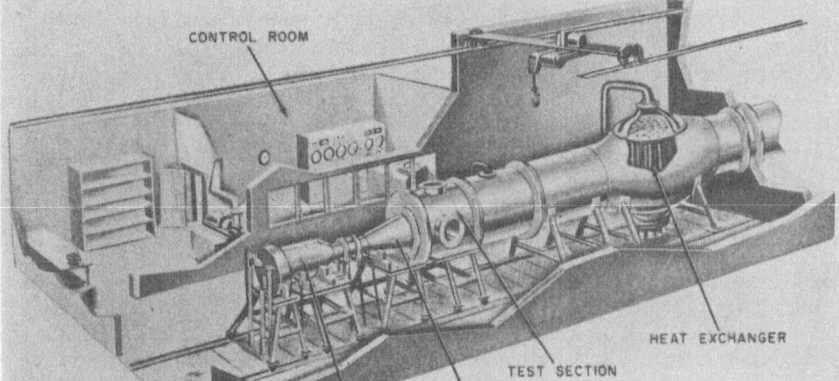

Fig. 22. Arc tunnel facility

or rectangular geometry (only the axially symmetric test section is shown in Fig. 22), will provide between 1 and 2 megawatts of power to the air, and will be capable of running in excess of 10 minutes. The test flow MACH number will be between 5 and 8, and the atmospheric velocity simulation in terms of test gas stagnation enthalpy will be in excess of 20,000 ft./sec. The facility also will incorporate controls to allow the programmed variation of test conditions so that major portions of re-entry trajectories can be simulated in their true time scales.

Conclusions

It was not intended and not possible that this paper be all inclusive in its discussion of current high temperature re-entry aerodynamic problems and their experimental study. Rather, it has been attempted to outline a rational approach to the subject of re-entry aerothermodynamics considering both the types of problems requiring study and the capabilities of currently available experimental tools. It is believed that both the shock tunnel and air arc facility have been shown to be useful in the study of re-entry problems and that both are at present complementary tools. It is also believed that the air arc facilities, primarily because of their long test time capabilities, have the greatest potential for future growth in this area.

Acknowledgement

The work reported upon in this paper was supported by the United States Air Force Ballistic Missile Division under Contract AF 04 (647) 269.

References

1. L. Lees, Laminar Heat Transfer Over Blunt-Nosed Bodies at Hypersonic Flight Speeds. Jet Propulsion 26, 259 (1956).
2. J. A. Fay and F. R. Riddell, Theory of Stagnation Point Heat Transfer in Dissociated Air. J. Aero/Space Sci. 25, 73 (1958).
3. S. M. Scala, Hypersonic Heat Transfer to Catalytic Surfaces. J. Aero/Space Sci. 25, 273 (1958).
4. S. M. Scala and G. W. Sutton, The Two-Phase Hypersonic Laminar Boundary Layer—A Study of Surface Melting. Proc. Heat Transfer and Fluid Mech. Inst.; University of California, Berkeley, Calif.; June, 1958.
5. A Large Air Arc for Hypersonic Wind Tunnels. G.E. MSVD Doc. No. 59SD6; 1 April, 1959.
6. R. W. Perry and W. N. MacDermott, Development of the Spark-Heated, Hypervelocity, Blowdown Tunnel—Hotshot. AEDC-TR-58-6, ASTIA Doc. No. AD 157 138; June, 1958.
7. W. R. Warren, The Design and Performance of the General Electric Six Inch Shock Tunnel Facility. Proc. 1st Shock Tube Symp., USAF Spec. Weap. Center, SWR-TM-57-2; February, 1957.
8. A. Hertzberg, W. E. Smith, H. S. Glick and W. Squire, Modifications of the Shock Tube for the Generation of Hypersonic Flow. AEDC-TN-55-15; March, 1955.
9. A. J. Vitale, E. M. Kaegi, N. S. Diaconis and W. R. Warren, Results from Aerodynamic Studies of Blunt Bodies in Hypersonic Flows of Partially Dissociated Air. Proc. Heat Transfer and Fluid Mech. Inst.; University of California, Berkeley, Calif.; June, 1958.
10. H. Maecker, Ein Lichtbogen für hohe Leistungen. Z. Physik 129, 108 (1951).
11. R. Weiss, Untersuchung des Plasmastrahles, der aus einem Hochleistungsbogen austritt. Z. Physik 138, 170 (1954).
12. J. H. McGinn and H. Wachman, Development and Investigation of Fluid Stabilized Arcs. G.E. MSVD Doc. No. 57SD680; August, 1957.
13. A. Ferri and P. A. Libby, A New Technique for Investigating Heat Transfer and Surface Phenomena Under Hypersonic Flow Conditions. J. Aero/Space Sci. 24, 264 (1957).
14. C. J. Harris and E. M. Kaegi, The Application of Pressure and Force Transducers in Shock Tunnel Aerodynamic Studies. Proc. 3rd Shock Tube Symp., USAF Spec. Weap. Center; March, 1959.
15. R. J. Vidal, Model Instrumentation Techniques for Heat Transfer and Force Measurements in a Hypersonic Shock Tunnel. WADC TN 56-315; February, 1957.
16. P. H. Rose and W. I. Stark, Stagnation Point Heat Transfer Measurements in Dissociated Air. J. Aero/Space Sci. 25, 86 (1958).
17. A. Seigal, Naval Ordnance Laboratory, Private Communication.
18. F. C. Foshag and N. S. Diaconis, The 2500 KW AC AIr Arc. Proc. Soc. Exper. Stress Analysis, Washington, D.C.; May, 1959.

Interplanetary Navigation

By

Itiro Sinra[1]

(With 5 Figures)

(Received July 21, 1959)

Abstract — Zusammenfassung — Résumé

Interplanetary Navigation. In interplanetary flight, the minimization of its duration will appear as one of the most important factors. Adoption of the secant path has naturally a good effect in shortening the duration, but it is necessary to know its extent beforehand. By using the secant path, the curvilinear distance becomes short, evidently, but this has the disadvantage of shrinking the heliocentric speed of the vehicle.

The way to get a general perspective for duration is given and the existence of an optimum secant angle for each initial velocity is clearly indicated. Also the degree of gain in duration is discussed.

Interplanetarische Navigation. Als einer der wichtigsten Faktoren beim interplanetarischen Flug wird die Verringerung der Flugdauer gelten. Die Annahme der Sekanten-Bahn hat natürlich eine gute Auswirkung auf die Kürzung der Flugdauer, doch ist es notwendig, vorher ihre Ausdehnung zu kennen. Bei Benutzung der Sekanten-Bahn wird die kurvenlineare Entfernung gering. Dies hat aber den Nachteil, daß die heliozentrische Geschwindigkeit des Fahrzeuges geringer wird.

Der Weg, um eine allgemeine Perspektive für die Flugdauer zu erhalten, wird angegeben und das Vorhandensein eines günstigsten Sekantenwinkels für jede Anfangsgeschwindigkeit eindeutig aufgezeigt. Ebenso wird das Ausmaß der Abkürzung der Flugdauer diskutiert.

Navigation interplanétaire. La réduction de la durée des vols interplanétaires apparaîtra comme une de leurs caractéristiques les plus souhaitables. L'utilisation de la trajectoire sécante correspond à une faible distance curviligne, qui tend à réduire la durée, mais elle a comme désavantage une réduction de la vitesse héliocentrique.

On indique la façon d'obtenir une estimation générale de la durée et l'existence d'un angle sécant optimum est clairement mise en évidence. Le gain possible sur la durée est discuté.

I. Introduction

Although there exist several papers which discuss interplanetary orbits, they seem to fail to take account of the initial conditions, in other words, circumstances existing at departure from the Earth.

For real interplanetary flight, the problem of duration will come out as one of the most essential factors. In the case of manned flight, it naturally determines the required amount of provisions and daily necessaries, and gives important

---

[1] The Faculty of Engineering, Meiji University, Kanda Surugadai, Chiyoda-ku, Tokyo, Japan.

effects on the payload of the spaceship. Even for the unmanned probe, the duration will exert influence on the magnitude and reliability of power source etc.

Historically the HOHMANN orbits are too well announced and it becomes the popular impression that it would take about eight and half months for the journey to Mars and about five months to reach Venus. It looks like this long duration had some effects on increasing the utopian nature of interplanetary travel.

In this manner, the problem of duration becomes the most indispensable one. The duration has naturally intimate relation with the initial energy of space ship. And henceforth we will analyze the interplanetary flight in laying emphasis on its duration.

II. Preliminary Considerations

Instead of the HOHMANN orbit we will discuss the secant path, but in order to simplify as much as possible the complicated circumstances, we make the following assumptions (as assumed by almost all papers):

(1) The planets move in circular orbits around the Sun at rest.

(2) The planes of the planetary orbit lie in the plane of the ecliptic.

(3) The spaceship describes at first a hyperbolic path around the Earth in the ecliptic plane and thereafter enters into orbit around the Sun.

(4) The spaceship makes free flight around the Sun ignoring the attraction of the planets.

III. The Transit Phase

The spaceship goes through some boundary which separates the domain between the geocentric and heliocentric regions. We can't draw a definite line for this boundary and so we proceed as follows:

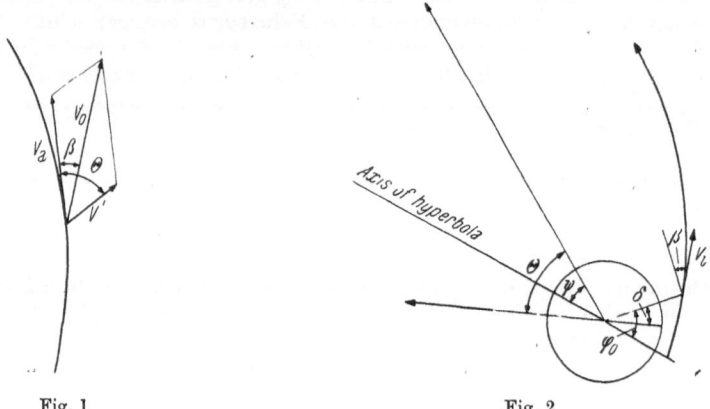

Fig. 1   Fig. 2

(1) Find the velocity vector when the radial distance from the Earth becomes infinity.

(2) As for the initial conditions of motion around the Sun, we take the velocity obtained in (1).

In these way, we denote the various velocities and angles as in Fig. 1.

Where

v_a is the linear velocity of the Earth in its circular orbit;

v' is the residual velocity of the spaceship when it enters in the heliocentric region;

v_0 is the linear velocity of the spaceship at the initial moment of heliocentric motion;

θ is the angle between the direction of forward movement of the Earth and the residual velocity of the spaceship going out of the geocentric region;

β is the angle between the direction of forward movement of the Earth and the initial velocity of heliocentric motion of the spaceship.

Evidently there exist following relations between them:

$$v'^2 = v_a^2 + v_0^2 - 2v_a\,v_0\,\cos\beta \tag{1}$$

$$\tan\theta = v_0\,\sin\beta/(v_0\,\cos\beta - v_a). \tag{2}$$

Also from the motion in the geocentric region, we obtain the following relations:

$$v_i^2 - v_E^2 = v'^2 \tag{3}$$

denoting v_i as the initial linear velocity of the spaceship leaving the Earth's atmosphere and v_E as the escape velocity of the Earth at that place.

Analysis of the hyperbola in the geocentric field gives the following equations:

$$\tan\varphi_0 = \frac{x_0\,\sin\beta'\,\cos\beta'}{x_0\,\cos^2\beta' - 1} \tag{4}$$

$$\tan\psi = \sqrt{x_0\,(x_0 - 2)}\;\cos\beta' \tag{5}$$

$$\theta = \psi + \varphi_0 - \delta \tag{6}$$

where

β' is the angle between the direction of velocity v_i and the horizon at that place;

ψ is the angle between the axis of hyperbola and the direction of its asymptote;

δ is the direction angle of the launching point measured from the line of motion of the Earth;

and

$$x_0 = 2v_i^2/v_E^2.$$

(4) and (5) may be derived from the expression of the eccentricity e being written as

$$e^2 = 1 + x_0\,(x_0 - 2)\;\cos^2\beta'.$$

IV. Calculation of the Duration

Here we consider only the duration in the heliocentric region.

In the field of universal gravitation, using polar coordinates, we can write

$$v^2 - \frac{2\mu}{r} = 2E \tag{7}$$

$$v^2 = \dot{r}^2 + r^2\dot{\theta}^2 = \dot{r}^2 + \frac{h^2}{r^2} \tag{8}$$

in which μ is the constant of gravitational field.

Furthermore:

$$2E = v_0^2 - \frac{2\mu}{r_0} = \frac{\mu}{r_0}\,(x - 2) \tag{9}$$

$$x = \frac{r_0\,v_0^2}{\mu} \tag{10}$$

$$h^2 = \mu\,r_0\,x\,\cos^2\beta \tag{11}$$

where β indicates the same angle as Fig. 1 and r_0 is the distance between the Sun and the Earth.

Combining all these equations, we obtain

$$\dot{r}^2 = \frac{2\mu}{r} - \frac{\mu r_0 x \cos^2 \beta}{r^2} + \frac{\mu}{r_0}(x-2) =$$
$$= \frac{\mu}{r_0 r^2}\left\{(x-2)r^2 + 2r_0 r - r_0^2 x \cos^2 \beta\right\}. \tag{12}$$

From which:

$$\frac{dr}{dt} = \pm \frac{1}{r}\sqrt{\frac{\mu}{r_0}\left\{(x-2)r^2 + 2r_0 r - r_0^2 x \cos^2 \beta\right\}} \tag{13}$$

obviously, the $(+)$ sign corresponds to the case when r increases with time and the $(-)$ sign to the case when r decreases with time.

From (13)

$$t = \pm \sqrt{\frac{r_0}{\mu}} \int_{r_0}^{r_1} \frac{r\,dr}{\sqrt{(x-2)r^2 + 2r_0 r - r_0^2 x \cos^2 \beta}}. \tag{14}$$

This integral separates into three cases.
(1) The case when $x < 2$
then

$$t = \pm \sqrt{\frac{r_0}{\mu(2-x)}} \int_{r_0}^{r_1} \frac{r\,dr}{\sqrt{\dfrac{2r_0}{2-x}r - r^2 - \dfrac{r_0^2 x \cos^2 \beta}{2-x}}} =$$

$$= \pm \sqrt{\frac{r_0}{\mu(2-x)}} \int_{Z_0}^{Z_1} \frac{\left(Z + \dfrac{r_0}{2-x}\right)dZ}{\sqrt{A^2 - Z^2}} =$$

$$= \pm \sqrt{\frac{r_0}{\mu(2-x)}} \left[-\sqrt{A^2 - Z^2} + \frac{r_0}{2-x}\sin^{-1}\frac{Z}{A}\right]_{Z_0}^{Z_1} =$$

$$= \pm \sqrt{\frac{r_0}{\mu(2-x)}} \left\{\sqrt{A^2 - Z_0^2} - \sqrt{A^2 - Z_1^2} + \frac{r_0}{2-x}\left(\sin^{-1}\frac{Z_1}{A} - \sin^{-1}\frac{Z_0}{A}\right)\right\}$$

where

$$Z = r - \frac{r_0}{2-x}, \qquad A = \frac{r_0 e}{2-x} \left.\begin{array}{c} \\ \\ \end{array}\right\}$$
$$e^2 = 1 - x(2-x)\cos^2 \beta. \tag{15}$$

Reversing to r

$$t = \pm \sqrt{\frac{r_0^3}{\mu(2-x)^3}}\left\{(t_1 - t_0) - e(\cos t_1 - \cos t_0)\right\} \tag{16}$$

where

$$t_1 = \sin^{-1}\frac{r_1(2-x) - r_0}{r_0 e}$$
$$t_0 = \sin^{-1}\frac{1-x}{e}. \left.\begin{array}{c} \\ \\ \\ \end{array}\right\} \tag{17}$$

(2) The case when $x = 2$

then

$$t = \pm \sqrt{\frac{1}{2\mu}} \int_{r_0}^{r_1} \frac{r\,dr}{\sqrt{r - r_0 \cos^2 \beta}}$$

$$= \pm \sqrt{\frac{1}{2\mu}} \int_{Z_0}^{Z_1} \frac{(Z + r_0 \cos^2 \beta)\,dZ}{\sqrt{Z}}$$

$$= \pm \sqrt{\frac{1}{2\mu}} \left[\frac{2}{3} \sqrt{Z^3} + 2r_0 \cos^2 \beta \cdot \sqrt{Z} \right]_{Z_0}^{Z_1}$$

$$= \pm \frac{1}{3} \sqrt{\frac{2}{\mu}} \left\{ \sqrt{Z_1} \, (Z_1 + 3 r_0 \cos^2 \beta) - \sqrt{Z_0} \, (Z_0 + 3 r_0 \cos^2 \beta) \right\}$$

where

$$Z = r - r_0 \cos^2 \beta.$$

Reversing to r

$$t = \pm \frac{1}{3} \sqrt{\frac{2}{\mu}} \left\{ \sqrt{r_1 - r_0 \cos^2 \beta} \, (r_1 + 2 r_0 \cos^2 \beta) - r_0^{\frac{3}{2}} \sqrt{1 - \cos^2 \beta} \, (1 + 2 \cos^2 \beta) \right\}. \quad (18)$$

(3) The case when $x > 2$.

then

$$t = \pm \sqrt{\frac{r_0}{\mu (x - 2)}} \int_{r_0}^{r_1} \frac{r\,dr}{\sqrt{r^2 + \frac{2 r_0}{x - 2} r - \frac{r_0^2 x}{x - 2} \cos^2 \beta}} =$$

$$= \pm \sqrt{\frac{r_0}{\mu (x - 2)}} \int_{Z_0}^{Z_1} \frac{\left(Z - \frac{r_0}{x - 2} \right) dZ}{\sqrt{Z^2 - B^2}} =$$

$$= \pm \sqrt{\frac{r_0}{\mu (x - 2)}} \left[\sqrt{Z^2 - B^2} - \frac{r_0}{x - 2} \cosh^{-1} \frac{Z}{B} \right]_{Z_0}^{Z_1} =$$

$$= \pm \sqrt{\frac{r_0}{\mu (x - 2)}} \left\{ \sqrt{Z_1^2 - B^2} - \sqrt{Z_0^2 - B^2} - \frac{r_0}{x - 2} \left(\cosh^{-1} \frac{Z_1}{B} - \cosh^{-1} \frac{Z_0}{B} \right) \right\}$$

where

$$Z = r + \frac{r_0}{x - 2}, \qquad \qquad B = \frac{r_0 e}{x - 2} \quad \left. \right\} \quad (19)$$

$$e^2 = 1 + x \, (x - 2) \cos^2 \beta.$$

Reversing to r

$$t = \pm \sqrt{\frac{r_0^3}{\mu (x - 2)^3}} \left\{ e \, (\sinh t_1' - \sinh t_0') - (t_1' - t_0') \right\} \quad (20)$$

where

$$t_1' = \cosh^{-1} \frac{r_1 (x - 2) + r_0}{r_0 e} \quad \left. \right\}$$

$$t_0' = \cosh^{-1} \frac{x - 1}{e}. \quad \left. \right\} \quad (21)$$

Given the destination planet, r_1/r_0 can be determined, for example:

$$\text{for Mars} \qquad r_1/r_0 = 1.524$$
$$\text{for Venus} \qquad r_1/r_0 = 0.723$$

and the duration t may be calculated from the adequate equation of (16), (18) and (20), starting from the fixed values of x and β.

V. The Minimization of the Duration

The curves of the duration of the flight to Mars were given in Fig. 3 as an example.

As it would be easily foreseen, the duration decreases rapidly as the angle β increases for the constant value of x. This results from the fact that the distance to the destination becomes shorter.

But our great concern is about the behavior when v' and θ are varied, particularly the problem of minimization of the duration by changing θ is quite interesting.

To solve this problem, we transform eqs. (1) and (2) into the forms

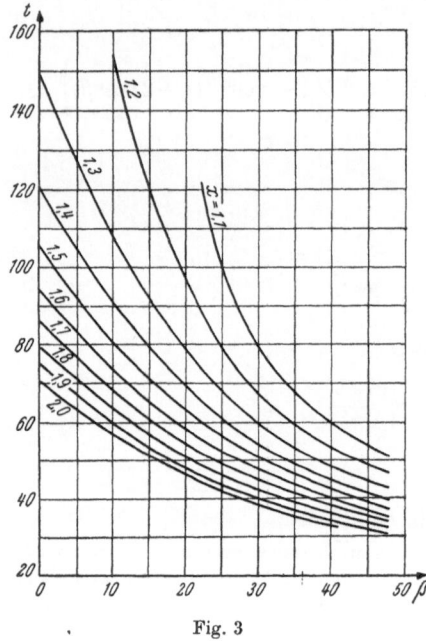

Fig. 3

$$x' = 1 + x - 2\sqrt{x}\cos\beta \qquad (22)$$

$$\tan\theta = \frac{\sqrt{x}\sin\beta}{\sqrt{x}\cos\beta - 1} \qquad (23)$$

where

$$x' = r_0\, v'^2/\mu \equiv v'^2/v_a^2 \qquad (24)$$

or in another equivalent form

$$\left. \begin{array}{c} x = 1 + x' + 2\sqrt{x'}\cos\theta \\[2mm] \tan\beta = \dfrac{\sqrt{x'}\sin\theta}{1 + \sqrt{x'}\cos\theta} \end{array} \right\} \qquad (25)$$

And after changing the variables (x, β) into (x', θ), we would be able to calculate the duration in function of θ for each constant value of x'.

In the case to determine the duration for some definite set of (x', θ), the above said method must be followed. But relying on it will not give a clear image of our object, that is, the minimization of duration.

Therefore a somewhat roundabout way of attack was taken as follows:

(1) From Fig. 3, the curves are drawn which indicate x as function of β in keeping t constant. This is demonstrated by Fig. 4.

(2) Draw the curves which express x as function of β for each constant value of x' by means of eq. (22).

(3) To these curves add the next group of curves which express x as function of β for each constant value of θ, by means of eq. (23).

(4) Then put these two groups of curves which are demonstrated in Fig. 5 upon Fig. 4. As an example, the dotted line curve was written in Fig. 4 for the case $x' = 0.04$.

(5) Determine the point from the given values of the set of x' and θ in Fig. 5, and read the required duration from the curve-group of Fig. 4.

In this way, it becomes evident that there exists some definite θ which gives the minimum of the duration for each constant value of x'. x' being linearly connected to $v_i{}^2$, as can be seen from (3), there is a most optimum oblique angle θ for each initial speed v_i. For example, the most optimum θ is about 22° for $x' = 0.04$ and 33° for $x' = 0.09$. This angle shortens the duration from 113 days to 107 days for the case of $x' = 0.04$, 113 days being the duration when θ is zero.

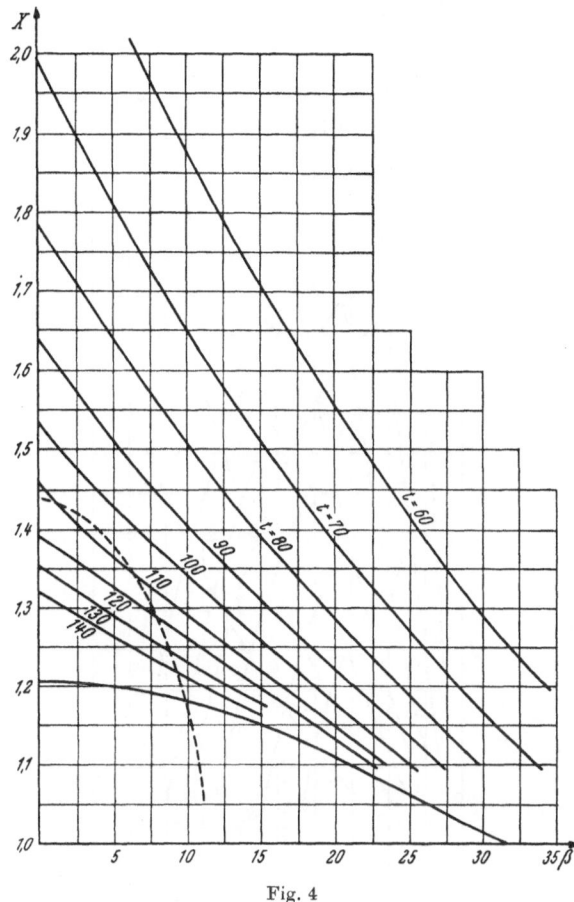

Fig. 4

VI. The Restraint for the Initial Condition

Obviously, the initial condition must be one which gives rise to the orbit that reaches the planet's orbit.

By this circumstance, there exists the restraint for the value of the initial condition.

We treat separately the case of outer planet and inner planet.

(1) For the outer planet, the necessary condition is that the aphelion-radius must be equal or longer than the distance to the planet from the Sun.

Aphelion-radius r_a is given by

$$r_a = \frac{r_0 (1 + e)}{2 - x} \tag{26}$$

x and e being the same as (10) and (15).

Therefore we can write

$$\frac{r_0 (1 + e)}{2 - x} \geq r_1$$

or

$$1 + \sqrt{1 - x (2 - x) \cos^2 \beta} \geq \frac{r_1}{r_0} (2 - x). \tag{27}$$

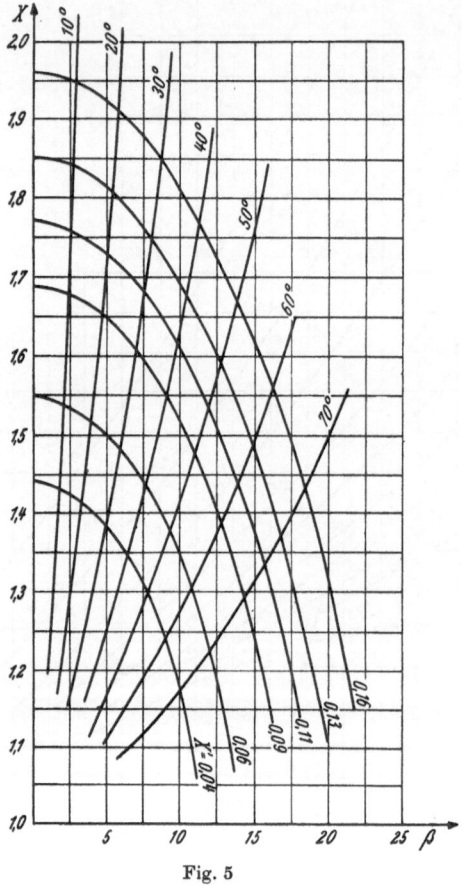

Fig. 5

From which

$$1 - x (2 - x) \cos^2 \beta \geq n^2 (2 - x)^2 - 2 n (2 - x) + 1. \tag{28}$$

Putting

$$n = r_1/r_0 > 1 \tag{29}$$

eq. (28) reduces to the expression

$$x \geq \frac{2 n (n - 1)}{n^2 - \cos^2 \beta}. \tag{30}$$

This is indicated in Fig. 4 by hatching.

(2) For the inner planet, the necessary condition is that the perihelion-radius must be equal to or shorter than the distance to the planet from the Sun.

Perihelion-radius r_p is given by

$$r_p = \frac{r_0 (1 - e)}{2 - x} .\tag{31}$$

Therefore we can write

$$\frac{r_0 (1 - e)}{2 - x} \leqq r_1$$

or

$$1 - \sqrt{1 - x (2 - x) \cos^2 \beta} \leq \frac{r_1}{r_0} (2 - x).\tag{32}$$

From which

$$1 - x (2 - x) \cos^2 \beta \geq 1 - 2 n (2 - x) + n^2 (2 - x)^2\tag{33}$$

putting

$$n = r_1/r_0 < 1\tag{34}$$

eq. (33) reduces to the expression

$$x (\cos^2 \beta - n^2) \leq 2 n (1 - n).\tag{35}$$

VII. On the Hohmann Orbits

The HOHMANN orbits are given by the condition that $\beta = 0$ and aphelion-radius or perihelion-radius is the same as the distance to the planet from the Sun.

Therefore in that case

$$e = \pm (x - 1)\tag{36}$$

and

$$r_a = \frac{r_0 \{1 + (x - 1)\}}{2 - x} = \frac{r_0 x}{2 - x} = r_1$$

$$x = \frac{2n}{n + 1}, \qquad n = \frac{r_1}{r_0} > 1 .\tag{37}$$

Also

$$r_p = \frac{r_0 \{1 - (1 - x)\}}{2 - x} = \frac{r_0 x}{2 - x} = r_1$$

$$x = \frac{2n}{1 + n}, \qquad n = \frac{r_1}{r_0} < 1 .\tag{38}$$

Putting these values into (16) and (17), we obtain

$$t_1 = \sin^{-1} \frac{n (2 - x) - 1}{x - 1} = \sin^{-1} \frac{\frac{2n}{n + 1} - 1}{\frac{2n}{n + 1} - 1} = \sin^{-1} (1) = \frac{\pi}{2} \text{ (when } n > 1)$$

$$t_1 = \sin^{-1} \frac{n (2 - x) - 1}{1 - x} = \sin^{-1} (-1) = -\frac{\pi}{2} \text{ (when } n < 1)$$

$$t_0 = \sin^{-1} \frac{1 - x}{x - 1} = \sin^{-1} (-1) = -\frac{\pi}{2} \text{ (when } n > 1)$$

$$t_0 = \sin^{-1} \frac{1 - x}{1 - x} = \sin^{-1} (1) = \frac{\pi}{2} \text{ (when } n < 1)$$

and

$$t = \sqrt{\frac{r_0^3}{\mu}} \left(\frac{n + 1}{2}\right)^{\frac{3}{2}} \pi .\tag{39}$$

Since $\sqrt{r_0^3/\mu}$ is equal to 58.12 (d), this gives 259 (d) for Mars $(n = 1.524)$ as is well known.

VIII. Conclusion

As is stated in Section II, this paper stands on many simplifying assumptions and it will be unsatisfactory for the exact calculation of the interplanetary journey. But by this analysis it became evident that there exists the optimum secant angle for each initial velocity v_i or x' which gives rise to the minimization of the duration. However this gain in the duration is not so large in contrary to general expectations. Therefore we must endeavour to realize the large initial velocity and increase x' as much as possible.

References

1. M. VERTREGT, Interplanetary Orbit. J. Brit. Interplan. Soc. 16, No. 6, 326 (1958).
2. J. COPELA ND, Interplanetary Trajectories under Low Thrust Radial Acceleration J. Amer. Rocket Soc. 29, No. 4, 267 (1959).

The Human Eye in Space
(Physiologic Aspect)

By

Hubertus Strughold[1]

(With 1 Figure)

(Received August 27, 1959)

Abstract — Zusammenfassung — Résumé

The Human Eye in Space (Physiologic Aspect). Visible radiation or light as it is found in nearby space is discussed from a physiologic point of view, with emphasis upon the luminance of the sky and the illumination from the sun. The difference between the darkness of a moonless sky on earth and the blackness of the sky in space is analysed. The visual appearance of the moon and earth to an orbiting astronaut and the strange spatial light distribution is examined with regard to his orientation in a weightless state. Attention is given to a possible hazard in space in the form of a retinal burn resulting from observing the sun with unprotected eyes *(helioscotoma retinae)*. Then the illumination from the sun is studied for the whole range from Mercury to Pluto, which may justify a subdivision of space into photic zones, namely an euphotic zone (favorable to space operations and to life on planets) surrounded by a hyperphotic and hypophotic region. Finally, in the weakly illuminated region far beyond Pluto (about 3 times its distance from the sun), colour discrimination becomes difficult for an astronaut and from here on the colorless dim-lighted world of interstellar space with its black, star-studded sky begins.

Das menschliche Auge im Weltraum (Physiologischer Aspekt). Sichtbare Strahlung oder Licht, wie es im nahen Weltraum gefunden wird, wird vom physiologischen Standpunkt diskutiert, mit Betonung der Helligkeit des Himmels und der Beleuchtung von der Sonne. Der Unterschied zwischen der Dunkelheit eines mondlosen Himmels auf der Erde und der Schwärze des Himmels im Weltraum wird analysiert. Die sichtbare Erscheinung der Erde und des Mondes für einen im Weltraum befindlichen Astronauten und die ungewohnte Lichtverteilung werden mit Berücksichtigung seiner Orientierung im gewichtslosen Zustand überprüft. Aufmerksamkeit ist einer möglichen Gefahr von Netzhautverbrennung gewidmet, die durch das Beobachten der Sonne mit ungeschützten Augen entstehen könnte *(helioscotoma retinae)*. Die Beleuchtung von der Sonne für den ganzen Planetenbereich von Merkur bis Pluto wird besprochen. Dies mag eine Unterteilung des Raumes in photische Zonen rechtfertigen: eine euphotische Zone, die Raumunternehmungen und Leben auf den Planeten begünstigt und die umgeben ist von einer hyperphotischen und hypophotischen Region. In dem schwach beleuchteten Gebiet weit jenseits von Pluto (in etwa 3mal dessen Entfernung von der Sonne) wird die Farbunterscheidung für einen Astronauten sehr schwierig und von hier beginnt die farblose, schwach beleuchtete Welt des interstellaren Raumes mit seinem schwarzen, sternbesäten Himmel.

---

[1] M.D., Ph.D., Professor of Space Medicine and Advisor for Research, USAF Aerospace Medical Center, Brooks Air Force Base, San Antonio, Texas, U.S.A.

L'oeil humain dans l'espace (aspect physiologique). Les radiations visibles ou la lumière dans l'espace proche sont étudiées du point de vue physiologique, avec l'accent sur la luminosité du ciel et la lumière émanant du soleil. La différence entre les ténèbres d'un ciel sans lune sur terre et d'un ciel dans l'espace est analysée. L'apparence visuelle de la lune et de la terre à un astronaute en orbite et l'étrange distribution de la lumière dans l'espace sont examinées par rapport à son orientation dans un état de gravité nulle. L'attention est attirée à un danger possible dans l'espace sous forme d'une brulure de la rétine resultant de l'observation du soleil sans protection des yeux (*helioscotoma retinae*). Puis l'illumination solaire est étudiée pour l'intervalle total de Mercure à Pluton, pouvant justifier une subdivision de l'espace en zones photiques, notamment une zone euphotique (favorable aux opérations dans l'espace et à la vie sur les planètes) entourée par des régions hyperphotique et hypophotique. Finalement, dans la région faiblement illuminée bien au delà de Pluton (à peu près 3 fois sa distance du soleil), l'appréciation des couleurs devient difficile pour un astronaute et à partir de cette limite commence le monde glauque de l'espace interstellaire avec son ciel noir parsemé d'étoiles.

The visual world encountered by an astronaut in space will be exotic; different in many respects, indeed, from the one seen from the earth's surface. Under the condition of weightlessness, vision is the only sensory means for orientation in space. Except for radio communication in the deep silence of space, vision offers the only sensory contact with the outside world. What the human eye sees in space and how its function may be affected by the optical scenery is, therefore, of tremendous importance to the astronaut; in addition, it is of scientific and general human interest.

In the following an attempt will be made to analyze the light conditions, i.e., visible radiation, in space considered essentially from a physiologic point of view (the standpoint of human physiology).

In terms of physiologic optics we are dealing essentially with the section from about 3,800 to 7,800 Ångström in the electromagnetic spectrum, to which the pigment of the human retina is specifically attuned.

First, we shall discuss this *visible radiation* as it is encountered in circumterrestrial or *nearby space*, with emphasis upon the brightness or *luminance of the sky* and the *illuminance from the sun*. In the second part of the paper we shall examine *solar illuminance* for the *whole distance from Mercury to Pluton*.

For a better understanding of light in space, it might be useful to contrast the conditions found there with those observed on the earth's surface and in the depth of the oceans.

Beginning with the latter, the deep regions of the *oceans* are permanently dark. It will be interesting to recall, later in this discussion, that in this lightless or aphotic abyss, numerous fish are found; some with light-producing (bioluminescent) organs and with eyes, and other fish with vestigual non-functioning eyes. These blind deep-sea fish sense their environment by means of mechanoreceptors and chemoreceptors only.

When a person in a bathysphere ascends from these depths of the ocean, the first slight traces of light perceptible by the human eye, appear about 500 to 600 meters below sea level (W. Beebe [3]; Jacques Piccard [20]); they are of bluish color and shift with decreasing depth to green and greenish-yellow in the sub-surface regions. One meter below the surface, illuminance from the sun is about 10,000 lumens per square meter or lux (lx)[1]. But even that close to the surface a diver perceives no clear evidence of a sun.

---

[1] For photometric terms and units see Annex.

Emerging from the hydrosphere to the bottom of the atmosphere, we see—during a cloudless noon—the sun, high in the sky, its rim blurred by an aureole which blends into a dome-shaped sky of bluish light. The aureole is caused by indirect sunlight, reflected by ice crystals in the higher atmosphere, and the blue sky light is indirect sunlight, scattered by the air molecules and fine dust. The photometric brightness, or luminance, of the blue sky, about 25° from zenith, as seen from sea level, is about 1,600 nit[1] (H. Haber [11]). Because of this veil of scattered light the stars remain invisible, and the moon is barely discernible. The illuminance from the sun is, roughly, 108,000 lx (= 10,000 foot candles), the average value at noon during sunshine at sea level at middle latitudes in summer (H. Haber [11]; E. O. Hulburt [13]).

When man now ascends in a rocket vehicle, he will find that the sky gradually becomes darker and the sun brighter because of the rarification of the air and the resulting disappearance of light-scattering. This has been observed in high altitude balloon flight by J. Piccard [19], A. W. Stevens and O. A. Anderson [1], D. G. Simons [23], and M. D. Ross and M. L. Lewis [22]. At 30 km (18 mi) the *luminance of the sky* decreases to 30 nit, and at 160 km (100 mi), it is only about 3×10^{-5} nit (R. Tousey and E. O. Hulburt [26]; H. Haber [11]). Against this low field brightness the stars are visible all the time. Because of the absence of a reflecting and scattering medium, the sun shines now without an aureole, as a luminous disk on a dark background. And, the color of the sun (and of the stars) should be more whitish, because no blue rays are scattered out by an atmospheric medium. *Solar illuminance* increases from its maximal sea level value of about 108,000 lx to about 140,000 lx at the top of the atmosphere (13,600 ft-c, according to H. Haber [11]; 12,700 ft-c, according to F. S. Johnson [14]; 14,000 ft-c at 30 km, according to R. B. Toolin and V. J. Stakutis [25]. This extra-atmospheric value is called the solar illuminance constant, which of course is not a constant in the true sense of the word.

Such is the basic difference between the atmospheric and extra-atmospheric photic environment in nearby space during the day; a bright, blue sky with a bright sun, on the one hand, and a deep, dark sky with a still brighter sun on the other.

The *darkness of space* is not the same as that in a moonless, clear sky on earth at midnight. In the latter case the luminance of the dark sky between visible stars, or the background luminance, is greater. In the first place this is caused by *night airglow* in the upper atmosphere—a faint, diffuse light, emitted by atomic oxygen, nitrogen and sodium brought into excited states by solar ultraviolet rays (D. Barbier [2]).

In the order of decreasing relative intensities the total night light of a terrestrial clear, moonless sky, comprises airglow, starlight, zodiacal light (sunlight reflected or scattered from micrometeorites and dust particles), (Van de Hulst [29]), and galactic light (Nebulae), all direct and scattered by the earth's atmosphere (K. S. Mitra [18]). The scattered component gives the background luminance a bluish shade. The total luminance is in the order of 10^{-4} nit [11]. The dominant light source is airglow. In the sky in space this light source, of course, is absent.

The total light encountered in the dark sky of space includes therefore direct starlight, zodiacal light, and galactic light only. The dominant light source is the stars. The total luminance of the dark space sky is in the order of 10^{-5} nit, by a factor of ten lower than that of the terrestrial sky at night.

In the night sky, as seen from earth, the stars are embedded in the mild luminance of the airglow, which mollifies the contrast between them and the

---

[1] See footnote 1, page 716.

surrounding darkness. With airglow absent in space, zodiacal light remains the
main background luminance, at least up to the distance of Jupiter (J. M. Levitt
[15]). Against the darker background in space the stars should appear brighter
by contrast, and actually they are brighter by about 30 per cent; this is the attenuation of light when travelling vertically through the atmosphere [16]. For the
same reason more stars should be visible from above the atmosphere than from
the earth's surface. And, of course, they would not twinkle because no atmospheric turbulence interferes. This has been observed by D. G. Simons during his
balloon flight up to 30 km in 1957 [23].

So much about the light conditions an astronaut orbiting around the earth
would experience in altitudes from 200 to 800 km. This is the region to which
manned satellite flight will be confined because of the Van Allen Radiation
Belt, beginning at 800 km [28].

When the moon is in sight from an orbiting space vehicle, the sun-illuminated
portion would be brighter by about 30 per cent than seen from the earth, because
the atmosphere no longer interferes. Brighter, also, should be the moon's portion
not directly illuminated by the sun. This luminosity is "earthlight on the moon,"
which is actually indirect sunlight reflected from the earth.

With this we arrive at the earth itself as a source of light. Thirty-six per
cent of the solar light falling upon this planetary body *in toto* is reflected or
scattered back into space. This causes the earth to appear as an illuminated
celestial body with an *albedo* value about five times as high as that of the moon
(0.07). Numerous photographs of the earth have already been made from rockets
at considerable altitudes (C. T. Holliday [12]; J. G. Vaeth [27]). The color of
the sunlight reflected or scattered back from the earth's atmosphere is bluish
white; a conclusion which has been made from spectrographic studies of the
"earthlight" on the moon. A certain area of the earth would also show to the
orbiting astronaut the moonlight on the earth, just as we see from earth the
earthlight on the moon's dark areas.

The physical optical conditions as described for the upper atmosphere and
nearby space are of greatest interest from the standpoint of Space Medicine,
or bioastronautics. They pose psychologic, physiologic, and medical problems.

The appearance of the earth as a light source in the photic environment of
nearby space causes a strange situation, in that it is bright below and dark above,
as seen from an orbiting space vehicle. This is the reverse of the situation on the
earth's surface, which appears generally—except in winter—in dark (green or
brownish) colors, with a bright dome of skylight above. As seen from space in
the vicinity of the earth, the strange spatial distribution of light and darkness
is of greatest interest from the standpoint of *orientation in space*, especially
since the eye is the only sense organ on which this function depends in any space
operation. This is so because the astronaut is weightless, and under this condition
the mechanoreceptors, such as the otolith organ, the pressoreceptors of the
skin, etc., cannot provide any information concerning his position and movement
in space. This is in contrast to the life of deep-sea fish with non-functioning eyes
which depend in this respect entirely on their mechanoreceptors, as already
mentioned.

The low field brightness of the sky, combined with an intensive illumination
from the sun, represents a strange optical situation found on earth only under
artifical conditions; e. g., theatrical stage lighting. Everything that is exposed
to sunlight—outside and inside the cabin—appears extremely bright; everything
in the shadow is dark. Light and shadow dominate the scenery. This *photoscotic
condition* poses interesting problems in the field of *contrast vision* and *retinal*

adaptation, and requires special attention in *human engineering of the space cabin* (P. Cibis [7]; H. Rose [21]).

The observation of the light sources, themselves, must be considered. Beginning with the weaker one, the sun-illuminated portion of the earth which may produce, at the plane of the eye, an illuminance value of more than 20,000 lux at satellite distances below the Van Allen Radiation Belt (H. J. Merrill [17]), might easily cause a dazzling glare (as described by T. C. McDonald [16]. with regard to higher atmospheric altitudes), especially when the orbiting astronaut emerges out of the shadow of the earth. (The cone of the earth's shadow reaches about 1,385,000 km, or 859,000 mi, into space.) It should be added that for the first time a bird's eye view of the Aurorae Polares will be possible from a polar satellite. And it might be interesting to learn whether or not astronauts will be able to perceive the so-called Gegenschein. or counterglow—a faint luminosity far above the earth's atmosphere opposite the sun, the cause of which is still a matter of dispute.

Fig. 1. Retinal burn (see arrow) acquired by the author on 17 April 1912, in Europe, during the observation of a solar eclipse, with an insufficiently smoked glass

Special medical attention must be given to possible *hazards to the eye* from looking too long into the sun. Retinal damages, with which we have to reckon in this respect, are actually heat effects caused by visible and the near infrared ray focused by the lens upon a small area within the *fovea retinae* (A. Birch-Hirschfield [4]; F. H. Verhoeff et al. [30]; H. G. Clamann [8];.. and D. G. Cogan [9]). They occur frequently on earth when a *solar eclipse* is observed with an insufficiently smoked glass. The result may be a *retinitis solaris* and in severe cases a thermal coagulation necrosis of the retinal tissue or a *retinal burn*. Fig. 1 shows a scar of such a retinal burn, which I acquired when I observed the total solar eclipse on the 17th of April 1912 in Europe with my right eye insufficiently protected. This photograph, made more than 40 years later, shows that such retinal lesions are usually irreparable. The subjective symptom is a small blind area or *scotoma* in the visual field, which is called eclipse blindness (*scotoma helieclipticum*).

The critical exposure time for the development of eclipse blindness is estimated to be one minute or less (F. C. Cordes [10]). Outside the atmosphere the danger of such retinal lesions associated with visual defects, which generally might be called *helioscotoma*, is of course greater, because the intensity of solar radiation is about 50 per cent higher (2 cal cm^{-2} min^{-1} in space against maximally 1.4 cal cm^{-2} min^{-1} on earth). Furthermore, because of the dark sky, the eye—when turned to the sun—is not adjusted by pupillary constriction to such an intensive illumination and is, therefore caught by a blitz-like surprise out of the darkness.

From data available in the literature about similar effects on rabbits produced by atomic flashes (V. A. Byrnes, D. V. L. Brown, H. W. Rose, and P. A. Cibis [6]), it can be estimated than an exposure time in the order of a quarter of a minute or less to the solar radiation in space, at the earth's distance, might be sufficient to cause retinal burn. Caution in this respect, therefore, is indicated, and protection of the eye by means of strong absorbing glasses or electronic devices must be considered. Also, a retractable, light-scattering visor attached to the helmet, serving as a kind of blue sky simulator, may be useful to an astronaut on the moon. Where, in the remote regions of the solar system, the sun loses its retina-burning power is difficult to extrapolate, because we deal with living tissue which possesses reactive capabilities. It might be somewhere beyond Jupiter. Even beyond Saturn, according to H. Rose [21], solar illuminance is intense enough to cause a dazzling glare.

With this, we are already in the second part of this paper, namely, to examine what the light conditions are in *deep space*, including the whole *interplanetary range from Mercury to Pluto*.

In such an attempt the illumination from the sun is the factor that interests us most because it is subjected to considerable variations with increasing planetary distances, in contrast to the brightness of the sky, which is everywhere dark, and may become a shade darker in the extrajovian space because of the disappearance of the zodiacal background light. As mentioned before, *solar illuminance* above the earth's atmosphere at the earth's mean solar distance amounts to about 140,000 lx (solar illuminance constant). According to the inverse square law, in the region of Venus this value increases to 268,000 lx, and at Mercury's distance to 938,000 lx; it decreases at Mars' distance to 60,000 lx, at Jupiter's distance to 5,200 lx, and at the mean orbital distance of Pluto to 90 lx (Table I).

Table I

Planet	Mean Solar Distance 10^6 km 10^6 mi	Mean Solar Distance in Astron. Units (A.U.)	Solar Illuminance Lux
Mercury	57.8 35.9	0.38	938,000
Venus	108.1 67.2	0.72	268,000
Earth................	149.5 92.9	1.00	140,000
Mars	227.7 141.5	1.52	60,000
Jupiter	777.8 483.3	5.20	5,200
Saturn	1426.2 886.2	9.54	1,500
Uranus	2869.1 1728.8	19.19	380
Neptune	4495.7 2793.5	30.07	150
Pluto	5900.0 3670.0	39.50	90

These tremendous variations in solar illuminance suggest a subdivision of the space of the solar system into *photic zones*. We might not go too far in speaking of a *euphotic belt*, which is the zone favorable to space operations and may include some 100 million km on both sides of the earth's orbital distance; this zone is surrounded by a *hyperphotic* and *hypophotic zone*.

The *euphotic belt*, or we might also call it biophotic belt, together with the euthermal belt, liquid water belt and oxygen belt, is an important component in the concept of a *general life-favoring zone* or *ecosphere* in the planetary system (H. STRUGHOLD [24]). The photic component in this relatively favorable ecologic belt is not only important from the standpoint of vision but also with regard to the *utilization of light for photosynthesis* in a space vehicle and with regard to the question of *life on other celestial bodies*, particularly on Mars which has a transparent atmosphere.

We get some idea about solar illumination within the planetary system by comparing the *apparent size of the sun* as seen at the distances of the various planets.

To an observer on Mercury the diameter of the solar disk would appear more than twice as large as seen from the earth. As seen from Mars, the sun would have a considerably smaller apparent dimension than our Moon. At the distance of Jupiter, the sun's diameter is only one-fifth as large as seen from the earth; and at the distance of Pluto, the sun would not appear much larger than the evening star, Venus, appears to us on earth.

And, yet, the illuminance from the sun at the mean distance of Pluto is still 90 lx; this is considerably above the threshold for reading, which is about 20 lx, and also above the threshold for color vision. Below 10 lx, color discrimination becomes difficult. Solar illuminance decreases to this value in the region about three times the distance of Pluto, or about 18 billion km (or more than 10 billion miles) from the Sun. Here, then, begins the colorless world of interstellar space, as far as it is related to the illuminating power of the sun. And the sun, itself, as seen with the eyes of a space traveller, joins the ranks of the multitude of the other stars of our galaxy.

I wish to express my appreciation to Dr. OSKAR L. RITTER for valuable discussions.

Annex

Photometric Terms and Units

Illuminance: Luminous flux incident on unit area. Unit: Lux, foot candle.
Lux (lx), synonym metercandle: An illuminance of one lumen per square meter.
Foot Candle (ft-c): Illuminance of one lumen per square foot.
Lumen (lm): The luminous flux emitted through a unit solid angle (one steradian) from a point source of one candela.
Luminance (photometric brightness): Luminous intensity of any surface in a given direction per unit projected area of the surface viewed from that direction. Unit: nit.
Nit (nt): A luminance of one candela per square meter.
Candela (cd): Unit of luminous intensity. New defined and internationally accepted candle.

References

1. H. G. ARMSTRONG, The Medical Aspects of the National Geographic Society-U.S. Army Air Corps Stratosphere Expedition of November 11, 1935. J. Aviat. Med. **36**, 55 (1935).
2. D. BARBIER, The Airglow. Vistas in Astronomy, Vol. 2, p. 929. London-New York: Pergamon Press, 1956.
3. W. BEEBE, Half Mile Down. New York: Harcourt, Brace, Comp. Inc., 1934.
4. A. BIRCH-HIRSCHFIELD, Die Wirkung der strahlenden Energie auf das Auge. Erg. allg. Path. path. Anat. 16, 603 (1914).
5. K. BUETTNER and H. HABER, The Aeropause. Science **115**, 656 (1952).
6. V. A. BYRNES, D. V. L. BROWN, H. W. ROSE and P. A. CIBIS, Choreoretinal Burns Produced by Atomic Flash. A. M. A. Arch. Ophthalm. **53**, 351 (1955).

7. P. A. CIBIS, Visual Adaptation Applicable to Visual Problems in Flight at Increasing Altitudes. Chapter 21, in: Physics and Medicine of the Upper Atmosphere, Edited by S. WHITE and O. O. BENSON, JR. Albuquerque, New Mexico: University of New Mexico Press, 1952.
8. H. G. CLAMANN, Netzhautschädigungen bei Fliegern. Z. Luftfahrtmedizin 2, 314 (1938).
9. D. G. COGAN, Lesions of the Eye from Radiant Energy. J. Amer. Med. Ass. 142, 145 (1950).
10. F. C. CORDES, Eclipse Retinitis. Amer. J. Ophthalm. 31, 101 (1948).
11. H. HABER, Manned Flight at the Borders of Space. J. Amer. Rocket Soc. 22, 269 (1952).
12. C. T. HOLLIDAY, Seeing the Earth from 80 Miles Up. National Geographic Mag. 98, 511 (1950).
13. E. O. HULBURT, Physical Characteristics of the Upper Atmosphere of the Earth. Chapter 3, in: Physics and Medicine of the Upper Atmosphere, Edited by S. WHITE and O. O. BENSON, JR. Albuquerque, New Mexiko: University of New Mexico Press, 1952.
14. F. S. JOHNSON, The Solar Constant. J. Meteorol. 11, 431 (1954).
15. J. M. LEVITT, Target of Tomorrow. New York: Fleet Publishing Corp., 1959.
16. T. C. McDONALD, Changing Concepts in Aviation Medicine. J. Aviat. Med. 26, 463 (1955).
17. H. J. MERRILL, Satellite Tracking by Electronic Optical Instrumentation. Chapter 3, in: Scientific Uses of Earth Satellites, Edited by J. A. VAN ALLEN. Ann Arbor, Michigan: The University of Michigan Press, 1958.
18. S. K. MITRA, The Upper Atmosphere. Calcutta: The Royal Asiatic Society of Bengal, 1947.
19. JEAN PICCARD, Sky, Earth and Sea. New York: Oxford University Press, 1956.
20. JACQUES PICCARD, Le bathyscaphe et les plongées du Trieste, 1953—1957. Lausanne, 1958.
21. H. ROSE, Perception and Reaction Time. Chapter 18, in: Physics and Medicine of the Atmosphere and Space, Edited by O. O. BENSON, JR., and H. STRUGHOLD. New York: J. Wiley, 1959.
22. M. D. ROSS and M. L. LEWIS, To 76,000 Feet by Strato-Lab Balloon. National Geographic Mag. 91, 269 (1957).
23. D. G. SIMONS, Pilot Reactions during Manhigh II Balloon Flight. J. Aviat. Med. 29, 1 (1958); Observations in High Altitude, Sealed Cabin Balloon Flight. Chapter 9, in: Man in Space, Edited by K. F. GANTZ. New York: Duell, Sloan & Pearce, Inc.
24. H. STRUGHOLD, The Ecosphere of the Sun. J. Aviat. Med. 26, 323 (1955).
25. R. B. TOOLIN and V. J. STAKUTIS, Visual Albedo and Total Solar Illumination as a Function of Altitude. Bull. Amer. Meteorol. Soc. 39, 543 (1959).
26. R. TOUSEY and E. O. HULBURT, Brightness and Polarization of the Day-Light Sky at Various Altitudes above Sea Level. J. Opt. Soc. Amer. 37, 78 (1947).
27. J. G. VAETH, 200 Miles Up. Newark: The Ronald Press Co., 1955.
28. J. VAN ALLEN, The Great Radiation Belt. Chapter 1, in: Physics and Medicine of the Atmosphere and Space, Edited by O. O. BENSON, JR., and H. STRUGHOLD. New York: J. Wiley, 1959.
29. H. C. VAN DE HULST, The Zodiacal Light. Vistas in Astronomy, Vol. 2, p. 998. London-New York: Pergamon Press, 1956.
30. F. H. VERHOEFF, L. BELL and C. B. WALKER, Pathological Effects of Radiant Energy on the Eye. Proc. Amer. Acad. Arts Sci. 1916, 51.

On the Apparent Motion of an Earth's Artificial Satellite

By

J. J. de Orús[1], AAE

(With 2 Figures)

(Received June 8, 1959)

Abstract — Zusammenfassung — Résumé

On the Apparent Motion of an Earth's Artificial Satellite. It is known that in radio observations of an artificial satellite there is the problem to be solved, starting from the ephemerides, of the time when the satellite is at its minimum distance from a determined Earth station. Experimentally the said time is shown by the maximum variation of the displacement of the frequencies of the signals perceived (Doppler effect), and by means of its repeated registration the "observation minus calculus" can be found, which supplies the corrections to the orbital elements of the satellite. Though such radio observations do not achieve the precision of the visual ones, their independence of meteorological and ionospherical conditions which assure their registration et each transit has rapidly enforced them.

The calculation of the time when the satellite is at a minimum distance from a determined Earth station is considerably simplified owing to two favourable circumstances: the slight oblateness of the terrestrial globe and the little relation between the periods of revolution of the satellite and of rotation of the Earth ($\leqslant 1/16$). As we shall see in this brief report, we can obtain the said time by means of a fairly convergent development, which solves our interesting problem relative to the apparent motion of an Earth's artificial satellite.

Über die scheinbare Bewegung eines künstlichen Erdsatelliten. Bekanntlich muß bei Radiobeobachtungen eines künstlichen Satelliten das Problem gelöst werden, von den Ephemeriden ausgehend die Zeit zu finden, wann sich der Satellit im kleinsten Abstand von einer bestimmten Beobachtungsstation auf der Erde befindet. Es läßt sich zeigen, daß der genannte Zeitpunkt experimentell durch die maximale Variation der Frequenzverschiebung der empfangenen Signale (Doppler-Effekt) erhalten werden kann. Durch ihre wiederholte Registrierung kann man die Beobachtungsfehler und die Korrektionen der Bahnelemente des Satelliten ableiten. Obwohl solche Radiobeobachtungen nicht die Genauigkeit visueller Beobachtungen erreichen, hat ihre Unabhängigkeit von meteorologischen und ionosphärischen Bedingungen, welche ihre Registrierung bei jedem Bahndurchgang garantieren, sie rasch an Bedeutung gewinnen lassen.

Die Beobachtungen der Zeit des kleinsten Satellitenabstandes von einer bestimmten Erdstation wird durch zwei günstige Umstände beträchtlich vereinfacht: die geringe Abplattung der Erde und das kleine Verhältnis zwischen den Umlaufperioden des Satelliten und der Rotation der Erde ($\leqslant 1/16$). Wie aus der kurzen, hier vorliegenden Mitteilung ersichtlich ist, kann man den erwähnten Zeitpunkt mit Hilfe einer ziemlich konvergenten Entwicklung erhalten, wodurch das Problem hinsichtlich der scheinbaren Bewegung eines künstlichen Erdsatelliten gelöst ist.

---

[1] Fabra Observatory, Barcelona, Spain.

Sur le mouvement apparent d'un satellite artificiel. Dans les observations radio-électriques d'un satellite artificiel se pose le problème de déterminer à partir des éphémérides l'époque où la distance à une station d'observation est minimum. La détermination expérimentale de cette époque se fait par le maximum de variation du déplacement de la fréquence par effet Doppler. La répétition du procédé et l'application du calcul du minimum des erreurs d'observation fournit les corrections aux paramètres de l'orbite. Les observations radioélectriques n'ont pas la précision des observations visuelles. Cependant leur insensibilité aux conditions météorologiques et ionosphériques les ont rapidement imposées.

Le calcul de l'époque de la distance minimum est simplifié par le fait que l'aplatissement de la terre est faible et que la période de révolution du satellite est petite comparée à la rotation de la terre ($\leqslant 1/16$). Ceci permet de résoudre le problème par un développement bien convergent.

I. Initial Conditions

Let us assume that we have the ephemerides of the artificial satellite which supply, every 4 minutes and for each time t_0, the geographical coordinates,

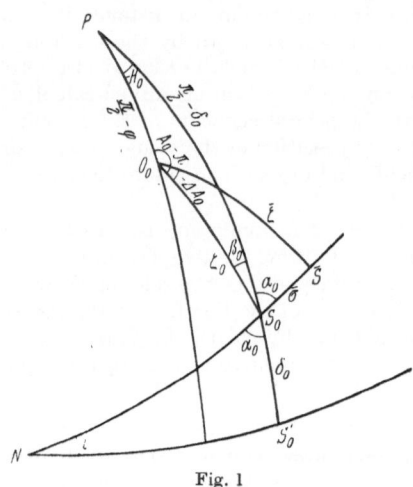

Fig. 1

longitude and latitude (to tenths of degrees), and the altitude h_0 (to ten kms.) from the point at which a vertical meets the satellite over the said point[1]. Let us assume, moreover, that some equatorial elements of the satellite are known, although these are only approximate. Let R be the radius of the Earth (6371 kms.); in each time t_0 the distance r_0 of the satellite from the centre will be $r_0 = R + h_0$ and, assuming a spherical Earth, its declination δ_0 will be $\delta_0 = \varphi_0$.[2]

Let λ and φ be the geographical coordinates of the station O. With $H_0 = \lambda_0 - \lambda$ denoting the hour angle of the satellite S_0 in the time t_0 (positive longitudes towards the west), the geocentric zenithal distance ζ_0 and the azimuth A_0 in the said time t_0 will be given by the known relations (Fig. 1):

$$
\left.
\begin{aligned}
&\mathrm{cosin}\, \zeta_0 = \sin \varphi \sin \delta_0 + \cosin \varphi \cosin \delta_0 \cos H_0 \\
&\sin \zeta_0 \sin A_0 = \mathrm{cosin}\, \delta_0 \sin H_0 \\
&\sin \zeta_0 \cos A_0 = -\cosin \varphi \sin \delta_0 + \sin \varphi \cosin \delta_0 \cos H_0 \, .
\end{aligned}
\right\} \tag{1}
$$

If p_0 is the horizontal parallax of the satellite in the time t_0, this will be on the horizon of the station for $\zeta_0 \leqslant \dfrac{\pi}{2} - p_0$, i.e. if it agrees:

$$
\cos \zeta_0 \geqslant \sin p_0 = \frac{R}{R + h_0} \, . \tag{2}
$$

---

[1] Telegrammes from U.S.N.A.C.-I.G.Y.

[2] With a greater approximation it would be:

$$
r_0 = R_e \, (1 - \varepsilon \sin^2 \varphi_0) + h_0 \, .
$$

$\delta_0 = \varphi_0 - \varepsilon \dfrac{R_e}{r_0} \sin 2\varphi_0$ R_e = equatorial radius of the Earth (6378 kms.)

ε = terrestrial oblateness (1/297).

By a simple numerical or graphical calculation, depending on the ephemerides we can obtain in what time t_0, between the ones in the table, ζ_0 is minimum, having also the condition (2). Thus we shall obtain through formulae (1) some approximate values which we shall call also ζ_0 and A_0.

II. First Approximation

If the Earth did not revolve, the minimum distance of the satellite from the station would take place at a minimum zenithal distance, i.e. for the spherical normal through O to the orbit of the satellite (disregarding the small variation of its radius vector in ± 2 m). Let us denote with a straight line the values concerning such minimum distance. Being $\Delta A_0 = \overline{A} - A_0$, from the rectangular triangle $\Delta O_0 \overline{S} S_0$ (Fig. 1). the following equations are deduced:

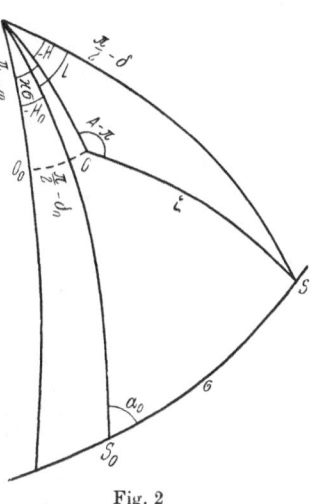

$$\left.\begin{aligned}
\tan \bar{\sigma} &= \cosin (a_0 + \beta_0) \tan \zeta_0 \\
\sin \overline{\zeta} &= \sin (a_0 + \beta_0) \sin \zeta_0 \\
\cotan \Delta A_0 &= -\tan (a_0 + \beta_0) \cosin \zeta_0
\end{aligned}\right\} \quad (3)$$

calculable through the auxiliar angles a_0 and β_0 given from the formulae:

$$\left.\begin{aligned}
\sin a_0 &= \frac{\cosin i}{\cosin \delta_0} \\
\sin \beta_0 &= -\frac{\sin A_0}{\cosin \delta_0} \cosin \varphi
\end{aligned}\right\} \quad (4)$$

(i = equatorial inclination of the orbit).

Let n_0 be the motion of the satellite in the time t_0:

$$n_0 = \frac{n\, a^2 \sqrt{1 - e^2}}{r_0^{\,2}} \quad (5)$$

Fig. 2

(n = mean motion, a — semi-major axis, e = eccentricity). If we assume, very approximately, a uniform movement in ± 2 m. we shall obtain[1]:

$$\bar{t} = t_0 + \frac{\bar{\sigma}}{n_0}. \quad (6)$$

From (3), (4), (5) and (6) it is easy to calculate the values $\overline{\zeta}$, $\overline{A}$, $\bar{\sigma}$ and $\bar{t}$ relative to the minimum distance.

III. Second Approximation

The earth really revolves; let m be its angular velocity. Being $k = \dfrac{m}{n_0}$, the angle that it rotates during the interval of time $t - t_0$ will be $m\,(t - t_0) = k\, n_0(t - t_0) = k\sigma$. In the time t is $H = H_0 + k\sigma - L$ (Fig. 2), and in the spherical triangle ΔOPS there is:

$$\cosin \zeta = \sin \varphi \sin \delta + \cosin \varphi \cosin \delta \cosin (H_0 + k\sigma - L)$$

---

[1] If n is taken instead of n_0 it is shown that an error could be committed of the order of $4\,e\,(t - t_0)$, too great for our purpose.

or in the triangle $\Delta S_0 PS$:

$$\sin \delta = \sin \delta_0 \cosin \sigma + \cosin \delta_0 \sin \sigma \cosin a_0$$

$$\cosin \delta \sin L = \sin \sigma \sin a_0$$

$$\cosin \delta \cosin L = \cosin \delta_0 \cosin \sigma - \sin \delta_0 \sin \sigma \cosin a_0 \; ;$$

also:

$$\cosin \zeta = \sin \varphi \, (\sin \delta_0 \cosin \sigma + \cosin \delta_0 \cosin a_0 \sin \sigma)$$
$$+ \cosin \varphi \, (\cosin \delta_0 \cosin \sigma - \sin \delta_0 \cosin a_0 \sin \sigma) \cosin (H_0 + k\sigma) \qquad (7)$$
$$+ \cosin \varphi \sin a_0 \sin \sigma \sin (H_0 + k\sigma) .$$

The expression (7), which supplies ζ in function of σ, is of the form:

$$\cosin \zeta = x \, (\sigma) + y \, (\sigma) \, u \, (k\sigma) + z \, (\sigma) \, v \, (k\sigma)$$

and being $\dfrac{du}{d\sigma} = -kv \; ; \; \dfrac{dv}{d\sigma} = ku,$

and designating hereafter with an accent derivation respecting the argument, the zenithal distance ζ will be the minimum if the following equation is agreed:

$$x' + y'u + z'v - k \, (yv - zu) = 0. \qquad (8)$$

Following a classic method in astronomy, we are going to solve eq. (8), transcending in σ, by means of the development in MCLAURIN's formula of the function $\sigma = \sigma \, (k)$. Deriving successively in (8), and being $x'' = -x, \; \ldots \; u'' = -u, \ldots,$ we obtain:

$$- [x + yu + zv + k \, (y'v - z'u)] \, \sigma'$$
$$+ [y'u' + z'v' - k \, (yv' - zu')] \, (\sigma + k\sigma') \qquad (9)$$
$$- (yv - zu) = 0 .$$

$$- [x' + y'u + z'v - k \, (yv - zu)] \, \sigma'^2$$
$$- [yu' + zv' + k \, (y'v' - z'u')] \, 2\,\sigma' \, (\sigma + k\sigma')$$
$$- (y'v - z'u) \, 2\,\sigma' - (yv' - zu') \, 2 \, (\sigma + k\sigma') \qquad (10)$$
$$- [x + yu + zv + k \, (y'v - z'u)] \, \sigma''$$
$$- [y'u + z'v - k \, (yv - zu)] \, (\sigma + k\sigma')^2$$
$$+ [y'u' + z'v' - k \, (yv' - zu')] \, (2\,\sigma' + k\,\sigma'') = 0 .$$

Making in the last ones $k = 0$ [1] and taking into account that:

$$\bar{x} + \overline{yu} + \overline{zv} = \cosin \bar{\zeta}$$
$$\overline{yv} - \overline{zu} = -(\overline{yu'} + \overline{zv'}) = \cosin \varphi \sin \bar{\zeta} \sin \bar{A}$$
$$\overline{y'u'} + \overline{z'v'} = -(\overline{y'v} - \overline{z'u}) = \cosin \varphi \cosin \bar{A}$$
$$\overline{yv'} - \overline{zu'} = \cosin \varphi \, (\cosin \varphi \cosin \bar{\zeta} + \sin \varphi \sin \bar{\zeta} \cosin \bar{A})$$

if, $\bar{\sigma} = \sigma \, (0)$, we remove $\bar{\sigma}' = \sigma' \, (0)$ in (9) and substitute in (10), neglecting terms of higher order, we obtain finally:

$$\sigma = \bar{\sigma} \left[1 + \frac{k \cosin \varphi}{\cosin \bar{\zeta}} \cosin \bar{A} + \left(\frac{k \cosin \varphi}{\cosin \bar{\zeta}} \right)^2 \cosin 2\bar{A} + \ldots \right]$$
$$- \sin \bar{\zeta} \left[\frac{k \cosin \varphi}{\cosin \bar{\zeta}} \sin \bar{A} + \left(\frac{k \cosin \varphi}{\cosin \bar{\zeta}} \right)^2 \sin 2\bar{A} + \ldots \right] \qquad (11)$$

a development which constitutes the scope of the present communication.

[1] Making $k = 0$ in (8) the formula (3) can be obtained, which gives $\bar{\sigma}$.

Given that in the rectangular triangle $\varDelta \, O_0 \overline{S} S_0$ (Fig. 1) is verified:

$$\left. \begin{array}{l} \overline{\sigma} \simeq - \sin \overline{\zeta} \tan \varDelta \, A_0 \\ \tan \overline{\zeta} = \tan \zeta_0 \cosin \varDelta \, A_0, \end{array} \right\}$$

if we limit ourselves to the lineal approximation of (11), the said solution takes the simple form:

$$\sigma = \overline{\sigma} - k \cosin \varphi \tan \zeta_0 \sin (A_0 + 2 \varDelta \, A_0), \tag{12}$$

a very practical formula which, in the majority of cases, supplies σ with an approximation of the order of the ephemerides $(0°,1)$. According to (12), with the corresponding approximation (2 sec.), the time t of the minimum distance of the satellite from the station will be given by the formula:

$$t = \overline{t} - \frac{m}{n_0{}^2} \cosin \varphi \tan \zeta_0 \sin (A_0 + 2 \varDelta \, A_0)^1 .$$

---

[1] The following approximate formula, deduced elementarily, is also very useful:

$$\zeta = \overline{\zeta} + \overline{\sigma} k \cosin \varphi \sin (A_0 + \varDelta \, A_0).$$

The Drag Brake Manned Satellite System

By

R. W. Detra[1], A. R. Kantrowitz[1], F. R. Riddell[1] and P. H. Rose[1]

(With 14 Figures)

(Received August 27, 1959)

Abstract — Zusammenfassung — Résumé

The Drag Brake Manned Satellite System. The manned satellite designs which have been most frequently proposed are extrapolations either of airplanes or of missile nose cones. This paper presents a new approach to the design of a manned satellite system. The central feature of this design is a variable area drag brake. For a given capsule and payload, the elements associated with attitude stabilization, recovery from orbit, re-entry, landing accuracy control and final landing of the system described comprise less than half the weight required for the same functions in presently envisioned retro-rocket systems.

Die Verwendung der Widerstandsbremse für bemannte Satelliten. Die bisher öffentlich diskutierten Entwürfe bemannter Satelliten fußen sämtlich auf entsprechend modifizierten Prinzipien des Flugzeugbaues oder der Konstruktion von Kegelnasen für Raketen. Der vorliegende Aufsatz betrachtet die Konstruktion solcher Flugkörper unter einem anderen Aspekt. Statt in der angedeuteten Richtung weiterzuarbeiten, haben wir bei der Entwicklung eines bemannten Satelliten einen gänzlich neuen Weg beschritten. Das Ergebnis dieser dreijährigen Bemühungen ist der Entwurf eines bemannten Satelliten, begründet auf der Idee einer Widerstandsbremse, in der Absicht, der bemannten Raumfahrt zu einem frühen Start zu verhelfen.

Das Problem der Wärmeentwicklung und der Abbremsung, die ein menschlicher Fahrgast während des Wiedereintauchens in die Erdatmosphäre überstehen muß, beherrschte bisher fast alle Diskussionen über bemannte Satelliten. Wir sind jedoch der Ansicht, daß die während des Interkontinentalen Ballistischen Raketenprogramms erzielten Fortschritte uns den Entwurf eines Flugkörpers ohne Überbetonung des Erhitzungsproblems erlauben. Ebenso haben raumfahrtmedizinische Untersuchungen ergeben, daß die beim Wiedereindringen bei konstantem Reibungswiderstand zu erwartende Verzögerung von 8 g kein ernsthaftes Problem darstellt. Infolgedessen scheint uns, daß das Wiedereindringproblem für die Konstruktion eines solchen Flugkörpers nicht ausschlaggebend sein sollte. Eine Anzahl anderer Aufgaben müssen bei der Entwicklung eines bemannten Satelliten gelöst werden. Erstens bedeutet die zu fordernde hohe Festigkeit struktureller Teile, die aus der Umhüllung der Trägerrakete herausragen, eine hohe Gewichtseinbuße. Andererseits werden große Oberflächen benötigt, um Widerstand und andere aerodynamische Kräfte während des Fluges durch die dünne obere Atmosphäre zu erzeugen. Dies führt zwangsläufig zur Verwendung zusammenfaltbarer Strukturen. Eine Vorrichtung zur Ermöglichung einer befriedigenden Landung muß ebenfalls im Entwurf miteinbegriffen werden. Schließlich leuchtet ein, daß eine Verringerung der Anzahl der zur Ausübung dieser verschiedenen Funktionen erforderlichen Einzelteile von dem so überaus wichtigen

---

[1] Avco Research and Advanced Development Division, 201 Lowell Street, Wilmington, Massachusetts, U.S.A.

Standpunkt der Zuverlässigkeit her stets von Vorteil sein wird. Die von uns vorge-schlagene Konstruktion eines bemannten Satelliten baut sich auf der zentralen Idee einer zusammenfaltbaren, leichten, schirmartigen Widerstandsbremse aus Edelstahl auf. Öffnen und Schließen der Vorrichtung während des Fluges bewirkt eine Wider-standsänderung von 20:1. Diese Widerstandsänderung verursacht eine entsprechende Änderung der Sinkgeschwindigkeit, d. h. der Zeit, die sich der Satellit auf einer Kreis-bahn halten kann. Volle Ausnutzung der Veränderlichkeit des Widerstandes gemäß dem jetzigen Programm erlaubt eine Landung innerhalb von ±150 naut. Meilen (±275 km) von einem vorgeschriebenen Punkt.

Die ausgefaltete Struktur hat eine niedrige Belastung von nur 1,5 lbs./ft.² (7,61 kg/m²). Infolgedessen kann der Flugkörper bereits hoch oben in der Atmosphäre abgebremst werden, während die entwickelte Wärme bei Temperaturen abgestrahlt wird, welche die im Turbinenbau heute üblichen nie überschreiten. Glücklicherweise ergibt die Widerstandsbremse gleichzeitig eine so niedrige Endgeschwindigkeit, daß zusätzliche Fallschirme überflüssig sind. In der Widerstandsbremse schlagen wir also eine einzige neue Vorrichtung vor, welche die Vielzahl der in anderen Entwürfen individuellen Systemen übertragenen Funktionen gleichzeitig erfüllt. Neben ihrer offensichtlich größeren Zuverlässigkeit erlaubt auch die vielfache Verwendbarkeit dieser einfachen Vorrichtung eine ganz entschiedene Gewichtseinsparung.

Le satellite piloté a frein aérodynamique. Les conceptions de satellite piloté qui ont été divulguées à ce jour, sont, toutes, des adaptations de techniques utilisées soit en aéronautique, soit dans le dessin des cônes d'engins. Cet article présente un nouveau point de vue quant au dessin de tels véhicules. Au lieu d'extrapoler une des techniques ci-dessus, nous avons décidé d'approcher le problème des satellites pilotés d'une manière originale. Trois ans d'efforts ont permis le dessin d'un satellite piloté basé sur un frein aérodynamique et permettront de réussir plus rapidement des vols inter-planétaires.

Le problème de l'échauffement lors de la ré-entrée ainsi que la décélération à la-quelle est soumis un passager humain, ont dominé la plupart des discussions relatives au satellite piloté. Cependant, nous pensons que les progrès accomplis lors du déve-loppement des ICBM nous ont permis de concevoir un véhicule sans accorder une importance exagérée à l'échauffement de ré-entrée et, de plus, les progrès faits en re-cherches aéro-médicales montrent clairement que les décélérations de *8 g.* rencontrées pendant la phase de ré-entrée avec freinage aérodynamique constant, ne posent pas de sérieux problèmes. Nous en déduisons que le problème de ré-entrée ne doit pas être le facteur le plus important lors du dessin du véhicule. Nombres d'autres problè-mes doivent être pris en considération lors de la conception d'un satellite piloté: Premièrement, la haute résistance exigée des éléments structuraux en saillie de l'en-veloppe du véhicule porteur conduit à une augmentation du poids de celui-ci. D'autre part, de larges surfaces sont nécessaires pour créer la résistance aérodynamique ou d'autres forces du même genre, dans l'atmosphère raréfiée rencontrée par le véhicule en orbite. Ces différentes conditions conduisent à des structures pliantes. Un méca-nisme permettant un atterissage satisfaisant doit également être inclus dans le dessin. Enfin, il est évident qu'une réduction du nombre d'éléments nécessaires à l'accomplisse-ment des diverses fonctions sera toujours avantageuse en ce qui concerne la sûreté d'opération du véhicule. Le satellite piloté que nous proposons est construit autour d'un frein aérodynamique, de forme rappelant un parapluie, pliable, léger, fait d'acier inoxidable. En orbite, une variation de la résistance de 20:1 est obtenue entre la posi-tion ouverte et fermée. Cette variation de la résistance provoque un changement cor-respondant dans l'angle de descente ou dans la durée de révolution orbitale. Le con-trôle de la résistance d'après un programme pré-établi, permet un atterrissage en un endroit choisi avec une précision de ±150 miles nautiques.

La structure une fois ouverte, à une très faible charge, seulement 1½ lbs/ft² (7,61 kg/m²) et, par conséquent, le véhicule décélère dans les couches supérieures de l'atmosphère en rayonnant la chaleur à des températures ne dépassant jamais celles rencontrées dans les turbines d'aujourd'hui. Par une heureuse coïncidence, le frein aérodynamique permet des vitesses finales suffisamment faibles pour rendre le parachute d'atterrissage inutile. Par conséquent, nous proposons un nouvel élément simple, le frein aérodynamique, remplissant un grand nombre de fonctions qui, dans d'autres conceptions, nécessitent différents systèmes. En plus de la sûreté d'opération, l'emploi de cet élément simple permet une importante réduction de poids.

Introduction

It seems clear that the earliest manned satellite should be designed for flights ranging from one orbit to perhaps as long as several days. To achieve such life-times the satellite's orbit must thus be high enough so that the deceleration due to aerodynamic drag is less than a critical value, which ranges between 10^{-4} and $10^{-6}g$'s. The lowest altitude at which such flights can be achieved has the following advantages over higher altitude orbits:

1. A given booster can place a larger payload in orbit.

2. Escape problems from slightly sub-orbital velocities are less severe than for higher altitudes.

3. Radiation belt problems are reduced by the use of minimum altitudes.

Thus it appears that the earliest manned satellite flights will be made at altitudes at which there is a small but appreciable air density.

Controlled flight, either manned or unmanned, at these altitudes and near orbital speeds, introduces new concepts which must be considered in the design philosophy of vehicles to be operated in this regime. Although the flight speeds are high, the air density at these altitudes is extremely low, resulting in very low dynamic pressures. Consequently, very light structures are adequate for producing the forces needed to interact with the highly rarefied atmosphere. Since the forces are small per unit area, structures must be large in order to generate useful control forces.

Using boosters now under development for launching the satellite, the maximum dynamic pressure during launch may reach values as high as 1000 lbs./sq. ft. This value is several orders of magnitude larger than the highest forces experienced at any other time in the flight. Consequently the large, light structures needed to interact with the upper atmosphere should not protrude from the launch vehicle envelope. In addition, extensive structures attached to a booster may impose severe penalties on the launch capabilities of the booster because of additional drag. Therefore the extremely light structures needed for the high altitudes and speeds should be foldable during the launch phase.

Manned re-entry into the Earth's atmosphere from satellite orbits will require small initial flight path angles because of limitations of deceleration. The deceleration problem has been faced in aeromedical research so that human limitations are now well established and have been shown to be far beyond what is needed for launch or any near tangential re-entry. The total heat transferred to a re-entering body increases with decreasing initial flight path angle. For example, the total heat which must be absorbed per unit area at the stagnation point during a tangential re-entry with satellite velocity, is four times larger than the same value for an ICBM type re-entry at the same velocity, nose radius, and ballistic drag parameter. Because the problem lies mainly in this large amount of total heat input, rather than in the rate of heat input, metal heat sinks are the least economical means of handling the heating problem. This is shown in Fig. 1. Even though ablation can show a sizeable performance improvement over the heat sink, a radiating heat shield can be shown to be the lightest possible device for dissipating the re-entry heat from shallow angle satellite re-entry. Properly designed, the same lightweight structure used for radiating heat away during re-entry can also provide control forces. In order to radiate the re-entry heat at equilibrium temperatures sufficiently low to allow the use of currently available materials, the re-entry body must be blunt. Blunt bodies can be designed for tangential satellite re-entry whose maximum equilibrium temperatures and stresses are well within the state

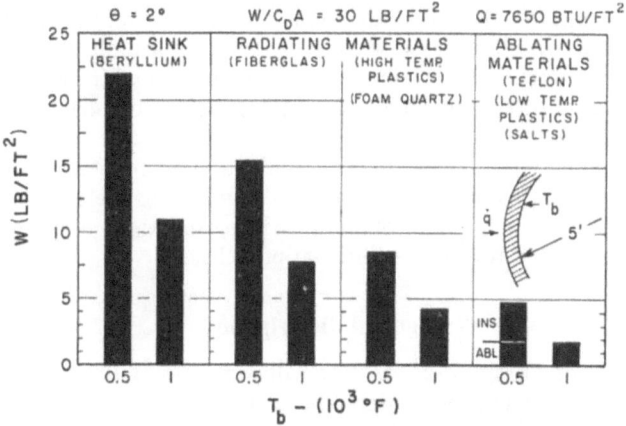

Fig. 1. A comparison of the performance of several materials for satellite re-entry heat shields

of the art of current materials technology. It has thus seemed to us that the re-entry problem should not at this time dominate the design of a manned satellite system.

The design we propose is built around the foldable, stainless steel, umbrella-like drag brake shown in Fig. 2. This structure is launched in a folded position

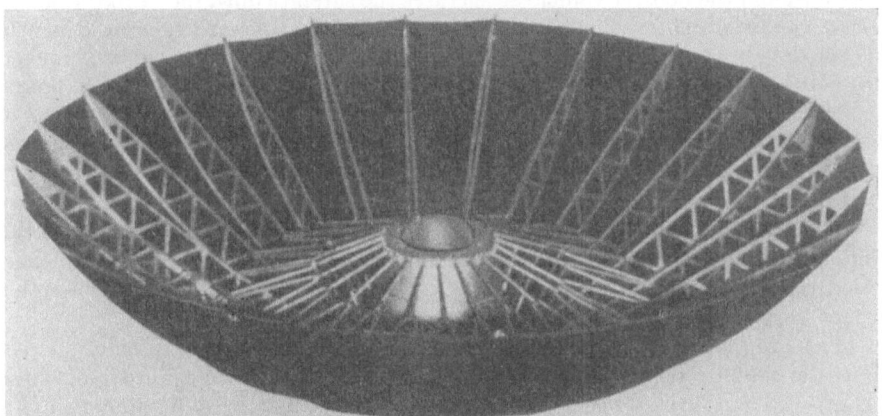

Fig. 2. Photograph of drag brake model in the re-entry position

resulting in an envelope very similar to an ogival missile nose. Fully opening the structure results in a 20 to 1 drag increase and a corresponding variation in orbital lifetime. Controlled variation of the vehicle drag monitored by an accelerometer and a clock permits landing at a pre-selected point with an accuracy of 150 miles. The extended structure gives a low surface loading ($W/C_p A = 1.5$ lbs./sq. ft.) so that the vehicle decelerates high (peak heating at 270,000 ft.; peak deceleration at 250,000 ft.) in the atmosphere. Radiating the heat away yields temperatures that are in all cases lower than standard turbine operating temperatures permitting the use of readily available stainless steels throughout the structure. It is a favorable coincidence that the same drag brake yields sufficiently low terminal velocity (45 ft./sec.) so that no additional parachutes are required. We thus present a design having a single new component, the drag brake, that performs functions which in other satellite designs are performed by a multiplicity of devices, each of which must work sequentially for a safe return. In addition to the obvious reliability advantage of the multiple use of this single component, a substantial saving in weight results.

Let us look further into the drag brake satellite design and performance, narrow our discussion by omitting problems of human engineering such as cabin ecology, details of emergency escape, and communications, and treat only those aspects which differ from other re-entry vehicle approaches.

I. General Description

A normal flight of a drag brake satellite will pursue the following chronological sequence. The human passenger enters the capsule through a central hatch at the top, while the drag brake is in the open position on its preflight stand. After the capsule hatch is secured, the drag brake will be folded into the launch configuration with the ribs vertical around the capsule and the skin folded inboard between the ribs (Fig. 3). A nose cover protects the folded ribs and the skin from the high dynamic pressures during launch. The entire vehicle is mounted onto the booster with a short transition section.

The structure is designed to withstand the aerodynamic heating, accelerations, and vibrations of the launch phase in the closed position. At staging, when the vehicle is essentially out of the dense part of the atmosphere, the nose cover is jettisoned and the booster vehicle accelerates the satellite until the proper velocity vector has been achieved. Then the satellite and booster are separated gently and the drag brake vehicle starts its journey in space. The satellite will be launched into a 105 nautical mile, near-circular orbit with a period of about $1^{1}/_{2}$ hours. This orbit allows a lifetime of two days in the low drag (or closed) position of the drag brake and also allows emergency recovery in approximately two hours.

At 105 nautical miles the small but distinct amount of residual atmosphere exerts aerodynamic forces on the vehicle. Because of the large extended structure of the brake, these forces are adequate for orientating the vehicle in orbit. the residual small amplitude oscillation having a period of several minutes. This period is short enough to maintain parallel flight path orientation to the Earth's surface throughout the orbital period of 90 minutes.

For a drag brake satellite, re-entry is made up of a number of distinct phases. First, the satellite descends from the orbital altitude to the region from which a controlled re-entry is made. After preselection of a desired landing point. a command can be issued either from a ground station or from within the capsule to start the controlled landing approach. The landing point control system now takes over. It automatically modulates the drag by a factor of 20, continuously varying the drag brake between the trail and full open positions, duplicating a desired drag-time history. This time program of acceleration is used to achieve the proper altitude and velocity vector at a point approximately 1000 miles from the landing, the so-called re-entry point.

From this point on we are assured of landing within ± 150 miles of the preselected landing point along the orbital track. No further functions have to be initiated or controlled. We merely depend upon the integrity of the static structure, locked in position, to withstand the dynamic pressure and skin temperature for which it is conservatively designed and ground tested.

During re-entry the skin temperature on the drag brake is above 1,000° for only $2\frac{1}{2}$ minutes during which time it will reach a momentary peak of 1200° F. Present day jet engine tail pipe materials have excellent strength properties at

← NOSE CONE

← DRAG BRAKE

← TRANSITION SECTION

← BOOSTER

Fig. 3. Schematic of drag brake vehicle in launch position. Figure shows photograph of folded drag brake model positioned for mounting on the transition section and covered by the protective nose cone

these temperatures for much longer periods of time. The peak temperature on the capsule skin is higher because only the front side is free to radiate. It reaches a maximum equilibrium temperature of less than 1500° F at the stagnation point, and it is above 1,000° for only about 4 minutes. The insulation needed around the capsule to prevent heat losses to space in orbit assures adequate thermal protection for the capsule and payload during re-entry.

A maximum deceleration of 8.2 g's occurs about 1 minute after peak heating. At this time the capsule skin temperature is already reduced to 1200° F and the drag brake skin temperature is below 1000° F. The deceleration remains above 4 g's for only 1 minute. As the vehicle structure cools, it decelerates and approaches near its terminal velocity at 200,000 ft. At this altitude terminal velocity is at a Mach number of about 2.0. The vehicle continues its descent with decreasing terminal velocity and achieves a final landing velocity of 46 ft./sec. A summary of the re-entry conditions is shown in Fig. 4. During the terminal descent, which

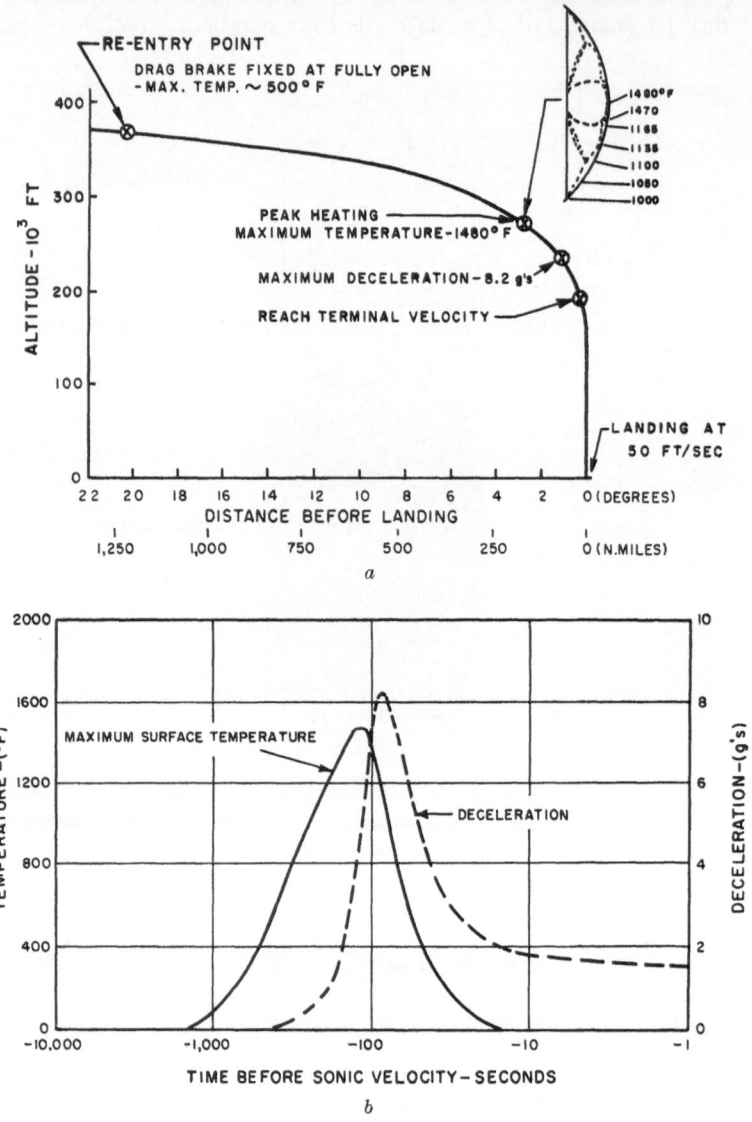

Fig. 4. Re-entry conditions. Upper figure shows relative position of characteristic phenomena along a typical re-entry trajectory. Lower figure indicates time relation between heating and loading during re-entry

takes approximately 15 minutes, the large metallic surface of the drag brake is an excellent radar target for locating and tracking the vehicle during the landing.

The complete re-entry requires only the proper operation of a simple mechanical folding structure and a control system to satisfy all the necessary functions of recovering a satellite vehicle. It accomplishes orientation in orbit, smooth deceleration, dissipation of atmospheric heating, and a safe landing on earth and in addition, due to its simplicity, represents a light weight for launch. The combination of these functions into a single structure meets the most important requirement for manned re-entry—reliability.

II. Detailed Description

A. Capsule

In addition to being an enclosure for the payload the function of the capsule is to provide the structural attachment points for the drag brake and its actuators (see Fig. 5). In the present preliminary design the basic capsule consists of a nearly

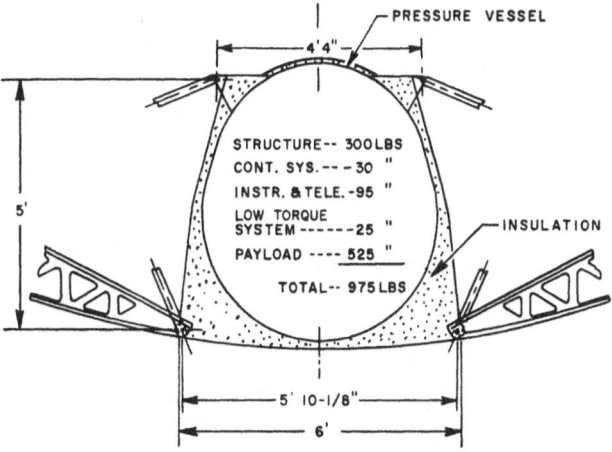

Fig. 5. Capsule schematic. Drawing indicates major dimensions, support and hinge points for the drag brake and general configuration of the payload cavity in the capsule

spherical pressure vessel in which the environment required by the passenger can be maintained. Around this pressure vessel is a layer of light weight insulation for temperature control. Over the layer of insulation is a thin metallic radiation shield whose surface is suitably coated to achieve the desired capsule temperature while the satellite is in orbit. The radiation shield also serves the function of dissipating the re-entry heating from the capsule. Integrated into the basic capsule is a structure which connects and supports the upper and lower structural rings on which the drag brake structural members are pivoted. For the present preliminary design, the weight of the capsule and the structural attachments is about 21% of the vehicle re-entry weight.

B. Drag Brake

The very low maximum dynamic pressure encountered by the drag brake re-entry vehicle permits the use of very light weight structures. (See Fig. 6.) The large size required of the fully opened drag brake dictates the use of thin materials in order to conserve weight. A similar requirement exists in the design of personnel parachutes but for the drag brake it is significantly changed in three respects:

1. The folding structure must have a reasonable strength weight ratio at re-entry temperatures.

2. Small diameter tension members cannot be used to suspend a payload below the principal drag area, or canopy, due to the heating problem and this precludes a conventional parachute type canopy held open by dynamic pressure.

3. Large porosity of the principal drag area seriously aggravates the heating problem during re-entry.

There are four principal components to the drag brake.

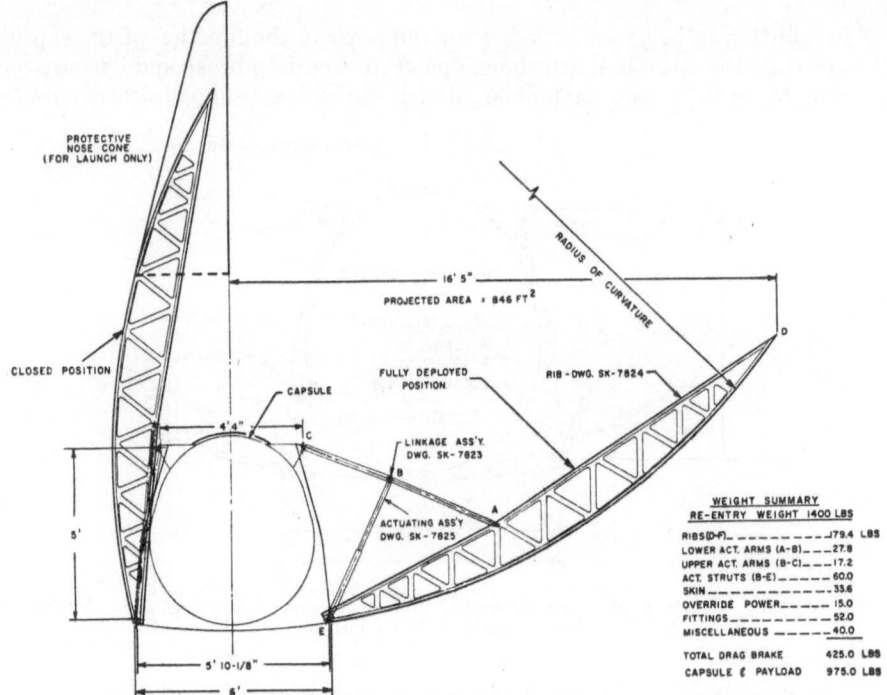

Fig. 6. Drag brake. Drawing shows drag brake rib, actuation linkage and lists typical weights for the various drag brake components

1. *Skin:* The skin performs the dual function of providing a large area to produce the high drag and of radiating the energy produced by aerodynamic heating. A thin skin of high emissivity can radiate energy in both direction, thereby achieving relatively low equilibrium temperatures. The drag brake skin is a fine wire mesh (400×400) attached to the ribs to provide tension strength for transferring the dynamic pressure loads to the ribs. In order to reduce the porosity of the skin strips of $1/2$ mil stainless steel are placed over the wire mesh, each one overlapping the adjacent one like shingles on a roof. The shingle strips are attached at each end to the ribs. They are not intended primarily as structural members, but they are capable of carrying the loads just as well as the primary tension structure, the wire mesh. Due to their higher modulus, the strips will assume the load first and pull taut. They are restrained from fluttering by being held against the wire mesh by the dynamic pressure.

2. *Ribs:* The ribs perform the function of concentrating the skin loads which are then carried to the capsule as concentrated compression loads through the lower and upper arms. They represent by far the largest fraction of the drag brake weight and consequently have received most attention in design. The ribs

are designed to minimum weight as bending members, similar to aircraft wing spars. They are made of 12 to 20 thousands stainless steel sheet with stamped flanges and cap strips and beaded diagonal members. A deep section rib is used to conserve weight.

3. *The Arms:* The lower and upper arms, pivoted for rotation in only one direction at the elbows, carry the loads from the ribs to the brackets around the top

Fig. 7. Full scale heating test facility. The photograph at the top shows the existing hot gas generator which uses six turbo jet engines installed as shown. The drawing below indicates how the existing facility is fitted with a test section shroud for making heating and load tests on the full scale drag brake

of the capsule. The arms are loaded as columns by concentrated compression loads. They are made of 2 inch diameter thin-walled stainless steel tube.

4. *Pneumatic Actuator Struts:* The actuator struts provide the drag modulation in orbit by means of a positive, two-way, push-pull force between the elbows of the arms and the fittings around the bottom of the capsule. For final deployment at the re-entry point, the actuator struts push the upper and lower arms over center into a locked position. It should be noted that the drag brake will operate satis-

factorily with only a small number of the 24 actuators operating. This multiplicity greatly improves the reliability of the actuation system.

For emergency one-shot opening the pneumatic actuator struts are equipped with a squid fired propellant charge which produces the necessary pressure to actuate the struts.

A major feature of the drag brake is that it can, in fact, be tested in full scale on the ground. The low maximum temperatures and loads encountered by a drag brake vehicle during re-entry from a satellite orbit, make full scale ground tests possible with existing test equipment. The exhaust gases from an aviation gas turbine engine provide a satisfactory medium in which full scale tests can be made. An existing facility large enough for this job is the All American Engineering Company's "Turbo-cat," shown in Fig. 7. Fitted with an appropriate test section shroud, its six turbo jet engines can generate gas at the correct temperature and pressure for statically testing the drag brake structure under elevated temperatures and loads simultaneously. Programmed controls on the engines make it possible to simulate the temperature and load variation encountered along the re-entry trajectory.

C. Landing Point Control System

The simplest sequence for a safe return to Earth by a drag brake equipped satellite is to actuate the drag brake to the full open and locked position. In this case the landing point depends upon the altitude, density and density variation and the accuracy is not very good. By introducing a simple control system for closing and opening the drag brake as required, the landing point can be made predictable to within ± 150 nautical miles along the earth track. Wind drift can cause a maximum of ± 20 miles error in a transverse direction.

The operation of a variable drag landing point control system depends on the fact that the descent rate or lifetime of decaying satellite orbits can be changed significantly by changes in the vehicle ballistic parameter, $W/C_D A$. For instance, from an altitude of about 90 nautical miles one can re-enter in as many as 15 orbits or in less than one orbit, depending on the drag brake position for a vehicle whose $W/C_D A$ can be varied from 1.5 to 30. This is shown in Fig. 8. It can be shown that a large amount of control over the range is available by drag variation. In fact, in the geometric mean drag brake position for a 20:1 drag variation, $W/C_D A = 7.0$ and at only 3000 nautical miles from the landing point, one can still alter the landing point by approximately plus 2000 or minus 1500 nautical miles. This effect is shown on Fig. 9. Drag variation is a very effective means of adjusting the range of a re-entering satellite.

A detailed control system philosophy which operates on the above stated principle to achieve a preselected landing point has been developed. The system depends on two measurements, time and acceleration, and the fact that the vehicle is initially inserted into a nearly circular orbit. The time serves to locate the vehicle with respect to the landing point. As long as the satellite is still in a near circular orbit, the orbital period is essentially invariant. The small ellipticities due to drag and drag variations make very small uncertainties in the location from time alone. Thus timing done from either the insertion point or from a later fix locates the satellite accurately. This relationship between time and location holds until the onset of large deceleration forces.

A nominal drag variation with time can be established for the controlled drag phase of an ideal re-entry. This nominal drag variation is stored in the control system. A body-mounted accelerometer measures the drag and compares it to the nominal value. The control system can then sense whether the drag is correct, too high, or too low and can produce signals which vary the drag brake position so as to eliminate the difference between the nominal and measured drag values.

This control system is capable of correcting the effect of ellipticity, altitude error at the start of the controlled phase, error in our knowledge of the upper atmosphere, and density variations of either the latitude or diurnal type. The result of an error analysis has fixed a density range from which the control system is able to achieve a landing point accuracy of ± 150 nautical miles in the orbital plane. This ability to achieve the proper landing point from a range of initial conditions is indicated by the "landing slot" shown in Fig. 10. During controlled phase the satellite flies within the "landing corridor," correcting the drag brake position in accordance with the control system dictates, until it achieves the proper velocity vector at the "re-entry point."

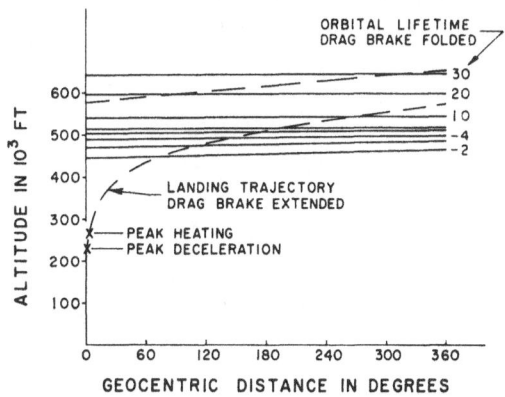

Fig. 8. Landing point control. The solid lines indicate the orbital history for the drag brake folded. Fully opening the drag brake increases the rate of descent and decreases the lifetime as depicted by the dashed line

From the re-entry point to final touchdown, no further functions are required. At the re-entry point the nose of the capsule has heated up to a temperature of about 500° F, and the vehicle is about 1000 nautical miles from the final landing point. The linear relationship between distance and time has been accurate to ± 15 nautical miles to this point.

In order to initiate a re-entry at a preselected landing point from any initial altitude, the control system works as follows: The first task is to de-

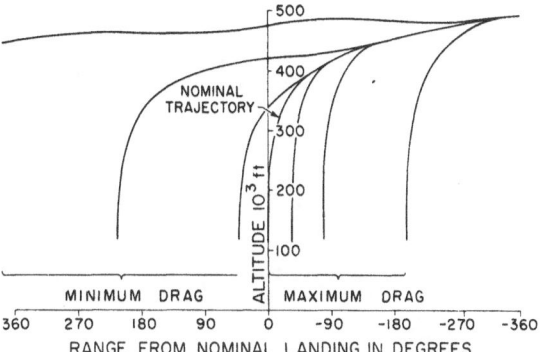

Fig. 9. Range control for variable drag recovery from a circular orbit. Nominal or ballistic drag parameter $W/C_D A = 7$ lbs/ft². Maximum drag is 20 times minimum drag

termine the relative position of the satellite at the time a landing is desired to the so-called landing slot. Because the landing slot is defined as a range of decelerations (at given drag brake positions) this is easily accomplished by the body-mounted accelerometer. A range of $1—10^{-4}g$'s is sufficient for this instrument. The drag brake vehicle must achieve the value of deceleration at the landing slot by adjusting the descent rate by opening or closing the drag brake. When the proper value of deceleration has been achieved the drag brake is closed and the descent rate is kept to a minimum until the proper position for initiation of automatic control is achieved. This proper position is defined by the landing point and the length of the landing corridor, i.e., it is a known time from any fixed point on the orbit. Note that this is the only information about the orbit which is necessary to effect a landing because the altitude (or better the density) is determined internally. The initial insertion by the booster vehicle guidance system is sufficiently accurate so that one can predict the orbital period very closely, which is equivalent to a very small uncertainty per orbit. Thus, it is only necessary to improve the insertion information by tracking after a number of orbits. This task is easily accomplished by the present Minitrack network without requiring the construction of any further tracking capability.

Once the automatic control has been actuated the satellite flies down the landing corridor, continually adjusting the drag brake position in such a manner

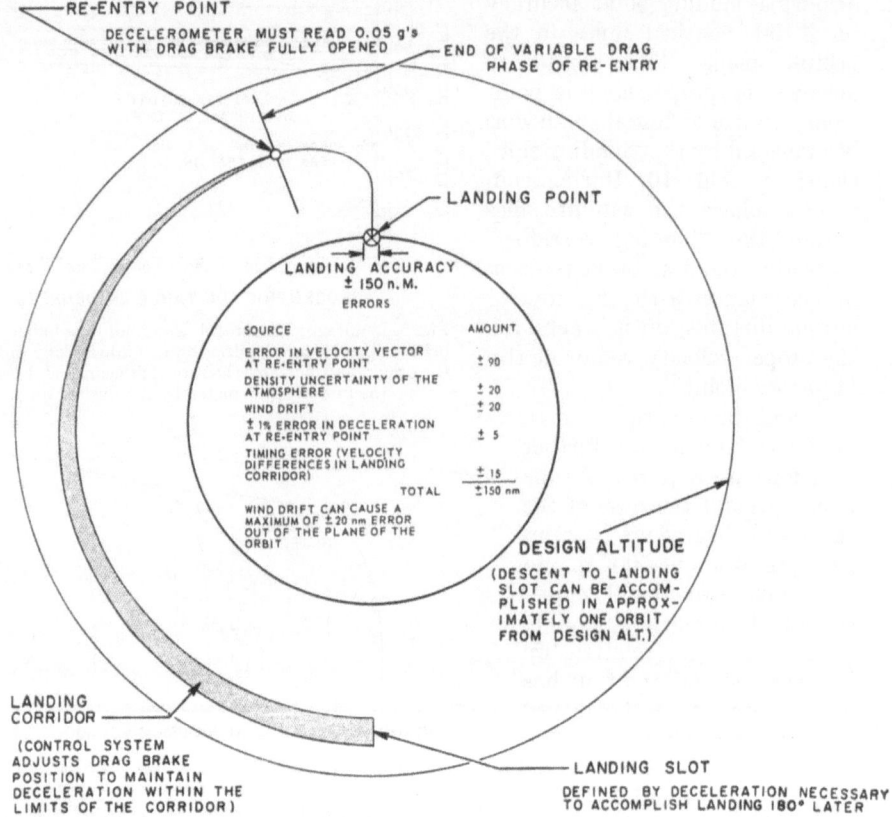

Fig. 10. Schematic of landing sequence

as to make the velocity vector approach the nominal one. At the re-entry point this process has been completed and the drag brake is opened fully and locked. From this point on, the landing error starts to accumulate. The major contributions to this error are density uncertainties and density variations below the altitude of the re-entry point, the error in the velocity vector at this point, the error in altitude at the re-entry point, and the wind drift.

Because of the high drag of the open drag brake, $W/C_D A$ 1.5, peak heating occurs at a very high altitude, and consequently the heat transfer rates and the radiative equilibrium temperatures are relatively low. At peak heating, reached at 270,000 ft. altitude, the deceleration is about 4 g. After peak heating the deceleration continues to increase and reaches 8.2 g at 233,000 ft. The time above 8 g is approximately 20 seconds. This time is well within the established limits of human endurance.

The relative time of occurrence of the peaks of heating and deceleration point out an interesting future possibility for drag brake vehicles. If at the instant of peak heating (i.e., 4 g) the acceleration is held constant by retracting the drag brake slightly, then the peak re-entry deceleration can be reduced to 4 g without any increase in maximum temperature. The peak heating temperature would merely be held for a somewhat longer time, increasing the total heat input. This does not require any additional weight for a radiating re-entry vehicle

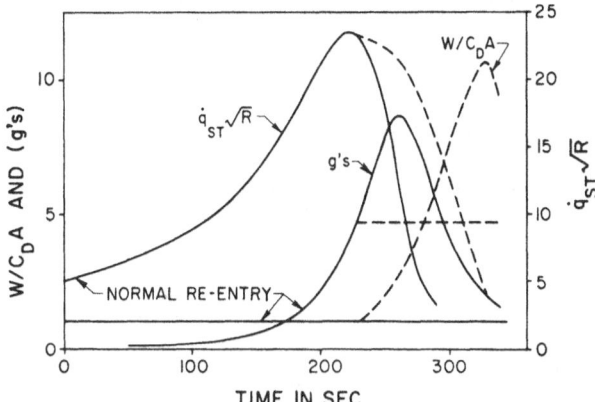

Fig. 11. Effect of drag variation on re-entry deceleration. Solid curves are for a normal re-entry. Dashed curves show $W/C_D A$ variation required to limit maximum deceleration to 4.6 g's and indicate also the resultant heat transfer rate

structure but is does require the actuation of a hot structure. Fig. 11 shows the re-entry decelerations, temperatures and corresponding $W/C_D A$ variation effected by this maneuver.

D. Provisions for Safe Return Under Emergency Conditions

There are several regions in a normal launch profile in which the possibility of booster failure must be considered. If booster failure occurs at these conditions then special conditions are imposed on the re-entry vehicle, if it is to be returned safely to the Earth. The first and most likely to be called-upon condition would be a ground launch escape. In the event of booster failure on the pad or shortly thereafter, solid propellant rockets burning for only one second, mounted in the protective nose cone, are fired and pull the satellite vehicle away from the booster. Aerodynamic stability is supplied during this phase. After a coast period the vehicle reaches its apogee where the rocket casings, nose cone, and transition piece are jettisoned and the drag brake is fully deployed by the emergency system. The vehicle then descends at terminal velocity and lands in the vicinity of the launch site.

The next most likely requirement for escape will be in the region of maximum dynamic pressure during launch. A similar sequence is followed for this emergency situation, however, the separation distance between the vehicle and booster escape rocket burnout is less. Since the booster flight path angle at this condition has departed considerably from the vertical, the dynamic pressure after the vehicle has decelerated and coasted to apogee is still aproximately 35 lbs./sq.ft. Since 35 lbs./sq.ft. is a greater load on the drag brake actuating mechanism than normal re-entry, this condition designs the actuator struts and pressure system.

The last emergency condition to be considered is the region of suborbital escape. After the dynamic pressure is reduced to a negligible amount but before the vehicle is in orbit, a massive escape system such as previously described will no longer be necessary. Therefore, about 80 % of the escape system weight can be saved by dropping it at staging rather than taking it into orbit. Once out of the dense atmosphere, re-entry is essentially the same as from orbital flight. However, more severe re-entry conditions both in maximum temperature and maximum deceleration exist if re-entry must be made just prior to achieving orbital velocity. Fig. 12 shows the maximum re-entry conditions as a function of booster velocity at emergency separation for launch into a 105 nautical mile

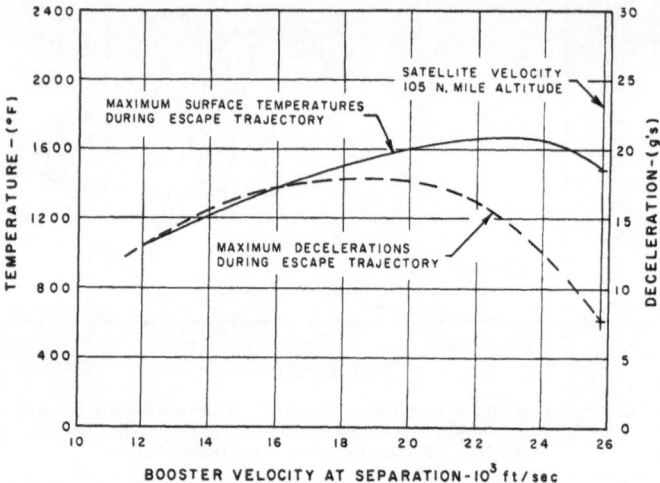

Fig. 12. Re-entry after emergency escape. Curves indicate the maximum temperature and deceleration during re-entry after escape as a function of booster velocity at separation along a launch trajectory into a 105 n.m. altitude orbit

altitude orbit. Both the maximum temperature and deceleration can be significantly higher than encountered during normal re-entry. For this reason the structure has been designed for re-entry from the most severe suborbital condition with a small safety factor over the yield stress. This means that under normal re-entry conditions the safety factor will be over three times the value designed for a suborbital escape.

E. Stability

The aerodynamic stability of the drag brake vehicle has been considered in the several flight regimes of the mission. After insertion into orbit, any residual oscillation induced by the vehicle booster combination is reduced to a minimum value. Below these tumbling rates the aerodynamic forces on the folded drag brake are greater than the inertial forces and the aerodynamic restoring moments will preserve the attitude in orbit. The natural period of this oscillation is about 3 minutes at the orbital altitude. Thus the drag brake will continue to re-align itself as the angle of its axis changes with respect to the Earth's axis throughout the orbit. In hypersonic, supersonic, and transonic flow, the large radius of curvature with the relatively low (forward) center of gravity gives a static stability margin many times that of a stable ballistic missile nose cone. Stability calculations have been made which indicate no tendency of the oscillation to grow after peak deceleration, such as exhibited by marginal re-entry shapes. Scale model drop tests from a 100 ft. tower have been made to check subsonic stability. These tests indicate static and dynamic stability, provided that the center of gravity is kept well below the capsule center.

III. Comparison with Retro-Rocket System

We wish now to compare the drag brake satellite recovery system with a much publicized competitive system, i.e., the retro-rocket system. This comparison will be made on the basis of weight, complexity and reliability, as well as the component functions which are important in reaching the desired goal with the highest confidence level.

Lifting re-entry vehicles, which are characteristic of today's aircraft or extrapolations thereof, have not been included in this comparison. They are more

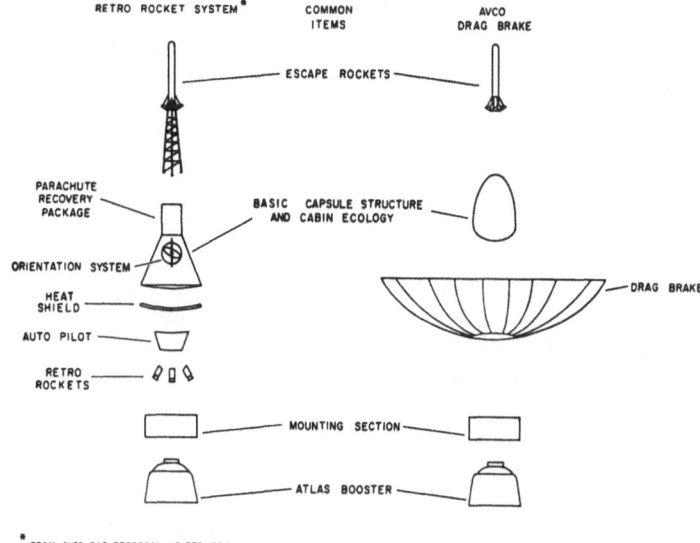

* FROM AVCO RAD PROPOSAL, 10 DEC 1958

Fig. 13. Functional comparison

complex and require more development and usually have a high ratio of weight in orbit to payload weight. This ratio has been estimated to be about four by proponents of lifting re-entry vehicles. Until the need and utility of lift during re-entry from a satellite orbit is justified in the light of this weight penalty, the role of lifting vehicles remains uncertain.

In this comparison we presume an automatic flight to the extent that the recovery could be made successfully even in the event of complete incapacitation of the occupant. Certainly this will be a necessary ground rule for the first series of manned orbital flights and will remain so until man's capabilities in a space environment are definitely and satisfactorily established. Even for more advanced systems of the future there seems to be a strong argument for building a fail-safe system, which includes the essential automatic control features discussed below.

Let us first make a functional comparison of the two manned satellite systems under consideration. For ease of comparison we have separated out the common components as shown in Fig. 13. In addition to the booster, the mounting section, basic capsule structure. and cabin ecology, the first stage escape system rockets are included in the list of common items. Although the escape system details may differ from one manned satellite system to another, the problems which must be solved are similar and hence the fundamental features will be common to all manned systems using the same booster. Separate provision for escape during the second stage firing. when the aerodynamic pressure is nearly non-existent and the booster motor can be shut down, must be provided as will be pointed out below.

Shown along the margins of Fig. 13 are the additional components which must provide for the two systems in order to carry out the functions necessary for a successful flight in orbit and return to the Earth's surface. For the retro-rocket system we mention first the retro-rockets. These rockets serve two functions in that in addition to providing the necessary impulse for recovering the satellite from orbit, they can be used for escape in the event of a booster failure during the second stage firing. The remaining components listed, each with a single function, are necessary for a successful recovery of the satellite.

Once in orbit and free of the booster any undesired residual angular momentum must be removed by the orientation system. After the desired orbital flight is completed but before the retro-rockets are fired, the orientation system must determine a direction in space either by memory or by making several external observations and orient the satellite and its attached retro-rockets. The resultant orientation must be such that retro-rocket impulse will produce the desired orbit perturbation and hence yield the correct range control. To insure correct alignment of the impulse vector it is necessary to hold the satellite in a fixed direction while the retro-rockets are firing. This requires an autopilot type of system.

A heat shield and associated insulation must be provided to protect the occupant and satellite structure from the intense heating encountered during re-entry into the Earth's atmosphere along the resultant ballistic trajectory. The high re-entry radiation equilibrium temperatures characteristic of retro-rocket satellite designs, i.e., ballistic drag parameter $W/C_D A = 30$—50 lb./ft.^{2}, preclude the use of any light weight radiation shield on the forward side of the vehicle. Either a metal heat sink or an ablation type heat shield must be used. Should a metal heat sink be used it is desirable to jettison it after the re-entry heating is over so that the stored heat in the shield will not continue to flow into the inhabitable capsule during the slow terminal descent. An ablation type heat shield exhibits several advantages over the heat sink in that its weight would be only about one-half that of the best heat sink, it stores less heat and hence would not have to be jettisoned, and its inherent lower thermal conductivity would reduce the insulation weight requirements (see Fig. 1).

Finally, to reduce the impact velocity at sea level to an acceptable level, a sequence of recovery parachutes must be deployed at the altitude where the terminal velocity is low enough for reasonable parachute opening loads.

The drag brake manned satellite system has fewer additional components over the common items than the retro-rocket system since it makes multiple use of its major component, the drag brake. Starting first with escape during the launch phase of the flight, a second stage escape system is provided to perform separation and escape from the booster in the event of failure after the major escape rockets have been jettisoned at booster staging. However, for a successful launch these rockets are not needed and are separated from the satellite when in orbit.

As discussed earlier the drag brake, in addition to providing a preferred orientation in orbit, serves all the functions necessary for successfully recovering a satellite from orbit. Programmed actuation of the drag brake enables recovery from orbit and effects landing point control, thus eliminating the need for retro-rockets and autopilot. The characteristic low ballistic drag parameter, $W/C_D A$, of a drag brake vehicle permits radiation cooling with readily available materials during re-entry heating. The low terminal velocity at sea level permits landing at ground level without the use of recovery parachutes. Hence, the drag brake, a single component, serves the function of four separate sub-systems of the retro-rocket satellite recovery system.

The five different sub-systems of the retro-rocket system, in addition to the common items, must function properly in a definite sequence in order to have a successful flight. For the drag brake system, on the other hand, only the drag brake and the initial orientation system must operate for a normal flight. A third sub-system, the second stage escape system, is used only in the event of an aborted

Common Items

Basic structure	167 lbs.
Inhabitable capsule	135
Escape system	695
Environmental equipment	70
Pilot support	46
Cockpit	17
Navigation aids	41
Communications equipment	212.0
Instruments	15
Recording equipment	24
Power supply	157
Programmer	9
Man and suit	210
	1798 lbs.

Retro-rocket System[1]		Avco Drag Brake	
1. Added structure	133 lbs.	1. 0	
2. Beryllium heat shield	398	2. 0	
3. Automatic orientation systems	137	3. Low torque system	25.5
Power	66	4. 0	
		5. 0	
4. Retro-rocket	165	6. Drag brake	425
5. Recovery systems	207	7. Second stage	
6. 0		Escape	30
7. 0		8. Drag brake control	
8. 0		system	20
	1106 lbs.		500 lbs.

[1] From Avco RAD proposal Dec. 10, 1958

Launch weight	2902 lbs.	2297 lbs.
Orbital weight	2207	1527
Re-entry weight	1914	1399

Fig. 14. Weight comparison

launch. This multiple use of a single component and the resultant simplicity suggests that the drag brake recoverable satellite will have a higher reliability in a shorter development time and can be developed for a smaller cost than the corresponding retro-rocket recovery system.

From the operational standpoint the drag brake system can claim some additional advantages over the retro-rocket system. As pointed out in the previous section, the drag brake effects the quoted landing point accuracy without detailed information about the orbit. In fact, an entire flight can be made without communication with the satellite after it is successfully launched into orbit. Without this communication landing point errors which are related to launching errors of a typical booster accumulate at a maximum rate of about 10 nautical miles per orbit. A simple one-way communication system which resets a timing device in the satellite once each orbit as it crosses a reference great circle on earth, can eliminate or minimize the above-mentioned landing point errors.

On the other hand, the landing point control of a retro-rocket system is very sensitive to uncertainties of the orbital altitude. For example, at an orbital altitude of 100 nautical miles the re-entry range sensitivity is of the order of 20 to 40 nautical miles per nautical mile uncertainty in orbital altitude. In order to execute an accurate recovery from orbit, it is clear the satellite control system must be provided with accurate orbit information so it can compensate for these uncertainties. Thus, the satellite must be tracked from the ground and the orbit accurately computed, and the pertinent information must be transmitted to the satellite. A short duration flight of one or two orbits, as the first manned flights will probably be, requires, therefore, the services of an expansive network of tracking and communication stations around the Earth so the orbit can be quickly and accurately determined. To obtain accurate landing point on earth the complete retro-rocket satellite recovery system is, by comparison to the drag brake system, complicated operationally and requires large amounts of equipment and manpower.

In addition to the obvious reliability advantage of a simpler system, substantial savings in weight result for the drag brake. Shown in Fig. 14 is a weight comparison for the two systems described above. These weights are based on a detailed engineering design effort and are representative of current state of the art of re-entry vehicle design. It is to be noted that for the same payload the drag brake vehicle shows a weight advantage of about 600 lbs over the retro-rocket system. Since the useful payload weight is the same in both cases this weight difference is exactly the weight of the equipment needed to recover the payload by the two methods. In this case the drag brake recovery equipment weight is less than half that of the retro-rocket vehicle. The corresponding ratios of weight in orbit to payload weight are 1.39 for the drag brake vehicle and 2.0 for the retro-rocket vehicle. The weight advantage of the drag brake vehicle can be used either for increasing the already generous design safety factor or for an increase in payload. Weight will always be a prime consideration in the design of space vehicles, at least until a major break-through in launching propulsion systems is achieved. The desirability and utility of a weight advantage for any satellite system will always be clear and distinct.

Conclusion

This paper presents a unique solution to the problem of controlled recovery of a manned satellite from orbit. The approach which has been used is an attempt to answer the specific question of how to make the best manned satellite system, starting from first principles rather than by making extrapolation according to the disciplines of airplane or missile nose cone technology. The result of this three-year effort has been the invention of a simple device called the drag brake, which could achieve early realization of manned space flight.

Acknowledgement

The authors wish to acknowledge the contributions of their associates at the Avco-Everett Research Laboratory who over the past several years have participated in the studies which led to the development of the drag brake concept. In the development of the engineering details of the concept, the Avco-Everett Research Laboratory has been aided by several of the other divisions of Avco Corporation; in particular, Lycoming Division. In addition the mechanical engineering and structural design of the drag brake was done by All American Engineering Co.

References

1. J. A. FAY and F. R. RIDDELL, Theory of Stagnation Point Heat Transfer in Dissociated Air. J. Aeronaut. Sci. **25**, 73 (1958).
2. N. H. KEMP and F. R. RIDDELL, Heat Transfer to Satellite Vehicles Re-entering the Atmosphere. Jet Propulsion **27**, 132 (1957).
3. P. H. ROSE and W. I. STARK, Stagnation Point Heat Transfer Measurements in Dissociated Air. J. Aeronaut. Sci. **25**, 86 (1958).
4. P. H. ROSE, R. F. PROBSTEIN and M. C. ADAMS, Turbulent Heat Transfer through a Highly Cooled Partially Dissociated Boundary Layer. J. Aeronaut. Sci. **25**, 751 (1958).
5. N. H. KEMP, P. H. ROSE and R. W. DETRA, Laminar Heat Transfer around Blunt Bodies in Dissociated Air. J. Aeronaut. Space Sciences **26**, 421 (1959).
6. B. KIVEL and K. BAILEY, Tables of Radiation from High Temperature Air. Avco-Everett Research Laboratory Research Report 21. 1, December 1957.
7. F. R. RIDDELL and J. D. TEARE, The Difference Between Satellite and Ballistic Missile Re-entry Problems. October 1958. Paper presented at 2nd Annual AFOSR-Astronautical Symposium, Denver, Colorado, April 1958.
8. H. A. BETHE and M. C. ADAMS, Theory for the Ablation of Glassy Materials. J. Aeronaut. Space Sciences **26**, 321 (1959).
9. S. FELDMAN, Hypersonic Gas Dynamic Charts for Equilibrium Air. Avco-Everett Research Laboratory Research Report 40. January 1957.
10. J. KECK, J. CAMM, B. KIVEL and T. WENTINK JR., Radiation from Hot Air. Ann. Physics **7**, 1 (1959).
11. S. GEORGIEV, H. HIDALGO and M. C. ADAMS, On Ablation for the Recovery of Satellites. Heat Transfer and Fluid Mechanics Inst. Preprints of Papers, Stanford, Cal.: Stanford University Press, 1959.
12. R. W. DETRA, F. R. RIDELL and P. H. ROSE, Controlled Recovery of Non-lifting Satellites. May 1959. Paper presented at American Rocket Society Controllable Satellites Conference, Cambridge, Massachusetts, April 30—May 1, 1959. ARS Rep. 784—59.
13. W. E. POWERS, S. GEORGIEV and M. C. ADAMS, An Experimental and Theoretical Study of Quartz Ablation at the Stagnation Point. 13. Avco-Everett Research Laboratory Research Report 57. June 1959.
14. R. W. DETRA, N. H. KEMP and F. R. RIDDELL, Addendum to 'Heat Transfer to Satellite Vehicles Re-entering the Atmosphere.' Jet Propulsion **27**, 1256 (1957).
15. F. R. RIDDELL and R. W. DETRA, Returning Alive from Space. Paper presented at the American Astronautical Society Meeting, January 1958.
16. M. C. ADAMS and R. F. PROBSTEIN, On the Validity of Continuum Theory for Satellite and Hypersonic Flight Problems at High Altitudes. Jet Propulsion **28**, 999 (1958).
17. J. E. HAYES, P. H. ROSE and W. E. VANDER VELDE, Analytical Study of a Drag Brake Control System for Hypersonic Vehicles. WADC Technical Note to be published.
18. R. C. HAKES, Some Fundamental Problems Associated with Injecting, Orbiting, and Recovering a Man from Orbit. Presented at IAS 27th Annual Meeting, New York, N. Y., January 1959.
19. R. L. PHILLIPS and C. B. COHEN, The Use of Drag Modulation to Reduce Deceleration Loads during Atmospheric Re-entry. Report GM-TR-0165-00352, Space Technology Laboratories, Los Angeles, California, April 9, 1858.

Space Power[1]

By

William W. T. Crane[2]

(With 7 Figures)

(*Received August 27, 1959*)

One of the greatest problems in space flight today is that of providing a reliable source of electrical power for auxiliary systems. Conventional batteries are not feasible for space missions lasting longer than a few days; and, in fact, only two systems developed to date can furnish a long-lived source of electrical energy in a light-weight package. These are solar cells and isotopic power.

As these competitive systems develop, their relative advantages and disadvantages will become apparent; but at present it is still impossible to predict the ultimate number of watts per pound available from each. This paper will describe recent research and test data which indicate so far that a safe, reliable and practical radioisotope-fueled auxiliary power supply can be built using existing technology.

The principles underlying isotopic power are simple. Heat is generated when the particulate and electromagnetic radiations emitted by a sealed quantity of radioisotopes are absorbed in the isotope and in the surrounding containment material. The heat is then partially converted into electricity, using a suitable energy conversion device, with the remainder of the heat being dissipated to the external environment. By using direct conversion systems, it has been possible to produce isotope-fueled generators with no moving parts whatsoever—a tremendous aid to reliability.

Radioisotopes provide self-contained sources of energy in relatively high power densities whose natural decay rates determine, in major part, the lifetime designed into the power plant. These power plants lend themselves especially to remote, unmanned application. Rugged and reliable, they hold strong promise of enabling man to probe the environment of the planets of this solar system and to prepare himself for the conditions to be encountered on his arrival.

Heat generation in isotopic power systems stems from containment of two major types of radioactive materials. The first is the waste material arising from the controlled fission of uranium and the second is alpha-emitting isotopes produced by irradiation of suitable target materials in nuclear reactors. Each type of source material has distinct advantages for specific missions, but for space probes into the new environments where one of the properties to be investigated is the electromagnetic radiation a distinct advantage is found in the use of alpha-emitters as the source of energy. After extensive examination of the alpha-emitters available, it has been determined that only four isotopes have merit as heat sour-

---

[1] This work was performed under contract AT (30-3)-217 to the Aircraft Reactor Branch of the United States Atomic Energy Commission.

[2] Martin Nuclear Division, Baltimore 3, Maryland, U.S.A.

ces. These are Curium-242, Polonium-210, Curium-244 and Plutonium-238. Potential availability and cost, plus the lack of hard gamma radiation, were the deciding factors in the selection of these four isotopes.

Two of these four isotopes, Curium-242 and Polonium-210, are in the short-life, high-power category, while Plutonium-238 has an 86.4-year half-life and will answer the problem of long-lived heat sources. Curium-244 with its 17.9-year half-life will serve as a source of power for this intermediate period of time.

Alpha-emitting isotopes which are feasible to use in power generating devices are listed in Table I.

Table I. *Specific Power of Various Isotopic Power Sources*

Isotope	Fuel Form	Practical Density (grams per cubic centimeter)	Compound (watts per cubic centimeter)
Polonium-210	Po	9.4 (solid)	90.0
Curium-242	Cm·8 Pt	15.6	256.0
Curium-244	Cm	7.0	20.0
Plutonium-238.................	PuC	12.5	6.9
Plutonium-238.................	Pu	16.0	9.3

The fuel forms given in Table I are considered the most practical from a standpoint of fabrication for these power packs and are listed with the densities. The last column gives the specific power or watts per cubic centimeter which can be fabricated today.

In the construction of a radioisotope heat source, two basic biological hazards must be considered. First, there is a direct radiation hazard from the penetrating radiation associated with alpha and beta decay, such as X-radiation from bremsstrahlung, alpha-neutron reactions, and gamma photons from decay products and fuel impurities. Second, there is a potential hazard from the possibility of the radionuclides dispersing into the biosphere and subsequent biological assimilation by humans through inhalation and ingestion.

The use of the alpha-emitters that are considered in Table I minimise the hazard of direct radiation. The amount of direct radiation associated with 100 thermal watts of these isotopes may be seen in Table II.

Table II.
Radiation Associated with 100 Thermal Watts of Various Isotopic Power Sources

Isotope	Fuel Form	Gamma Radiation (1 yard, rem per hour)	Neutron Radiation (1 yard, rem per hour)	Total Radiation (1 yard, rem per hour)
Po-210	Po	0.020	0.0065	0.0265
Cm-242	Cm·8 Pt	0.0043	0.0325	0.0368
Cm-244	Cm	0.0083	0.0045	0.0128
Pu-238	PuC	0.00001	0.00078	0.00079
Pu-238	Pu	0.00001	0.00053	0.00054

These figures are the radiation associated with an unshielded isotope heat source, considering those gamma radiations that will escape from the fuel block. It can readily be seen that the radiation hazard associated with these isotope heat sources is almost non-existent.

The concept of absolute containment is employed on these alpha-emitting isotopes to assure encapsulation of the radioisotope heat source under any conceivable condition, thus avoiding any possibility of biological accumulation to

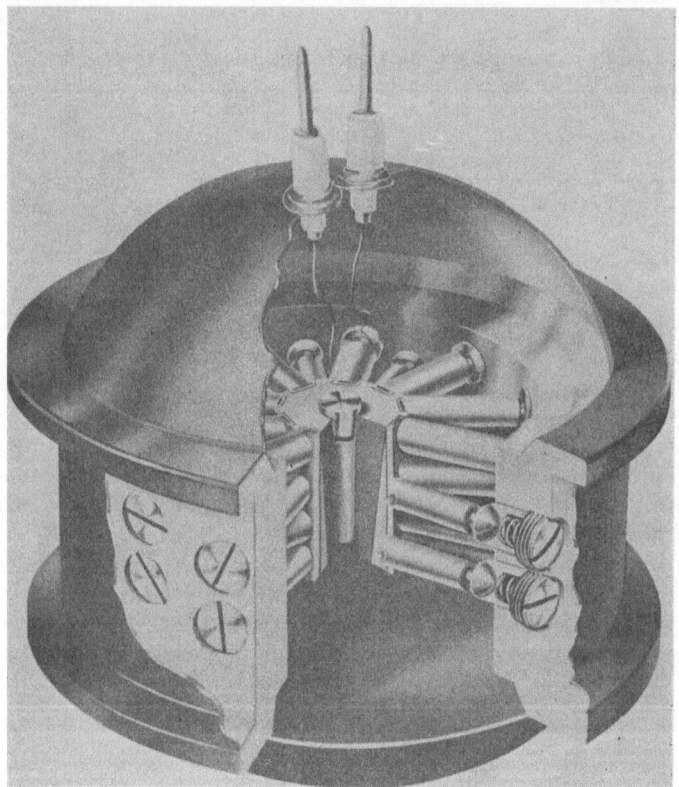

Fig. 1. Artist's concept of SNAP-III

plants or man. The source material is sealed in capsules of high-temperature, high-strength materials to assure safe containment under all anticipated conditions.

Since safety is the most important consideration in the design of an isotopic power heat source, extensive tests have been undertaken to evaluate various heat source designs. Cold simulated radioisotope fuels have been employed in these tests.

These tests have included both low and high velocity impact tests. The contained radioisotope heat source was impacted against targets such as granite, concrete and packed earth, at velocities varying from terminal velocity to three times terminal velocity. All specimens were heated to temperatures equivalent to impact temperature and fired from rocket sleds. All specimens that were fired at speeds equivalent to terminal velocities remained intact, although extensive deformation was encountered on impact with granite.

These same radioisotope heat sources have been subjected to missile failure tests such as exploding mixtures of liquid oxygen and kerosene. Though the specimens were subjected to very high transient temperatures and pressures from the detonation of this mixture they maintained their integrity in all tests. Even when fuel blocks were heated to more than 2000 degrees Fahrenheit and plunged directly into liquid oxygen (at almost 300 degrees below zero), the test specimens failed to support combustion and survived intact.

Additional tests are planned to evaluate the integrity of these heat sources under the high-temperature aerodynamic heating encountered on re-entry to the atmosphere. These will consist of plasma jet tests, using varied amounts of ablative material to determine what is needed to guarantee survival of these packages through the highest temperature that can possibly be encountered on re-entry to earth.

At the end of this extensive testing program we hope to establish clearly that an isotopic power device can be safely utilized in any environment, be it terrestrial, marine or space, with no danger to the people of earth.

A combination of a radioisotope heat source with a static energy conversion device was presented to the public by the United States Atomic Energy Commission in January 1959. This device, which was a "proof of principle", was built by the Nuclear Division of The Martin Company using a Minnesota Mining and Manufacturing

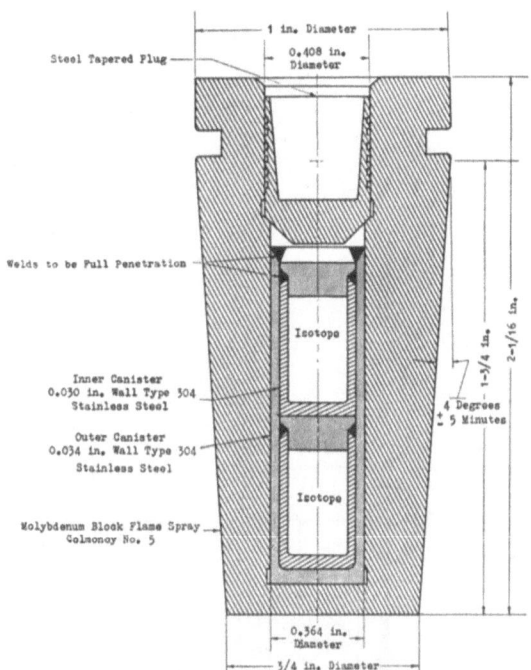

Fig. 2. Internal containment of SNAP-III

Company thermoelectric converter. An artist's concept of this SNAP-III-B (Secondary Nuclear Auxiliary Power) is shown in Fig. 1, and Fig. 2 shows the internal containment.

Polonium-210 was selected as the fuel for the generator because of its high specific alpha and low gamma activities, facilitating ease of handling and demonstration. Furthermore, it was readily available from the Mound Laboratories (operated for the AEC by the Monsanto Chemical Company).

Other isotopes show more promise for certain operational uses, however; and, as a matter of fact, the size and design of the SNAP-III-B is such that Curium-242 could be substituted for Polonium-210 in it with little difficulty. Although Curium-242 would extend the half-life of the unit only slightly (from 138 to 162 days), this isotope has other advantages, some of which have already been cited: Its gamma emission is even lower than that of Polonium-210; its power output per unit of volume in usable form is almost three times as great; (and the fact that it can be used in solid form makes it easier to contain safely at high temperatures than volatile polonium).

Plutonium-238 or Curium-244, of course, would increase the life of an operational unit to decades; but their lower power density would require some increase in the size of the generator.

The Polonium-210 for SNAP-III was supplied in encapsulated form. Each of the two stainless steel cylinders contained about 800 curies of activity, and was closed with a tapered plug. The plug was sealed by heliarc welding with 100% penetration. Each capsule was carefully tested for alpha activity and helium leakage. The two cylinders were inserted into a close-fitting third cannister, which was sealed and tested using the same techniques. The third cannister was enclosed in a molybdenum core which had been flame coated to prevent oxidation and sealed with a stainless steel tapered pin plug under helium pressure. The latter provides a thermal bond between the steel capsule and the molybdenum core. Closure with a pin plug obviates the need for a high-temperature seal which could affect the heliarc welds adversely. Tapering the molybdenum block insures good thermal contact with the hot junction ring of the thermoelectric converter.

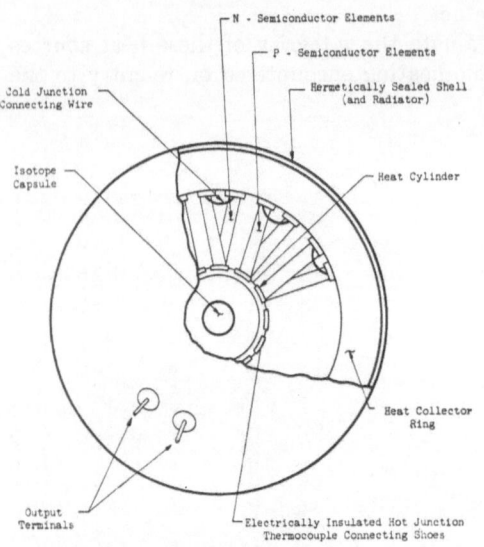

Fig. 3. Sectional view of SNAP-III prototype thermo-electric generator

Situated radially about the contained heat source are 27 couples of doped lead telluride elements supplied by Minnesota Mining and Manufacturing Company. These elements are doped with bismuth or sodium to provide N- and P-type semiconductors, with one N-type and P-type thermo-element defining a couple, and all connected in series. Each of the elements provides a parallel path for heat flow from the heat source to the container-radiator. Fig. 3 shows a top view of the converter.

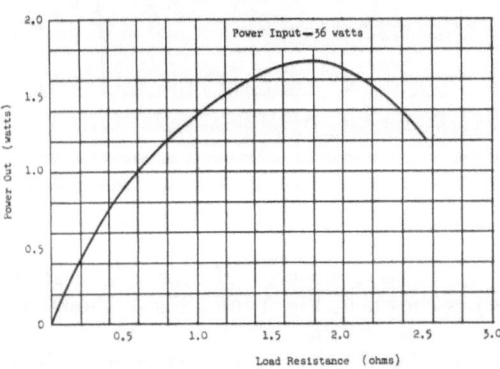

Fig. 4. Power out vs load resistance for fixed power input

Since the thermocouples must be electrically insulated from each other, mica sheet or a ceramic coating is used between the hot junction and the heat source. At the cold junction terminus an alumina coating process, called Martin Hard Coat, provides a thin, electrically insulating thermal path to the container-radiator. Each of these elements is spring-loaded at the cold junction side to insure positive, low resistance electrical contact at the hot and cold junction.

Table III lists many of the important performance characteristics of the SNAP-III-B unit. However, this device was constructed for "proof-of-principle" and little attempt was made to minimize weight. Simple design changes and optimal selection of materials in the container would reduce the weight from the present four pounds to less than three without seriously affecting power output or efficiency.

Table III. *SNAP-III-B Characteristics*

Dimensions	4.75 in. diameter; 5.5 in. high
Weight	4 lb
Source	1700 curies (0.38 **gm**) of Polonium-210
Thermal output	60 w
Electrical output	3.3 w to optimum load of 1.7 ohms
Efficiency	5.5 %
Total electrical power output over 280 days (approximately 2 half-lives)	10,000 w-hr
Conventional battery equivalent (over 2 half-lives)	160 lb of zinc-silver cells
Open-circuit voltage	5.0 v
Closed-circuit voltage	2.5 v to 1.7 ohm load
Thermoelectric elements	lead telluride: P-type doped with bismuth, N-type doped with sodium
Hot junction temperature	496° C
Cold junction temperature	107° C
Copper shell temperature	99° C
Dose rate at shell surface	500 mr/hr
Dose rate at 1 feet	50 mr/hr
Dose rate at 5 feet	1 mr/hr

Note: Performance data are for unit at beginning of first half-life of Polonium-210 source.

Fig. 4, a curve of power output as a function of load resistance on the generator, illustrates some of the interesting characteristics of a thermoelectric device. Though the power output has a fairly wide variation over the entire range of load resistance shown, under a ± 30 percent range of load resistance from a matched load, the power output changes only 15 percent.

It is noteworthy that in this generator the efficiency remains nearly constant with decreasing power output. The figure of merit of the doped lead telluride increases with decreasing temperature and counterbalances the decreasing CARNOT efficiency. The overall efficiency, therefore, tends to remain constant. Fig. 5 illustrates this point.

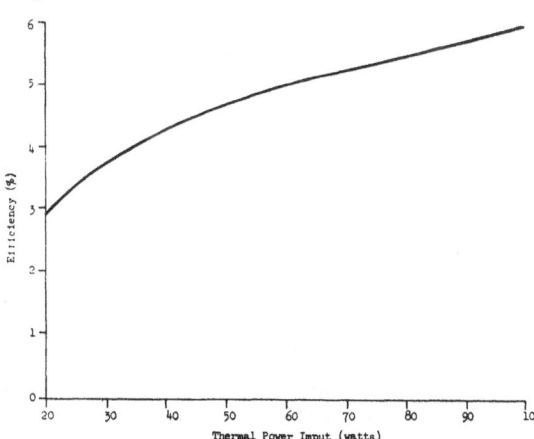

Fig. 5. Extrapolated experimental efficiency of SNAP-III-B vs thermal power input (space ambient conditions)

Fig. 6 shows the variation in power output with lifetime of the existing generator, extrapolated over projected lifetimes. Similar curves are extrapolated for other potential radioisotope fuels.

Constant power output may be obtained over a portion of the half-life of these devices by dumping the excess heat created early in life through a controlled paral-

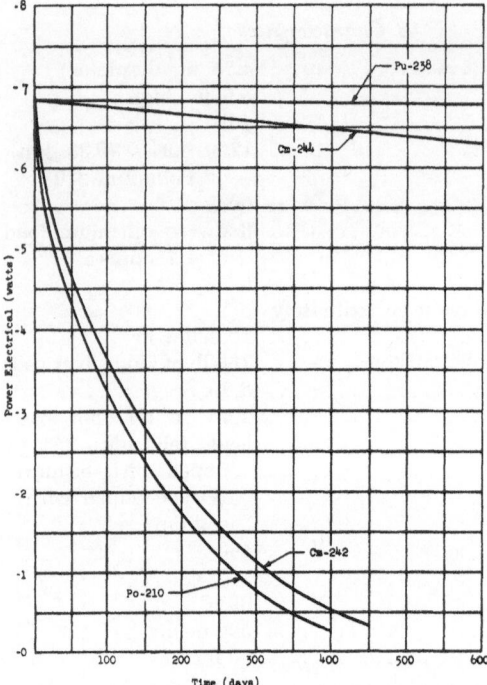

Fig. 6. Isotope electrical power watts versus time

lel heat path. One method under development controls power output by adjusting the gas pressure in an enclosure filled with a porous insulation. It has been determined that the conductivity of such an insulator varies almost linearly with the gas pressure enclosed. The gas pressure is controlled by a valving system requiring no power and which allows gas to escape to space as the cold junction temperature decreases.

A program to establish the ruggedness and reliability of these thermoelectric isotope generators has been in progress for some months. The Martin concept requires that reliability be designed into a product and then this reliability be established by testing. Besides the previously described tests to which the heat source and containment block were subjected, intensive vibration, acceleration and shock tests are being conducted at present on three complete generator systems.

The vibration tests that have been completed at the time of this paper may be seen in Table IV.

Table IV. *Thermoelectric Generator Vibration Tests*

Time (seconds)	Type Signal	G Level (root mean)
0-15	White Gaussian noise (band limit 15-1500 cycles per second)	2
15-30	None	0
30-32	White Gaussian noise	12.5
32-332	White Gaussian noise plus sinusoid sweeps from 15-500 cycles per second	7.05
332-632	White Gaussian noise plus sinusoid sweeps from 500-1500 cycles per second	9.5

These generators have been subjected to these vibrations twice in each of three planes. During the test some lowering of power output was noted and some noise has been observed during testing on the oscilliscope, but in all cases within five minutes after the tests ceased the power returned to its original value.

Accelerations up to 15 g have been placed on these generators for five minutes in each plane of symmetry with no effect.

These generators have been subjected to 50 g shock tests with a rise time of less than one millisecond in each of the three mutually orthogonal directions with all generators satisfactorily operating after these tests.

It is our intention to test these units until damage occurs, and to use this data to redesign an even more rugged power supply.

The long life and inherent reliability of isotopic power systems suggest numerous applications in the Age of Space. An instrumented satellite or space probe which has lost its means of transmitting data to earth stations is next to useless, yet the use of chemical batteries imposes severe limits on the lifetime of a transmitter in space. Solar cells offer one solution, as already suggested; but unless some mechanism is included to keep them oriented toward the sun their effectiveness is reduced. Thus, isotopic power may prove invaluable to satellite navigation aids, communications satellites and deep space probes.

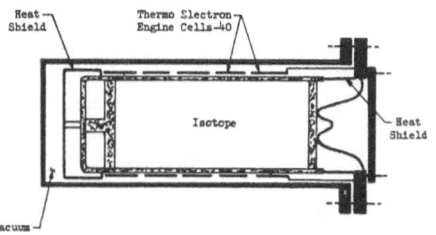

Fig. 7. Isotope powered thermo electron engine

Within the next few years man hopes to observe our neighboring planets by remote means as a prelude to true space exploration; and in the case of Venus, particularly, the only feasible source of auxiliary energy seems to lie in isotopic power. Using existing means of propulsion, a trip to Venus would last several months—far beyond the limits of a reasonable supply of chemical batteries. Once a soft landing of the instrumented payload had been achieved, solar cells would be rendered useless by the thick atmospheric blanket which covers Venus. A light, reliable, rugged isotopic power supply, on the other hand, could be expected to survive the journey and enable a transmitter to give us a continuing report of conditions observed and recorded by the instruments on the planet's surface. In order to stay within weight limitations, a "trickle charge" could be used with an energy storage system to build up every hour or every few hours to the level needed for such long distance transmission.

Extensive work is in progress on more advanced isotopic power systems to produce auxiliary power with markedly greater weight savings. Some of these advanced systems (Fig. 7) will use the thermoionic effect to convert the heat from the radioisotope fuel into electrical power. It is anticipated that safely packaged isotopic power supplies can be built during the next few years to produce eight watts per pound and to operate at significantly higher efficiencies than the five or six percent achieved by SNAP-III-B.

We believe that isotopes face a great future in the realm of Space Power.

Some Results of the Measurement of the Mass Spectrum of Positive Ions by the 3rd Artificial Earth Satellite

By

V. G. Istomin[1]

(With 11 Figures)

(Received August 27, 1959)

Abstract — Zusammenfassung — Résumé

Some Results of the Measurement of the Mass Spectrum of Positive Ions by the 3rd Artificial Earth Satellite. The radio frequency mass-spectrometer installed in the third Soviet satellite registered positive ions with mass values of 32, 30, 28, 18, 16 and 14 which were identified as single charged ions of molecular oxygen, nitrogen oxide, molecular nitrogen, atomic oxygen and atomic nitrogen respectively. The data obtained ranged from altitudes of 225 to 980 kilometers and latitudinal intervals of 27° to 65° north latitude. Certain patterns in the changing composition of the ionosphere with altitude and geographic latitude have been discovered.

The conclusions drawn from the analysis of the spectrum are: 1. that at these altitudes atomic oxygen ions predominate; 2. that the ionosphere composition changes with the altitude: the relative content (in relation to atomic oxygen) of heavy molecular ions decreases with an increase in altitude; but the relative content of the atomic nitrogen ion increases with increasing altitude; 3. that the ionosphere composition changes with the latitude: the relative atomic nitrogen ion content, as well as the content of ions of molecular nitrogen, molecular oxygen and nitrogen oxide, is greater in the northern part of the investigated latitude range, i.e. 55°—65°.

Einige Ergebnisse der Messungen des Massenspektrums positiver Ionen mit Hilfe des dritten künstlichen Erdsatelliten. Das in dem dritten sowjetischen Erdsatelliten eingebaute Radiofrequenz-Massenspektrometer registrierte positive Ionen mit den Massenwerten 32, 30, 28, 18, 16 und 14, die als einfach geladene Ionen von molekularem Sauerstoff, Stickoxyd, molekularem Stickstoff, atomarem Sauerstoff und atomarem Stickstoff identifiziert wurden. Die erhaltenen Daten beziehen sich auf Höhenbereiche von 225 bis 980 km und Breitenintervalle von 27 bis 65° n. B. Es wurden gewisse Strukturen wechselnder Zusammensetzung der Ionosphäre in Abhängigkeit von Höhe und geographischer Breite entdeckt.

Die aus der Analyse des Spektrums gezogenen Schlüsse sind: 1. In diesen Höhen herrschen Ionen atomaren Sauerstoffs vor. 2. Änderungen der Zusammensetzung der Ionosphäre mit der Höhe: Der relative Gehalt (im Verhältnis zum atomaren Sauerstoff) an Ionen schwerer Moleküle nimmt mit zunehmender Höhe ab; doch der relative Gehalt an Ionen des atomaren Stickstoffs wächst mit zunehmender Höhe. 3. Änderungen der Zusammensetzung der Ionosphäre mit der Breite: Der relative Gehalt an Ionen des atomaren Stickstoffs, ebenso an Ionen des molekularen Stickstoffs, molekularen Sauerstoffs und Stickoxyds, ist im nördlichen Teil des untersuchten Breitenbereiches, nämlich von 55° bis 65° n. B., größer.

---

[1] Institute of Applied Geophysics, Academy of Sciences of the U.S.S.R., Moscow, U.S.S.R.

Quelques résultats de mesure sur le spectre de masse d'ions positifs par le troisième satellite artificiel soviétique. Le spectrographe de masse à radio fréquence installé dans le troisième satellite soviétique a enregistré des ions positifs de masses 32, 30, 28, 18, 16 et 14 identifiés respectivement avec les ions à une seule charge d'oxygène moléculaire, oxyde d'azote, azote moléculaire, oxygène atomique et azote atomique. Les données ont été obtenues à des altitudes variant entre 225 et 980 km. et des latitudes variant entre 27° et 65° nord. Certaines configurations dans la composition variable de l'ionosphère ont été découvertes.

Les conclusions tirées de l'analyse spectrale sont: 1. Les ions d'oxygène atomique prédominent à ces altitudes, 2. la composition de l'ionosphère se modifie, le nombre relatif d'ions moléculaires lourds (en relation à l'oxygène atomique) décroissant avec l'altitude, le nombre relatif d'ions atomiques d'azote croissant avec l'altitude, 3. cette composition se modifie aussi en latitude: le nombre relatif d'ions d'azote atomique et le nombre d'ions moléculaires d'azote, d'oxygène et d'oxyde d'azote étant plus grand dans la partie septentrionale de l'intervalle de latitude exploré, notamment de 50° à 65°.

I. Introduction

A 7—5 cycle version of the Bennett type radio frequency mass-spectrometer was used in the third satellite for investigating the mass spectrum of the positive ions of the ionosphere. The design of the instrument, its basic parameters and measuring methods are described in work [1].

In accordance with the program of the instruments' operations in the satellite and the available power supply, the mass-spectrometer functioned from the 15th through the 25th May. A large volume of material—about 15,000 mass spectra at altitudes ranging from 225 to 980 kilometers—was received during that time. The measurements were made only in the northern hemisphere at intervals of latitude from 27° to 65° north latitude. Since the altitude and geographic latitude of the satellite in its orbital flight are in a definite way inter-related, and the time changes in this relation in the case of the third satellite, due to orbital regression, were sufficiently slow, the data on the ionosphere composition obtained in all revolutions represent approximately the same complex altitude-latitude cross sections of the atmosphere. Fig. 1 shows the orbit of the third satellite in altitude-geographic latitude co-ordinates for approximately the

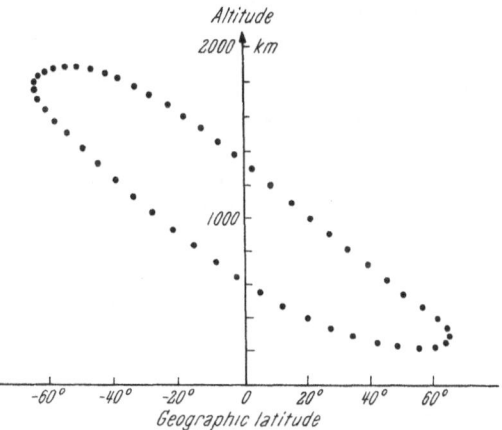

Fig. 1. Orbit of the 3rd satellite in the altitude-geographic latitude coordinates approximately the first 10 days of its existence

first 10 days of its existence. It may be seen that any value of the geographic latitude within the range of 65° is corresponded by only two values of the satellite's flight altitude and, conversely, any value of the flight altitude is corresponded by only two latitude values. This situation complicates the interpretation of the results obtained, and makes it difficult to determine the strictly altitudinal and strictly latitudinal relations of the atmosphere parameters under investigation.

It should be pointed out, in addition to the abovesaid, that all of the material received refere to daytime, that is what was obtained were the spectra of the atmosphere ions illuminated by the Sun. Most of the data were obtained during the morning hours, Moscow time, between 7:00 and 11:00 a.m. The local time will naturally be different, depending on the geographic longitude of the particular trajectory point of the satellite.

II. The Basic Characteristics of the Obtained Spectra

Characteristics of the mass spectra of the ions obtained by the third satellite, in addition to the basic (true) mass peaks, are also the so-called harmonic (false) peaks which make the deciphering of the data difficult. The harmonic peaks are present because the effective retarding potential of the mass-spectrometer was found to be greatly lowered for two reasons. Due to the fact that the mass-spectrometer is located on a moving object, all the atmospheric ions possess, in addition to the thermal speeds, a directional (regulated) speed equal to that of the satellite, in relation to the instrument. Because of this regulated speed $v = 8.10^5$ cm/sec, an ion with a mass number M and a charge $q = 4.8.10^{-10}$ CGSE will possess a definite energy equivalent to a certain amount of accelerating voltage,

$$\Delta V_v = 300 \frac{M \cdot m_0}{2q} v^2 \text{ volts} \tag{1}$$

where $m_0 = 1.67.10^{-24} g$ is the value equal to 1/16 of the atomic weight of oxygen. In the case of ions with a mass number 16, the speed 8.10^5 cm/sec is equivalent to an energy of 5.35 electron-volts, and in the case of ions with a mass number 30 it is equivalent to 10 electron-volts. In view of this, the retarding potential of the instrument with reference to equal-mass ions is found to be lowered by various values which are numerically equal to the ion energies expressed in ev (electron-volts). Besides, as an analysis of the obtained spectra shows, the satellite had a negative potential of several volts which was also conducive to an additional reduction of the effective retarding potential by a value which is the same for all atmospheric ions.

As a result of the reduction of the effective retarding potential brought about by these two causes, the sensitivity of the mass-spectrometer was found to be several times higher, more in connection with the heavy ions than with the light ones, and the mass resolution was accordingly reduced approximately by one half in comparison with what it was when the instrument was adjusted in the laboratory.

The instrument had been adjusted to the resolving power $R = \frac{M}{\Delta M} \approx 20$ in the region of mass numbers 20. Here, as usual, M represents the mass number of the peaks and ΔM its width at the base expressed in atomic mass units. In the obtained spectra of atmospheric ions the peak base resolution in the field of mass numbers 16—14 is equal to 7—10. That means that in the given experiment the mass peaks in the region of light ($M \sim 16$) masses were fully resolved if their mass numbers differed by approximately 2 or more atomic mass units. In the region of heavy masses ($M \sim 30$), the peaks whose mass number differed by 2 atomic units were not fully resolved. The aforesaid should be borne in mind when the experiment results are reviewed.

The dynamic range of the instrument was found to be considerably wider with respect to high intensities due to the availability of harmonic peaks in the spectra. As the light harmonic level of mass 16 was on the average equal to

0.1 of the basic peak amplitude, when peak 16 was intensive enough for the amplifier's low-sensitive output to become saturated (the current exceeded 10^{-8} amperes), its value could be estimated by the value of the light harmonic peak.

The mass numbers of all the peaks in the spectra, as had been expected [2], were found to have moved on the instrument's mass scale toward the light masses under the influence of the satellite's speed and its negative charge[1]. We repeat that in the radio frequency mass-spectrometer in use the mass scanning is realized by a change of the negative accelerating voltage according to the sawtooth law. In laboratory conditions the mass number of the peak is determined by the following formula:

$$M = \frac{V}{k} \qquad (2)$$

where M is the mass number of the peak in atomic mass units, V is the voltage value of the scanning in volts at the time the peak occurs, and k is the instrument's constant which is equal to 7.2 v/a.m.u. (volt/atomic mass units).

When the instrument is used on the satellite, the negative voltage value of scanning V will be increased by the value ΔV_v which is numerically equal to the electron-volt value of the corresponding ion moving at a speed of 8.10^5 cm/sec, as well as by the satellite's negative potential value φ. Formula (2) should in this case be rewritten as follows:

$$M = \frac{I}{k}(V + \Delta V_v + \varphi) = M_{dec.} + \Delta M_v + \Delta M_\varphi . \qquad (3)$$

Here $M_{dec.}$ represents the mass number of the peak determined by deciphering the entry of the scanning voltage at the time it occurred, ΔM_v is the correction taking into account the satellite's speed, and ΔM_φ taking its charge into account.

The value ΔV_v itself, as can be seen from formula (1), depends on the mass number of the ion, and the formula for determining the mass number of the peak may therefore be recorded as follows:

$$M = I/k \left(\frac{V + \varphi}{I - 300\, m_0\, v^2/2q\, k} \right) = \frac{M_{dec.} + \Delta M\varphi}{I - 300\, m_0\, v^2/2q\, k} \qquad (4)$$

or definitively, by substituting the numerical values of the magnitudes:

$$M = \frac{I}{k} \left(\frac{V + \varphi}{I - 0.334/k} \right) = \frac{M_{dec.} + \Delta M\varphi}{0.954} . \qquad (5)$$

As may be seen from formula (5), the shift of the mass peaks under the influence of the speed with regard to the instrument in use should be: 0.74 atomic mass units for mass 16 and 1.39 atomic mass units for mass 30.

III. Deciphering the Spectra

The major difficulty in deciphering the mass ion spectra was the separation of the basic (true) mass peaks from the harmonic (false) peaks. The nature of the spectrum registration is shown in Fig. 2 where copies of four photograms are reproduced. The first and second represent ion spectra in the range of the perigee along the low and high sensitivity channels of the mass-spectrometer amplifier,

---

[1] Ion peaks refered to contamination, which were recorded only during the first and third revolutions on May, 15 as, for instance, the peak 18^+—water H_2O^+ [5], suffered no such shift. This fact proved the correctness of contamination ion peaks' identity.

respectively, and the third and fourth are analogical recordings in the altitude range of about 350 km. The registrations were made on 23 May 1958, at about 9:00 a.m. Moscow time. The first two entries were made when the geographic latitude of the satellite's flight was 55° north latitude, and the second two entries when it was 64° north latitude. Conspicuous in the perigee range of ion spectra is plenty of peaks and a low resolution particularly in the region of large mass numbers.

The separation of the basic and harmonic peaks in the spectra was made by way of comparing their amplitudes and the nature of their changes according to the altitude of the satellite's flight and its orientation. The places of the harmonics in the spectra had been determined beforehand by laboratory experiments with reference to this particular instrument. Another criterion that serves to confirm the correctness of the identification of the basic and harmonic peaks is the fact that the relative width of the harmonic peaks is smaller than that of the basic peaks. That had also been established by laboratory experiments.

After the correction was made for the satellite's speed according to formula (5), the mass numbers of the peaks accepted as the basic (true) ones differed from the even integral values by some part of an atomic mass unit. That difference was explained by the negative charge of the satellite. For example, in the two days of 22 and 24 May (revolutions 96 and 122) the average differences between the parts of the revolution wherein the measurements were made amounted to 0.3 and 0.5 atomic mass unit,

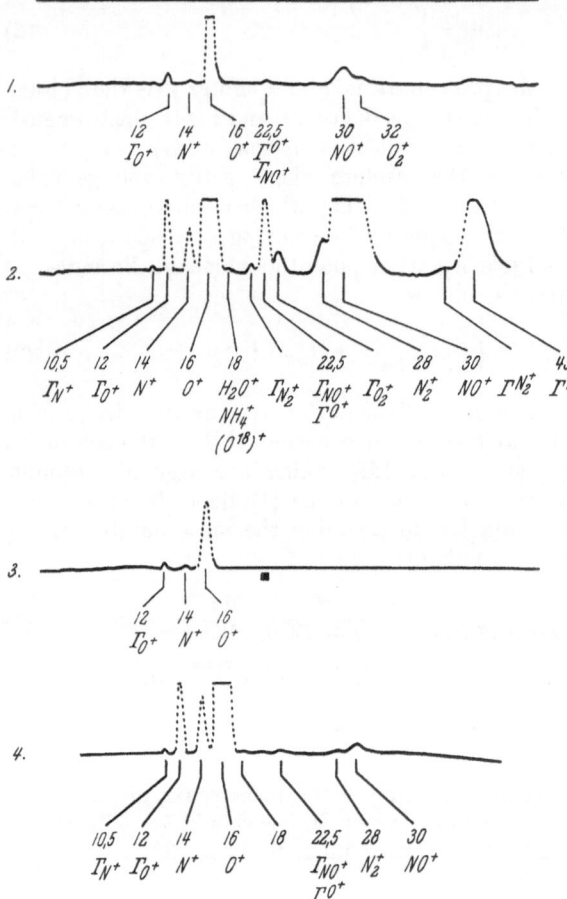

Fig. 2. Registration of ion mass spectra made on 23 May 1958, about 9:00 a.m. Moscow time. 1st and 2nd registrations in the low and high-sensitivity channels of the mass-spectrometer amplifier. Altitude 230 km, latitude 55°. 3rd and 4th registrations, same as above, altitude 350 km, latitude 64°, north latitude

respectively. Taking the value constant of the instrument as $k = 7.2$ v/a.m.u. (volt/atomic mass units), the satellite's negative potential will be equal to 2.2 and 3.6 volts.

Predominant in all the spectra, in point of intensity, is the peak whose mass number was close to 16 after the correction for the satellite's speed had been made. On this basis, the given peak was identified with the atomic oxygen peak O^+.

The second most intensive light peak, also present in the spectra at all altitudes, is the one with a mass number 14. It would be most natural to identify it with the atomic nitrogen peak N^+. The peaks with mass numbers 12 and 10.5 were identified with the light harmonic peaks 16 and 14, respectively.

Adjoining peak 16 in the field of heavy masses is the peak with a mass number 18 which is very weak but clearly detectable in many spectra. Its relative intensity has been determined as equal to $i_{18}/i_{16} = (0.15 \pm 0.05)$ %. This peak is not the harmonic of some basic peak present in the spectrum, and its occurrence can be explained only by the presence of a corresponding ionized molecule or atom in the atmosphere. The given ion can be identified either with the ion of water H_2O^+, or with the ion of another hydrogen compound, NH_4^+. In the light of such an identification, the presence of an ion with a mass number 18 should serve as an indirect indication of the presence of appreciable quantities of neutral or ionized hydrogen in the upper atmosphere. There is another and more reasonable possibility, however: the mass 18 peak should apparently be attributed to an isotopic ion of oxygen with an atomic weight 18 whose relative abundance is 0.2 %. The problem can be finally resolved by further accurate measurements.

In addition to the group of light peaks 18, 16 and 14 in the spectra obtained at different altitudes in the perigee area, there is a fairly conspicuous group of heavy peaks with mass numbers, 32, 30 and 28 which were also classified as basic (true) peaks. The most intensive among them is the peak with a mass number 30 which should be identified with the ion of the nitrogen oxide peak NO^+. The peaks with mass numbers 32 and 28 should naturally be attributed to the ions of molecular oxygen O_2^+ and molecular nitrogen N_2^+. Despite the fact that the given three heavy peaks have not been fully resolved, and the extreme ones, particularly peak 32, frequently combine with the average location peak 30, their presence in the spectra is undoubted and quite evident.

All the other peaks in the spectra, represented in Fig. 2, should be classified among the harmonic (false) peaks. Thus the group of three peaks, the most intensive of which is the one with a mass number 22.5, is represented as consisting of three light harmonic peaks 32, 30 and 28. In some spectra it is also possible to discern here the heavy harmonic of peak 16 with a mass number 22.8. The peaks with mass numbers 40 and 43 are the heavy harmonics of peaks 28 and 30.

All the peaks in the photogram recordings of the mass spectra shown in Fig. 2 are marked according to the mentioned identification. The basic (true) peaks are designated by appropriated chemical symbols, and the harmonic (false) peaks are indicated by the letter Γ with a particular index. For example, the light harmonic of atomic oxygen is shown as Γ_{O^+}, and the heavy harmonic of a nitrogen oxide ion as Γ^{NO^+}.

IV. The Patterns Discovered in the Changing Composition of the Ionosphere According to Altitude and Geographic Latitude

When reviewing the results cited in this section, one should bear in mind that all the data apply only to the intensities of the mass peaks measured in units of current. The peak intensities should naturally be connected with the relative and absolute concentrations of the appropriate ions but this connection may be quite complicated, and the question has not yet been finally elucidated. The difficulty involved in a comparison of the relative intensities of ionic peaks with a view to obtaining information regarding the concentration of the appropriate ions is that it is necessary to take into account the mass discrimination which must inevitably occur in the described experiment.

The discrimination of ions of various masses will occur, first of all, in the field of the satellite itself if it has no zero potential relative to the ambient plasma. The presence of a drawing field created by the first grids of the mass-spectrometer tube will bring about an additional discrimination of ions. Finally, in the analyzer itself the ions with various masses will be found in different

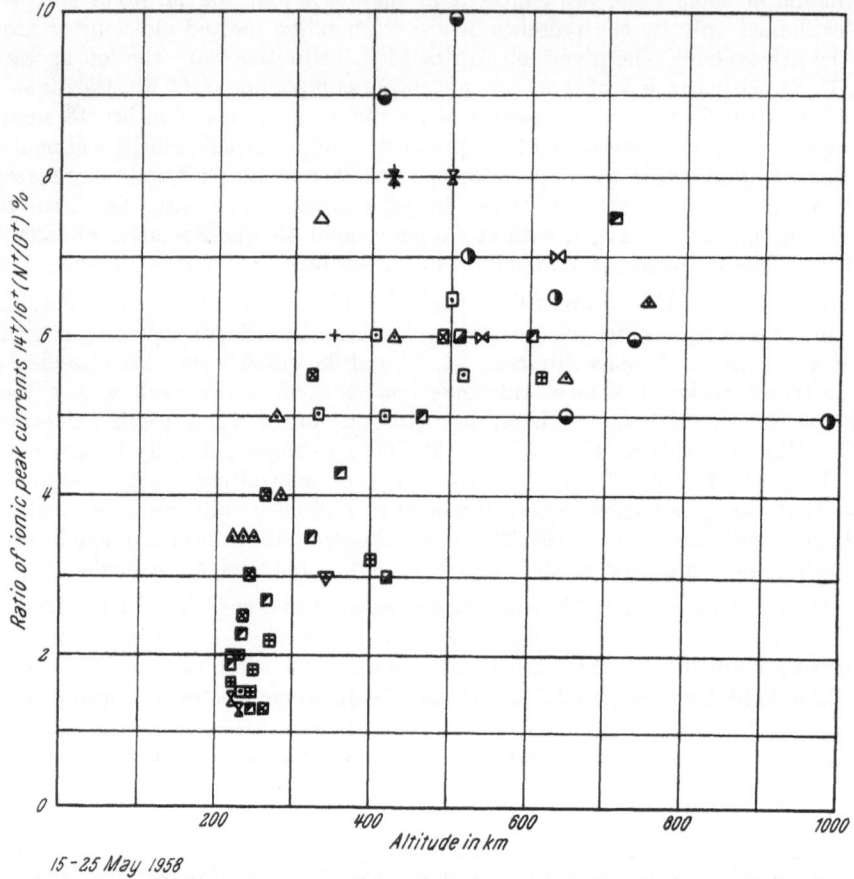

15-25 May 1958

Fig. 3. Changing concentration of atomic nitrogen ions in relation to atomic oxygen ions according to altitude (based on the data of the 12 revolutions between 18 and 24 May 1958)

states depending on the initial energy with which they have entered the analyzer. As was noted in section II, the initial energy of the ions with mass numbers 16 and 30 on the satellite under experimental conditions was approximately twice as high. In accordance with this, the level of the retarding potential with respect to these ions will also be different, and the peak intensity ratio will not equal the concentration ratio of the corresponding ions when they enter the analyzer[1].

----

[1] The first two factors—discrimination in electric fields—are conducive to a reduction of the relative intensities of the heavy ions as compared with the light ones, whereas the last factor operates in an opposite direction and may compensate for that reduction in some measure or even produce an inverse effect.

However, it should be borne in mind that the effect of mass discrimination, whatever the reasons for it, can be great only in the case of ions whose masses are considerable different from one another, and small in the case of ions close to each other in mass number. In this latter case the ratio of mass peak intensities is close to the ratio of concentrations of the given ions. For example, the ratio of the peak intensities of atomic nitrogen and atomic oxygen i_{14}/i_{16} or molecular nitrogen and nitrogen oxide i_{28}/i_{30} should be close to the ratio of their concentration in the ionosphere, whereas the ratio of the peak intensities of nitrogen oxide and atomic oxygen i_{30}/i_{16} is possibly considerably different from their relative concentrations.

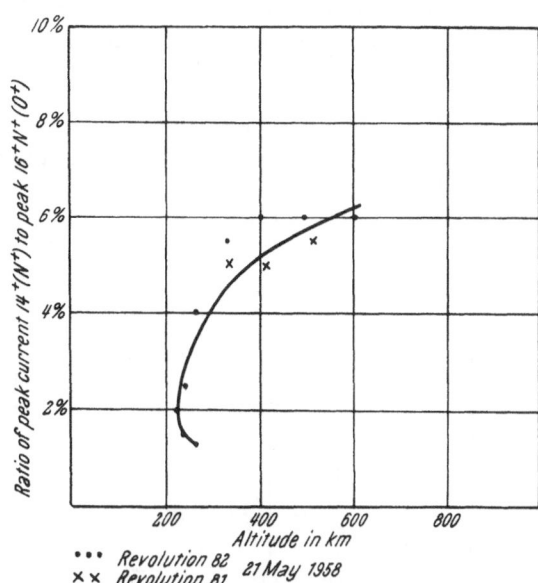

Fig. 4. Changing concentration of atomic nitrogen ions in relation to atomic oxygen ions according to altitude (based on the data of the 2 revolutions of 21 May 1958)

Both the absolute and relative intensities of the mass peaks in the spectra reveal considerable changes with time. Inasmuch as the functioning of the instrument was controlled and remained unchanged all the time, the changes in the spectra should be attributed to: (a) the changing orientation of the analyzer's inlet opening relative to the satellite's flight direction due to the rotation of the latter; (b) the changing coordinates of the satellite—altitude, geographic latitude and longitude, and (c) the changing external conditions, the factors affecting the ionization of the atmosphere.

The first group of changes connected with the satellite's rotation is made apparent by the revealed periodicity which coincides with the one established in the other experi-

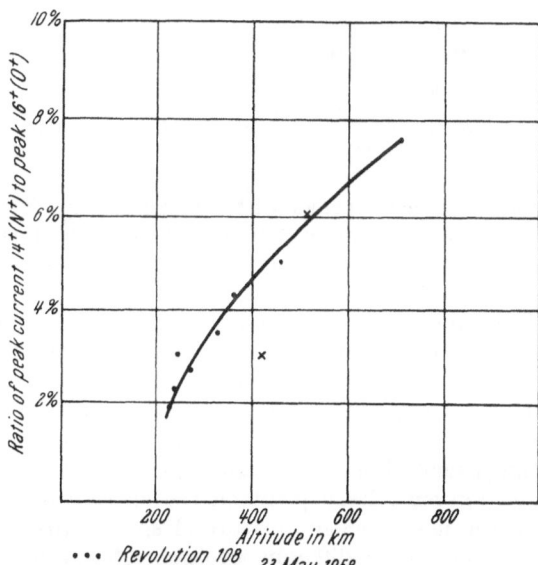

Fig. 5. Changing concentration of atomic nitrogen ions in relation to atomic oxygen ions according to altitude (according to the data of the 2 revolutions of 23 May 1958)

ments on the third satellite (see [3] and [4], for example). An examination of sufficiently lengthy recordings (on the order of tenths of a second) makes it possible to single out the groups of spectra obtained during the most favorable orientation of the analyzer's inlet opening in relation to the flight direction, namely when the angle between the normal to the inlet opening of the mass-spectrometer tube and the satellite's speed vector had a minimum value. A theoretical examination of this question shows that in these conditions the discrimination of ions with various masses should be down to a minimum. This is confirmed also by an analysis of the obtained recordings. The data on the processing of these particular spectra were used primarily in analyzing the composition of the ionosphere and its changes with altitude and geographic coordinates.

As reported earlier [5], the ion of atomic oxygen O^+ is the everpresent and predominant component of the ionosphere at all the investigated altitudes from 225 to 980 km. This makes it convenient to compare the intensities of all

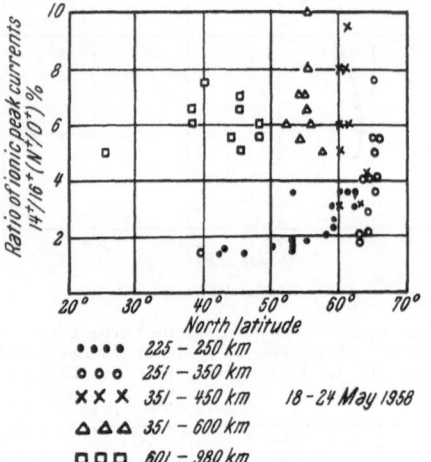

● ● ● ●   225 – 250 km
○ ○ ○   251 – 350 km
× × ×   351 – 450 km   18 – 24 May 1958
△ △ △   351 – 600 km
□ □ □   601 – 980 km

Fig. 6. Changing concentration of atomic nitrogen ions in relation to atomic oxygen ions according to geographic latitude (according to the data of the 12 revolutions between 18 and 24 May 1958)

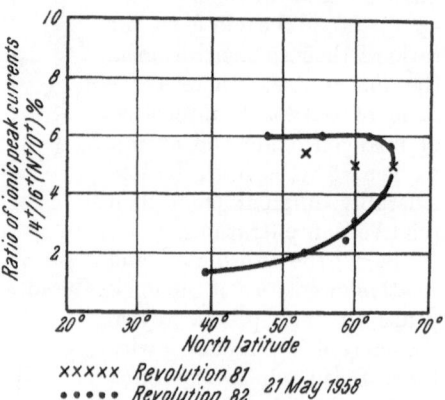

× × × × ×   Revolution 81   21 May 1958
● ● ● ● ●   Revolution 82

Fig. 7. Changing concentration of atomic nitrogen ions in relation to atomic oxygen ions according to geographic latitude (according to the data of the 2 revolutions of 21 May 1958)

the other mass peaks with the peak intensity of atomic oxygen. The second atomic component of the ionosphere is the nitrogen ion N^+. The ionic current (intensity) of the atomic nitrogen peak changes in relation to atomic oxygen, depending on altitude and geographic latitude, within the range of 1.3 to 8—10 %. The tendency to an increasing content of atomic nitrogen ions, also reported earlier [5], was confirmed with regard to all the processed spectra. This dependence is graphically represented in Fig. 3. Overlaid on the horizontal axis is the altitude in kilometers, and on the vertical axis the relative peak intensity of atomic nitrogen as a percentage of the atomic oxygen peak. As pointed out above, this magnitude is close to the concentration ratio of the ions of atomic nitrogen and oxygen. Every point on the chart was obtained as a result of averaging up the measurements of several spectra (from 5 to 20 spectra). The different signs refer to the different revolutions between the 18 and 24 May 1958. Attention is called to the fairly large dispersal of the points, amounting to ±3 % at some altitudes, but it can be seen that that dispersal is not occasioned by errors in the measurements—it reflects the actual changes in the composition of the ionosphere from day to day and from revolution to revolution. This can be seen on the charts of Figs. 4 and 5 which represent the same dependence of the relative concentration of atomic nitrogen ions on the altitude, but the data of one day are used for every chart. Evidently, the points corresponding to one revolution arrange themselves in a small spread on a smooth curve while the points obtained from the preceding or following revolution are already considerably different from the curve.

In examining the charts on Figs. 3—5 it should be borne in mind that, in view of the specific nature of the experiment on the satellite as pointed out above, they do not justify any conclusion on the purely altitudinal changes of the ionosphere composition as the changing altitude of the satellite's flight was paralleled by a change of its geographic latitude. The latitudinal relation of the ionosphere composition can be seen particularly distinctly between the altitudes of 225—350 km where the latitude change is considerable—from 25° to 65° north latitude.

The chart in Fig. 6 shows the data on the concentration ratio of atomic nitrogen and atomic oxygen ions in relation to the particular latitude. The

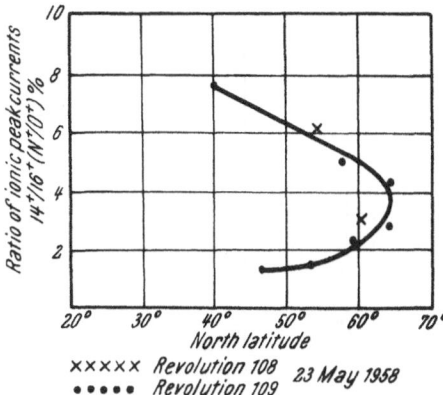

Fig. 8. Changing concentration of atomic nitrogen ions in relation to atomic oxygen ions according to geographic latitude (according to the data of the 2 revolutions of 23 May 1958)

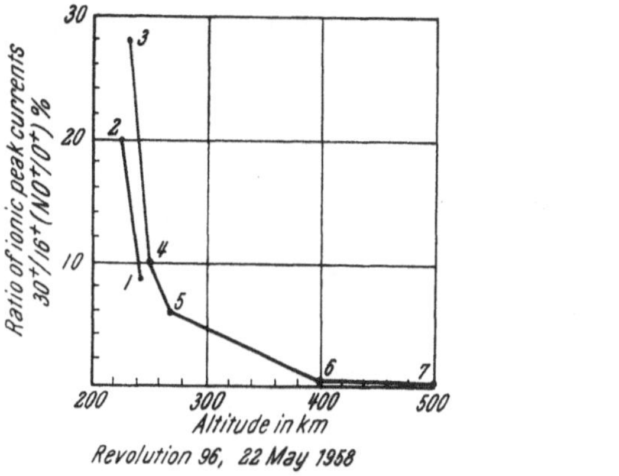

Fig. 9. The changing relative intensity of the ionic peak of nitrogen oxide according to altitude and geographic latitude (according to the data of one revolution of 22 May 1958). The points have been renumbered in the order of increasing geographic latitude: 1 and 2—south latitudes (before the perigee), and 3—7—north latitudes (after the perigee)

entire altitude range is divided into 5 intervals, and the points applying to the various intervals are designated by appropriate symbols. It is evident that in the altitude ranges of 225—250 km and 251—350 km the relative concentration of atomic nitrogen ions becomes considerably greater during the transition from the latitude range of 30°—50° to that of 55°—65° north latitude. In the altitude range of 351—450 km the latitudinal effect cannot be discerned due to the insignificant change of latitude within the mentioned range. As for the still higher altitudes, 451—600 km and 601—980 km, it is possible to conclude that the relative concentration of atomic nitrogen ions is either no longer related to the latitude or that this relation is considerably weaker but qualitatively the same as in the lower altitudes.

With regard to the spread of the points on the chart of Fig. 6, reference should be made to what has already been said about the chart of Fig. 3, that is that the point spread reflects in some measure the ionosphere composition from revolution to revolution and from day to day. This is confirmed by the charts of Figs. 7 and 8 each of which reflects the latitudinal change of the relative concentration of atomic nitrogen ions for one day. It is evident that all the points with a relatively small spread arrange themselves on a smooth curve.

The altitudinal relations of the relative concentrations of molecular ions of oxygen, nitrogen and nitrogen oxide are similar in nature. At the perigee altitude, the relative intensities of the corresponding ionic peaks are at a maximum, and they diminish fairly rapidly with altitude. The most intensive molecular ion is the nitrogen oxide ion. At an altitude of about 230 km the intensity of the nitrogen oxide ion peak is 25 %—35 % of the intensity of the atomic oxygen ion peak. The second molecular ion in point of intensity is the oxygen ion O_2^+. Its intensity at the perigee is about 7 %—12 % of that of atomic oxygen. The peak intensity of molecular nitrogen at the perigee is 1.5 %—3 %.

The characteristic altitude-relation curves of the relative intensities of the molecular peaks NO^+, O_2^+ and N_2^+ (in relation to O^+), according to the data of a single revolution of 22 May 1958 are shown in Figs. 9—11. The characteristic feature of these curves is that they all are clearly divisible into two branches:

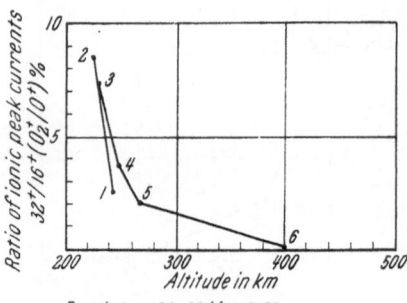

Revolution 96, 22 May 1958

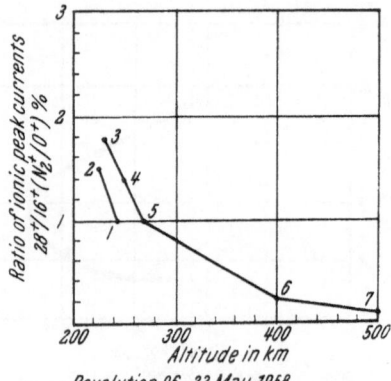

Revolution 96, 22 May 1958

Fig. 10. The changing relative intensity of the ionic peak of molecular oxygen according to altitude and geographic latitude (according to the data of one revolution of 22 May 1958). The points are numbered in the same way as in Fig. 9

Fig. 11. The changing relative intensity of the ionic peak of molecular nitrogen according to altitude and geographic latitude (according to the data of one revolution of 22 May 1958). The points are numbered in the same way as in Fig. 9

southern and northern. The points on the charts have been renumbered in the order of increasing geographic latitude: points 1, 2 apply to the southern latitudes (before the perigee), and points 3—7 to the northern latitudes (after the perigee). The latitude increases successively from point 1 to point 6. The last point, 7, corresponds to about the same latitude as point 4.

On all the charts of Figs. 9—11 the southern branch of the curve runs below the northern which points to the relation between the relative concentrations of molecular ions NO^+, O_2^+ and N_2^+ and the geographic latitude: the concentration of ions of nitrogen oxide, molecular oxygen and molecular nitrogen also increases with increasing latitude in relation to the concentration of atomic oxygen ions.

The molecular oxygen ion peak is traced to a maximum altitude of 400 km with the relative peak intensity $O_2^+/O^+ \sim 0.1$ %, and the nitrogen oxide and molecular nitrogen ion peaks to an altitude of 500 km. At altitudes of 400—500 km the peak intensities NO^+ and N_2^+ are about the same in relation to O^+ and equal to 0.2—0.1 %. Thus it is only above 500 km that the molecular ions are no longer found, and the ionosphere becomes purely atomic and oxygen-nitrogen, correct to 0.1 %.

V. Conclusions

1. The daytime measurements of the mass spectra of positive ions at altitudes of 225—980 km showed that atomic oxygen ions O^+ are predominant in that sphere. Registered also, in addition to the atomic oxygen ions, were ions with

mass numbers 14, atomic nitrogen N^+; 18, oxygen isotope $(O^{18})^+$; 28, molecular nitrogen N_2^+; 30, nitrogen oxide NO^+ and 32, molecular oxygen O_2^+.

2. It was found that the composition of the ionosphere in the investigated region changes with the altitude. The relative content (in relation to O^+) of heavy molecular ions NO^+, O_2^+ and N_2^+ diminishes with increasing altitude, and the relative content of the light ion N^+ increases with increasing altitude. At altitudes above 400 km the relative content of molecular oxygen ions is less than 10^{-3}, and above 500 km the content of molecular ions of nitrogen and nitrogen oxide is less than 10^{-3}.

3. It was found that the composition of the ionosphere depends on the geographic latitude. The relative content of atomic oxygen ions is considerable larger at altitudes of 225—350 km in the transition from the latitude range of 30°—50° north latitude to the latitude range of 55°—65° north latitude. The relative content of ions of molecular nitrogen, molecular oxygen and nitrogen oxide is also considerably larger in the latitude range of 55°—65° north latitudes compared to the latitudes further south.

Acknowledgment

The author expresses his gratitude to the head of the laboratory B. A. MIRTOV for his constant interest in the work and the discussion of the results, as well as to the laboratory staff members S. V. VASUKOV, A. A. PERNO and R. P. SHIRSHOV for their generous assistance in deciphering the telemetered recordings and processing the abundant experimental material.

References

1. V. G. ISTOMIN, Collected Works, Artificial Earth Satellites, ed. 3, 98 (1959).
2. B. A. MIRTOV and V. G. ISTOMIN, Uspekhi fisitcheskikh nauk (Progress of Physical Sciences) **63**, 1, 227 (1957).
3. S. SH. DOLGINOV, L. N. ZHUZGOV and N. V. PUSHKOV, Collected Works, Artificial Earth Satellites, ed. 2, 50 (1958).
4. V. V. MIKHNEVITCH, Collected Works, Artificial Earth Satellites, ed. 2, 26 (1958).
5. V. G. ISTOMIN, Collected Works, Artificial Earth Satellites, ed. 2, 32 (1958).

The Meteoritic Environment from Direct Measurements

By

Maurice Dubin[1]

(Received August 27, 1959)

Abstract — Zusammenfassung — Résumé

The Meteoritic Environment from Direct Measurements. Direct measurements of the distribution of micrometeorites in the vicinity of the Earth has been made recently from rockets, satellites and deep space probes. The principal methods involved detection of phenomena associated with the impact of the meteoroid on the surface of the vehicle, such as the extent of crater damage, the flash of the impact, and the impact-vibration. A review of the results obtained by the various investigators from these experiments is made. The impact-vibration method using crystal transducers has contributed the largest data sample. These measurements have been carried out on a number of rockets, alpha 1958, delta 1958, and on Pioneer I. The rate of influx of micrometeorites upon the Earth has been determined from this limited data sample. It has been found that daily variations may be greater than one order of magnitude. The measurements for particle masses from 10^{-9} to 10^{-10} grams indicated that the daily accretion rate of micrometeoritic particles upon the Earth is of the order of ten thousand tons per day. Finally, the requirements for future work in the direct measurement of the distribution of interplanetary material of astro-physical significance are reviewed, and the problems of astronautical interest relating to interplanetary matter are also presented.

Über die Meteoritenumgebung, ermittelt durch direkte Messungen. Direkte Messungen der Verteilung der Mikrometeoriten in der Umgebung der Erde sind vor kurzem mit Raketen, Satelliten und interstellaren Sonden ausgeführt worden. Die Hauptmethoden bezweckten die Wahrnehmung von Phänomenen, die mit dem Aufschlag von Meteoriten auf die Oberfläche des Flugkörpers zusammenhingen, wie Ausmaße des Kraterschadens, Aufblitzen des Aufschlages und Aufschlagvibration. Ein Überblick über die von verschiedenen Forschern mit diesen Experimenten erhaltenen Ergebnisse wird gegeben. Die Aufschlagvibrationsmethode, die Kristallüberträger benutzt, hat die größte Datensammlung eingebracht. Diese Messungen wurden mit einer Anzahl von Raketen — Alpha 1958, Delta 1958 und Pioneer I — durchgeführt. Die Größe des Einstromes von Mikrometeoriten auf die Erde wurde aus dieser begrenzten Datensammlung bestimmt. Es ergab sich dabei, daß die täglichen Variationen größer als eine Größenordnung sein können. Die Messungen von Teilchenmassen von 10^{-9} bis 10^{-10} g zeigten, daß die tägliche Zuwachsrate mikrometeoritischer Teilchen auf der Erde in der Größenordnung von 10000 Tonnen liegt. Schließlich werden die Bedürfnisse für zukünftige Arbeit an den direkten Messungen der Verteilung von interplanetarem Material, die von astro-physikalischer Bedeutung sind, erörtert und auch die astronautisch wichtigen Probleme bezüglich der interplanetaren Substanz dargestellt.

Evaluation de la situation météoritique par mesures directes. Des mesures directes de distribution de micrométéorites au voisinage de la Terre ont été faites récemment

---

[1] National Aeronautics and Space Administration, Washington 25, D.C., U.S.A.

par fusées, satellites et sondes lointaines. Les méthodes principales ont recours aux phénomènes d'impact à la surface du véhicule tels que: étendue du dommage, luminosité à l'impact, vibrations induites. La méthode des vibrations enregistrées par transducteurs à cristaux a fourni l'échantillon de données le plus important. De telles mesures ont été faites sur un certain nombre de fusées: alpha 1958, delta 1958, Pioneer I. Le taux de pénétration dans l'atmosphère terrestre a été déterminé à partir de cet échantillonnage limité. Les variations journalières ont été trouvées supérieures à un ordre de grandeur. Les mesures pour les masses particulaires comprises entre 10^{-9} et 10^{-10} grammes indiquent un taux d'accrétion de particules micrométéoritiques pour la Terre de l'ordre de dix mille tonnes par jour. Les exigences posées pour la poursuite des recherches par mesure directe de la distribution de matière interplanétaire intéressant l'astrophysique sont passées en revue et les problèmes relatifs à la présence de cette matière et d'intérêt astronautique sont indiqués.

The environment of vehicles in the interplanetary space and in satellite orbits contains debris of asteroidal and cometary origin. This solid component of interplanetary matter moving at heliocentric velocities may impact upon space vehicles with their very high energy densities. This individual amount of incident energy is by far the largest compared to incident radiations and other particles in space. For space flight, meteors represent one of the greatest hazards.

Direct measurements of the density of micrometeorites in the vicinity of the earth have recently been made from rockets, satellites, and space probes. Visual and radio measurements of meteors have been the major source of data concerning interplanetary material. From these measurements information regarding the concentration of interplanetary material, the mass distribution, and the distribution of orbits has been obtained. Meteorite falls and dust collections have been used to obtain samples of the extraterrestrial material. The great advances in rocketry during the past decade have supplied techniques for direct measurement of the debris in the interplanetary space. Only from direct measurements of the space density of meteoritic material and the impact effects upon a space vehicle may the astronautical designer accurately determine the hazards and design requirements for interplanetary space ships.

Such direct measurements, however, are difficult to carry out because the interplanetary material is distributed in a random manner. A large amount of this material is in the ecliptic and moves in direct orbits, and there exist high concentrations of material moving in cometary orbits also. The interplanetary space is large in volume, and the number density of particles is small enough to allow relatively safe travel through space, and small enough also to make the direct measure of meteoritic particles a fairly formidable problem. In order to obtain a statistical sampling of direct impacts, the sensor must be large in area and exposed to the space environment for a long time. Because micrometeorites of very small mass are relatively more numerous than the larger particles, this component of the interplanetary matter has been subjected to the first measurements.

The techniques for detecting micrometeorites involve a means of detecting an effect resulting from the impact on the space vehicles. The impact vibration on the skin of the vehicle detected with a crystal transducer has contributed the largest data sample. The flash of light from the impact has been monitored on one rocket flight, and a number of coils of fine wire have been exposed on two satellites to monitor discontinuities in the wire resulting from meteoritic impact craters.

Table I. Summary of Micrometeorite Measurements

Vehicle	Date	Area cm²	Time Exposed sec.	Area × Time cm² sec.	Impact Rate No. m⁻² sec.⁻¹	Sensitivity gm cm sec.⁻¹	Sensitivity Mass in gm
V-2	8 Dec 49	$\sim 10^4$	$\sim 10^2$	$\sim 10^6$	$\sim 6 \times 10^{-1}$	?	10^{-2} ?
Aerobee	17 Nov 55	75	$\sim 10^2$	7.5×10^2	1.8×10^2	—	10^{-15} ?
Aerobee	16 Jul 57	5×10^3	$\sim 10^2$	5×10^5	2.3×10^{-1}	2×10^{-3}	$\sim 10^{-9}$
Aerobee	16 Oct 57	5×10^3	$\sim 10^2$	5×10^5	6.4×10^{-1}	5×10^{-4}	$\sim 10^{-10}$
Cosmic rocket .	21 Feb 58	9×10^2	$\sim 3 \times 10^2$	$\sim 3 \times 10^4$	3.1×10^1	?	?
Vanguard	27 May 58	8×10^3	$\sim 10^3$	8×10^5	2.1×10^{-2}	$\sim 5 \times 10^{-3}$	$\sim 10^{-9}$
Alpha 1958	1 Feb 58	2.3×10^3	8×10^4	1.8×10^8	1.7×10^{-2}	2.5×10^{-3}	8×10^{-10}
Alpha 1958	1 Feb 58	1.1×10^1	6×10^6	7×10^7	0	—	$\sim 2 \times 10^{-9}$
Gamma 1958 ..	26 Mar 58	1.2×10^1	3.7×10^6	4.5×10^7	$(4 \times 10^{-4} ?)$	—	$\sim 2 \times 10^{-9}$
Delta 1958	15 May 58	3.4×10^3	10^8 ?	$\sim 10^{11}$?	2×10^1 1×10^1 1.7×10^{-3}	1×10^{-1}	$\sim 10^{-9}$
Pioneer I	10 Oct 58	3.8×10^2	1.1×10^5	4.2×10^7	4.0×10^{-3}	1.5×10^{-4}	$\sim 10^{-10}$

Table I is a summary of most of the information available to-date from direct measurements from rockets, satellites, and space probes. Data from a number of low altitude rocket flights have not been included. Additional data from Delta 1958, Mechta, and Explorer VI may soon be available.

For all the flights listed, vibration transducers or microphones were used as sensors except on the Aerobee of 17 Nov. 1955, on which a light flash detector was used, and the wire grid experiment on Alpha 1958 and Gamma 1958.

Of particular importance in evaluating this data is the product of the sensitive area and the time of exposure. Satellite vehicles are ideal for obtaining long time exposures. Delta 1958 with magnetic storage has an estimated area-time product of 10^{11} cm² sec. Alpha 1958 without data storage has the next largest area-time product of 2 10^8 cm² sec. and Pioneer I is next with 4×10^7 cm² sec. Although the area-time product for the wire grid experiment is comparable to that of Pioneer I, the sensitivity of the experiment was not high, and hence the total number of events sampled (about two) is inadequate for statistical validity. By the same reasoning, the data from most of the rocket flights cannot be weighted very strongly; these flights have mainly been tests for later satellite experiments.

The first experiment for micrometeorite detection was performed in 1949 on a V-2 by BOHN and NADING. A number of additional rocket flights were carried out beginning in 1955. BERG and MEREDITH performed the light flash experiment with rather interesting results; the sensitivity of the experiment has still not been accurately determined, although it appears possible that macromolecules were detected by this experiment. A number of additional rocket experiments leading to satellite experiments were carried out by the author with BOHN and ALEXANDER and by LAGOW. In the Soviet Union, NAZAROVA, KOMISSAROV, BOLOSKOV et al. investigated this problem.

An important part of the problem is the calibration of the detection system. In several cases hypervelocity calibration to velocities of 5 km/sec. were used to determine the mass of the impacting particles. The equipment on Alpha and Gamma 1958, and Pioneer I, was calibrated in this manner. Large daily variations have been found in the data from Alpha 1958, and this may explain the difference in the influx rate relative to Pioneer I.

The reports of the Soviet data from Delta have been puzzling since the early reports by NAZAROVA in Moscow and KRASSOVSKY in Amsterdam were several orders of magnitude higher than a later report by NAZAROVA and the lowest average impact rate by KOMISSAROV et alii. The impact rate results from the cosmic rocket appear rather high also. The hypervelocity calibration by KELLS indicates that the crystal transducer is sensitive to the impulse of the impact; the Soviet calibrations do not confirm this as indicated by sensitivities shown in Table I for Alpha 1958 and Delta 1958. Although the sensitivity in the mass column is equivalent, the momentum sensitivity differs by a factor of 40.

Thus, although the best data samples were obtained on Alpha 1958, Delta 1958, Pioneer I, and the Vanguard, there are still discrepancies that require clarification. The best data sample appears to be from Alpha 1958 for which 153 events were recorded. A diurnal variation may have been detected, and an extrapolation from the calibration and an assumed mass distribution indicate that the order of 10,000 tons of extraterrestrial material is falling on the earth daily. The presently available data samples are grossly inadequate. Much work has yet to be done, both on calibration of sensor system and data acquisition, using effective area-time exposures for mapping the distribution of matter in the interplanetary space.

Relativity Advances of the Perigee of Artificial Earth Satellites

By

Jesús Tharrats[1]

(With 4 Figures)

(Received August 27, 1959)

Abstract — Zusammenfassung — Résumé

Relativity Advances of the Perigee of Artificial Earth Satellites. The relativistic advance of the perigee of artificial Earth satellites is studied by means of the space-time structure introduced by the terrestrial sphere taking account of its density variation from centre to surface. The result gives in general a correction for SCHWARZSCHILD's point model of the order of thirty per cent more. The formulae obtained are applied to the computation of the relativistic advance of the perigee in the case of the principal artificial Earth satellites.

Relativistische Vorverlegung des Perigäums künstlicher Erdsatelliten. Die relativistische Vorverlegung des Perigäums künstlicher Erdsatelliten wird studiert unter Zuhilfenahme der Raum-Zeitstruktur, die durch die Erdkugel eingeführt wird, wobei ihre Dichteänderung vom Zentrum zur Oberfläche in Rechnung gesetzt wird. Das Resultat liefert im allgemeinen eine Korrektur für SCHWARZSCHILDs Punktmodell in der Größenordnung von 30 %. Die erhaltenen Formeln werden auf die Berechnung der relativistischen Verschiebung des Perigäums im Fall der hauptsächlichen künstlichen Erdsatelliten angewendet.

Avance relativiste du périgée des satellites artificiels. Cette avance est étudiée dans la structure spatio-temporelle induite par la sphère terrestre en tenant compte de la variation de densité du centre à la surface. Le résultat fournit une correction de l'ordre de trente % ou davantage sur le modèle ponctuel de SCHWARZSCHILD. Les formules obtenues sont appliquées au calcul de l'avance relativiste du périgée pour les principaux satellites artificiels.

I. Introduction

In this communication the effects of the General Relativity on the artificial satellites shall be discussed. To-day, the interior of our planet is known with sufficient accuracy, and, at least, essentially there is no reason to refuse the corrections which such structure introduces on a point-shaped model, chiefly if one remembers that these satellites circulate near the terrestrial surface (Fig. 1).

The results that were obtained concerning the advance of the perigee in the case of a material non-homogeneous sphere establish a correction to be done at SCHWARZSCHILD's point-shaped model, as it was utilized in [1] and [2] and it is very convenient to have correct formulae in order to determine with the greatest accuracy these secular perturbations of the satellites.

---

[1] Seminary of Theoretical Physics, University of Barcelona, Spain.

II. Shape of the Earth

The distribution of densities $\varrho\,(r)$ from the centre of the Earth up to the crust has been determined by seismic informations and, in order to consider the confidence that can be allowed, to introduce this function in the formulae which follow, we will quickly indicate how they have been attained.

Seismic observations give for velocity of P waves (longitudinal waves) a monotonous increase from 7.75 km/sec. to 13.64 km/sec. for depths from 33 to 2898 km from the surface level. These velocities for S waves (transversal waves) go from 4.35 to 7.30 km/sec. At 2898 km depth there is the Mohorovičić discontinuity and the rates of the velocities change abruptly; P waves velocity changes from 8.10 to 11.31 km/sec at depths from 2.898 to 6371 km.

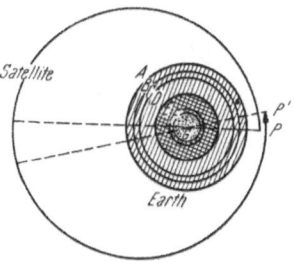

Fig. 1. Advance (PP') of the perigee

There are three equations to compute the density ϱ and two Lamé's coefficients $\lambda,\ \mu$ in each point of the terrestrial sphere: in fact, because the velocities v_P and v_S of P and S waves are known, we have:

$$v_P = \sqrt{\frac{\lambda + 2\mu}{\varrho}}, \quad v_S = \sqrt{\frac{\mu}{\varrho}}.$$

On the other hand, the compressibility module $\dfrac{1}{\varrho}\dfrac{d\varrho}{dp}$ (where p is pressure) equals $\dfrac{1}{\lambda + \dfrac{2}{3}\mu}$, and because $\dfrac{dp}{dr} = -\dfrac{GM}{r^2}\varrho$, we have:

$$\frac{d\varrho}{dr} = \frac{d\varrho}{dp}\frac{dp}{dr} - \frac{\varrho}{\lambda + \dfrac{2}{3}\mu}\frac{GM\varrho}{r^2}$$

and considering Lamé's coefficients in function of the propagation velocities we finally have the Williamson-Adams formula [3]:

$$\frac{d\varrho}{dr} = -\frac{GM}{r^2} \cdot \frac{\varrho}{v_P{}^2 - \dfrac{4}{3} v_S{}^2}. \qquad (1)$$

From this we can obtain through a numerical integration the function $\varrho\,(r)$. The results can be seen in the paper published by Jacobs [4].

For our computations, it is convenient to express the function

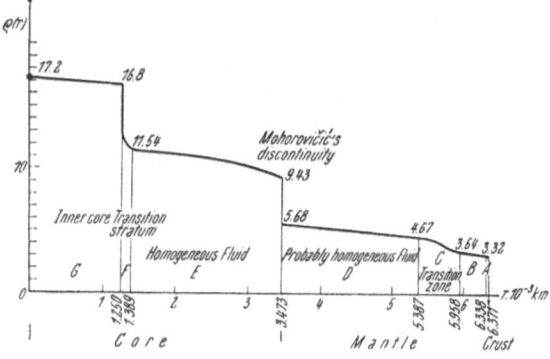

Fig. 2. Density of different regions of the Earth according to distance from the centre. Bullen's zones A, B, ..., G correspond to our different forms of $\varrho(r)$: eq.(2)

$\varrho\,(r)$ in intervals of simple analytic functions that can be obtained by dividing the terrestrial radius in the same zones A, B,... G of Bullen [5]. Fig. 2 shows the density form of these zones and the analytic form of $\varrho\,(r)$ for each of them is clearly seen. The following results are in agreement with the results of Bullen [4]:

1. Zone A : $\varrho_A(r) \sim 8.65 - 0.84\,r$ $6.338 < r \leqslant 6.371$
2. Zone B : $\varrho_B(r) = 8.65 - 0.84\,r$ $5.958 < r < 6.338$
3. Zone C^1 : $\varrho_C(r) = -49.78 + 20.88\,r - 2r^2$ $5.387 < r < 5.958$
4. Zone D : $\varrho_D(r) = 7.52 - 0.53\,r$ $3.473 < r < 5.387$ (2)
5. Zone E : $\varrho_E(r) = 11.965 - 0.532\,r - 0.125\,r^2$ $1.389 < r < 3.473$
6. Zone F : $\varrho_F(r) = 11.4 + 3.8\exp[-17.8\,(r-1.25)]$ $1.250 < r < 1.389$
7. Zone G : $\varrho_G(r) = 17.2 - 0.32\,r$ $0 < r < 1.250$

$$(r \text{ in thousands km.})$$

If we compare the results of densities as given in [4] with those deduced from the above formulae, we shall see that, with an error less than two per cent (probably less than the experimental errors themselves), these simple analytic formulae can be taken.

As it shall be immediately seen, it is important to find the k-order moments which can be deduced from a simple integration:

$$M_k = 4\pi \int_0^R dr\,r^k\,\varrho(r) = 4\pi\left\{\int_0^{1.250} dr\,r^k\,\varrho_G(r) + \ldots + \int_{6.338}^{6.371} dr\,r^k\,\varrho_A(r)\right\}$$

$$\tag{3}$$

$$M_2 = 4\pi\int_0^R dr\,r^2\,\varrho(r) = M = \text{whole mass}.$$

In order to operate on magnitudes without dimensions, we shall introduce, instead of the moments, the numerals which characterize the inner structure of the sphere. If ζ_k is an intermediate radius: $0 < \zeta_k \leqslant R$ we have:

$$M_k = (\zeta_k)^{k-2}\,4\pi\int_0^R dr\,r^2\,\varrho(r) = (\zeta_k)^{k-2}\,M$$

or, taking:

$$\mu_k = \frac{\zeta_k}{R}\,;\quad \{0 < \mu_k \leqslant 1\}$$

the k-moments have this form:

$$M_k = (\mu_k)^{k-2}\,M\,R^{k-2}\,. \tag{4}$$

The problem is reduced to the determination of these numerals μ_k for the terrestrial structure. For instance, for an homogeneous sphere it is:

$$\mu_k = \left(\frac{3}{k+1}\right)^{\frac{1}{k-2}}\quad\text{or}\quad M_k = \frac{3}{k+1}\,M\,R^{k-2}\,.$$

Formulae (2) and (3) give Table I.

Table I

$(\mu_{2k})^{2k-2}$	μ_2	$(\mu_4)^2$	$(\mu_6)^4$	$(\mu_8)^6$	$(\mu_{10})^8$	. . .
Earth	1.000	0.546	0.332	0.219	0.199	. . .
Homogeneous sphere...	1.000	0.600	0.429	0.333	0.273	. . .

---

[1] Actually, the most suitable approximation is (as shown in Fig. 2) a cubic: $\varrho_C(r) = 4.33 - 0.2\,u - 0.02\,u\,(u-1) - 0.0033\,u\,(u-1)\,(u-2)$; $[u = 10\,(r-5.671)]$; in the calculations a quadratic form is taken for sake of simplicity.

III. The Relativistic Earth

The case of the Earth considered as punctual to study the advance of the perigee of artificial satellites has been largely treated [1, 2]. This case being very well known, the space-time structure is given by SCHWARZSCHILD's formula:

$$ds^2 = -\frac{dr^2}{1 - \frac{2M}{r}} - r^2(d\theta^2 + \sin^2\theta\, d\Phi^2) + \left(1 - \frac{2M}{r}\right)dt^2 \tag{5}$$

(M = mass of the Earth; G = Gravitational constant = 1, c = speed of light = 1). To see how an internal structure of the Earth modifies this quadratic form it is needed to reconsider the explanations with the tensor of tensions T_k^i. The problem was solved by SCHWARZSCHILD for a sphere with a homogeneous fluid [6], and the more general cases have been solved by BRILLOUIN [7], DE DONDER [8], and HAAG [9, 10]. But it is easier to resolve the problem following the method of WHITEHEAD [11], as SYNGE has already done. The fundamental quadratic form, in this theory, is determined by the moments M_k of the distribution of densities:

$$ds^2 = -(1 + 2A)dr^2 - (1 + 2B)r^2(d\theta^2 + \sin^2\theta\, d\Phi^2) + 4C\, dr\, dt +$$
$$+ \left(1 - \frac{2M}{r}\right)dt^2 \tag{6}$$

$$A = \frac{M_2}{r} - \frac{2}{3}\frac{M_4}{r^3} = \frac{M}{r}\left\{1 - \frac{2}{3}\left(\frac{R}{r}\right)^2\mu_4^2\right\} \quad (R = \text{radius of the Earth}) \quad (G = c = 1)$$

$$B = \frac{1}{3}\frac{M_4}{r^3} = \frac{1}{3}\frac{R^2}{r^3}\mu_4^2$$

$$C = \frac{1}{r}\left[M - \sum_{n=0}^{\infty} \frac{M_{2n+4}}{(2n+1)(2n+3)r^{2n+2}}\right] = \frac{M}{r}\left[1 - \sum_{0}^{\infty}\left(\frac{R}{r}\right)^{2n+2}\frac{(\mu_{2n+4})^{2n+2}}{(2n+1)(2n+3)}\right].$$

Then, for the Earth's model, by replacing the values of Table I and considering that it suffices to expand up to the μ_{10} term, we obtain:

$$A = \frac{M}{r}\left|1 - 0.364\left(\frac{R}{r}\right)^2\right|$$

$$B = \frac{0.182}{r}\left(\frac{R}{r}\right)^2$$

$$C = \frac{M}{r}\left\{1 - \left[0.182\left(\frac{R}{r}\right)^2 + 0.022\left(\frac{R}{r}\right)^4 + 0.006\left(\frac{R}{r}\right)^6 + 0.003\left(\frac{R}{r}\right)^8 + o\right]\right\}$$
$$(o < 0.001).$$

IV. Advance of the Perigee

The advance of the perigee in the punctual scheme, as deduced from eq. (5) is:

$$\left[\frac{d\tilde{\omega}}{dt}\right]_S = \frac{24\,\pi^3 a^2}{c^2 P^2 (1 - e^2)} \tag{7}$$

(a = semi-major axis, P = period, e = eccentricity), or introducing KEPLER's constant: $K = 4\pi\frac{a^3}{P^2} = GM$

$$\left|\frac{d\tilde{\omega}}{dt}\right|_S = \frac{1}{1 - e^2}\left(\frac{19732}{a_{km}}\right)^{\frac{5}{2}} (''/\text{year}). \tag{7'}$$

If $\bar{Q}$, $\bar{q}$ are the apogee and the perigee altitudes, in R units, we have:

$$\left(\frac{d\tilde{\omega}}{dt}\right)_S = 95.48\,\frac{e^{\frac{5}{2}}}{1-e^2}\,\frac{1}{(\bar{Q}-\bar{q})^{\frac{5}{2}}}\;(''/\text{year})\,. \tag{7''}$$

Fig. 3 shows the advance of the perigee due to a point mass plotted against $\bar{Q}-\bar{q}$ for several of the eccentricities. Forbidden zone has been obtained approximately with following considerations:

$$a-c=a\,(1-e)=R+q>R\,,\qquad Q-q=2c=2\,e\,a$$

then

$$\bar{Q}-\bar{q}>\frac{2e}{1-e}\,.$$

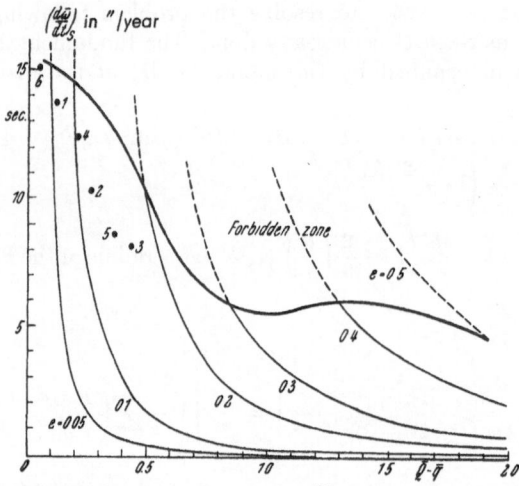

Fig. 3. Advance of the perigee due to a point mass. $\left(\dfrac{d\tilde{\omega}}{dt}\right)_s$ is plotted against $\bar{Q}-\bar{q}$ (Apogee altitude—Perigee altitude, in R units) at various eccentricities. $1, 2, \ldots, 6$ are the satellites as in Table II

For a material sphere, formula (6) gives [12]:

$$\frac{d\tilde{\omega}}{dt}=\left(\frac{d\tilde{\omega}}{dt}\right)_S\left\{1+\frac{M_4}{6\,M a^2}\,\frac{4+e}{(1-e^2)^2}\right\} \tag{8}$$

or by introducing the numerals μ:

$$\frac{d\tilde{\omega}}{dt}=\left(\frac{d\tilde{\omega}}{dt}\right)_S\left\{1+\frac{(\mu_4)^2}{6}\left(\frac{R}{a}\right)^2\frac{4+e}{(1-e^2)^2}\right\}. \tag{8'}$$

The correction is, then, important when the magnitude $\left(\dfrac{R}{a}\right)^2$ is considered, because the satellite runs near the terrestrial surface.

In our case, for the Earth, (8') gives:

$$\frac{d\tilde{\omega}}{dt}=\left(\frac{d\tilde{\omega}}{dt}\right)_S\left\{1+0.091\left(\frac{R}{a}\right)^2\frac{4+e}{(1-e^2)^2}\right\}. \tag{9}$$

Putting $a-c=R+q=R\,(1+\bar{q})$, one obtains:

$$\frac{d\tilde{\omega}}{dt}\Big/\left(\frac{d\tilde{\omega}}{dt}\right)_S=1+0.091\,\frac{4+e}{(1+\bar{q})^2}\left(\frac{1-e}{1-e^2}\right)^2. \tag{9'}$$

Fig. 4 shows the relationship between the real advance and that due to a point mass as a function of perigee altitude (in R units) and the eccentricity.

Finally, Table II is given for the advance of the perigee, in seconds per year for the main satellites launched up to this date.

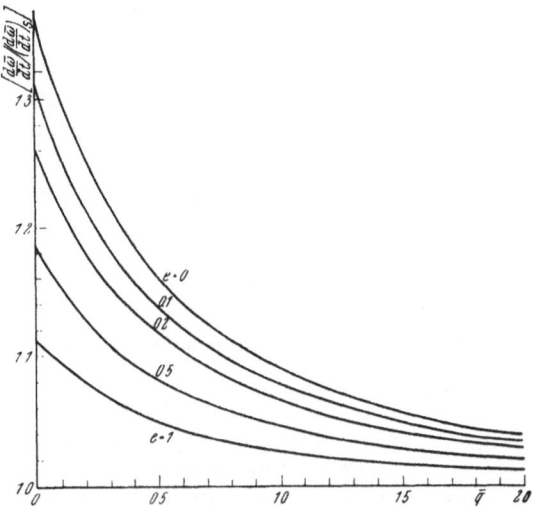

Fig. 4. $\dfrac{d\widetilde{\omega}}{dt}\Big/\left(\dfrac{d\widetilde{\omega}}{dt}\right)_s$ plotted against $\bar{q}$ (altitude of the perigee in R units) at various eccentricities

Table II

Satellite			e	$\bar{Q}$	$-$	$\left(\dfrac{d\widetilde{\omega}}{dt}\right)_s$	$\dfrac{d\widetilde{\omega}}{dt}$
1	1957 α 2	(Sputnik 1)	0.052	0.149	0.036	13″66	17″61
2	1958 α	(Explorer 1)	0.141	0.398	0.058	10″31	13″35
3	1958 β 2	(Vanguard 1)	0.191	0.622	0.103	8″10	10″35
4	1958 δ 2	(Sputnik 3)	0.111	0.296	0.036	12″39	15″88
5	1959 α 1	(Vanguard α 1)	0.166	0.520	0.119	8″69	10″85
6	1959 γ	(Discoverer 2)	0.008	0.055	0.038	14″93	19″98

References

1. E. Esclangon, C. r. acad. sci. Paris **226**, 23 (1948).
2. H. Krause, Proceedings of the VII International Astronautical Congress, Rome, 1956, p. 570-575. Roma: Associazione Italiana Razzi, 1956.
3. E. D. Williamson and L. H. Adams, J. Wash. Acad. Sci. **13**, 413 (1923).
4. J. A. Jacobs, Handbuch der Physik, vol. XLVII (Geoph. I). Berlin–Göttingen–Heidelberg: Springer, 1956.
5. K. E. Bullen, An Introduction to Theory of Seismology. Cambridge: University Press, 1953.
6. K. Schwarzschild, Sitz. Ber. Akad. Wiss. Berlin (1916) 424.
7. M. Brillouin, C. r. acad. sci. Paris **174**, 1585 (1922).
8. Th. De Donder, La Gravifique Einsteinienne. Paris: Gauthier-Villars, 1921.
9. J. Haag, C. r. acad. sci. Paris **176**, 658 (1923).
10. J. Haag, Le problème de Schwarzschild. Mem. Sc. Math. XLVI. Paris: Gauthier-Villars, 1931.
11. A. N. Whitehead, The Principle of Relativity. Cambridge: University Press, 1922.
12. J. L. Synge, Proc. Roy. Soc. London **211**, 303 (1952).

The Reactions of Terrestrial Microorganisms to Simulated Martian Conditions

By

Irving Davis[1] and **John D. Fulton[2]**

(With 4 Figures)

(*Received August 27, 1959*)

Abstract — Zusammenfassung — Résumé

The Reactions of Terrestrial Microorganisms to Simulated Martian Conditions. The existence of life on the planet Mars has been hypothesized by several authors. It is generally stated that the Martian planet would probably be the most likely planet in our solar system, other than Earth, capable of supporting "life". The nature of such possible Martian life has been suggested to be similar to certain simple forms of terrestrial life. The first attempt to test the validity of this assumption by means of laboratory experimentation was previously reported from this laboratory. These early studies demonstrated the ability of certain of the indigenous bacteria of soil to survive and multiply during exposure to an environment closely simulating that presumed to be present on the planet Mars.

Recent experimentation has been directed toward an understanding of the relation of temperature, moisture, and other factors to the survival and multiplication of homogeneous cultures of bacteria in soil under these simulated Martian environmental conditions.

Several generalities regarding the simulated Martian environment may be stated.

1. Soil bacteria, selectively adapted to a simulated Martian environment, survive and multiply.

2. Sporeforming bacteria appear to have a higher rate of cell multiplication than strictly vegetative cells.

3. The factor of temperature cycling ($+25°$ C, diurnal, to minus $25°$ C, nocturnal) appears to yield greater numbers of viable cells than room temperature conditions.

4. Recoverable moisture in the soil-bacteria specimens in the Mars jars held under conditions of temperature cycling appears consistently less than that found under room temperature conditions.

The practical applications of such studies are discussed.

Die Reaktionen irdischer Mikroorganismen gegen simulierte Mars-Umweltsbedingungen. Von verschiedenen Autoren wurde die Hypothese einer Existenz von Leben auf dem Planeten Mars aufgestellt. Die Annahme ist allgemein, daß dieser Planet in unserem Sonnensystem derjenige ist, der neben der Erde wahrscheinlich am ehesten imstande ist, Leben zu beherbergen. Es wurde vermutet, daß die Natur

[1] Captain, U.S.A.F. (M.S.C.); Dr. DAVIS is the Assistant Chief, Department of Microbiology-Cellular Biology, School of Aviation Medicine, U. S. Air Force, Brooks Air Force Base, Texas, U.S.A.

[2] Lt. Colonel, U.S.A.F. (M.S.C.); Dr. FULTON, formerly Chief, Department of Microbiology-Cellular Biology, School of Aviation Medicine, U. S. Air Force, Brooks Air Force Base, Texas, presently is assigned to the U. S. Air Force Arctic Aeromedical Laboratory, Ladd Air Force Base, Alaska, U.S.A.

eines solchen möglichen Lebens auf dem Mars ähnlich der gewisser einfacher irdischer Lebensformen ist. Über den ersten Versuch, die Gültigkeit dieser Annahme mit Hilfe von Laboratoriumsversuchen zu prüfen, wurde schon früher aus unserem Laboratorium berichtet. Diese früheren Studien zeigten die Fähigkeit gewisser im Boden ansässiger Bakterien, zu überleben und sich zu vermehren, wenn sie einer Umgebung ausgesetzt wurden, die der auf dem Mars vermuteten äußerst ähnlich war.

Versuche aus jüngster Zeit richteten sich auf ein Verständnis der Beziehungen von Temperatur, Feuchtigkeit und anderen Faktoren zu der Überlebensfähigkeit und Vermehrung homogener Bakterienkulturen in Böden unter solchen Bedingungen einer simulierten Mars-Umwelt.

Einige Allgemeinheiten bezüglich der simulierten Marsbedingungen können angegeben werden.

1. Bodenbakterien, die in ausgewählter Weise einer simulierten Mars-Umwelt angepaßt sind, überleben und vermehren sich.

2. Sporenbildende Bakterien scheinen eine höhere Zellvermehrungsgeschwindigkeit zu besitzen als ausgesprochen vegetative Zellen.

3. Der Faktor eines Temperaturzyklus (+25° C bei Tag bis — 25° C bei Nacht) scheint größere Zahlen lebensfähiger Zellen zu liefern als Bedingungen bei Zimmertemperatur.

4. Regenerierbare Feuchtigkeit in den Proben von Bodenbakterien in „Mars-Kulturgläsern", die unter den Bedingungen des Temperaturzyklus gehalten werden, scheint durchwegs geringer zu sein als bei Zimmertemperatur.

Es werden die praktischen Anwendungen derartiger Studien erörtert.

Les réactions de micro-organismes terrestres à un milieu martien simulé. L'existence d'une forme de vie sur la planète Mars a fait l'objet de plusieurs hypothèses. L'avis général est que, en dehors de la Terre, Mars est la planète pour laquelle l'hypothèse d'une forme de vie est la plus vraisemblable. Il a été suggéré que cette forme rappellerait les formes les plus élémentaires de vie sur la Terre. Les premières tentatives de vérification expérimentale ont été conduites dans le laboratoire des auteurs. Elles ont montré la capacité de certaines bactéries du sol de survivre et de se multiplier dans un milieu simulant celui que l'on pense exister sur Mars.

Les expériences récentes ont cherché à comprendre les relations entre température, humidité et d'autres facteurs avec la survie et la multiplication de cultures homogènes dans ces conditions simulées. Les résultats généraux suivants peuvent être avancés:

1. Les bactéries du sol, adaptées sélectivement au milieu martien simulé, survivent et se multiplient.

2. Les bactéries à spores semblent avoir un taux de multiplication cellulaire plus élevé que les cellules strictement végétatives.

3. Le cyclage de la température (+25° C diurne à —25° C nocturne) semble conduire à un nombre supérieur de cellules viables que les conditions ambiantes normales.

4. L'humidité récupérable dans les spécimens de bactéries du sol contenus dans les bocaux martiens dans les conditions de cyclage de la température semble systématiquement inférieure à celle trouvée dans les conditions normales.

Les applications pratiques de ces études sont discutées.

The feasibility of extended flight in space has become more imminent in recent years and has increased interest in studies of the planets as possible environments for life. The existence of life on the planet Mars has been hypothesized by several authors [1—9]. It is generally stated that the Martian planet would probably be the most likely planet in our solar system, other than Earth, capable of supporting "life". The nature of such possible Martian life has been suggested to be similar to certain simple forms of terrestrial life. The first known attempt to test the validity of this assumption by means of laboratory experimentation was reported from our laboratory [10]. These continuing studies on the survival and multiplication of certain terrestrial forms of life under a simulated Martian environment will be elaborated upon in this communication.

Considerable information is available regarding the environmental conditions present, or hypothesized to be present, on the planet Mars. The more comprehensive reference texts are those written by the French astronomer, DE VAUCOULEURS [11, 12], and that edited by the American astronomer, KUIPER [13]. Much of this information has been abstracted and commented upon by STRUGHOLD [1]. These references provide us with the basic data of the Martian environment. It must be noted, however, that the qualitation and quantitation of these several environmental factors are controversial and, as such, are subject to change depending on the latest scientific observations. At first consideration it might appear that most, if not all, of these environmental conditions would be sufficiently extreme to preclude survival and multiplication of terrestrial life. However, several authors [1, 8, 9, 11, 13] have pointed out many terrestrial forms of life existing under extremes of natural terrestrial conditions which are similar in certain respects to certain of the existing Martian conditions. These terrestrial forms include the bacteria and algae found in hot springs, and the lichens, algae and even higher botanical forms growing under very hot and desert conditions, in very cold climates, and at relatively high altitudes. Indeed, TIKHOV [8, 9] has compared spectroscopic patterns obtained from high altitude and/or cold adapted vegetation with those patterns obtained from the Martian green areas and suggested certain similarities between the two patterns.

The original stimulus for laboratory experimentation in this new field of science, Astrobiology, was provided by Dr. STRUGHOLD in a "tentative outline of the functional specifications for an ecological Mars simulator" [14]. The environmental conditions considered included soil, water, atmosphere, temperature, and radiations. Each of these factors, with the exception of radiation, can be readily reproduced under laboratory conditions. Certain types of radiation, such as infrared and ultraviolet, can be produced without difficulty; cosmic radiation, which is considered by some to be a factor on Mars, at this time can not be produced in the laboratory. The gravitational force found on Mars, of course, can not be simulated in a laboratory at the Earth's surface.

Studies from our laboratory have employed the environmental factors shown in Table I. It is realized that those limits established for our experimental con-

Table I. *Simulated Martian Environment*

Factors	Simulated by
Atmospheric pressure	65 mm Hg.
Moisture Content	Approx. 1 %
Atmospheric composition	Commercial nitrogen gas
Soil type	Red sandstone, lava soil
Temperature range	+ 25 to — 25° C
	Diurnal Nocturnal

ditions are not precisely those established by the Martian specialists but it is felt that these approach the true conditions closely enough to be valid.

▶ These simulated conditions of the Martian environment are produced in modified Brewer anaerobic jars (Fig. 1) which are referred to as "Mars jars." The jars containing the microbial specimens are evacuated and flushed with nitrogen gas several times to remove residual atmospheric oxygen. They are then filled with commercial nitrogen to 65 mm pressure. The nocturnal-diurnal temperature range is simulated by holding the Mars jars at room temperature (approx. 25° C.) during the daylight hours and then placing them in the freezer (approx. minus 25° C) during the night. Soil types are simulated by employing red sandstone which is relatively high in iron oxide, and red and black lava soil obtained from the colored marl hills of Arizona. Moisture content of the soil is established by controlled desiccation of the soil and subsequent addition of water to the desired level. No attempts to simulate specific types of radiation have been employed in the studies completed thus far.

The first experiments to study the survival and growth of microorganisms exposed to simulated Martian environment were performed in our laboratory

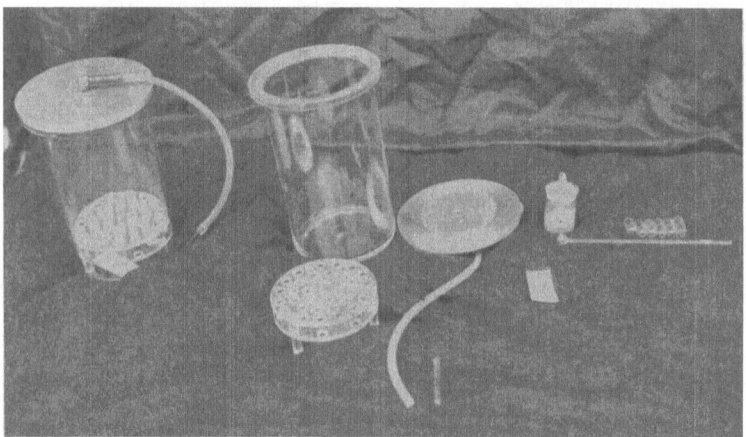

Fig. 1. "Mars Jars" and its component parts

[10, 15, 16]. In these studies sterile soil samples were inoculated with unsterilized portions of the same soils. The indigenous soil microflora included spore and vegetative forms of aerobic and anaerobic bacteria, yeasts, fungi, and actinomycetes. The initial microbial populations were determined and the remaining samples exposed to the simulated Martian environment in the Mars jars. Periodically, the jars were opened and multiple aliquots of the soil were removed and analyzed by standard laboratory techniques. Statistical analyses for significance were performed.

In these studies, the microbial population reproportioned itself in time with a gradual loss of the obligate aerobic forms (both bacteria and fungi) with a proportional increase in the anaerobic and facultative anaerobic forms. The greater percentage of the survivors were found to be divisible into three rather specific morphologic forms as follows: 1) highly pleomorphic, non-sporeforming organisms tentatively identified as *cornyebacteria*, 2) slender, short Gram-negative rods which as yet have not been classified, and 3) sporeforming organisms showing the morphologic characteristics of *clostridia*. The first two forms are facultative anaerobes, while the latter form is an obligate anaerobe. An increase in the total numbers of the surviving microorganisms also was observed. This would indicate that these forms were capable not only of surviving but also of multiplying.

More recent studies from this laboratory [17, 18] have been directed toward an understanding of the biological role of each of the several simulated Martian

environmental factors. In these experiments pure cultures were employed of
several bacterial species isolated from soil samples previously maintained in a
simulated Martian environment. Cultures of each species were admixed with
separate sterile soil specimens. Aliquots were measured into capsules and the
latter were placed into the Mars jars (Fig. 1). In addition to the several factors
simulating the Martian environment that were held constant, the design of each
experiment reflected the specific Martian environmental factor, or a combination
of these factors that were the variables. The effect of temperature cycling on the
survival and multiplication of selected facultative anaerobic bacteria subjected
to a simulated Martian environment will be discussed.

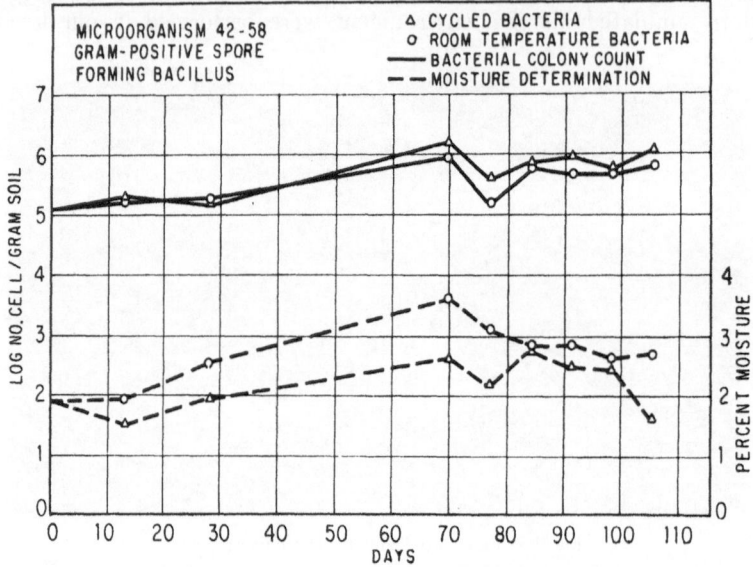

Fig. 2. The effect of the temperature factor in a simulated Martian environment on the total viable
cells and recoverable moisture in soil-bacteria specimens exposed in "Mars Jars." Bacterial species:
Gram-positive sporeforming bacillus

Some of the data obtained using a Gram-positive sporeforming bacillus are
shown in Fig. 2. The two solid upper curves represent log number of viable cells
per gram of soil. Data from the Mars jars that were cycled for temperature are
indicated by the triangles. Data from those jars retained at room temperature
are noted by the circles. Percent moisture recovered from the soil-bacteria specimens is shown with the two broken-line lower curves. It may be noted that during
the exposure time, the total viable cell counts of both cycled and room temperature specimens increased tenfold ($p < .001$). However, the bacteria in the cycled
jars showed significantly higher viable cell counts during the experimental period
($p < .01$). This is probably a reflection of the greater average rate of cell multiplication produced under the temperature cycling conditions simulating the
Martian environment compared to specimens held at room temperature. The
recoverable moisture curves in Fig. 2 show that the specimens which were cycled
for temperature to simulate this Martian environmental factor contained significantly less moisture on the average than those specimens which were held at
room temperature ($p < .01$).

An identical protocol using a Gram-positive non-sporeforming bacillus was
carried out. Some of the results are shown in Fig. 3. A significant drop in total

viable cells occurred initially ($p < .01$). From the fortieth day of incubation through the experimental time, there was no net change in total viable cells at either of

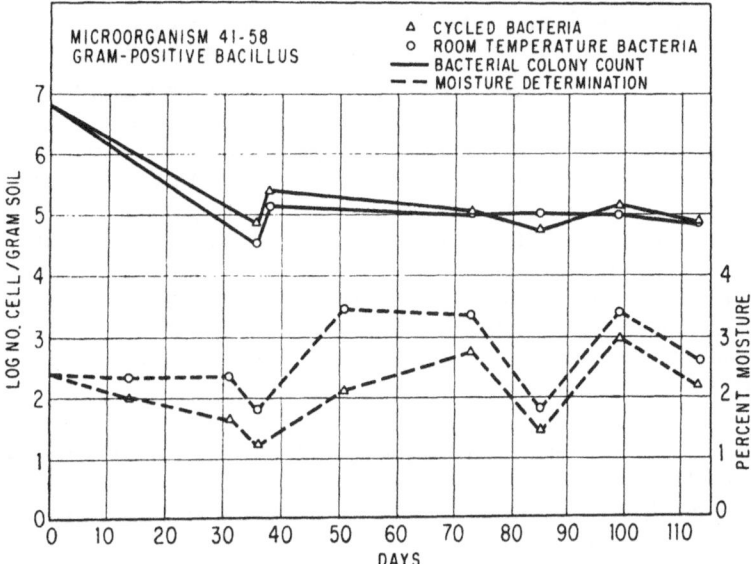

Fig. 3. The effect of the temperature factor in a simulated Martian environment on the total viable cells and recoverable moisture in soil-bacteria specimens exposed in "Mars Jars." Bacterial species: Gram-positive non-sporeforming bacillus

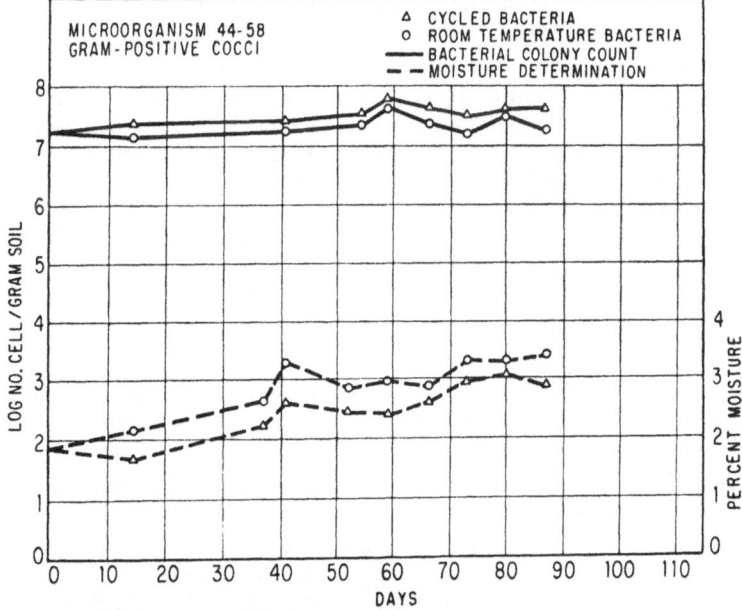

Fig. 4. The effect of the temperature factor in a simulated Martian environment on the total viable cells and recoverable moisture in soil-bacteria specimens exposed in "Mars Jars." Bacterial species: Gram-positive *coccus*

the two temperature conditions. It is reasonable to suppose that this maintenance of a constant level of viable cells is a reflection of dying cells as well as proliferating ones. In addition, it will be noted that the observation made with the previous bacterial species holds true for this organism also, that is, the average percent recoverable moisture of the specimens that were cycled for temperature during the simulation of the Martian environment was significantly less than those values obtained from the specimens that remained at room temperature ($p < .01$).

The third bacterial species that will be commented on was a Gram-positive *coccus*. As shown in Fig. 4, total viable cells were significantly higher ($p < .001$) during the time that the specimens were exposed to the simulated Martian environment that included the cycled temperature factor than when similar specimens were held in this simulated environment but at room temperature. In addition, an indication of increasing total viable cells with exposure time is shown in these two curves. Percent recoverable moisture, as found with the previous species discussed, was shown to be significantly less ($p < .01$) in those specimens maintained under cycled temperatures than in those held at room temperature. Also significant was the fact that percent recoverable moisture increased linearly with time ($p < .001$).

The results presented permit certain generalities regarding the survival and multiplication of terrestrial microorganisms exposed to a simulated Martian environment.

1. Soil bacteria, selectively adapted to a simulated Martian environment, survive and multiply.

2. Sporeforming bacteria appear to have a higher rate of cell multiplication than strictly vegetative cells.

3. A simulated Martian environment including temperature cycling, an environmental characteristic on Mars, appears to yield greater numbers of viable cells than a simulated Martian environment under room temperature conditions.

4. Recoverable moisture from specimens in a simulated Martian environment including temperature cycling appears consistently less than that found in specimens maintained in the simulated environment held at room temperature. Having summarized some of our laboratory findings, a few remarks concerning the applied aspects of such studies seem appropriate. One of our ultimate goals is to determine if simple forms of terrestrial life can not only survive and multiply but also contribute some product or perform some function required to support man in specific extraterrestrial environmental situations.

Man's survival on extraterrestrial bases will require a tremendous effort in logistic support. The greater the distance from a possible base of supply, the greater will be the logistic problems involved. As a consequence, it is obvious that the development of a functional small-scale, self-sustaining "ecosphere" will be mandatory for long-term operations of an extraterrestrial observational or exploratory base.

Another recently developed area of interest to which our studies can contribute concerns the possibility of contamination of extraterrestrial bodies with simple forms of terrestrial life, such as might occur as a result of the impactment of contaminated rockets on the extraterrestrial bodies. Studies such as those described herein can contribute information concerning the possibilities of such contamination and establish the requirements and need for rocket decontamination procedures.

Finally, it is suggested that studies on the physiology, genetics, and taxonomy of microorganisms under simulated extraterrestrial environments may contribute to astrobiology, space medicine, and science in general.

References

1. H. STRUGHOLD, The Green and Red Planet. Albuquerque: The University of New Mexico Press, 1953.
2. P. MOORE, The Worlds Around Us. New York: Abelard Schuman, 1956.
3. R. S. RICHARDSON, Exploring Mars. New York: McGraw-Hill, Inc., 1954.
4. I. M. LEVITT, A Space Traveler's Guide to Mars. New York: Henry Holt and Co., 1956.
5. K. HEUER, Men of Other Planets. New York: Pellegrini and Cudahy, 1951.
6. H. S. JONES, Life on Other Worlds, 2nd revised ed. New York: Macmillan Co., 1954.
7. F. A. PEREIRA, Introduçao a Astrobiologia. São Paulo: Sociedade Interplanetaria Brasileira, 1958.
8. G. A. TIKHOV, Astrobotany. Alma-Ata: Academy of Sciences of the Kasakh S.S.R., 1949.
9. G. A. TIKHOV, Is Life Possible on Other Planets? J. Brit. Astronom. Assoc. **65**, 193 (1955).
10. J. A. KOOISTRA, JR., R. B. MITCHELL, and H. STRUGHOLD, The Behavior of Microorganisms under Simulated Martian Environmental Conditions. Publ. Astronom. Soc. Pacif. **70**, 64 (1958).
11. G. DE VAUCOULEURS, Physics of the Planet Mars, Revised edition. London: Faber and Faber, 1953.
12. G. DE VAUCOULEURS, The Planet Mars. Translated by T. A. MOORE. London: Faber and Faber, 1950.
13. G. P. KUIPER, The Atmospheres of the Earth and Planets. Chicago: University of Chicago Press, 1951.
14. H. STRUGHOLD, Personal Communication.
15. J. A. KOOISTRA, JR., Unpublished Data.
16. J. D. FULTON, Survival of Terrestrial Micro-organisms under Simulated Martian Conditions. In: Physics and Medicine of the Atmosphere and Space. New York: John Wiley, 1959.
17. I. DAVIS, C. E. CRAFT, and J. D. FULTON, The Relation of Temperature and Moisture to the Survival and Multiplication of Selected Bacteria Subjected to a Simulated Martian Environment. Tex. Rep. Biol. Med. **17**, 198 (1959).
18. I. DAVIS, J. D. FULTON, and C. E. CRAFT, Unpublished Data.

The Corpuscular Rays from the Sun

By

C. E. Andersen[1]

(*Received August 27, 1959*)

Abstract — Zusammenfassung — Résumé

The Corpuscular Rays from the Sun. According to the theory predicted by the author at the VIth I.A.F. Congress in Copenhagen 1955, the sun expels material occasionally as surge eruptions with speeds which may be greater than the minimum escape velocity and these solar plasma masses proceed as narrow streams which may hit the Earth's atmosphere near the magnetic poles with a particle energy and flux of the same order of magnitude as the charged particles in the beam from a cyclotron and hence being fatally dangerous due to electromagnetic radiation of secondary nature. Now experience has unveiled charged particles with the same energy near the magnetic equatorial plane of the Earth, being presumably deflected particles of the sun plasma streams, which get long lifetimes near the Earth. Comparison between theory and experience is given.

Die korpuskularen Strahlen von der Sonne. Gemäß einer von dem Verfasser beim VI. I.A.F.-Kongreß 1955 in Kopenhagen entwickelten Theorie schleudert die Sonne gelegentlich Material als Stoßeruptionen mit Geschwindigkeiten aus, die größer sein können als die Fluchtgeschwindigkeit. Diese solaren Plasmamassen pflanzen sich als eng begrenzte Ströme fort, welche die Erdatmosphäre in der Nähe der magnetischen Pole mit einer Teilchenenergie und einem Fluß derselben Größenordnung treffen können, wie sie die geladenen Partikel in einem Teilchenstrom aus einem Zyklotron besitzen. Infolgedessen müssen sie infolge sekundärer elektromagnetischer Strahlung von tödlicher Gefahr sein. Nunmehr hat die experimentelle Erfahrung geladene Teilchen mit der gleichen Energie in der Nähe der Ebene des magnetischen Erdäquators gefunden, die wahrscheinlich abgelenkte Partikel der solaren Plasmaströme sind und lange Lebensdauer in der Nähe der Erde besitzen. Es wird ein Vergleich zwischen der Theorie und der experimentellen Erfahrung gezogen.

Le rayonnement corpusculaire du soleil. Suivant la théorie avancée par l'auteur au VIe Congrès de la F.I.A., le soleil émet occasionnellement des particules avec une vitesse supérieure à la vitesse de libération. Ces courants de plasma peuvent pénétrer l'atmosphère terrestre au voisinage des pôles magnétiques avec une énergie et un flux comparables à ceux d'un cyclotron. Ils constituent un danger en raison du rayonnement secondaire. Des particules de même énergie ont maintenant été observées dans le voisinage du plan équatorial. Ce sont vraisemblablement des particules déviées provenant des courants de plasma, qui ont une durée de vie considérable au voisinage de la terre. On fait une comparaison entre théorie et expérience.

During the International Astronautical Congress, held in Copenhagen 1955, a lecture was given, called: "The Great Danger: the Corpuscular Rays from the Sun", and an 11-page hectographed paper was distributed among the participants.

---

[1] Römersgade 7⁴, Copenhagen K, Denmark.

During the congress in Amsterdam last year preliminary experimental results were given about an intense corpuscular radiation in the neighborhood of the Earth.

Later it turned out that this radiation was mainly concentrated near the equatorial plane of the earth, while the predicted radiation should occur mainly near the geomagnetic poles. But the qualitative composition of the rays, say the particle types and their energy distribution, are essentially the same. Hence, the conclusion is near, that the particles are the same, only appearing in two phases of existence.

As both a theory and experience have been given it might perhaps seem superfluous to proceed with the theme. But on the other hand it is of interest to see, if there is consistency between the predicted theory and experience, though post experience refinements of the theory surely will give a more precise and detailed picture.

Moreover, some sides of the original theory have not yet been confirmed. They could easily be investigated, since the space vehicles have come to existence.

The theory was, that the sun gives off corpuscular rays with velocities greater than the gravity-escape velocity. There are presumably two types of emission, one characterized by plasma which proceeds directly away from the sun as self-focussing material, thought stretching out and being intrinsically accelerated to long narrow streams. Another emission is presumably rather diffuse, at least not concentrated in narrow streams. It consists of spread material with generally smaller kinetic energies.

The first form of emission is associated with solar flares and hence sunspot areas. The other form of emission is not associated with sunspots. This form of emisison is presumably the dominating one, though not giving rise to sudden and violent effects. Curiously enough, it comes from only one side of the sun. It has been traced through 200 sun revolutions and has a phase time of 26.88 days, a rather exact period, as against to the sun spot-revolution period which varies with the sun-latitude.

Both forms of corpuscle radiation give rise to aurora phenomenae, though not alike.

The aurora rays proceed to a certain lower limit in the atmosphere. From the path length (g cm^{-2}) the energy of protons can easily be calculated and appears to be of the order of a mev. But this is only valid for particles with straight paths. As the particles are spiralling along the magnetic force lines, their real path lengths are much longer and the energies consequently greater.

From the luminosity effect of the aurorae, especially when observed from the ground directly in the ray directions, that is the small area with maximum luminosity, one may calculate a minimum of the stream energy per sq. cm per sec. I calculated it from certain observation data in the literature. In extreme cases it might amount to 3.10^{-3} cal. per sec. per cm^2. But as only a little part of the energy is transformed to visual light energy, the real value is many times greater.

From the total energy of the stream and the energy per particle, the total number of particles was calculated. With 0.5 mev per particle and 0.3 cal per cm^2 per sec the number of particles in the stream hitting the earth atmosphere might be as much as 10^{13}. This number, quoted from my first lecture, must be reduced by perhaps: 2 powers of ten, because the paths of the particles in the atmosphere are unknown, due to their spiralling paths so that the particle energies are much greater.

Nevertheless the number is very great, almost of the same order as the particle beam from a cyclotron.

As the material is plasma, protons, alpha-particles and electrons, all having presumably the same kinetic energies—even the electrons are easily accelerated, due to their lightness, and hence will have much greater speeds than the spiralling protons and other nuclei—they will be very dangerous. The charged nuclei can give rise to coulomb excitation of rocket shells of aluminium, wich has many excitation levels between 0.5 and 1 mev. Particularly the bremsstrahlung from the electrons will be dangerous, because their deceleration gives X-rays exactly as in a hospital.

If e.g. 1 $^0/_{00}$ of the kinetic energy of the stream—0.3 cal. per cm^2 per sec.— is converted to X-rays, which pass the rocket hull, this is 3.10^{-4} cal or more than 10^4 erg. per cm^2 per sec. As 1 röntgen is 83 ergs absorbed in 1 g of air, it is more than 100 röntgens per sec., if absorbed uniformly in a layer of about 1 cm or water or flesh.

This and other results were communicated to the participants of the Copenhagen Congress, but none seemed to be frightened by that.

The conclusion was drawn that it would be fatally dangerous to be hit by the concentrated streams of plasma-materials from the sun. But the chance of being hit would be small, particularly near the Earth, when being near the equatorial plane, when the sun had no sunspots just facing the earth and when turning the calm side against the Earth (see later). The existence of the belt radiation was not foreseen.

As mentioned above the sun emits corpuscular radiation associated with the sunspot areas, but also from other areas. In 1948—51 only 15% of the aurorae and geomagnetic effects observed were associated with the sunspot areas.

Now it is a curious experience, done by Dr. JOHANNES OLSEN, the leader of the Danish Geomagnetic Research in Greenland through about 35 years, that the sun has a "friendly" side and an "unfriendly" side (apart from the sunspot areas).

While the lifetimes of the sunspots are generally a month ore some months, very seldom more than a year, dr. OLSEN could trace the corpuscle-emitting side of the sun through all the period 1926—41. It was followed through circ. 200 sun-revolution periods. The revolution periods of the sunspots are varying with the latitude. But the revolution period of the sun's corpuscle emitting side turned out to be constant, being 26.88 days[1]. This is a very interesting astrophysical discovery.—But it is also of some interest in connection with space travel.

The radiation in question has presumably smaller particle energies than the energies of the particles in the streams. But as the particles are very much spread out, the total radiation particle number may well be much greater than the number of particles in the streams, and they seem to occur everywhere, when the sun turns its dangerous side against us. Space travellers will be hit not only near the Earth, but also far away from the Earth. Hence this radiation might perhaps be dangerous. At least it must be investigated.

Now it should be possible to prove both the stream particle theory and the one-face emission theory directly from artificial moons and planets. This must be among the foremost research objects.

The belt with similar radiation, concentrated near the Earth's magnetic equatorial plane, found by VAN ALLEN, seems to have the same particle nature as the predicted radiation from the sun, and is probably due to a capture phenomenon.—Now the problem arises: Does the belt radiation originate from the

---

[1] See: Terrestrial Magnetism and Atmospheric Electricity, p. 123—134, 1948.

stream corpuscular radiation from the sun or from the emission from a certain area of the sun with emission all times, and not associated with sunspots, flares etc. ?

It will be of great interest to find out, if the belt radiation shows a regular variation with the 26.88 day period though dampened and phase-shifted. If this may be proved to be the case, it should theoretically be possible to deduce the lifetimes of the belt radiation particles.

In practice, it may be difficult, even if the lifetime of the belt particles should be as short as a sun-revolution period, because the stream radiation will interfere. May be this is the dominating belt-radiation producing factor. If that is the case, the belt radiation will, if lifetimes are short, vary rather violently. At least there seems to be a possibility of the proof: Does the belt radiation arise mainly from the one or the other source of corpuscular radiation from the sun ?

Experience is more reliable than any predicted theory. Sometimes the experiences turn out to be very difficult, in other cases they give unexpected clear results. It is my hope that experimenters will take up these and connected problems.

A Study of Hypersonic Ablation

By

Sinclaire M. Scala[1]

(With 21 Figures)

(Received August 27, 1959)

Abstract — Zusammenfassung — Résumé

A Study of Hypersonic Ablation. The successful design of a hypersonic vehicle for manned re-entry requires "a priori" knowledge of the behavior of high temperature materials subjected to a severe aerothermochemical environment. The three general classes of materials currently being contemplated for high temperature application to re-entry vehicles include metals, ceramics and plastics. Metals and the crystalline forms of certain ceramics are typical crystalline solids. Graphitic materials and the general category of ceramic oxides, carbides and nitrides are typical amorphous solids. The plastics are essentially thermosetting resins, which are sometimes reinforced with organic or refractory fibers in varying proportions and orientations.

The response of a given material in each of the broad classes listed above will depend critically on the rate of application, magnitude, duration and nature of the heating load. That is, the response of a material to a thermal load during hypersonic re-entry is not so much dependent on the total integrated heat transfer during the time of flight as it is on the precise combination of the instantaneous enthalpy and pressure of the chemically reacting gas which envelopes the vehicle as it re-enters.

If the coefficient of thermal conductivity of an oxidation resistant material is sufficiently high relative to the rate of heat transfer, then equilibration of the heat transfer with the heat sink capacity of the material occurs, and the temperature of the material rises with time. If the surface temperature remains below the softening point or below the vaporization temperature during the entire heating cycle, this type of material retains its structural integrity and ablation does not occur. Most metals can, in fact, behave in this fashion. At higher heating rates, the temperature near the surface of the material rapidly rises to the point where thermal degradation of one form or another begins.

For a certain class of refractory materials, the ablation process will usually consist of melting followed by vaporization from the interface between the gas and liquid phase boundary layers. The fraction of solid that enters the gas phase during the ablation process comprises the most effective utilization of the heat absorbing and heat blocking action of the entire mass transfer process. When the fractional part of the solid which gasifies, called Γ the gasification ratio, approaches unity, the process is called sublimation. When the gasification ratio is zero, the phenomenon is one of pure melting. The value of the gasification ratio is not a fixed constant for a given material but is a function of the material and its environment. That is, Γ depends critically on the enthalpy, pressure and chemistry of the environment in which the material performs. Quartz-like crystalline and amorphous refractories are typical of this class of materials.

---

[1] Manager, High Altitude Aerodynamics, Space Sciences Laboratory, Missile and Space Vehicle Department, General Electric Company, Philadelphia 24, Pennsylvania, U.S.A.

Other refractory materials, such as graphite, do not melt but undergo heterogeneous chemical reactions. The rate of mass loss from the surface of this type of material will depend on an oxidation process, which is rate-controlled at low surface temperatures, but rapidly becomes diffusion-controlled as the surface temperature rises. In this type of ablation process, the chemical heat release during the oxidation process and the blocking action of the mass transfer process due to the thickening of the boundary layer will depend critically on the heterogeneous reaction rates. When it is assumed that the rates of reaction are infinitely fast, then the equilibrium constants of the various combustion reactions will yield sufficient constraints to make the solution to the problem determinate.

In considering the various classes of plastic materials, thermal degradation of a reinforced plastic involves pyrolysis of the solid phase, which results in the formation of gaseous products such as methane, ethylene, acetylene and hydrogen, leaving a solid residue behind in the form of a non-uniform cross-linked char sponge. The problem is also complicated by combustion reactions which occur between the gaseous hydrocarbons and the atomic and molecular oxygen present in dissociated air. Another class of thermoplastic resins decomposes by means of a process called depolymerization. In this case, a complex hydrocarbon polymer undergoes chemical reactions leading to the formation of monomer. This monomer can subsequently undergo combustion reactions with the oxygen in the gas which flows around the vehicle, resulting in an increase in heat transfer to the ablating plastic. Typical of this type of material is Teflon.

In this study, the analysis was confined to the region of most intense heating, namely the forward stagnation point. The multicomponent hypersonic laminar boundary layer equations were derived and solutions were obtained by considering chemical as well as gas dynamic factors. The diffusion and counterdiffusion of the various chemical components including oxygen and the products of vaporization or combustion were evaluated and microscopic chemical compatibility conditions were satisfied at the surface. The blocking action due to boundary layer thickening was included, as well as the heat release due to the chemical reactions, for each of the categories of ablating materials discussed above, e. g.

a) The ablation of a quartz-like refractory vaporizing oxide,
b) the ablation of a graphite-like refractory material,
c) the ablation of a reinforced plastic, and
d) the ablation of a thermoplastic resin.

The dependence of the "effective heat of ablation", (which is the amount of heat blocked or absorbed per pound of material ablated) upon enthalpy, pressure and chemistry was determined for the entire hypersonic flight regime.

For purposes of further understanding the behavior of these materials, a systems comparison was also performed in which each of the materials was employed as a thermal shield for a re-entry vehicle for a variety of re-entry satellite trajectories.

Untersuchung der Hyperschall-Ablation. Der erfolgreiche Entwurf eines Hyperschall-Fahrzeuges „für bemannten Wiedereintritt" benötigt a priori Wissen über das Verhalten von Hochtemperatur-Materialien, die einer strengen aerothermochemischen Umgebung ausgesetzt sind.

Drei allgemeine Klassen von Materialien, für die gegenwärtig für hohe Temperatur eine Anwendung in Wiedereintrittsfahrzeugen erwogen wird, schließen Metalle, Tonwaren und plastische Materialien ein. Metalle und die kristalline Form bestimmter Tonwaren sind typische kristalline feste Körper. Graphitmaterialien und die allgemeine Kategorie der Tonwarenoxyde, Carbide und Nitride sind typische amorphe feste Körper. Plastische Materialien sind Kunstharze, die manchmal mit organischen oder refraktären Fasern in verschiedenen Proportionen und Richtungen verstärkt sind.

Die Eignung eines gegebenen Materials in jeder der oben angeführten umfangreichen Klassen hängt von dem Maßstab der Anwendung, Größe, Dauer und Natur der Wärmeaufnahme ab. Das heißt, die Eignung eines Materials für eine thermische Beladung während des Hyperschall-Wiedereintritts hängt nicht so sehr von der totalen integrierten Wärmeübertragung während der Flugzeit ab, wie von der genauen Kombination der augenblicklichen Enthalpie und dem Druck eines chemisch reagierenden Gases, das das Fahrzeug beim Wiedereintreten einhüllt.

Wenn der Koeffizient der thermischen Leitfähigkeit eines gegen Oxydation widerstandsfähigen Materials relativ zur Rate der Wärmeübertragung groß genug ist, dann tritt Gleichgewicht zwischen der Wärmeübertragung und dem Wärmesenkvermögen des Materials ein, und die Temperatur des Materials erhöht sich mit der Zeit. Wenn die Oberflächentemperatur unter dem Erweichungspunkt oder unter der Verdampfungstemperatur während des ganzen Erhitzungszyklus bleibt, dann wird diese Art von Material ihre strukturelle Integrität behalten und Ablation (Schwund) wird nicht eintreten. Die meisten Metalle werden sich in der Tat auf diese Weise verhalten.

Bei höherer Erhitzung steigt die Temperatur in der Nähe der Oberfläche des Materials schnell bis zu dem Punkt an, wo thermische Degradation der einen oder anderen Form eintritt.

Für eine bestimmte Klasse refraktärer Materialien wird der Schwundprozeß gewöhnlich aus Schmelzen und nachfolgendem Verdampfen der Grenzschichten zwischen der gasförmigen und flüssigen Phase bestehen. Der Teil des festen Körpers, der während des Schwundprozesses in die Gasphase eintritt, stellt die beste Nutzbarmachung der hitzeabsorbierenden und hitzeblockierenden Wirkung während des ganzen Massetransferprozesses dar. Wenn der Bruchteil des festen Körpers, der vergast wird, Γ-Vergasungsverhältnis genannt, eins wird, wird der Prozeß als Sublimation bezeichnet. Wenn der Vergasungsquotient Null ist, tritt die Erscheinung des reinen Schmelzens auf. Der Wert des Vergasungsquotienten ist für ein gegebenes Material keine Konstante, sondern eine Funktion des Materials und seiner Umgebung. Das heißt, Γ hängt von Enthalpie, Druck und Chemie der Umgebung ab, in der sich das Material befindet. Quarzähnliche, kristalline und amorphe refraktäre Substanzen sind typisch für diese Klasse von Materialien.

Andere refraktäre Materialien wie Graphit schmelzen nicht, sondern erleiden heterogene chemische Reaktionen. Der Maßstab des Masseverlustes der Oberfläche dieser Art von Material wird von einem Oxydationsprozeß abhängen, der bei niedrigen Oberflächentemperaturen geschwindigkeitskontrolliert ist und schnell diffusionskontrolliert wird, wenn die Oberflächentemperatur steigt. In dieser Art von Ablationsprozeß wird die Freisetzung chemischer Wärme während des Oxydationsprozesses und der blockierenden Aktion des Massetransferprozesses, der durch eine Verdickung der Grenzschicht hervorgerufen wird, kritisch von den Geschwindigkeiten der heterogenen Reaktionen abhängen. Wenn angenommen wird, daß die Geschwindigkeiten der Reaktionen unendlich schnell sind, werden die Gleichgewichtskonstanten der verschiedenen Verbrennungsreaktionen genügend Zwang erzielen, um die Lösung für das Problem bestimmbar zu machen.

Zieht man verschiedene Klassen plastischer Materialien in Erwägung, findet man, daß thermische Degradation eines verstärkten plastischen Materials Pyrolyse der festen Phase mit sich bringt, die in der Bildung von Gasprodukten wie Methan, Äthylen, Acetylen und Wasserstoff besteht und einen festen Rückstand in Form eines uneinheitlichen, verkohlten Schwammes hinterläßt. Das Problem wird ebenso durch Verbrennungsreaktionen kompliziert, die zwischen den gasförmigen Kohlenwasserstoffen und dem atomaren und molekularen Sauerstoff, der sich in dissoziierter Luft befindet, auftreten. Eine andere Klasse plastischer Materialien zersetzt sich durch Polymerisation. Hier erleidet ein komplexes Kohlenwasserstoff-Polymeres chemische Reaktionen, die zur Bildung eines Monomeren führen. Dieses Monomere wird dann Verbrennungsreaktionen mit dem Sauerstoff durchmachen, der sich in dem das Fahrzeug umhüllenden Gas befindet. Dies äußert sich in einer Zunahme der Wärmeübertragung auf das sich zersetzende plastische Material. Typisch für diese Art von Material ist Teflon.

In der vorliegenden Untersuchung beschränkt sich die Analyse auf das Gebiet der intensivsten Erhitzung. Die aus vielen Bestandteilen bestehenden laminierten Hyperschall-Grenzschicht-Gleichungen wurden abgeleitet und Lösungen wurden durch Betrachtung chemischer und gasdynamischer Faktoren erhalten. Die Diffusion und Gegendiffusion der verschiedenen chemischen Komponenten, einschließlich des Sauerstoffs und der Verdunstungs- und Verbrennungsprodukte, wurden ausgewertet und die mikroskopisch-chemischen Verträglichkeitsbedingungen waren an der Oberfläche ausreichend. Die blockierende Aktion, die auf Grenzschichtverdickung beruht, und die Hitzefreisetzung durch chemische Reaktionen wurden für jede Kategorie der oben diskutierten zersetzenden Materialien einbezogen:

a) Die Ablation eines quarzähnlichen, refraktären, verdampfenden Oxydes;

b) die Ablation eines graphitähnlichen, refraktären Materials;

c) die Ablation eines verstärkten plastischen Materials;

d) die Ablation eines *thermoplastic*, kunstharzartigen plastischen Materials.

Die Abhängigkeit der „wirksamen Ablationshitze" (die der Betrag der blockierten oder absorbierten Wärme pro Pfund zersetzten Materials ist) von Enthalpie, Druck und Chemie wurde für das ganze Hyperschallflug-System bestimmt.

Zwecks weiteren Verstehens des Verhaltens dieser Materialien wurde ein Systemvergleich angeführt, in dem jedes Material als thermischer Schutz für Wiedereintrittsfahrzeuge für eine Reihe von Wiedereintritts-Satellitenbahnen verwendet wurde.

Etude de l'ablation en régime hypersonique. La conception heureuse d'un véhicule hypersonique destiné à la ré-entrée avec équipage demande une connaissance préalable du comportement des matériaux soumis à des conditions aérothermochimiques sévères. Les métaux, les céramiques et les plastiques sont les trois classes de matériaux envisagés actuellement. Les métaux et les formes cristallines de certaines céramiques sont des solides cristallins typiques. Les graphites et la catégorie générale des oxydes, nitrures et carbures céramiques sont des solides typiquement amorphes. Les plastiques sont essentiellement des résines thermodurcissables, quelquefois renforcées en diverses proportions et orientations par des fibres organiques ou réfractaires.

Dans chacune de ces classes le comportement du matériau dépend essentiellement de l'intensité, de la variation, de la durée et de la nature de la charge thermique. Autrement dit, le comportement ne dépend pas tellement de la chaleur totale reçue que de la combinaison exacte de l'enthalpie instantanée et de la pression du gaz chimiquement actif qui entoure le véhicule durant la phase de ré-entrée.

Si le coefficient de conductivité thermique d'un matériau résistant à l'oxydation est assez élevé relativement au taux de transfert de chaleur un équilibre s'établit entre le transfert et la capacité d'absorption et la température du matériau augmente avec le temps. Si la température en surface reste inférieure au point de ramollissement ou à la température de vaporisation durant tout le cycle d'échauffement, l'intégrité structurelle est maintenue et il n'y a pas d'ablation. La plus part des métaux se comportent en fait de cette manière. Pour des taux d'échauffement plus grands la température de surface monte rapidement au point où l'une ou l'autre forme de dégradation thermique commence à se manifester.

Pour certaines classes de matériaux réfractaires le processus d'ablation consiste usuellement en une fusion suivie de vaporisation à l'interface gazeuse-liquide de la couche limite. La fraction solide qui passe en phase gazeuse est la plus utile pour s'opposer à l'absorption. Quand la fraction solide Γ, qui passe à l'état gazeux est nulle, le processus est de pure fusion. Γ n'est pas une caractéristique fixe d'un matériau mais dépend aussi de l'ambiance, et surtout de l'enthalpie, pression et caractère chimique du milieu ambiant. Les réfractaires amorphes et cristallins du genre du quartz sont typiques dans cette classe de matériaux.

D'autres matériaux réfractaires, tels que le graphite, ne subissent pas de fusion mais des réactions chimiques hétérogènes. Le taux de perte de masse de ces matériaux dépend d'un processus d'oxydation limité par cinétique chimique aux faibles températures, puis par diffusion quand la température de surface augmente. Dans ce type d'ablation, la chaleur d'oxydation et le blocage thermique par transfert de masse, dû à l'épaississement de la couche limite, dépendent de façon critique des taux de réaction hétérogènes. Si ces taux de réaction sont supposés infinis, les constantes d'équilibre des réactions fournissent assez de liaisons pour rendre le problème déterminé.

Pour ce qui est des matériaux plastiques, la dégradation thermique d'un plastique renforcé comporte une pyrolyse de la phase solide avec formation de produits gazeux tels que le méthane, l'éthylène, l'acétylène et l'hydrogène, laissant un résidu solide sous forme d'une matrice spongieuse. Le problème est compliqué par les réactions de combustion des hydrocarbones gazeux dans l'oxygène atomique et moléculaire présents dans l'atmosphère dissociée. Une autre classe de plastiques thermodurcissables se décompose par dépolymérisation. Le monomère résultant de la décomposition réagit avec l'oxygène pour accroître le taux de transfert de chaleur vers le plastique en ablation. Un matériau typique de ce genre est le Teflon.

Dans cette étude l'analyse a été confinée à la région d'échauffement la plus intense, le point de stagnation. Les équations de la couche limite hypersonique à composantes multiples ont été établies et leur solution prend en considération les phénomènes chimiques aussi bien que dynamiques. La diffusion et contre-diffusion des divers composants chimiques, comprenant l'oxygène et les produits de vaporisation et de combustion ont été évaluées et les conditions de compatibilité chimique satisfaites à la surface. Le blocage par épaississement de la couche limite et la chaleur dégagée par les réactions chimiques ont été inclus pour chacune des catégories mentionnées précédemment:

    a) ablation d'un oxyde réfractaire du genre quartz,
    b) ablation d'un matériau réfractaire du genre graphite,
    c) ablation d'un plastique renforcé,
    d) ablation d'une résine thermodurcissante.

La dépendance de la "chaleur effective d'ablation" ou quantité de chaleur bloquée ou absorbée par unité de poids du matériau vis à vis de l'enthalpie, pression et nature chimique du milieu ambiant a été déterminée pour l'entièreté du régime hypersonique.

En vue de mieux saisir le comportement de ces matériaux une comparaison a été faite pour diverses trajectoires de ré-entrée, chaque matériau étant supposé utilisé comme protection thermique.

Nomenclature

Symbols

$\mathfrak{A}$	defined in eq. (25)	$h = \sum_i C_i h_i$	static enthalpy of gaseous mixture
A	area		
Alt.	altitude		
B	effective collision frequency	ΔH_{vap}	enthalpy of vaporization
		$j_i = \varrho_i \vec{V_i}$	diffusion flux of species i
C_D	drag coefficient	$\vec{J}$	mechanical equivalent of heat
C_i	mass fraction of species i		
C_{p_i}	frozen specific heat of species i at constant pressure	k	specific reaction rate
		K	frozen thermal conductivity of gaseous mixture
$\overline{C}_p = \sum_i C_i C_{P_i}$	mean frozen gas specific heat of mixture	K_{p_i}	equilibrium constant of species i
C_s	specific heat of solid	$Le = \dfrac{\varrho \overline{C}_p \mathfrak{D}_{ij}}{K}$	frozen Lewis number
$\mathfrak{D}_{ij}$	binary diffusion coefficient		
$D_i{}^T$	thermal diffusion coefficient of species i	$Le^T = \dfrac{\overline{C}_p D_i{}^T}{K}$	frozen thermal Lewis number
E	activation energy	$l = \varrho \mu / \varrho_w \mu_w$	
f	similarity stream function	$\dot{m}_w = \varrho_w v_w$	interphase mass transfer
f_η	dimensionless velocity	M_i	molecular weight of species i
$h_i = \int_{T_{\text{ref}}}^{T} C_{P_i} dT + \Delta h_{f_i}{}^0$	enthalpy of species i including chemical	$\overline{M} = \sum_i X_i M_i$	mean molecular weight of gaseous mixture
$\Delta h_{f_i}{}^0$	heat of formation of species i evaluated at T_{ref}	n	number of chemical species

P_i — partial pressure of species i

$P = \sum_i P_i$ — local static pressure

$Pr = \dfrac{\bar{C}_p \mu}{K}$ — frozen PRANDTL number

Q — heat transfer

$\tilde{Q} = K \dfrac{\partial T}{\partial y} - \Sigma \varrho_i \vec{V_i} h_i$ — energy function

$\dfrac{\triangle \tilde{Q}}{\triangle \dot{m}}$ — "effectiveness quotient", unit change in energy function per unit mass transfer

Q^* — effective heat of ablation

r — radius of cross-section of body

$\Re$ — universal gas constant

$R = \dfrac{\Re}{\bar{M}}$ — specific gas constant

R_B — nose radius of body

T — temperature

T_0 — temperature at interior of solid

t — time

u — x component of velocity

v — y component of velocity

$\vec{V_i}$ — diffusion velocity of species i

$\vec{v_i}$ — absolute velocity of species i

V_∞ — flight speed

$\dot{w}_i$ — chemical source term, mass rate of production of species i by chemical reaction per unit volume per unit time

W — weight

x, y, r — body oriented coordinate system

$\dot{y}$ — linear rate of surface recession

α — thermal diffusivity

α_{vap} — vaporization coefficient

Γ — gasification ratio

δ — symbol denoting change in temperature due to endothermic reaction in condensed phase

ε — emissivity

η, ξ — similarity variables

λ — mean free path

$\theta = \dfrac{T}{T_e}$ — dimensionless temperature

Θ — EINSTEIN function

μ — viscosity

Ξ — second gasification ratio

ϱ — density

σ — STEFAN-BOLTZMANN constant

σ_{ij} — collision diameter

Ψ — stream function

Subscripts

aero. — aerodynamic
A — air atoms
cal. — calorimeter
char comb. — char combustion
e — outer edge of boundary layer
eq. — equilibrium
(g) — gaseous species
g — gas cap, shock layer
i — i^{th} species
k — vaporizing species
l — liquid phase
M — air molecules

p — pyrolysis
rad. — radiation
(s) — solid phase
s — stagnation point, solid phase
tot. — total
$v. p.$ — virgin plastic
w — wall, interface
o — sea level standard
∞ — upstream of shock, edge of boundary layer
η — denotes differentiation with respect to η

Introduction

Astronomers and astrophysicists have for many years applied the term "ablation" to the process of thermal erosion and disintegration which meteors experience when entering the earth's atmosphere. Considerable literature has appeared during the last thirty-six years, including the contributions of LINDE-MANN and DOBSON [1], SPARROW [2], ÖPIK [3, 4] and THOMAS [5], which deals with meteor observations and theories relating the luminosity of a meteor trail to the mass ablation rate. It is of interest to note that in the above theoretical studies of meteor ablation, it was postulated that the environment was primarily one of free molecule flow, and heat transfer to the surface occurred by means of thermal accomodation between high speed incident air particles and surface molecules.

Recent interest in the response of materials to severe heating stems from the analyses of the re-entry problem by ALLEN and EGGERS [6], GAZLEY [7], and others who considered the thermal protection of a space vehicle which re-enters the earth's atmosphere at hypersonic speed. The major implication of these studies was that the severity of the re-entry problem could be minimized by utilizing a vehicle having a blunt nose (large R_B), a small value of the ballistic parameter $W/C_D A$, and a shallow entry angle from the horizontal. A transient heat con-

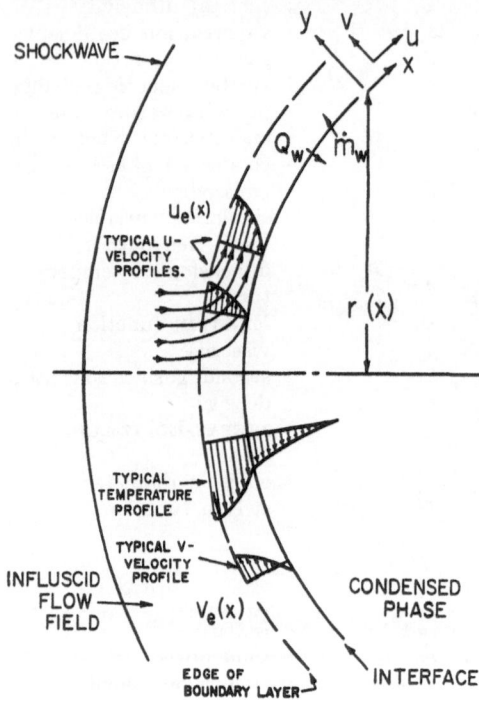

Fig. 1. Coordinate system and profiles for hypersonic stagnation point boundary layer

duction analysis shows readily that if the thermal diffusivity of a material is sufficiently high relative to the heating rate to which it is exposed, then the material merely absorbs the convective and net radiative heat transfer, and transmits the heat into the interior by conduction. However, during hypersonic re-entry, a blunt-nosed space vehicle generates a curved detached shock wave (see Fig. 1) which results in equilibrium gas cap temperatures of the order of 13,000° R, and stagnation pressures of the order of 100 atmospheres. Thus, significant heating occurs in the continuum regime, and if one utilizes the hypersonic heat transfer rates predicted by ROMIG [8], LEES [9], or FAY and RIDDELL [10], one finds that during re-entry, the surface temperature of a vehicle can rise rapidly to a level where thermal degradation occurs.

Although the initial definition of ablation connoted catastrophic failure, in contemporary usage the aerothermodynamicist prefers to use the word ablation to denote the more desirable situation in which the skin of a hypersonic re-entry vehicle degrades thermally, but does not fail structurally.

The earliest rigorous study of melting ablation under the conditions of hypersonic re-entry is due to SUTTON and SCALA [11], in which a glass-reinforced plastic was analyzed for a typical hypersonic flight condition. A later study by SCALA and SUTTON [12] considered the melting and vaporization of a glassy material having a relatively low viscosity in the liquid phase.

Other recent theoretical studies of the blunt body re-entry ablation problem include the work on the melting of a glassy material by BETHE and ADAMS [13], the study of sublimation by DENNISON and DOOLEY [14, 15], the study of teflon pyrolysis by GEORGIEV, HIDALGO and ADAMS [16], the studies of melting by LEES [17], LEW and FANUCCI [18], and ROBERTS [19], the analyses of melting and vaporization and combustion by SCALA [20, 21, 22] and the study of vaporization by SCALA and VIDALE [23]. In this paper, the author will present a unified picture of the mechanism of thermal degradation for four different classes of ablating materials, including melting and vaporization, pyrolysis, depolymerization and combustion ablation.

It has now become common practice to use the symbol $Q*$ to represent the "effective heat of ablation", which may be defined as the heat blocked or absorbed per unit mass loss from the ablating material. The quantity $Q*$ represents the complex interaction between a material and its environment, and consequently is not a constant property of a given material, but rather its magnitude depends on both the material properties and the environmental conditions to which the material is exposed during flight.

The "effective heat of ablation" may be defined by the ratio:

$$Q* = \frac{\text{heat transfer to a non-ablating calorimeter}}{\text{total mass ablated}} \tag{1}$$

where the calorimeter is assumed to have the same surface temperature, emissivity and catalytic efficiency [24] as the ablating material. The total effectiveness of the material is then due to the separate contributions of:

a) heat blocking due to boundary layer thickening caused by mass transfer

b) heat absorption within the material, including phase changes

c) heat absorption or release due to heterogeneous and homogeneous reactions between the injected species and dissociated air

d) heat blocking due to convection in the liquid phase (if one exists).

The magnitude of each of these terms will of course be different for different classes of materials. However, some specific comments can be made at this point with regard to (a). The contribution of this heat blocking term is usually of major significance, and may vary from several hundred BTU to thousands of BTU's per pound of injected gas. For hypersonic flight, the precise magnitude of this term has been established rigorously by the author [25] for the injection of air molecules into the dissociated hypersonic laminar boundary layer. The heat blocking term may be conveniently expressed by the quotient $\left(\frac{\Delta \widetilde{Q}}{\Delta \dot{m}}\right)_w$, which represents the unit change in the energy function $\widetilde{Q}$ per unit mass transfer. This function was correlated by the formula [25]:

$$\left(\frac{\Delta \widetilde{Q}}{\Delta \dot{m}}\right)_w = \frac{30}{\overline{M}_w}[240 + 0.48\,(h_s - h_w)], \frac{\text{BTU}}{\text{lb.}} \tag{2}$$

where $\overline{M}_w$ is the mean molecular weight of the gas at the surface and the enthalpies h_s and h_w are expressed in BTU/lb.

Note that this form of the equation may also be approximately valid for predicting the effects of the mass transfer of gases other than air, since the mean molecular weight of the gas at the wall appears in the denominator. This equation predicts that the injection of light-weight gases acts to increase the "effectiveness quotient", but only insofar as it alters the mean molecular weight of the gas at the wall from that of dissociated air. This is clearly a non-linear effect, since at high mass transfer rates, $\overline{M}_w$ is affected more strongly by foreign species than at low mass transfer rates.

In passing, it is sufficient to observe that the enhanced "effectiveness quotient" produced by light-weight gases is due to the larger heat capacity and larger diffusion coefficients of gases having a low molecular weight.

In fps units, the specific heat of a diatomic molecule is given by (see, for example [26]):

$$C_{p_i} = \left[\frac{7}{2} + \frac{\left(\frac{\Theta}{T} \right)^2 e^{\Theta/T}}{[e^{\Theta/T} - 1]^2} \right] \frac{\Re}{M_i J}, \frac{\text{BTU}}{\text{lb.}°\text{R}} \tag{3}$$

while the binary diffusion coefficient is given by (see, for example, [26]):

$$\mathfrak{D}_{ij} = 1.1712 \times 10^{-6} \, T^{3/2} \left(\frac{M_i + M_j}{M_i \cdot M_j} \right)^{1/2} \sigma_{ij}^{-2}, \frac{\text{ft.}^2}{\text{sec.}}. \tag{4}$$

Since both of these coefficients vary inversely with the molecular weight, one may expect the lighter gas species to diffuse further out from the wall than the heavier species originating at the wall, and also absorb more energy per unit mass than the heavier species. Consequently, one expects $\left(\frac{\Delta \widetilde{Q}}{\Delta \dot{m}} \right)_w$ to increase as $\overline{M}_w$ decreases.

As noted, in this study of the chemically reacting hypersonic laminar boundary layer, the analysis is confined to the forward stagnation point. Furthermore, for ballistic re-entry vehicles the major heating occurs in the continuum regime where the mean free path $\lambda_e \ll R_B$. Thus, the shock layer is a distinct region, separate from the boundary layer, and viscous merging [27] occurs at too high an altitude to affect the total heat transfer significantly.

In this case, the "influscid" flow field (absence of molecular transport fluxes, evidenced by phenomena such as conduction, viscous dissipation and molecular diffusion—i.e. the author's generalization of the word "inviscid") acts to produce a quasi-Newtonian pressure distribution over the body. This result enormously simplifies the determination of the boundary conditions at the outer edge of the viscous boundary layer. For example, the stagnation point velocity gradient is given simply by:

$$\left(\frac{du_e}{dx} \right)_s = \frac{1}{R_B} \sqrt{\frac{2 (P_e - P_\infty)}{\varrho_e}}. \tag{5}$$

The solution to the ablation problem then consists of the determination of the interphase mass transfer $\dot{m}_w$, the ablation rate $\dot{m}_{tot}$, the surface temperature T_w, and the effective heat of ablation Q^* for a material of given shape (characterized by a nose radius R_B) when it is subjected to the environmental conditions corresponding to hypersonic re-entry. It is noted that in treating a particular material, the surface temperature T_w, at which ablation occurs, is not known a priori, but is a variable which must be determined by the application of the appropriate conservation laws during the analysis of the ablation problem. That is, although the assumption of a constant "ablation temperature" is a useful expedient, it is an artificial concept which masks the effects of stagnation pressure.

Conservation Laws

Upon assuming the applicability of the modified Fick's law for diffusion, i. e.

$$\underset{\rightarrow}{j_i} = \varrho_i \underset{\rightarrow}{V_i} = -\varrho \mathfrak{D}_{ij} \nabla C_i - \frac{D_i^T}{T} \nabla T \tag{6}$$

the boundary layer equations become in the steady state for a body-oriented coordinate system, see Fig. 1:
global conservation of mass:

$$\frac{\partial}{\partial x} (\varrho u r) + \frac{\partial}{\partial y} (\varrho v r) = 0; \tag{7}$$

conservation of species i:

$$\varrho\left(u\frac{\partial C_i}{\partial x} + v\frac{\partial C_i}{\partial y}\right) = \frac{\partial}{\partial y}\left(\varrho\mathfrak{D}_{ij}\frac{\partial C_i}{\partial y} + \frac{D_iT}{T}\frac{\partial T}{\partial y}\right) + \dot{w}_i ; \tag{8}$$

conservation of momentum:

$$\varrho\left(u\frac{\partial u}{\partial x} + v\frac{\partial u}{\partial y}\right) = -\frac{\partial P}{\partial x} + \frac{\partial}{\partial y}\left(\mu\frac{\partial u}{\partial y}\right) ; \tag{9}$$

$$\frac{\partial P}{\partial y} = 0 \tag{10}$$

conservation of energy:

$$\varrho\bar{C}_p\left(u\frac{\partial T}{\partial x} + v\frac{\partial T}{\partial y}\right) = u\frac{\partial P}{\partial x} + \mu\left(\frac{\partial u}{\partial y}\right)^2 + \frac{\partial}{\partial y}\left(K\frac{\partial T}{\partial y}\right)$$

$$+ \sum_i\left[C_{p_i}\left(\varrho\mathfrak{D}_{ij}\frac{\partial C_i}{\partial y} + \frac{D_iT}{T}\frac{\partial T}{\partial y}\right)\right]\frac{\partial T}{\partial y} - \sum_i\dot{w}_i\,h_i . \tag{11}$$

It is noted in passing that if n denotes the number of chemical species present in the gas phase, then the order of the above system of non-linear differential equations is $2n+3$, since $n-1$ equations for the conservation of species are required.

Upon introducing a stream function defined by:

$$\varrho\,ur = \frac{\partial\Psi}{\partial y} ; \quad \varrho\,vr = -\frac{\partial\Psi}{\partial x} , \tag{12}$$

eq. (7) is automatically satisfied, and then applying the MANGLER-DORODNITSYN transformation defined by [9, 10, 24]:

$$\eta = \frac{\varrho_e u_e}{\sqrt{2\,\xi}}\int\limits_0^y\frac{\varrho}{\varrho_e}r\,dy, \tag{13}$$

$$\xi = \int\limits_0^x\varrho_w\mu_w u_e\,r^2\,dx \tag{14}$$

where
$$\Psi = \sqrt{2\xi}\,f(\eta), \tag{15}$$

the boundary layer equations are reduced to ordinary differential equations at the forward stagnation point. These may be written:

conservation of species i:

$$\left[\frac{l}{Pr}\left(Le\,C_{i_\eta} + Le^T\frac{\theta_\eta}{\theta}\right)\right]_\eta + f\,C_{i_\eta} + \frac{\dot{w}_i}{2\varrho\left(\dfrac{du_e}{dx}\right)_s} = 0 ; \tag{16}$$

conservation of momentum:

$$(l\,f_{\eta\eta})_\eta + f\,f_{\eta\eta} + {}^1\!/_2\left(\frac{\varrho_e}{\varrho} - f_\eta{}^2\right) = 0 ; \tag{17}$$

conservation of energy:

$$\left(\frac{\bar{C}_p l}{Pr}\theta_\eta\right)_\eta + \bar{C}_p f\theta_\eta + \frac{l}{Pr}\sum_i\left[C_{p_i}\left(Le\,C_{i_\eta} + Le^T\frac{\theta_\eta}{\theta}\right)\right]\theta_\eta$$

$$- \frac{\sum\limits_i\dot{w}_i\,h_i}{2\varrho\,T_e\left(\dfrac{du_e}{dx}\right)_s} = 0. \tag{18}$$

This system of differential equations requires $2n+3$ boundary conditions which may be written:
At the surface:

$$f_\eta(0) = f_{\eta w} = \frac{u_w}{u_e} \tag{19}$$

$$f(0) = f_w = -\dot{m}_w \left[2 \varrho_w \mu_w \left(\frac{du_e}{dx} \right)_s \right]^{-1/2} \tag{20}$$

$$\theta(0) = \theta_w = \frac{T_w}{T_e}. \tag{21}$$

One also requires $n-1$ boundary conditions of the form:

$$C_i(0) = C_{i_w}. \tag{22}$$

At the outer edge of the boundary layer, the variables approach their edge values asymptotically, e. g.

$$\lim_{\eta \to \infty} f'_\eta = 1.0; \qquad \lim_{\eta \to \infty} \theta = 1.0; \qquad \lim_{\eta \to \infty} C_i = C_{i_e}. \tag{23}$$

The above system of differential equations is coupled by the density (which is common to all the equations) and the temperature and gas composition which determine the thermodynamic properties and transport coefficients of the gas mixture. Once the gas phase boundary layer profiles including velocity u, temperature T and composition, i.e. the mass fractions C_i, have been established, the problem may be considered solved, since the gradients of velocity, temperature and composition may be utilized to determine all other physical quantities of interest.

In order to obtain numerical solutions to the system of eqs. (16), (17) and (18) subject to the boundary conditions (19), (20), (21), (22) and (23), one must usually resort to a high speed electronic computer. However, in the ensuing work, upon introducing certain assumptions it will be found possible to utilize the eigenvalues of a previous analysis [25], and thus new integrations are avoided.

It is first assumed that gas phase reactions are frozen (i.e. $\dot{w}_i = 0$), and in order to compensate for zero homogeneous rates, it is also assumed that the surface reaction rates are infinite which promotes a composition which is in local equilibrium at the surface. The implications of this pair of assumptions have been considered in detail by Lees [9], Fay and Riddell [10], and the author [21, 24]. Upon further assuming that thermal diffusion can be neglected, and that the Lewis numbers of all species are equal to the Lewis number for air molecules diffusing through air atoms and using the method of [23], one obtains for the diffusion flux of the ith species:

$$\underset{\to w}{j_i} = \left(\varrho_i \underset{\to}{V_i} \right)_w = -\mathfrak{A} \left(C_{i_w} - C_{i_e} \right) \tag{24}$$

where:

$$\mathfrak{A} = \frac{\left(\dfrac{Le}{Pr} \right)_w \theta_{\eta w} \sqrt{2 \varrho_w \mu_w \left(\dfrac{du_e}{dx} \right)_s}}{\theta_w - \overline{C}_p / \overline{C}_{p_w}} \tag{25}$$

and as noted, the eigenvalue $\theta_{\eta w}$ which is the non-dimensional temperature gradient, may be found tabulated in [25]. Since $\theta_{\eta w}$ depends on altitude, flight speed, surface temperature and mass transfer rate as do the other terms in eq. (25), $\mathfrak{A}$ has a functional dependence which can be written symbolically.

$$\mathfrak{A} = \mathfrak{A} \, (\text{Alt.}, \, V_{\infty}, \, T_{w}, \, \dot{m}_{w}). \tag{26}$$

By definition, the absolute mass flux of species i at the surface may be written:

$$(\dot{m}_i)_w = (\varrho_i \, v_i)_w = [\varrho_i \, (v + V_i)]_w . \tag{27}$$

Upon introducing eq. (24) into eq. (27), the latter becomes:

$$\dot{m}_{i_w} = C_{i_w} \, \dot{m}_w + \mathfrak{A} \, (C_{i_e} - C_{i_w}) \tag{28}$$

and since $\sum_i C_i = 1.0$, one also obtains:

$$\sum_i \mathfrak{A} \, (C_{i_e} - C_{i_w}) = 0. \tag{29}$$

Eq. (28) will be utilized at the surface in order to satisfy the microscopic chemical and physical compatibility conditions demanded by stoichiometric considerations, as well as impermeability relations.

Having considered the mass transfer equations briefly, let us now turn our attention toward heat transfer. At the stagnation point, the absolute heat transfer at a chemically reacting surface, in the presence of mass transfer, is given by:

$$[-Q_w] = \left[K \, \frac{\partial T}{\partial y} - \sum_i \varrho_i \, \overrightarrow{V_i} \, h_i - \dot{m} \, h + Q_{\text{rad.}} \right]_w \tag{30}$$

where the minus sign indicates heat transfer to the wall.

It is convenient to introduce the energy transfer function $\widetilde{Q}_w$ which is the sum of the conduction and diffusion terms:

$$\widetilde{Q}_w = \left(K \, \frac{\partial T}{\partial y} - \sum_i \varrho_i \, \overrightarrow{V_i} \, h_i \right)_w . \tag{31}$$

In this case, the effect of mass transfer can be expressed by the TAYLOR expansion:

$$\widetilde{Q}_w = (\widetilde{Q}_w)_{\dot{m}_w = 0} - \left(\frac{\Delta \widetilde{Q}}{\Delta \dot{m}} \right)_w \dot{m}_w \tag{32}$$

where the first term on the right hand side is synonymous with the aerodynamic heat transfer in the absence of mass transfer.

$\widetilde{Q}_w$ can be written in terms of the similarity variables as follows:

$$\widetilde{Q}_w = \sqrt{2 \varrho_w \mu_w \left(\frac{d u_e}{d x} \right)_s} \left(\frac{\overline{C}_p}{Pr} \right)_w \theta_{\eta_w} T_e + \mathfrak{A} \sum_i (C_{i_w} - C_{i_e}) \, h_{i_w} \tag{33}$$

where $\mathfrak{A}$ is given in eq. (25). Since an increase in mass transfer acts to decrease θ_{η_w}, the direct effect of mass transfer is to decrease $\widetilde{Q}_w$. The indirect effect of mass transfer is to alter the composition of the gas at the wall and hence depending on the algebraic sign of the heats of formation of the injected species, the second term on the right hand side of eq. (33) can either increase or decrease with mass transfer.

When air is injected into the boundary layer, eq. (2) indicates the large decrease in $\widetilde{Q}_w$ which one can expect at high flight speeds, as a consequence of mass transfer.

If the injected gas is chemically inert, its concentration builds up at the surface and drives out the air species, and thus alters the gas composition at the wall. If the injected gas is reactive, nitrogen molecules (which are chemically inert) are driven away from the surface, while atomic and molecular oxygen are consumed at the surface in combustion reactions. In either of these latter two cases, it is not necessarily true that eq. (2) applies exactly even if one assumes that the injected species have precisely the same transport properties as air molecules. Actually, of course, $\left(\dfrac{\Delta \widetilde{Q}}{\Delta \dot{m}}\right)_w$ can be readily evaluated from its definition during the solution to the problem. Let us therefore proceed with the analysis of the four different types of ablating materials.

Melting and Vaporization Ablation

It is well known that the recovered fragments of metallic meteorites represent only a small fraction of the original size of the meteor. This is due to the fact that when a metallic material is subjected to severe heating, it undergoes catastrophic failure by melting, in which only a negligible quantity of material vaporizes. This type of failure occurs because the liquid phase viscosity of a metal is very low, and hence the molten metal will be driven away from the stagnation region before appreciable vaporization can begin.

Apparently, however, stony meteors fare somewhat better than the metallic variety since molten stone has a high viscosity in the liquid phase which tends to promote vaporization shielding, a phenomenon already analyzed by the earliest workers in meteor theory, (e.g. Öpik).

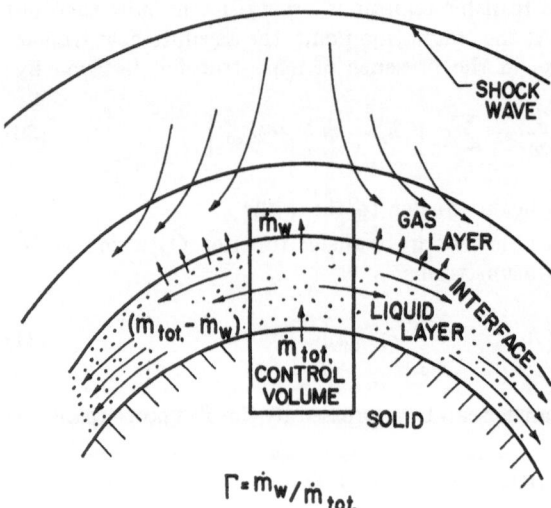

Fig. 2. Control volume for energy balance for refractory oxide

In this section, we will consider the ablation of a quartz-like refractory oxide material which melts and vaporizes, as it flows downstream away from the forward stagnation point, under the influence of aerodynamic and body forces. Fig. 2 shows a schematic representation of the melting and vaporization process at the stagnation point. For the type of material considered, it is noted that there need be no sharp phase change as the material is heated, rather, upon the application of heat the solid softens and enters a molten state.

The typical endothermic vaporization reactions that can occur at the interface between the liquid and the gas phase include:

$$SiO_2\,(l) \rightleftharpoons SiO\,(g) + {}^1/_2\,O_2 \tag{34}$$

$$SiO_2\,(l) \rightleftharpoons SiO_2\,(g). \tag{35}$$

The vaporization of this type of material has been studied by the author in great detail [21, 22] in which attention was focussed on the behavior of the gas phase boundary layer and the vaporization process itself. It was found that the total heat blocked or absorbed during the ablation process includes three coupled contributions, which are:

a) heat blocking due to boundary layer thickening caused by mass transfer,
b) heat absorption due to gasification,
c) convection of heat in the liquid phase.

If the symbol Γ is used to denote the fractional part of the ablating solid which enters the gas phase, i.e.

$$\Gamma = \frac{\text{vaporization rate}}{\text{ablation rate}} = \frac{\dot{m}_w}{\dot{m}_{\text{tot.}}} \tag{36}$$

then the conservation of mass and energy yield:

$$Q^* = \frac{Q_{\text{cal.}}}{\dot{m}_{\text{tot.}}} = \Gamma \left[\left(\frac{\Delta \widetilde{Q}}{\Delta \dot{m}} \right)_w + (h_w - h(s)) + \frac{1-\Gamma}{\Gamma} (h_l - h(s)) \right] \tag{37}$$

where the heat transfer to a calorimeter is given by:

$$Q_{\text{cal.}} = Q_{\text{aero.}} + Q_{\text{rad.}} \tag{38}$$

and the net radiant heat transfer may be written:

$$Q_{\text{rad.}} = \sigma (\varepsilon_g T_g^4 - \varepsilon_w T_w^4) . \tag{39}$$

Eq. (37), which at a given flight condition is essentially an equation in the surface temperature, cannot be evaluated until auxiliary relations are introduced to relate the composition of the gas at the surface to the surface temperature. That is, the vaporization rate $\dot{m}_w$ and the enthalpy of the gas at the interface h_w depend on both the composition of the gas and the temperature. Assuming eq. (34) to be the dominant reaction, results in a total of four unknown mass fractions at the surface. C_{O_w}, $C_{O_{2w}}$, C_{SiO_w} and $C_{N_{2w}}$, since nitrogen atoms will have recombined at most surface temperatures of interest. Since $\sum C_i = 1.0$ (which is a form of DALTON's law), only three additional relations are required.

Upon assuming a diffusion-controlled vaporization process, one has the equilibrium constant:

$$K_{P_{SiO}} = P_{SiO} (P_{O_2})^{1/2} = \frac{C_{SiO}}{M_{SiO}} \left(\frac{C_{O_2}}{M_{O_2}} \right)^{1/2} (\varrho R T)^{3/2} \tag{40}$$

which is a known function of temperature. In addition, one has the equilibrium constant for the oxygen system:

$$K_{P_O} = \frac{(P_O)^2}{P_{O_2}} = \left(\frac{C_O}{M_O} \right)^2 \left(\frac{M_{O_2}}{C_{O_2}} \right) \varrho R T \tag{41}$$

which is also a known function of temperature.

Finally, upon noting that the interface is impermeable to the element nitrogen, one may write:

$$(\dot{m}_N + \dot{m}_{N_2})_w = 0 \tag{42}$$

which becomes:

$$C_{N_{2w}} = \frac{-\mathfrak{A} (C_N + C_{N_2})_e}{(\dot{m}_w - \mathfrak{A})} \tag{43}$$

where stoichiometry requires that the interphase mass transfer be given by:

$$\dot{m}_w = \left(1 + \frac{M_{O_2}}{2 M_{SiO}}\right) \dot{m}_{SiO\,w} \tag{44}$$

and the mass transfer of silicon oxide is given by:

$$\dot{m}_{SiO\,w} = C_{SiO\,w} (\dot{m}_w - \mathfrak{A}) . \tag{45}$$

Upon utilizing the enthalpy data shown in Fig. 3 and the four auxiliary equations, results were obtained [21] for the full range of hypersonic flight conditions, and are shown in Fig. 4.

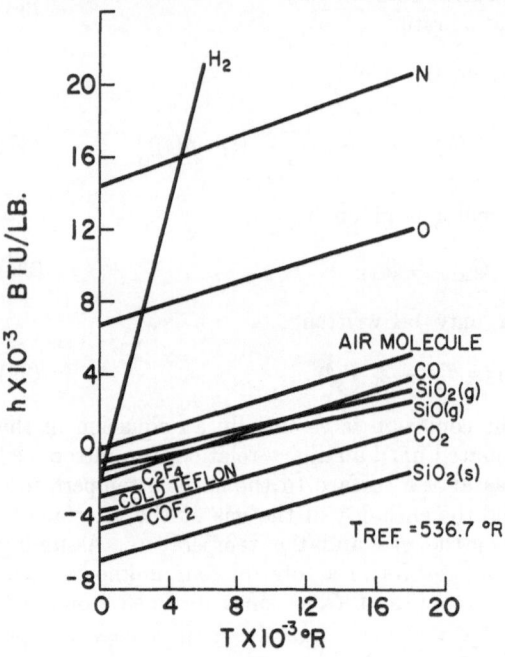

Fig. 3. Enthalpy including heat of formation of species in the boundary layer

It is first remarked that the procedure of treating Γ as an independent parameter has the effect of separating the effects of stagnation enthalpy from stagnation pressure. Thus, the upper curve shown in Fig. 4, which represents the case of pure sublimation, indicates that when $\Gamma = 1.0$, the effective heat of ablation is independent of stagnation pressure. It is also seen that $(Q^*)_{\Gamma=1.0}$ increases almost linearly with stagnation enthalpy $\left(h_s \cong \frac{V_\infty^2}{2gJ}\right)$. Finally, it was found that at constant flight speed (stagnation enthalpy) Q^* decreases linearly as the gasification ratio departs from unity. That is,

$$(Q^*)_{\Gamma \neq 1.0} = (Q^*)_{\Gamma=1.0} -$$
$$- (1 - \Gamma) \frac{\Delta Q^*}{\Delta \Gamma} \tag{46}$$

where $(Q^*)_{\Gamma=1.0}$ and $\frac{\Delta Q^*}{\Delta \Gamma}$ are both shown in Fig. 4 and differ from each other by a constant, ($\sim$heat of fusion). Thus, it is quite clear that Q^* is a minimum for pure melting ($\Gamma=0$) and what is required at this point is a knowledge of the behavior of Γ.

In the author's early analysis of a refractory oxide, it was discovered that the surface temperature T_w correlated directly with the logarithm of the stagnation pressure, and since the viscosity of a glassy material decreases exponentially with increasing temperature, it was anticipated that the value of Γ was not constant but was a decreasing function of stagnation pressure. The value of Γ for given flight conditions can, of course, be obtained by matching the shear stress, pressure gradient, interphase mass transfer, interphase heat transfer, interface temperature and interface velocity at the interface between the gas and liquid phase boundary layers [12]. This matching procedure was in fact, performed by FANUCCI [28] who utilized an integral analysis for the liquid phase and the author's unpublished work for the gas phase. It was found that Γ decreased with stagnation pressure at constant flight speed (stagnation enthalpy), and increased with flight speed at constant pressure, see Fig. 5. The magnitude of the ordinate in this figure depends on the precise transport properties utilized for the liquid phase. The value of Γ is large when the PRANDTL number of the

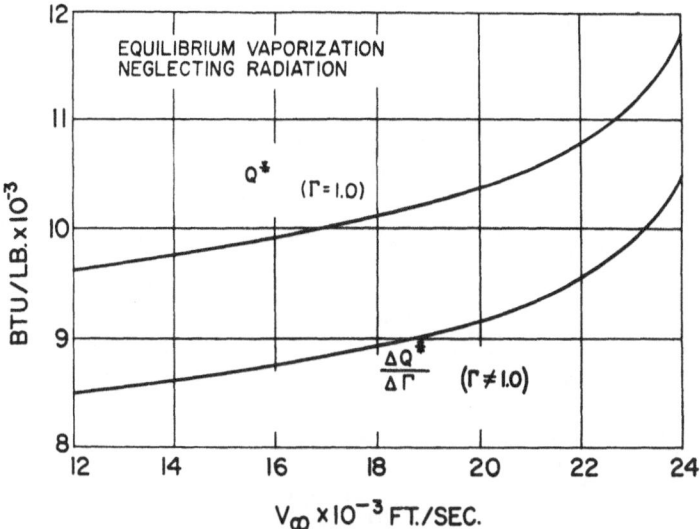

Fig. 4. Correlation of effective heat of vaporization with flight speed

liquid is large, i.e., when the viscosity is high and the thermal conductivity is low. Further, deceleration of the vehicle also tends to promote an increase

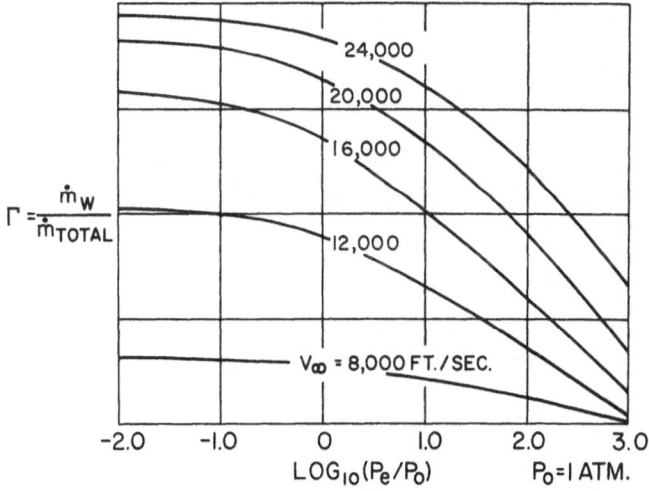

Fig. 5. Dependence of the gasification ratio upon stagnation pressure and flight speed

in Γ [28]. For discussion purposes, the maximum value of the ordinate in Fig. 5 can be taken as 0.4 when typical liquid properties are utilized for molten quartz and deceleration is neglected.

When the data of Figs. 4 and 5 are combined, one obtains the results shown in Fig. 6. Here it is clearly seen that the effective heat of ablation drops significantly at the higher pressures and rises with stagnation enthalpy in much the same manner as does the gasification ratio. Since $Q_{cal.}$, the numerator in eq. (37), is dependent on the magnitude of the surface temperature, it is not sufficient to merely present curves of Q^*. Consequently, the magnitude of the surface

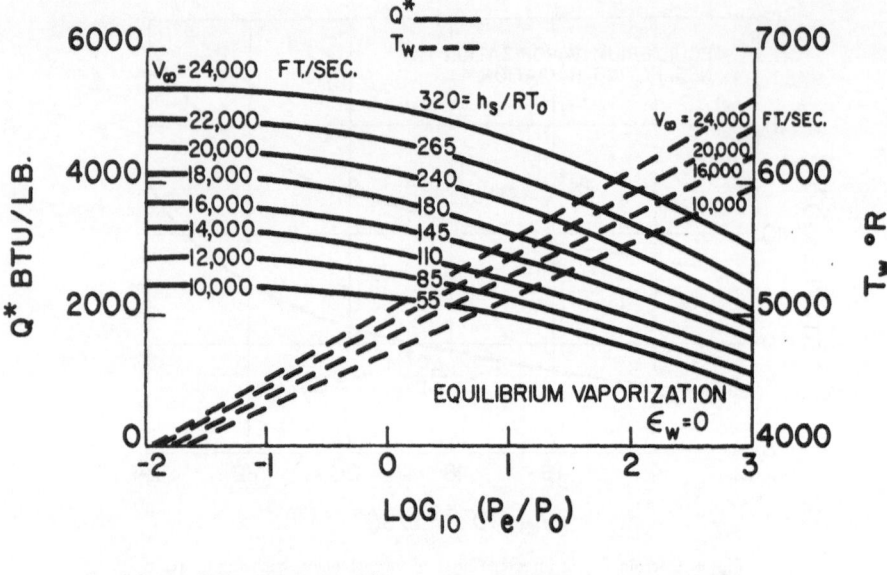

R=53.3 FT. LB/LB.-°R T_0=520°R P_0=1 ATM.

Fig. 6. Variation of heat of ablation and surface temperature of refractory oxide with stagnation pressure and flight speed

temperature is also indicated in Fig. 6 for the full range of hypersonic flight conditions.

Utilizing Fig. 6, one can simply determine the quasi-steady ablation rate for any flight trajectory by dividing the instantaneous calorimeter heat transfer

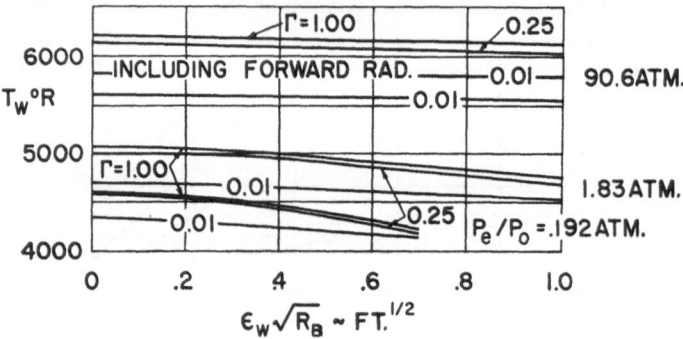

Fig. 7. Surface temperature vs surface emissivity factor for equilibrium vaporization of SiO_2

by the instantaneous value of Q^*. This will be done for a variety of satellite re-entry trajectories in a later section. It is also noted that although calculated for zero wall emissivity ($\varepsilon_w = 0$), the data of Fig. 6 are also approximately correct for a value of the surface emissivity parameter in the range $0 \leq \varepsilon_w \sqrt{R_B} \leq 0.5$ ft.$^{1/2}$, because, as is shown in Fig. 7, the surface temperature decreases only slightly with increasing $\varepsilon_w \sqrt{R_B}$ [22].

Surface Combustion

In the light of the preceeding dicussion, one tends to think of ablation as a process in which a large heat transfer rate causes the thermal degradation of a particular material. In the case of graphitic or carbonaceous materials, however, it is found that mass transfer occurs primarily by means of a diffusion-controlled oxidation process, and not as a consequence of heat transfer. Thus, during hypersonic flight, the net heat transfer into the surface tends to be modified from the calorimeter heat transfer rate, to the extent that exothermic combustion reactions occur in the boundary layer.

In an earlier study, the author considered the stagnation point ablation of graphite during hypersonic flight [20], however, it was assumed that carbon monoxide was the sole product of oxidation, and further, no estimates were given of the result-ing heat transfer. In this treatment, it will be assumed that the products of oxidation can include both CO and CO_2, and it will be shown that this more realistic treatment intro-duces an important effect on heat and mass transfer at low surface temperatures.

During surface com-bustion, the overall rate of reaction depends on the slowest and hence rate-con-trolling step, in a sequence of five steps. These include

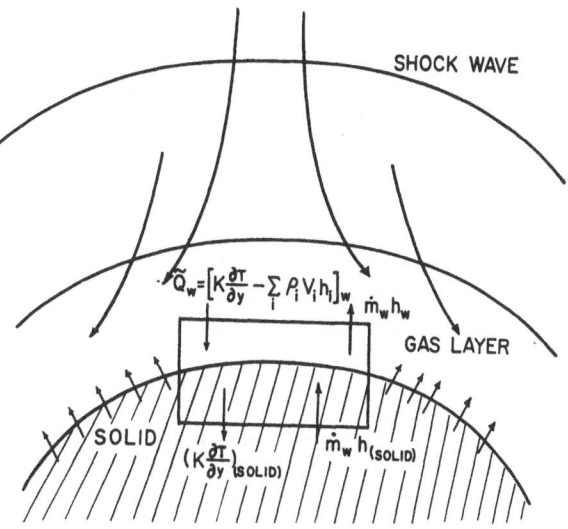

Fig. 8. Control volume for SiO_2 energy balance for combustible solid

the convection and diffusion of reactants to the surface, physical and chemical adsorption, chemical reaction, desorption of the combustion products and counter-diffusion and convection of these products into the stream. At very low surface temperatures, $T_w < 1000°$ R, the slowest step is usually the rate of chemical reaction, however, at higher temperatures the diffusion process is considerably slower than the rate of reaction, which means that, for most surface temperatures of interest, the combustion process is diffusion-controlled. We will present an analysis of the diffusion-controlled regime in which the following assumptions are made:

a) The gas phase composition can be approximated by a six component mixture consisting of oxygen and nitrogen atoms, oxygen and nitrogen molecules, carbon monoxide and carbon dioxide.

b) Gas phase reactions are assumed frozen, and nitrogen molecules are chemically inert at the carbon surface.

c) At surface temperatures below 5000° R, nitrogen atoms recombine at the surface, and do not form permanent compounds with carbon, e.g., no cyanogen is formed [29].

d) The composition of the gas at the surface may be determined by utilizing the local equilibrium constraint.

e) The LEWIS and PRANDTL numbers of the gas are taken equal to those for dissociated air.

Here again, it is anticipated that the assumption of local equilibrium at the surface compensates for the assumption of zero rates for the homogeneous reactions. Furthermore, since all of the reactions are confined to the surface, the question of which species tends to form first, carbon monoxide or dioxide is left unanswered [30].

In Fig. 8, we have indicated schematically a control volume suitable for the mass and energy balance at the surface of the ablating graphite. The conservation of mass requires that the interphase mass transfer be equal to the ablation rate, i.e.

$$\dot{m}_w = \dot{m}_{O(s)_w} \tag{47}$$

while the conservation of energy reveals that the heat conducted into the interior of the graphite is given by:

$$\left[K \frac{\partial T}{\partial y} \right]_{s,w} = Q_{cal.} - \left[\left(\frac{\Delta \widetilde{Q}}{\Delta \dot{m}} \right)_w + h_w - h(s) \right] \dot{m}_w \tag{48}$$

where $Q_{cal.}$ is given by eq. (38).

Since the instantaneous combustion rate $\dot{m}_w$ is small, one does not expect the heat conducted into the interior $\left(K \frac{\partial T}{\partial y} \right)_{s,w}$ to differ appreciably from $Q_{cal.}$, i.e. the graphite behaves like a heat sink material which also undergoes oxidation. Thus, the concept of an effective heat of ablation does not appear to be meaningful for graphite, since the heat which is absorbed does not produce a phase change, assuming, of course, that the surface temperature remains below the sublimation temperature (the usual case). Furthermore, since the injected species have a large negative heat of formation (see Table I or Fig. 3) the term in square brackets on the right hand side of eq. (48) may be negative. Thus, the chemical heat release can actually overwhelm the heat blocking term due to boundary layer thickening caused by mass transfer.

Table I. *Heats of Formation* ($T_{ref.} = 536.7° R$)

Species	$\Delta h_f°$ BTU/lb.	Reference
CO	—1770	[1]
CO_2	—3850	[1]
CF_4	—4459	[3]
COF_2	—4226	[3]
C_2F_4 (g)	—2736	[3]
C_2H_2	3753	[2]
CH_4	—2013	[2]
C_2H_4	810	[2]
SiO (g)	—1147	[4]
SiO_2 (s)	—6233	[4]
Teflon (s)	—3582	[3]

1. J. S. Gordon, Thermodynamics of High Temperature Gas Mixtures. Wright Air Development Center, ASTIA # 110735, 1957.
2. S. Glasstone, Textbook of Physical Chemistry, 2nd Ed. Princeton, N.J.: Van Nostrand Co., 1946.
3. H. Friedman and A. Myerson, Private communication.
4. G. Vidale, Private communication.

In order to evaluate the term in brackets on the right hand side of eq. (48), it is necessary to relate the gaseous composition to the surface temperature. As is generally true for injected foreign species, carbon monoxide and dioxide are present in maximum concentration at the surface, and vanish asymptotically at the outer edge of the boundary layer. Thus, although the boundary layer contains six gaseous species, only a maximum of four are present at the edge of the boundary layer, namely the four primary components of dissociated air e.g. O, O_2, N and N_2.

At the surface, oxygen atoms and oxygen molecules are consumed in combustion reactions, nitrogen atoms fully recombine, and hence only three species are present in appreciable concentration, e.g. N_2, CO and CO_2. Thus, in addition to an expression for the interphase mass transfer $\dot{m}_w$, one requires only three compatibility relations at the surface, in order to relate the three unknown mass fractions $C_{N_{2w}}$, C_{CO_w} and $C_{CO_{2w}}$ to the surface temperature T_w.

Now, by definition, the interphase mass transfer can be written:

$$\dot{m}_w = (\dot{m}_O + \dot{m}_{O_2} + \dot{m}_N + \dot{m}_{N_2} + \dot{m}_{CO} + \dot{m}_{CO_2})_w. \tag{49}$$

However, the impermeability of the surface to the element nitrogen requires that eq. (42) and (43) hold. Furthermore, combustion stoichiometry requires that:

$$\left(\frac{4}{7}\dot{m}_{CO} + \frac{8}{11}\dot{m}_{CO_2}\right)_w = -(\dot{m}_O + \dot{m}_{O_2})_w \tag{50}$$

since the following reactions can all occur simultaneously at the surface:

$$C(s) + O \rightleftharpoons CO \tag{51}$$

$$2C(s) + O_2 \rightleftharpoons 2CO \tag{52}$$

$$C(s) + 2O \rightleftharpoons CO_2 \tag{53}$$

$$C(s) + O_2 \rightleftharpoons CO_2 \tag{54}$$

Thus, it can be shown that eq. (49) reduces to:

$$\dot{m}_w = -\frac{\mathfrak{A}\left[\frac{3}{7}C_{CO_w} + \frac{3}{11}C_{CO_{2w}}\right]}{\left[1 - \frac{3}{7}C_{CO_w} - \frac{3}{11}C_{CO_{2w}}\right]}. \tag{55}$$

Another relationship is simply:

$$\sum_i C_{i_w} = 1.0. \tag{56}$$

Finally, the local equilibrium constraint implies that the gaseous carbon monoxide and dioxide must be in equilibrium with the solid phase. This is given by the reaction:

$$C(s) + CO_2 \rightleftharpoons 2CO \tag{57}$$

$$K_{P_{CO}} = \frac{(P_{CO})^2}{P_{CO_2}} = \left(\frac{C_{CO}}{M_{CO}}\right)^2\left(\frac{M_{CO_2}}{C_{CO_2}}\right)\varrho\,RT \tag{58}$$

where the partial pressures are expressed in atmospheres; the equilibrium constant $K_{P_{CO}}$ is a known function of surface temperature (see Table II).

Table II. *Equilibrium Constants*, $\log_{10} K_P = a - \dfrac{b}{T\,(°R)}$

Reference	Reaction	Equilibrium constant (P is in atm.)	a	$b \times 10^{-4}$
[1]	$O_2 \rightleftarrows 2\,O$	$K_P = \dfrac{(P_O)^2}{P_{O_2}}$	6.80	4.68
[1]	$N_2 \rightleftarrows 2\,N$	$K_P = \dfrac{(P_N)^2}{P_{N_2}}$	7.00	9.00
[2]	$SiO_2\ (1) \rightleftarrows SiO\ (g) + \tfrac{1}{2}\,O_2$	$K_P = P_{SiO}\,(P_{O_2})^{1/2}$	13.90	7.326
[1]	$2\,CO \rightleftarrows CO_2 + C\ (s)$	$K_P = \dfrac{P_{CO_2}}{(P_{CO})^2}$	−8.80	−1.503
[3]	$C_2F_4\ (s) \rightleftarrows C_2F_4\ (g)$	$K_P = P_{C_2F_4}$	10.49	1.664

1. B. LEWIS and G. VON ELBE, Combustion, Flames and Explosions of Gases. New York: Academic Press, 1951.
2. G. VIDALE, Private communication.
3. J. C. SIEGLE and L. T. MUUS, Pyrolysis of Polytetrafluoroethylene. Address to American Chemical Society Meeting (September 1956).

Since the equilibrium constant $K_{P_{CO}}$ increases exponentially with temperature, at low surface temperatures, $T_w \cong 1000°\,R$, eq. (58) indicates that the primary combustion product is CO_2. In the temperature range given by $1000°\,R < T_w < 3000°\,R$, there is a gradual transition from the dioxide to the monoxide, where the location of the knee of the curve is pressure dependent. Finally, at surface temperatures above 3000 °R the primary combustion product is CO. It should be noted that in the above discussion, we have made no distinction between the oxidation of graphite and the oxidation of carbon. While these two distinct materials have a different thermal diffusivity α, and hence conduct heat at a different rate, in the diffusion controlled regime they will have essentially the same thermochemical behavior and hence in predicting heat and mass transfer at the surface the two terms may be used interchangeably.

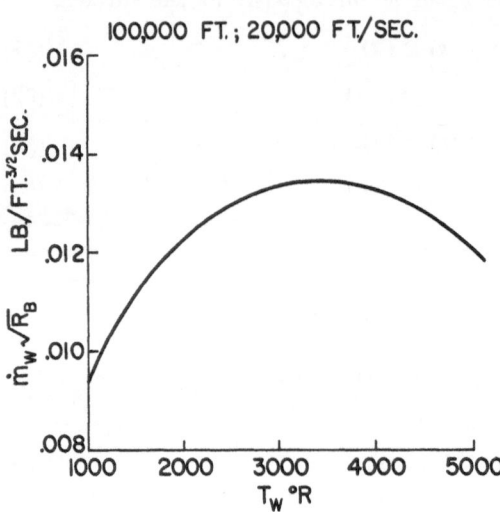

100,000 FT.; 20,000 FT/SEC.

Fig. 9. Mass transfer during diffusion-controlled graphite combustion

Typical results of the calculations, at an altitude of 100,000 ft. and a flight speed of 20,000 ft./sec., are shown in Figs. 9 and 10 for the diffusion-controlled regime. It is seen that at low surface temperatures, the ablation rate is low because two oxygen atoms are required to remove one carbon from the surface lattice to form CO_2. At the higher surface temperatures there is a gradual increase in ablation rate because as the shift from CO_2 to CO takes place two oxygen atoms can now remove two carbon atoms from the surface. One other effect can be discerned in Fig. 9, and that is at the highest surface temperatures the ablation rate begins to drop since the mean density of the gas in the boundary layer decreases with surface temperature.

Thus, for the theoretical model considered here, it was found that the ablation rate initially rises with temperature, and in the vicinity of 3000° R reaches a peak value, which is approximately fifty percent larger than that obtained at a surface temperature of 1000° R, and then decreases with surface temperature beyond that point.

The variation of the products of combustion and the ablation rate with surface temperature will also have an effect on the heat conducted into the interior of the graphite. The dependence of the heat transfer upon surface temperature neglecting radiative heat transfer is shown in Fig. 10 for the same environmental condition as Fig. 9. It is seen that the heat conducted into the interior decreases with increasing surface temperature, as does the aero-

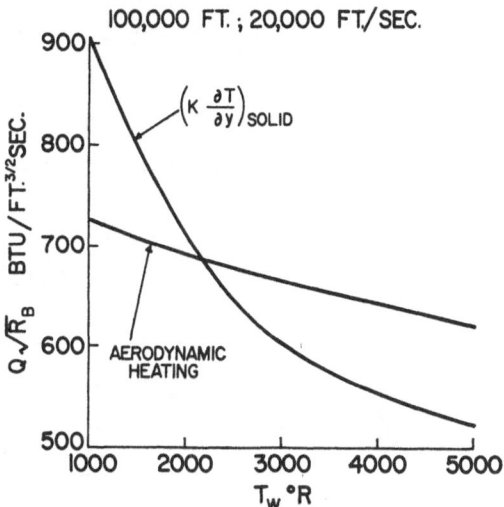

Fig. 10. Heat transfer during diffusion-controlled graphite combustion

dynamic heat transfer to a non-ablating surface. However, at a low surface temperature, the exothermicity of the chemical reactions given by eqs. (53) and (54) exceeds the contribution to heat blocking produced by mass transfer, since the heat release is high (see Table I) and the mass transfer is low. Hence, at 1000° R, the heat conducted into the interior of an ablating graphite surface exceeds the aerodynamic heat transfer by approximately twenty-five percent. The theory also predicts that the heat transferred into the graphite surface decreases rapidly with increasing surface temperature. This decrease is due not only to decreased conduction from the gas, but also to the combined effects of the decreased exothermicity of the chemical reaction at the surface and the increased mass transfer into the boundary layer as carbon monoxide becomes the chief combustion product. Whether or not the heat conducted into the interior of the graphite can actually fall below the aerodynamic heating, in the absence of re-radiation from the surface, as the theory predicts for the higher surface temperatures, must await experimental verification or perhaps the completion of a more refined theoretical analysis which the author is currently pursuing.

Calculations were also performed for the entire range of hypersonic flight conditions of interest, and the ablation rate and heat transfer rate (neglecting radiation) are shown in Figs. 11 and 12 respectively, for a range of flight speeds from 12,000 ft./sec. to 24,000 ft./sec., a range of altitudes from 10,000 ft. to 240,000 ft., and a range of surface temperatures from 1000° R to 5000° R. Although radiation has been neglected in Fig. 12, the heat transfer can be easily corrected for finite surface emissivity as well as gas cap radiation by writing:

$$\left[\left(K\frac{\partial T}{\partial y}\right)_{s,w}\right]_{\varepsilon_w \neq 0} = \left[\left(K\frac{\partial T}{\partial y}\right)_{s,w}\right]_{\varepsilon_w = 0} + Q_{\text{rad.}} \qquad (59)$$

where $Q_{rad.}$ is given in eq. (39), and the emissivity of the solid is given in Table III.

It is interesting to note that Fig. 11 indicates that although flight speed (stagnation enthalpy) and surface temperature (shifting chemical equilibrium) effects do exist in a hypersonic environment, the most important single independ-

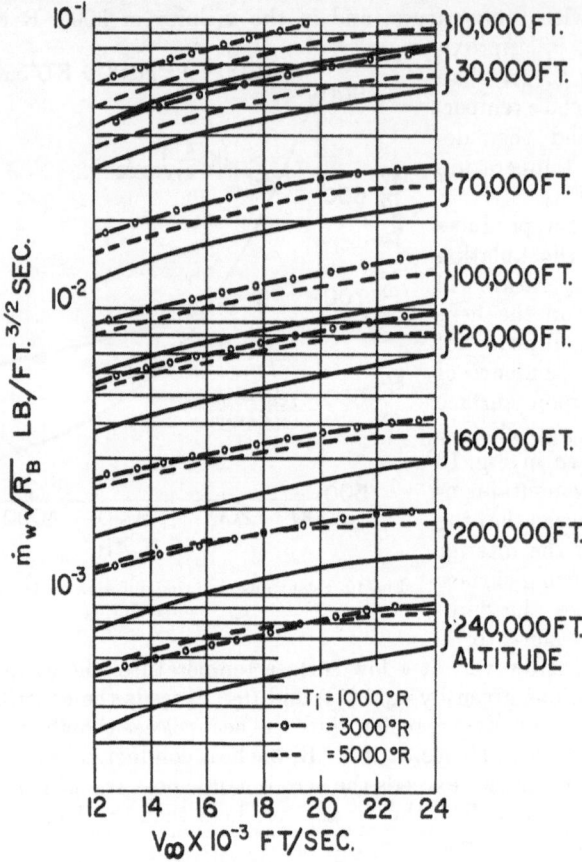

Fig. 11. Mass transfer for carbon combustion

ent variable is the stagnation pressure. This result was already shown by the author in an earlier study [20], since upon neglecting the surface temperature effect in [20] one has:

$$\dot{m}_w \cong 0.1 \sqrt{2\, \varrho_w \mu_w \left(\frac{d u_e}{d x}\right)_s} \tag{60}$$

and then introducing eq. (5), as well as assuming a simple viscosity law at the surface, i.e. $\mu_w \sim T_w$, one can readily obtain:

$$\dot{m}_w \sqrt{R_B} \cong 5 \times 10^{-6} \left(\frac{P_e^3}{\varrho_e}\right)^{1/4}, \frac{\text{lb.}}{\text{ft.}^{3/2}\,\text{sec.}} \tag{61}$$

where P_e is the stagnation pressure expressed in lb./ft.², and ϱ_e is the stagnation density in lb./ft.³. This equation clearly shows the marked dependence of the combustion rate upon the stagnation pressure.

Although one can use the theoretical predictions of Figs. 11 and 12 directly, it may be desirable to have approximate correlation formulae for these data which can be used for systematic studies of re-entry vehicle performance. Therefore, the author has obtained the following equation for the stagnation point ablation rate

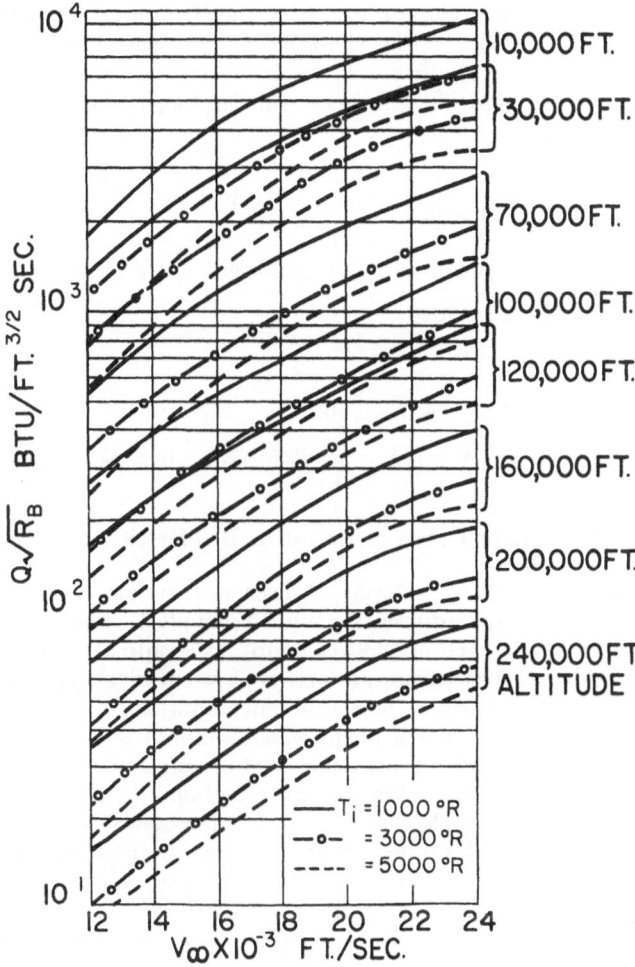

Fig. 12. Heat transfer into solid carbon

of a graphite surface, as a function of altitude, flight speed and surface temperature.

$$\dot{m}_w \sqrt{R_B} \cong c\,(10)^a (V_\infty)^b, \quad \frac{\text{lb.}}{\text{ft.}^{3/2}\,\text{sec.}} \tag{62}$$

where $\dot{m}_w$ is the ablation rate in lb./ft.^{2} sec. R_B is the nose radius in ft., V_∞ is the flight speed in ft./sec.; the exponents a and b are given by:

$$a = -\,(4.6759 + 9.1600 \times 10^{-6}\,\text{Alt.}) \tag{63}$$

$$b = 0.875 \tag{64}$$

while the coefficient c is given by:

$$c = 0.508 + 1.4725 \times 10^{-4}\,T_w \tag{65}$$

when T_w falls between 1000° and 3000° R, and the coefficient c becomes:

$$c = 1.17 - 1.045 \times 10^{-6}\,\text{Alt.} - (7.41 \times 10^{-5} - 2.85 + 10^{-10}\,\text{Alt.})\,T_w\,. \qquad (66)$$

when T_w exceeds 3000° R. In the above expressions, Alt. is the altitude expressed in ft., and T_w is the surface temperature in °R. Substitution shows that the above formula represents the data of Fig. 11 to within ± 15 percent error over the full range of interest.

In addition, one finds that the heat transfer (neglecting radiation) can be expressed in the form:

$$\left[\left(K\,\frac{\partial T}{\partial y}\right)_{s,w}\right]_{\varepsilon_w = 0} \sqrt{R_B} \cong c\,(10)^a\,(V_\infty)^b,\,\frac{\text{BTU}}{\text{ft.}^{3/2}\,\text{sec.}} \qquad (67)$$

where the exponents a and b and the coefficient c are given by:

$$a = -\,(7.32 + 8.7763 \times 10^{-6}\,\text{Alt.}) \qquad (68)$$

$$b = 2.5645 \qquad (69)$$

$$c = 1.828 - 4.156 \times 10^{-4}\,T_w + 4.097 \times 10^{-8}\,T_w^2 \qquad (70)$$

in which the fps system is again utilized for the physical dimensions.

The Thermal Degradation of a Reinforced Plastic

From the point of view of material fabrication, there are two major classes of plastic materials, e.g. the thermosetting and thermoplastic resins, and it appears that certain materials in each of these categories are quite attractive (see [31] to [36]), when considering the thermal protection of re-entering hypersonic vehicles. In this section, we will consider the thermal degradation of a reinforced plastic, which consists of a thermosetting resin, which is reinforced with either organic or refractory fibers in varying proportions and orientations. (A typical thermoplastic resin will be treated in the following section.)

Thermosetting resins include the following: aminos, epoxies, phenolics, polyesters and silicones. During the molding of these resins, at suitable pressures and temperatures, an irreversible chemical reaction called cross-linking occurs, during which the resin molecules form interconnecting chemical bonds, so that the final molded shape is essentially a three-dimensional super-molecule. Then, when the cured virgin plastic is exposed to severe heating, a form of thermal degradation known as pyrolysis occurs. This produces a partial "un-zipping" of the "chains" and results in the formation of high and low molecular weight gaseous fragments, such as hydrogen, carbon monoxide, ethylene, methane, acetylene and various free radicals which diffuse to the surface. A solid residue is left behind in the form of a rigid, non-uniform, cross-linked, carbonaceous structure, which contracts. As shrinkage stresses build up in the pyrolizing (carbonizing) layer, strain limits are rapidly exceeded and fissures and voids develop, so that the ultimate appearance of the surface of the pyrolyzing plastic is that of a char sponge. (See Fig. 13.)

As noted above, a large variety of different reinforcing fibers, including asbestos, glass, nylon and quartz have been suggested as a means of improving the strength and ablation characteristics of the resin, and these have been tested by SUTTON [31], and GRUNTFEST and SHENKER [34], and others. The precise contribution of a given reinforcing fiber to the overall behavior of the material will obviously depend on physicochemical properties such as the following:

a) the chemical composition of the fibers

b) the vapor pressure of the various chemical species released as a function of temperature

c) the viscosity of the molten layer, if one forms

d) the emissivity of the molten layer.

As is indicated in Fig. 13, the problem is further complicated by combustion reactions which occur either in the gas phase or at the surface of the material. The gaseous hydrocarbons produced during pyrolysis react chemically with the atomic and molecular oxygen present in the boundary layer. Hence, depending on the amount of reactive gas injected, some of the oxygen may diffuse to the surface and consume the char layer in heterogeneous combustion reactions of a nature discussed in the preceding section.

Consider the control volume, shown in Fig. 14, which is attached to the outer surface and which extends through the char layer, and the region undergoing pyrolysis, into the virgin plastic. Although the thickness of the char layer grows with time, it is assumed that the region undergoing pyrolysis is a thin zone of essentially constant thickness, and hence the problem may be treated by means of a quasi-steady ablation analysis.

It is noted that the determination of the ablation rate of a reinforced plastic differs from the case of pure sublimation, since the gasification process occurs in several distinct steps including pyrolysis and combustion. Thus, even though a liquid layer may not be involved, (i.e. the reinforcement may not be a glassy fiber), a certain fraction $(1 - \varGamma - \varXi)$, of plastic does not gasify, but remains behind as a char layer.

The gasification process, which occurs in the carbonizing zone immediately below the char layer, is a temperature-sensitive reaction in depth, and the pyrolyzing material obeys conservation laws quite similar to eq. (8) and (11) for the gas phase, which include $n-1$ equations for the local rate of pyrolysis:

$$\frac{\partial \varrho_i}{\partial t} + \nabla \cdot (\varrho_i \overrightarrow{v_i}) = \dot{w}_i \tag{71}$$

and the conservation of energy within the pyrolyzing material:

$$\varrho \, C_p \left(\frac{\partial T}{\partial t} + \overrightarrow{v} \cdot \nabla T \right) = \nabla \cdot (K \nabla T) - \sum_i \dot{w}_i \, h_i . \tag{72}$$

Each of the chemical source terms $\dot{w}_i$ obeys a kinetic equation which introduces rate constants having an increasing exponential temperature dependence. If these equations are first solved, one may then determine the mass of gas which flows through the interface between the char layer and the carbonizing plastic, by evaluating the summed integral

$$\dot{m}_i = \sum_p \int_0^\infty \dot{w}_p \, dy \tag{73}$$

which includes the contribution of all pyrolysis reactions. In this expression, $y = 0$ is at the interface, and the positive y axis is directed into the virgin plastic.

Unfortunately, at the present time, there is an incomplete understanding of the various pyrolysis mechanisms and hence the local reaction rates $\dot{w}_i$ cannot be predicted with great accuracy for most interesting complex hydrocarbon resins. Thus, no straightforward means exists for determining the interphase mass transfer as a function of the surface temperature, and furthermore, the gas composition at the surface is also not a predictable function of surface temperature.

We will therefore follow the author's procedure [37] of making a qualitative estimate of the behavior of a given reinforced plastic. Referring to Fig. 14 then, the conservation of mass requires:

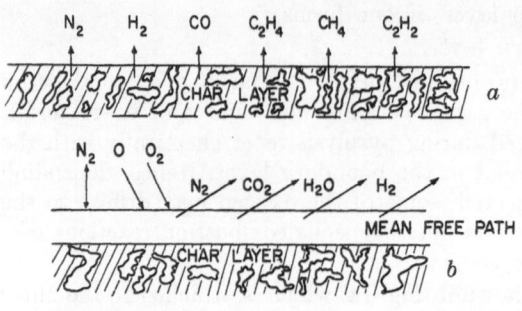

$$\dot{m}_w = \dot{m}_i + \dot{m}_{\text{char}\atop\text{comb.}} = (\Gamma + \varXi)\,\dot{m}_p$$

$$(74)$$

Fig. 13. a) Pyrolysis under vacuum conditions
b) Pyrolysis and combustion during hypersonic flight

where it is apparent that $(\Gamma + \varXi)$ cannot exceed unity in the steady state, and further, in vacuo $\varXi$ vanishes identically. An energy balance yields:

$$Q^* = \frac{Q_{\text{cal}}}{\dot{m}_p} = (\Gamma + \varXi)\left[\left(\frac{\varDelta\widetilde{Q}}{\varDelta\dot{m}}\right)_w + h_w\right] - h_{vp} \qquad (75)$$

where h_{vp} is the enthalpy of the virgin plastic and h_w is the enthalpy of the gaseous mixture present at the surface of the char layer, and $\dot{m}_p$ is the mass of solid which is pyrolyzed but not necessarily removed from the surface.

In an attempt to evaluate the gas composition at the surface of the char layer, which ultimately leads to the determination of both $\left(\frac{\varDelta\widetilde{Q}}{\varDelta\dot{m}}\right)_w$ and h_w, the maximum mass flux of oxygen to the surface was first evaluated by assuming that the element oxygen was completely consumed at the surface. Eq. (28) shows that this assumption implies:

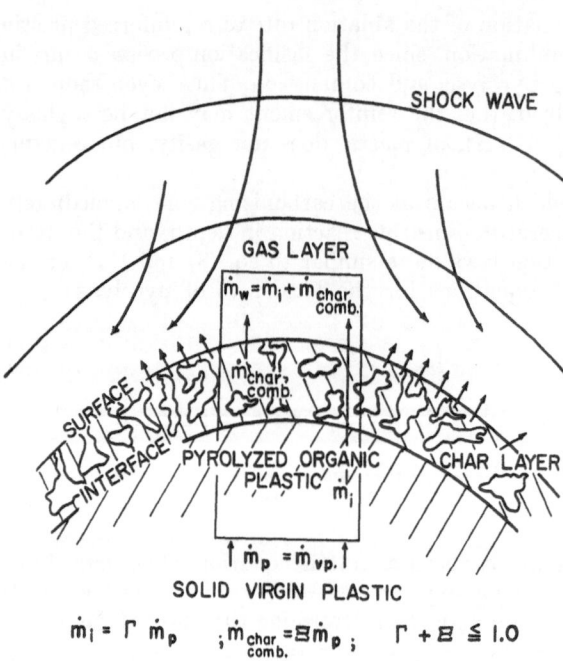

Fig. 14. Control volume for mass and energy balance for reinforced plastic

$$(\dot{m}_O + \dot{m}_{O_2})_w = \mathfrak{A}\,(C_O + C_{O_2})_e. \qquad (76)$$

Upon assuming the formation of gaseous dioxide at low surface temperatures and gaseous monoxide combustion products at higher temperatures, one obtains the results shown in Fig. 15. The mass transfer of the element oxygen given in this figure may be used to estimate the extent of exothermic diffusion-controlled oxidation reactions for arbitrary reinforced plastics.

Examination of the terms in eq. (75) leads one to conclude that the magnitude of the effective heat of ablation increases with stagnation enthalpy, and decreases with stagnation pressure. In general, one anticipates that a typical reinforced plastic will have the qualitative behavior shown in Fig. 16, where the maximum value of the ordinate depends on the choice of resin and the nature of the re-

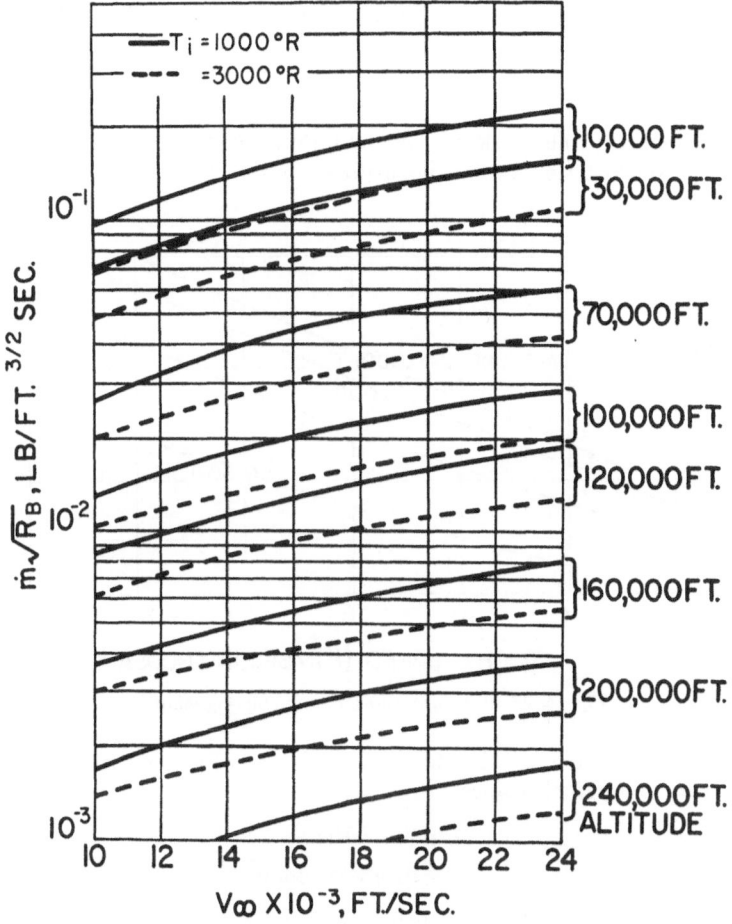

Fig. 15. Diffusion-controlled mass transfer of oxygen

inforcement utilized. For some of these materials, the stagnation pressure effect is not as pronounced as is indicated in Fig. 16, and thus, for certain materials, the curves of $Q*$ tend to coalesce, thus diminishing the variation of $Q*$ with altitude.

The Rate Controlled Pyrolysis of a Thermoplastic Resin

The general category of thermoplastic resins includes acrylics, cellulosics, fluorocarbons, nylon, polyethylene, and the vinyls. In this section, we will consider the ablation characteristics of polytetrafluoroethylene, which is a fluorocarbon commonly known as Teflon[1].

Although it would seem that a simple energy balance can be utilized to predict [16] the thermal response of Teflon, a more complete understanding requires that the chemical kinetic and gas dynamic aspects of the problem be considered as well as the purely thermodynamic factors.

---

[1] Dupont trademark.

The experimental work of Madorsky et al. [38], and Siegle and Muus [39], indicates that the depolymerization of Teflon polymer is first order with respect to the weight of solid. This can be expressed:

$$\frac{\partial W_p}{\partial t} = - k W_p \tag{77}$$

and leads directly to the result:

$$\dot{w}_p = - k \varrho. \tag{78}$$

Upon examining the work of [38] and [39], Friedman [40] recommended that the specific reaction rate k be utilized in the Arrhenius form:

$$k = B e^{- E/RT} \tag{79}$$

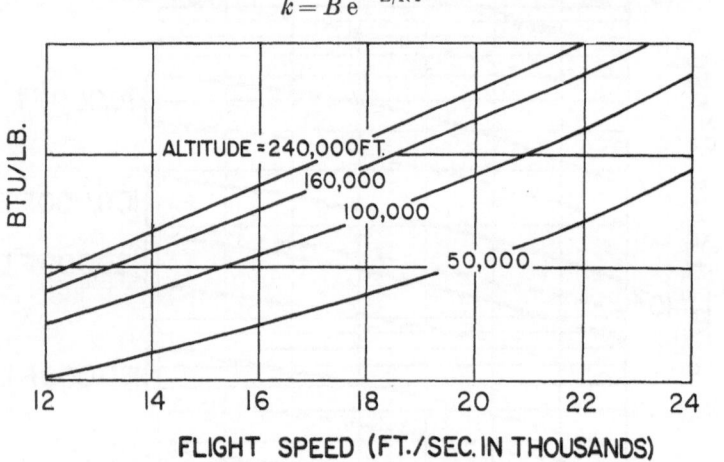

Fig. 16. Effective heat absorption of ablating material

where B is the effective collision frequency and E is the activation energy, with the following numerical values:

$$B = 9.4 \times 10^{18} \text{ sec.}^{-1}$$

$$E = 81.4 \text{ kcal./mole.} \tag{80}$$

It is noted that Knuth [41] recently investigated the behavior of Teflon for Couette flow over a flat plate utilizing a relation of the form of eq. (79).

We will now assume that the ablation of Teflon occurs as a quasi-steady depolymerization process, at the forward stagnation point, in which Teflon polymer $(C_2F_4)_n$ undergoes a rate-controlled pyrolysis reaction in which the solid gasifies to the monomer C_2F_4. The linear rate at which the surface recedes is $\dot{y}$, and is assumed to be a constant for given environmental conditions. Inspection shows that eq. (71) need not be utilized, and further, eq. (72) becomes:

$$\frac{d}{dy}\left(K \frac{dT}{dy} \right) + \varrho \dot{y} C_s \frac{dT}{dy} = \varrho B (h_m - h_p) e^{-E/RT} \tag{81}$$

where the non-homogeneous term represents the heat absorbed within the material during the endothermic depolymerization process.

Eq. (81) is a second order non-linear differential equation with the following boundary conditions:

At the surface: $y = 0$, $T = T_w$. $\tag{82}$

Within the solid: $y = \infty$, $T = T_o$. $\tag{83}$

Upon assuming that the Teflon has a constant thermal diffusivity α, and that the density of the polymer ϱ remains constant, one finds that the first order solution for the temperature distribution within the ablating solid is given by:

where

$$T = T_o + (T_w - T_o - \delta)\, e^{-\frac{y}{a} y} + \delta\, e^{-\frac{E}{R T_w}\left(1 - \frac{T_o}{T_w}\right)\frac{\dot y}{a} y} \qquad (84)$$

$$\delta = \frac{h_m - h_p}{C_s \dfrac{E}{R T_w}\left(1 - \dfrac{T_o}{T_w}\right)} \qquad (85)$$

Upon introducing eqs. (78), (79) and the linearized form of eq. (84) into eq. (73) and integrating, one obtains:

$$\dot y^2 = -\frac{a\, B\, e^{-E/R T_w}}{\dfrac{E}{R T_w}\left[\left(1 - \dfrac{T_o + \delta}{T_w}\right) + \dfrac{E\delta}{R T_w^2}\left(1 - \dfrac{T_o}{T_w}\right)\right]} \qquad (86)$$

which differs from the result obtained by Sutton [42] who had assumed $\delta = 0$. It is also noted that the above relation between ablation rate (i.e., $\dot m_w = \varrho\, \dot y$) and the surface temperature will predict a much smaller rate of mass transfer at a given value of T_w, than one would obtain from the assumption that the equilibrium vapor pressure is the driving force for the interphase mass transfer, (since $\alpha_{vap} = O(10^{-4})$).

Friedman and Myerson [43] recently considered the rates of reaction between the gaseous products of decomposition of Teflon and the products of dissociated air at high temperature. They concluded that the maximum heat transfer from the boundary layer to the solid due to combustion would occur if:

a) Teflon vaporized solely as C_2F_4
b) C_2F_4 was oxidized to COF_2. CF_4 and CO_2.
where it is noted that the chemical reaction

$$CO_2 + CF_4 \rightleftharpoons 2\, COF_2 \qquad (87)$$

has a heat of reaction which is only about seven percent of the heat liberated in the combustion reaction

$$C_2F_4 + O_2 \rightleftharpoons COF_2. \qquad (88)$$

Therefore eq. (88) will be assumed to be the dominant surface reaction. with negligible error in the final result.

An energy balance at the surface of the Teflon yields:

$$Q^* - \frac{Q_{cal}}{\dot m_w} = \left[\left(\frac{\varDelta\tilde Q}{\varDelta \dot m}\right)_w + (h_w - h(s))\right] \qquad (89)$$

which is essentially an equation in surface temperature and gas composition. As in the earlier sections. a means must be found for evaluating the composition of the gas at the surface. since as already noted several times, both $\left(\frac{\varDelta\tilde Q}{\varDelta \dot m}\right)_w$ and h_w depend on the composition of the gas at the surface, at a given surface temperature.

For the combustion model assumed above, there are a total of six chemical species in the boundary layer including O, O_2, N, N_2, C_2F_4 and COF_2. It is readily found that insufficient oxygen diffuses to the surface to consume all of the mono-mer and hence the species present at the wall include only N_2, C_2F_4 and the combustion product COF_2. The mass fractions of O, O_2 and N are virtually zero, since it may be assumed that the element oxygen is completely consumed and nitrogen atoms recombine completely at the surface. Since eq. (86) represents the relation between the interphase mass transfer and the surface temperature, only three additional relationships are required to determine the three unknown mass fractions $C_{N_{2w}}$, $C_{C_2F_{4w}}$ and $C_{COF_{2w}}$. The impermeability of the surface to the element nitrogen requires that eqs. (42) and (43) must hold. The complete consumption of the element oxygen at the surface in the combustion reaction given in eq. (88) requires that eq. (76) must hold.

Finally, combustion stoichiometry requires that

$$\dot{m}_{COF_{2w}} = \frac{132}{32}(\dot{m}_O + \dot{m}_{O_2})_w \qquad (90)$$

which upon utilizing eq. (28) can also be expressed as:

$$C_{COF_{2w}} = \frac{-0.9688\,\mathfrak{A}}{(\dot{m}_w - \mathfrak{A})} \qquad (91)$$

since $C_{COF_{2e}} = 0$ and $(C_O + C_{O_2})_e \cong 0.2348$.

Upon utilizing the thermodynamic data of Fig. 3 and Tables I and III, the

Table III. *Physical and Thermodynamic Properties*

Material	Emissivity		Thermal diffusivity (ft²/hr.)		Melting point (°R)	Normal boiling point (°R)
Carbon	0.80	[4]	0.12	[7]	6880 [1]	8051 [1]
Copper	0.10 (polished) 0.70 (oxidized)	[1]	4.353	[2]	2440 [1]	4696 [1]
Graphite	0.80	[4]	1.03 (1000° R) 0.37 (5000° R)	[6]	6790 [1]	8000 [1]
Iron	0.35—0.41	[1]	0.785	[2]	3254 [1]	5891 [1]
Nickel	0.40 (polished) 0.90 (oxidized	[1]	0.900	[2]	3110 [1]	5711 [1]
Quartz	0.10 (clear) 0.50 (opaque)	[5]	0.033	[5]	3011 [1]	5171 [1]
Teflon	0.30	[3]	0.0035	[8]	1080 [9]	1586 [10]

1. Handbook of Physics and Chemistry, 40th Ed. Cleveland, Ohio: Chemical Rubber Plant Co., 1958—1959.
2. Metals Handbook, ASM, 1947.
3. T. RIETHOF, Private communication.
4. O. SIMPSON and R. THORN, Spectral Emissivities of Graphite and Carbon. Argonne National Lab.
5. M. CAMAC, AVCO Everett Research Lab. RR 38, 1958.
6. N. RASOR and J. MCCLELLAND, Thermal Properties of Materials. ASTIA # AD 118144, 1957.
7. L. LOCH and J. SLYH, The Technology and Fabrication of Graphite. Amer. Inst. Chem. Engrs., 1954.
8. H. C. DUUS, Ind. Engng. Chemistry 47, 1445 (1955).
9. A. LEVITT and A. WONG, A Survey of the Elevated Temperature Properties of Teflon and KEL-F. Watertown Arsenal Lab., August 1956.
10. J. C. SIEGLE and L. T. MUUS, Pyrolysis of Polytetrafluoroethylene. Address to American Chemical Society Meeting (September 1956).

equations which represent the physicochemical constraints and the interphase mass transfer, can be introduced into eq. (89) which can be solved for the surface

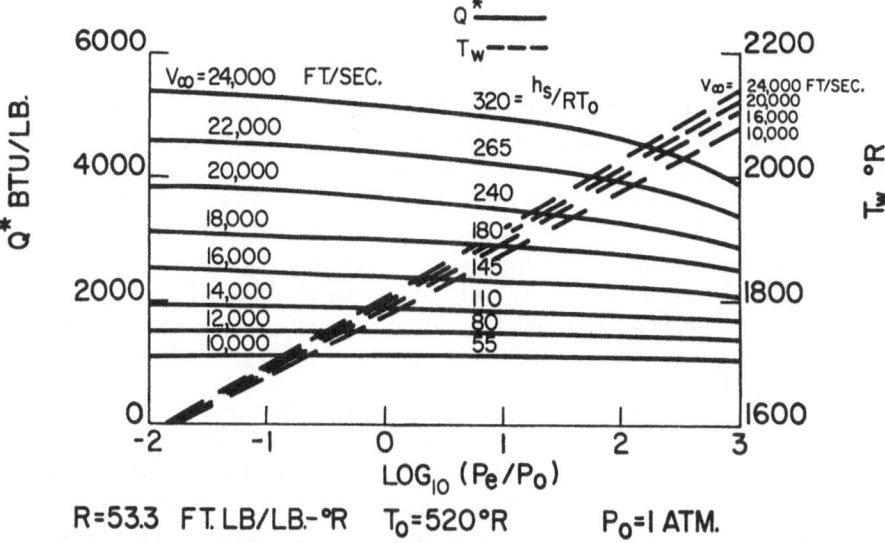

Fig. 17. Variation of heat of ablation and surface temperature of Teflon with stagnation pressure and flight speed

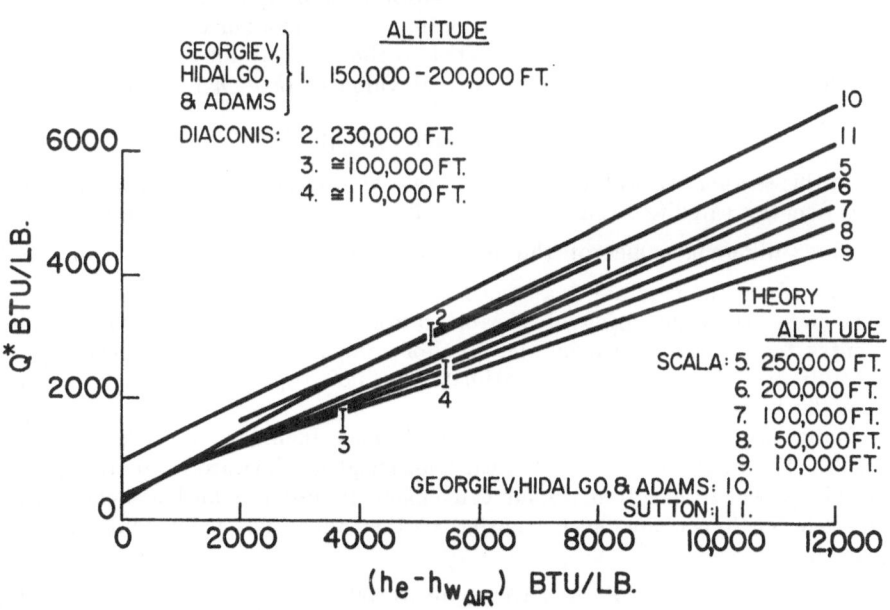

Fig. 18. Effective heat of ablation for Teflon

temperature. Once the surface temperature is determined, all other quantities of interest follow by quadrature.

From earlier work on Teflon, [16], [42], it was anticipated that the effective heat of ablation would correlate with stagnation enthalpy. However, in this analysis, the surface temperature was not assumed to be a constant, as in [16] and [42], and consequently, it was found that the surface temperature correlated as a linearly increasing function of the logarithm of the stagnation pressure, in a manner remarkably similar to quartz, and consequently Q^* appears to be stagnation pressure dependent.

Upon comparing Figs. 6 and 17, it is seen that quartz and Teflon seem to have quite similar behavior, although of course, the quartz surface operates "hot", while the Teflon operates relatively "cool" and hence is exposed to a higher value of $Q_{cal.}$. Upon cross-plotting the data of Fig. 17 versus the quantity $(h_e - h_{w_{Air}})$, one obtains the results shown in Fig. 18, and in addition, the most recent theoretical and experimental studies of Teflon ablation are also shown for purposes of comparison. These include the experimental and theoretical work of [16], the theoretical study which appeared in [42], and the experimental work of DIACONIS [44]. Note that the author's theoretical prediction concerning the stagnation pressure (altitude) effect seems to be partially confirmed by DIACONIS. Thus, one concludes that:

a) at constant altitude, Q^* increases with stagnation enthalpy

b) at constant stagnation enthalpy, Q^* increases with altitude.

One observation will now be made with regard to $h_{w_{Air}}$ which represents the enthalpy of the gas at the surface temperature of the ablating plastic, but treated as if the gas is air. It was found during the calculations that the composition of the gas at the wall consisted of approximately eighty percent by mass of unreacted monomer, ten percent nitrogen molecules and ten percent COF_2, over most of the range of flight conditions of interest. Therefore, the actual value of h_w was found to be essentially constant and equal to $- 2250$ BTU/lb., due to the large negative heat of formation of the monomer. This means that the curves in Fig. 18 must be translated to the right by an amount $\Delta h \cong 2600$ BTU/lb. if one wishes to have the actual value of $(h_e - h_w)$ as the abscissa.

Re-entry Satellite Ablation

In this section, we will determine the rate of ablation at the stagnation point of a re-entry satellite utilizing the four types of materials considered earlier, for three different values of the ballistic parameter $\dfrac{W}{C_D A}$, e.g. 50, 100, and 200 lb./ft.². If the satellite is assumed to be orbiting initially at an altitude of 900,000 ft., then the firing of suitable retrorockets will produce the satellite re-entry trajectories shown in Fig. 19. For comparison's sake, the velocity at the initiation of re-entry is approximately 24,000 ft./sec. at an angle of $2\frac{1}{2}°$ from the horizontal.

Many previous investigators have performed boundary layer calculations (8), (9), (10), in order to calculate heat transfer at the forward stagnation point. In this paper, it has been found convenient to use the author's correlation equation for hypersonic heat transfer at the stagnation point [45] which is given by:

$$Q_{\text{aero.}} \sqrt{R_B} = 1.10 \ (10.0)^a \ (0.100 \ V_\infty)^b, \ \frac{\text{BTU}}{\text{ft.}^{3/2} \text{sec.}} \tag{92}$$

where the fps system is used for the physical dimensions:

$Q_{\text{aero.}}$ = aerodynamic heat transfer rate, $\dfrac{\text{BTU}}{\text{ft.}^2 \text{sec.}}$

R_B = nose radius, ft.

V_∞ = flight speed, ft./sec.

and the exponents are given by

$$a = -(0.9689 + 3.888 \times 10^{-5}\, T_w)\,(5.626 + 9.84 \times 10^{-6}\,\text{Alt.})\tag{93}$$

$$b = (0.9793 + 2.5953 \times 10^{-5}\, T_w)\,(2.830 + 3.00 \times 10^{-7}\,\text{Alt.})\tag{94}$$

where the surface temperature T_w is expressed in °R, and Alt. is the altitude expressed in ft.

When one utilizes eq. (92) and the data of Figs. 6, 11, 16 and 18, for the three trajectories of Fig. 19, one obtains the instantaneous ablation rates shown in

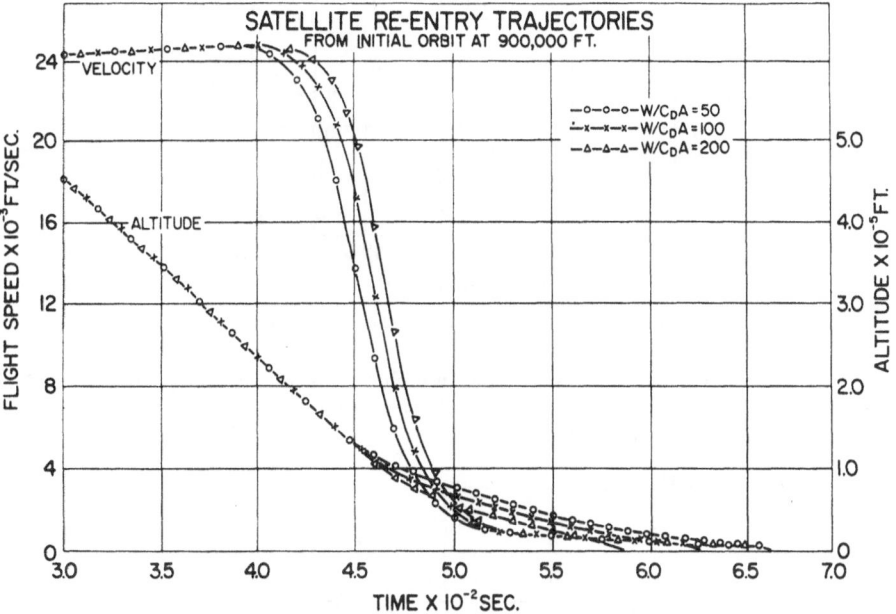

Fig. 19. Satellite re-entry trajectories

Fig. 20, for a satellite with a nose radius of six feet, and a $W/C_p A$ of 200 lb./ft. ^{2}.

The emissivity of quartz was taken as $\varepsilon_w = 0.1$, and it is noted that the author has already shown [45] that for this type of material which vaporizes at a high surface temperature, the re-radation from the material can be large and the ablation rate is consequently reduced.

The emissivity of the graphite was taken as $\varepsilon_w = 0.8$. It is interesting to observe that the peak ablation rate for graphite lags in time behind the other three materials, since it is in phase with the stagnation pressure, which lags the peak heat transfer rate which is the chief cause of thermal degradation for the other three materials.

The effective heat of ablation of the reinforced plastic was assumed to be given by the relation

$$Q^* = 1500 + \frac{1}{2}\, h_e,\ \frac{\text{BTU}}{\text{lb.}}\tag{95}$$

which is representative of reinforced plastics which do not exhibit a pronounced dependence of the effective heat of ablation upon the stagnation pressure. The results are seen to compare favorably with the quartz-like material.

Finally, in Fig. 20 one observes that Teflon has the same characteristic response to the heating cycle which other ablating materials evidence, however, the magnitude of the ablation rate is considerably higher because this material

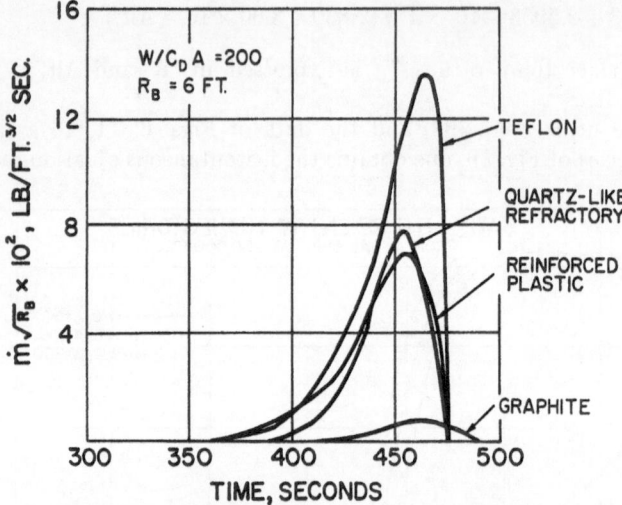

Fig. 20. Rate of ablation during satellite re-entry

operates at a low surface temperature which yields a high $Q_{cal.}$. Furthermore, Teflon has a low heat of pyrolysis which is further reduced by the combustion reactions which produce COF_2.

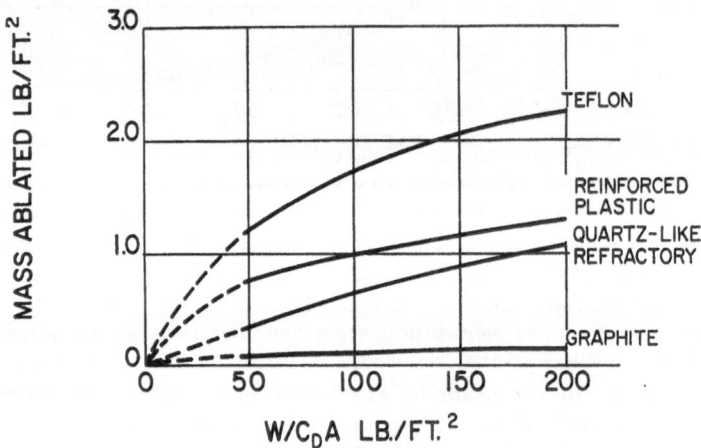

Fig. 21. Total mass ablated at stagnation point of re-entry satellite

When the area under each of the curves of Fig. 20 is obtained by integration, one obtains the total mass of ablating material required during re-entry. These integrated results are shown in Fig. 21. The materials are seen to rank in the following order, as far as minimum ablation is concerned:

1. Graphite,
2. Quartz-like refractory,
3. Reinforced Plastic,
4. Teflon.

It must be noted of course that heat can penetrate into the interior of these materials, particularly if the ablation rate is low; the mere fact that the mass of material ablated is small does not indicate that the material is always best for all applications. For example, graphite will undoubtedly yield the most stable shape during hypersonic re-entry, but unless it is backed up by a light-weight thermal insulator, the pay load will be deleteriously affected. Again, the reinforced plastic has a higher ablation rate but its thermal shielding (or self-insulating) characteristics can be notably better than graphite. Thus, although we have been primarily concerned with attempting to present a clear picture of the various types of ablation phenomena, one must also remember to include transient heat transfer effects within the unaffected material, particularly when the re-entry time is long, (as in manned satellite re-entry).

Conclusions

In this paper, we have indicated both quantitatively and qualitatively the type of behavior one may anticipate from four different classes of ablating materials during hypersonic flight.

The study indicates that for most materials, Q^*, the effective heat of ablation increases almost linearly with stagnation enthalpy and tends to decrease with an increase in stagnation pressure.

A unified treatment has been presented in which it was shown that by having a knowledge of certain minimum physicochemical data, the magnitude of the effective heat of ablation and/or the heat and mass transfer could be predicted theoretically for the full range of hypersonic flight conditions of interest.

It was shown that the specific area of greatest uncertainty was in the kinetics of condensed phase heterogeneous reactions. In addition it would appear that more exact theoretical solutions are required to establish the magnitudes of the diffusion fluxes in multicomponent gaseous mixtures containing light weight chemically reacting species.

Finally, analysis indicates that there is probably no single best ablating material but, rather the selection clearly depends on the type of application intended.

Acknowledgements

The author wishes to acknowledge the many helpful discussions held with his colleagues Dr. G. VIDALE, Dr. H. FRIEDMAN, Dr. M. LINEVSKY, Dr. W. ZINMAN and Dr. A. MYERSON of the Physical Chemistry Operation and Dr. I. GRUNTFEST of the Material Studies Operation of the General Electric Co. Space Sciences Laboratory.

The author wishes to acknowledge the assistance of Mr. LEON GILBERT and Miss EILEEN FOXHILL in performing the calculations and drawing the figures.

The numerical calculations were programmed for digital computation on an IBM 704 computer by Messers PAUL GORDON and J. ROYER.

This investigation is based on work supported by the United States Air Force, Ballistic Missiles Division Contract No. AF 04(647)-269.

References

1. F. A. LINDEMANN and G. M. B. DOBSON, Theory of Meteors and Density and Temperature of the Outer Atmosphere to Which it Leads. Proc. Roy. Soc. London, Ser. A **102**, 411 (1923).
2. C. M. SPARROW, Physical Theory of Meteors. Astrophysic. J. **63**, 90 (1926).
3. E. ÖPIK, Atomic Collisions and Radiation of Meteors. Acta Comm. Univ. Tartuensis **26**, No. 2 (1933).

4. E. ÖPIK, Researches on the Physical Theory of Meteor Phenomena. Tartu Observatory Publ. **29**, No. 5 (1937).
5. R. N. THOMAS, Heat Transfer and the Ablation Process in Meteors. Meteors, pp. 1—7. London-New York: Pergamon Press, 1955.
6. H. J. ALLEN and A. J. EGGERS, A Study of the Motion and Aerodynamic Heating of Missiles Entering the Earth's Atmosphere at High Supersonic Speeds. NACA RM A53D 28 (1953).
7. C. GAZLEY, JR., Heat Transfer Aspects of the Atmospheric Re-entry of Long Range Ballistic Missiles. Rand Corp. R-273 (1954).
8. M. ROMIG, Stagnation Point Heat Transfer for Hypersonic Flight. Jet Propulsion **26**, 1098 (1956). Addendum **27**, 1255 (1957).
9. L. LEES, Laminar Heat Transfer over Blunt Nosed Bodies at Hypersonic Flight Speeds. Jet Propulsion **26**, 259 (1956).
10. J. FAY and F. RIDDELL, Theory of Stagnation Point Heat Transfer in Dissociated Air. J. Aeronaut. Sci. **25**, 73 (1958).
11. G. W. SUTTON and S. M. SCALA, The Two-Phase Laminar Boundary Layer—A Study of Melting Ablation. Rand Corporation Report Mass Transfer Cooling for Hypersonic Flight, No. S-51 (June 1957).
12. S. M. SCALA and G. W. SUTTON, The Two-Phase Hypersonic Laminar Boundary Layer—A Study of Surface Melting. Heat Transfer and Fluid Mechanics Institute, pp. 231—240 (June 1958).
13. H. A. BETHE and M. C. ADAMS, A Theory for the Ablation of Glassy Materials. AVCO Research Laboratory RR 38 (November 1958).
14. M. R. DENNISON and D. A. DOOLEY, Combustion in the Laminar Boundary Layer of Chemically Active Sublimators. Publication No. U-110, Aeronutronic Systems, Inc. (September 1957).
15. M. R. DENNISON, Combustion in the Turbulent Boundary Layer of Chemically Active Sublimators. Publication No. U-166, Aeronutronic Systems, Inc. (March 1958).
16. S. GEORGIEV, H. HIDALGO, and M. C. ADAMS, On Ablation for the Recovery of Satellites. Heat Transfer and Fluid Mechanics Institute, pp. 171—180 (June 1959).
17. L. LEES, Similarity Parameters for Surface Melting of a Blunt Nosed Body in a High Velocity Gas Stream. J. Amer. Rocket Soc. **29**, 345 (1959).
18. J. FANUCCI and H. LEW, Effect of Mass Transfer and Body Forces on Two-Phase Boundary Layers. General Electric Co., M.S.V.D., R.M. No. 35, T.I.S. Document No. R59SD380 (April 1959).
19. L. ROBERTS, A Theoretical Study of Stagnation-Point Ablation. NACA TN 4392 (September 1958).
20. S. M. SCALA, Surface Combustion in Dissociated Air. Jet Propulsion **28**, 340 (1958).
21. S. M. SCALA, Sublimation in a Hypersonic Environment. General Electric Co., M.S.V.D., R.M. No. 20, T.I.S. Document No. R 58SD289 (October 1958). Also, J. Aero/Space Sci. **27**, No. 1 (1960).
22. S. M. SCALA, Vaporization of a Refractory Oxide During Hypersonic Flight. Heat Transfer and Fluid Mechanics Institute, pp. 181—192 (June 1959).
23. S. M. SCALA and G. L. VIDALE, Vaporization Processes in the Hypersonic Laminar Boundary Layer. General Electric Co., M.S.V.D., R.M. No. 14, T.I.S. Document No. R59SD323 (January 1959). To appear in: Internat. J. Heat and Mass Transfer **1**, No. 1 (1960).
24. S. M. SCALA, Hypersonic Stagnation Point Heat Transfer to Surfaces Having Finite Catalytic Efficiency. Proceedings of the Third U.S. National Congress of Applied Mechanics, pp. 799—806 (June 1958).
25. S. M. SCALA, Transpiration Cooling in the Hypersonic Laminar Boundary Layer. (To be published.)
26. S. M. SCALA and C. BAULKNIGHT, Transport and Thermodynamic Properties in a Hypersonic Laminar Boundary Layer, Part I—Properties of the Pure Species. J. Amer. Rocket Soc. **29**, 39 (1959).
27. S. M. SCALA, Hypersonic Viscous Shock Layer. J. Amer. Rocket Soc. **29**, 520 (1959).

28. J. Fanucci, Private communication.
29. W. Zinman, Private communication.
30. M. Gerstein and K. P. Coffin, Combustion of Solid Fuels. High Speed Aerodynamics and Jet Propulsion, Vol. II. Princeton, N. J.: University Press, 1956.
31. G. W. Sutton, The Ablation of Reinforced Plastics in Supersonic Flow. General Electric Co., M.S.V.D., R.M. No. 4, T.I.S. Document No. R57SD644 (July 1957).
32. I. J. Gruntfest, Developments in Plastics for High Temperature Service (Above 1000° F). A.S.M.E. Paper No. 59-MD-1 (May 1959).
33. Thermal Protection of Structural, Propulsion, and Temperature Sensitive Materials for Hypersonic and Space Flight. Chicago Midway Laboratories Data Report CML-DR-M152-2 (October 1958).
34. I. J. Gruntfest and L. H. Shenker, Behavior of Reinforced Plastics at Very High Temperatures, Part I. Modern Plastics **35**, 155 (1958).
35. I. J. Gruntfest, L. H. Shenker, and V. N. Saffire, Behavior of Reinforced Plastics at Very High Temperatures, Part II. Modern Plastics **36**, 137 (1959).
36. M. Dank, R. A. Nelson, W. H. Sutton, and W. R. Sheridan, Water Stabilized Arc Tests on Non-Metallic Materials. J. Electrochem. Soc. **106**, 4, 319 (1959).
37. S. M. Scala, The Thermal Degradation of Reinforced Plastics During Hypersonic Re-entry. General Electric Co., M.S.V.D., T.I.S. Document No. R59SD401 (July 1959).
38. S. L. Madorsky, V. E. Hart, S. Straus, and V. A. Sedlak, Thermal Degradation of Tetrafluoroethylene and Hydrofluoroethylene Polymers in a Vacuum. J. Res. Nat. Bur. Standards **51**, 327 (1953).
39. J. C. Siegle and L. T. Muus, Pyrolysis of Polytetrafluoroethylene. Address to American Chemical Society Meeting (September 1956).
40. H. Friedman, The Mechanism of Polytetrafluoroethylene Pyrolysis. General Electric Co., M.S.V.D., T.I.S. No. R59SD385 (June 1959); Address to American Chemical Society Meeting (September 1959).
41. E. L. Knuth, Compressible Couette Flow with Diffusion of a Reactive Gas from a Decomposing Wall. Heat Transfer and Fluid Mechanics Institute, pp. 104—113 (June 1958).
42. L. Steg, Materials for Re-entry Heat Protection of Satellites. Presented at ARS Semi Annual Meeting (June 1959), Preprint No. 836—59.
43. H. Friedman and A. Myerson, Private communication.
44. N. Diaconis, J. Fanucci, and G. W. Sutton, The Heat Protection Potential of Several Ablation Materials for Satellite and Ballistic Re-entry into the Earth's Atmosphere. 4th Symposium on Ballistic Missile and Space Technology, Space Technology Laboratories, Cal. (August 1959).
45. S. M. Scala, The Thermal Protection of a Re-entry Satellite. General Electric Co., M.S.V.D., T.I.S. Document No. R59SD336 (March 1959).

On the Directing of Intense Photonic Beams by Means of Electron Gas Mirrors[1]

By

Eugen Sänger[2]

(With 9 Figures)

On last year's I.A.F.-Congress at Amsterdam [1], I had the opportunity of pointing out that one way to produce the high radiation intensities of about 10^9 cal cm^{-2} sec^{-1}, such as are required for photonic beam propulsion or for weapons beams, might exist in the black body radiation of heavy plasmas having some 10^5 centi-degrees temperature.

This kind of radiation is at the outset directed towards all directions of space, thus creating the problem of how to concentrate and align them into a parallely directed beam, i.e. to effect the transformation [5] of Fig. 1. Fig. 1 shows the fundamental energy-transformations of jet-propulsion physics.

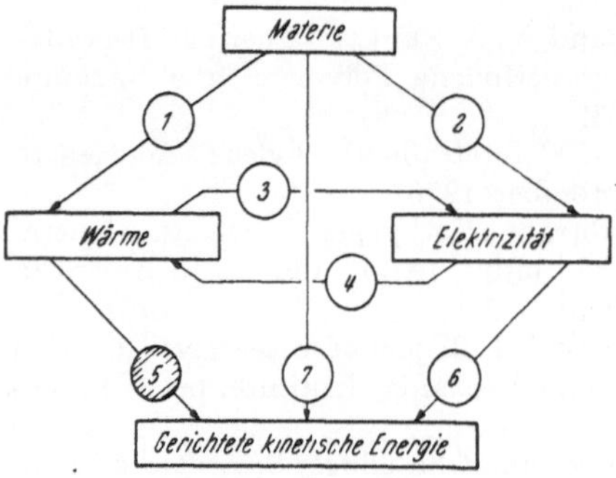

Fig. 1. Energy transformations due to jet propulsion
(Materie = matter, Wärme = heat, Elektrizität = = electricity, Gerichtete kinetische Energie = = directed kinetical energy)

As photons cannot be influenced either by electric or by magnetic fields, reflection by an electronic gas may be considered.

With our thermal rockets we can technically control heat transfers up to about 10^3 cal cm^{-2} sec^{-1} to the walls exposed to fire. With regard to the radiation intensity mentioned above, the reflectors here required thus must at least have a reflection degree of roughly

$$R = 1 - 10^{-6}$$

and that within the shortwave radiation range between approximately 10^{-4} cm and 10^{-7} cm of wave length.

The elementary process of a photon reflection within an electron gas is the elastic collision of a single photon against an electron, known as the COMPTON effect.

Fig. 2 shows the reflection degree R of the COMPTON collision, i.e. the ratio of photon energy after, as against prior to, collision, as a function of wave length λ of the photon, as well as of the angle of deflection ϑ of the photon.

---

[1] Abbreviated version. For a full account of these investigations see Astronaut. Acta **5**, 266—286 (1959).

[2] Forschungsinstitut für Physik der Strahlantriebe e. V., Stuttgart-Flughafen, Bundesrepublik Deutschland.

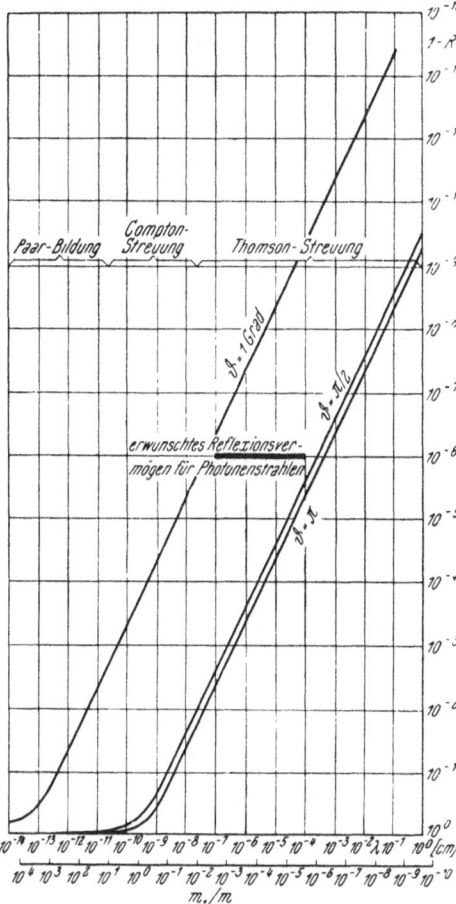

Fig. 2. Reflection degrees R at the elastic collision of a photon having an intrinsic mass at rest (mass of the electron) (Paar-Bildung = pairing, COMPTON-Streuung = = COMPTON dispersion, THOMSON-Streuung = = THOMSON dispersion, Grad = degree, erwünschtes Reflexionsvermögen für Photonenstrahlen = reflective power desired for photonic beams)

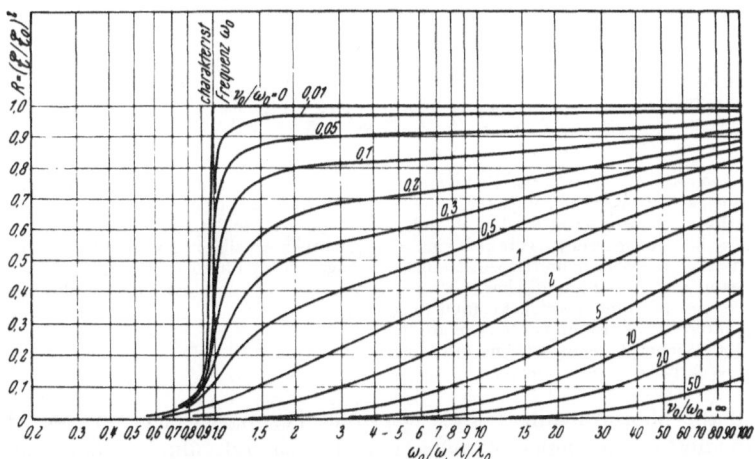

Fig. 3. Reflective power R of a gas layer, homogeneously ionized, against the relative frequencies ω_0/ω, or wave lengths λ/λ_0, with the relative number of collisions ν_0/ω_0 as parameter (Charakteristische Frequenz = characteristic frequency)

The reflection degree shows a strong increase with increasing wave lengths; however, it also shows an increase with decreasing angles of deflection ϑ and, with a small angle of deflection, it lies within the range of the reflection degrees here required.

With a large degree of reflection, i.e. a good reflectivity, a reversion of the photonic beam is thus possible only by way of very numerous individual col-

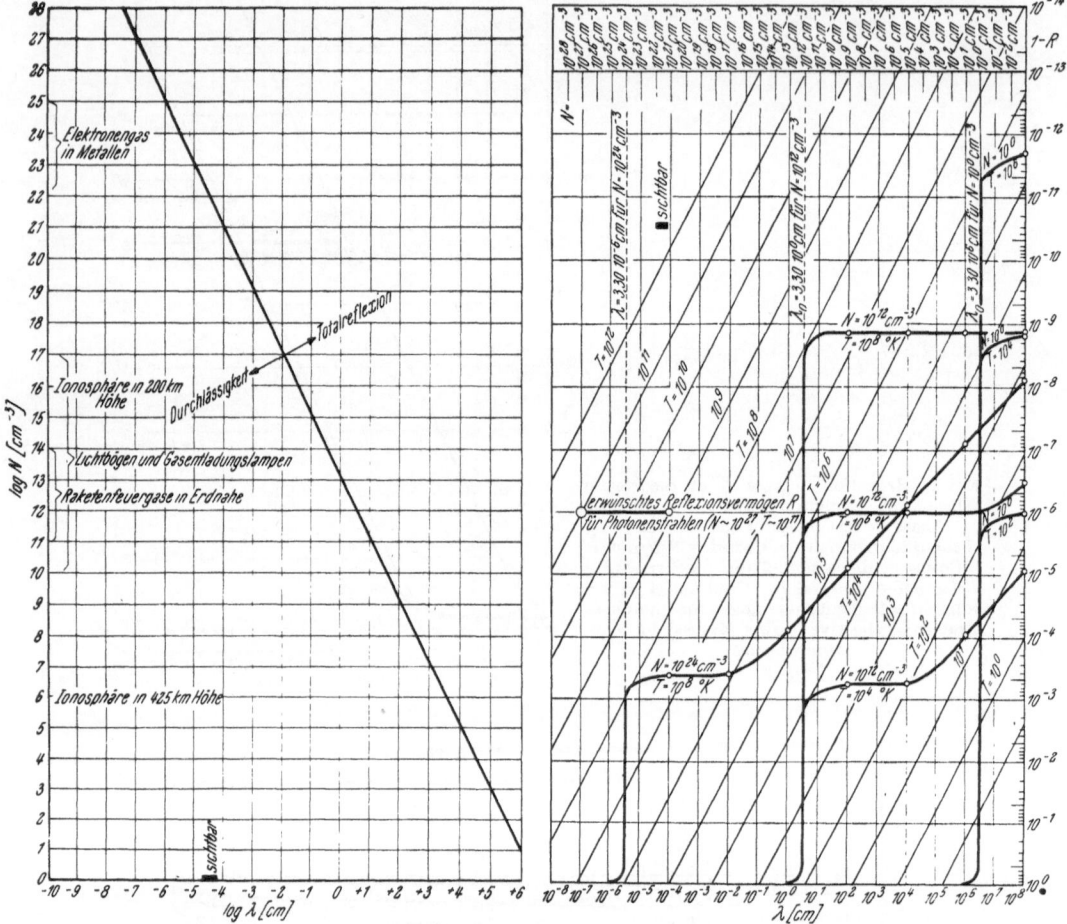

Fig.4. Total reflection on electron gas (Elektronengas in Metallen = electron gas in metals, Ionosphäre in 200 km Höhe = ionosphere at 200 km altitude, Durchlässigkeit = transparency, Totalreflexion = = total reflection, Lichtbögen und Gasentladungslampen = electric arcs and gas discharge lamps, Raketenfeuergase in Erdnähe = rocket combustion gases within terrestrial proximity, Ionosphäre in 425 km Höhe = ionosphere at 425 km altitude, sichtbar = visible range)

Fig. 5. Theoretical reflective power R of singly fully ionized plasmas of varying temperatures T [°K] and electron density N [cm^{-3}] (Sichtbar = visible, erwünschtes Reflexionsvermögen R fur Photonenstrahlen = reflective power R desired for photonic beams)

lisions of the photon against single electrons, i.e. in the electron gas. Thereby, the individual small angles of deflection of each individual collision may add up to a deviation of the beam reflected by ultimately 180 degrees.

Fig. 3 correspondingly shows the reflection degree of a homogeneous, ionized, gas layer plotted against the wave length, with a vertically incident radiation, and for the characteristic optical frequencies ω_0 and ν_0 of the gas layer. Thereby, ω_0 is defined merely through the electron density within the gas layer, while ν_0 is defined by the number of collisions which an electron is subjected to by other particles per second, within the gas layer—this number of collisions being related to electrical resistance.

Again, the reflecting power R for small wave lengths is seen to be very small

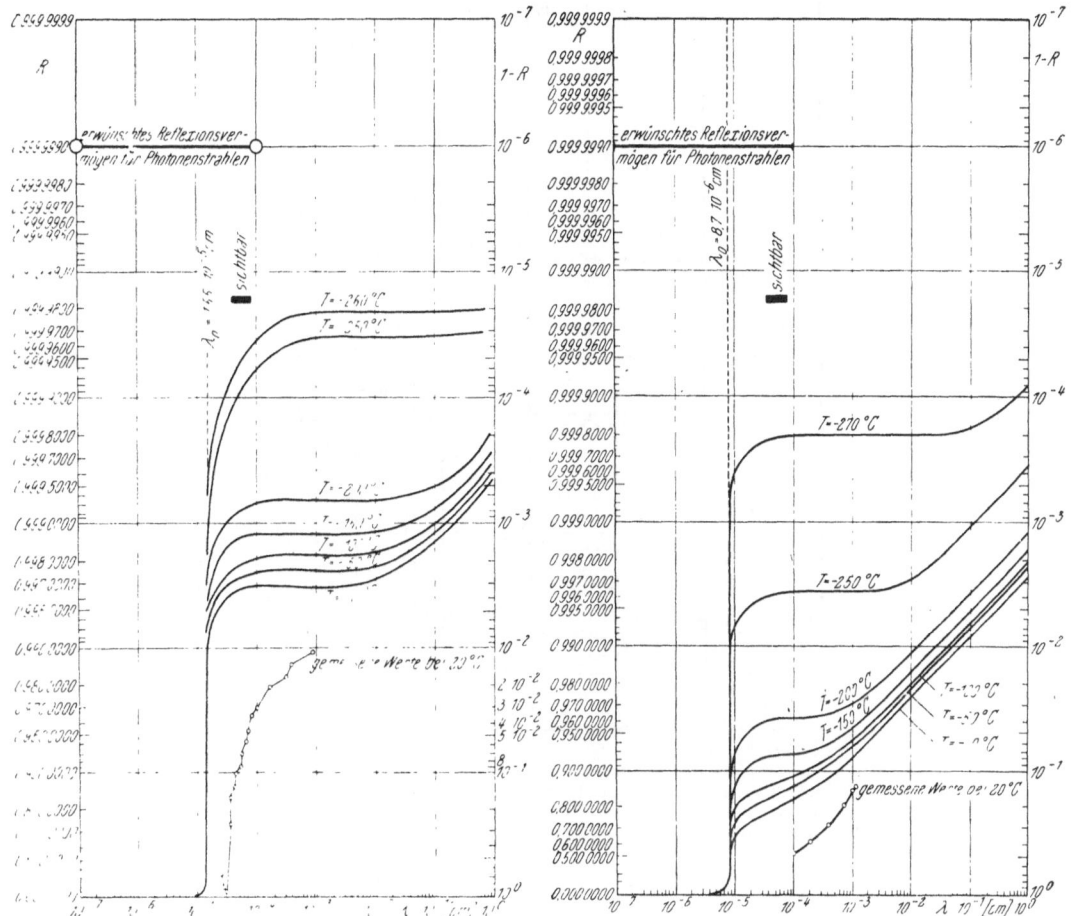

Fig. 6. Theoretical reflective power R of silver at low temperatures
(Erwunschtes Reflexionsvermogen fur Photonenstrahlen = reflective power desired for photonic beams, sichtbar = visible, gemessene Werte bei 20° C = values measured at 20° C)

Fig. 7. Theoretical reflective power R of tin at low temperatures
(Erwünschtes Reflexionsvermögen für Photonenstrahlen = reflective power desired for photonic beams, sichtbar = visible, gemessene Werte bei 20° C = values measured at 20° C)

above the characteristic frequency ω_0—the plasma thus virtually being transparent. Above the critical wave λ_0, however the reflecting power is seen to increase the more, the smaller ν_0—i.e. the greater the electrical conductivity.

With a vanishing ν_0, that is with vanishing electron collisions, or with an infinite electrical conductivity the reflecting power becomes unity, i.e. radiation incident vertically into the gas layer is totally reflected.

Fig. 4 shows the functional relationship of the critical wave length λ_0—where this total reflection of virtually ideal conductors begins—with the electron density N within the gas.

Fig. 5 shows the course of the reflecting power, the reflectivity R, plotted against the wave length, for singly fully ionized plasmas of various temperatures ($T = 10^2$, 10^4, 10^6, and 10^8 °K) and of various electron densities ($N = 10^0$, 10^{12},

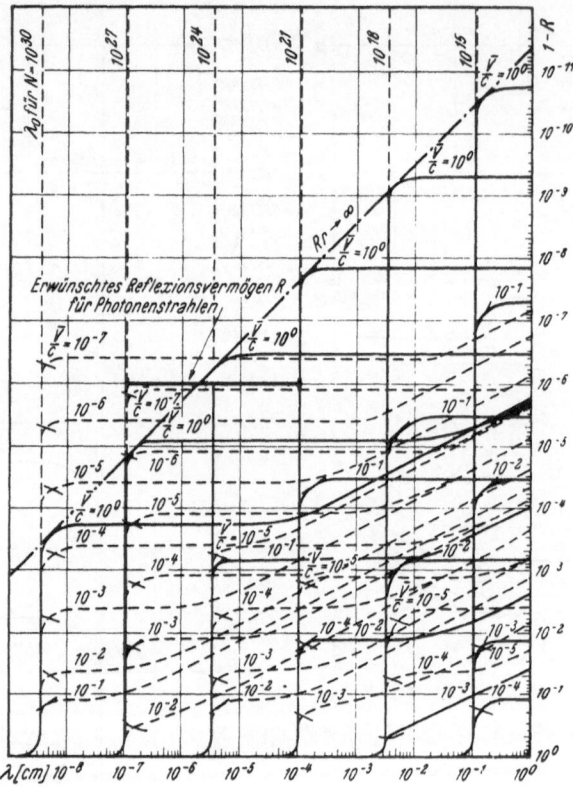

Fig. 8. Theoretical reflective power R of pure electron gas, not degenerated, in thermal equilibrium, of various mean electron velocities $\bar{v}/c$ and electron density N [cm^{-3}]
(Erwünschtes Reflexionsvermögen R für Photonenstrahlen = reflective power R desired for photonic beams)

and 10^{24} cm^{-3}). The characteristic behavior of the curves on Figs. 2 and 3, with its reflecting power increasing with greater wave lengths can again be recognized.

At the same time, the region of the reflecting power of $(1-R) = 10^{-6}$ between $\lambda = 10^{-4}$ and 10^{-7} cm as is desired for photonic beams, has been entered on the diagram. By extrapolating the positions of those distinct salients of the curves, where the reflecting power suddenly vanishes against shorter wave lengths, we find that for a singly fully ionized plasma the reflecting power required would be possible at an electron density of about $N = 10^{+27}$ cm^{-3} and a temperature of roughly $T = 10^{11}$ °K.

Fig. 6 shows the application of the reflection theory underlying Fig. 3 to silver known as a good speculum metal. According to this theory, the specular properties of silver totally disappear in the ultraviolet region below 1550 Å, while above this wave length the reflecting power is the better, the lower the temperature of the metal—i.e., the more its electrical conductivity is improving.

The decrease of metal temperature into the region of supraconductivity —here not entered any more—i.e., the region of virtually infinite metallic conductivity, does not yield any improvement of the reflecting power, according

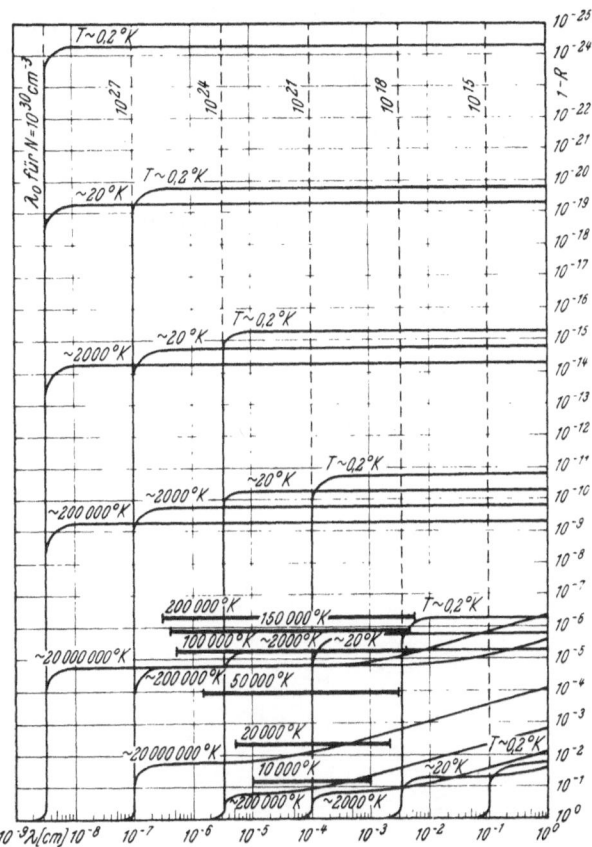

Fig. 9. Reflection on pure, thermal, degenerated electron gas

to hitherto experimental experience. But even if such an improvement could be contrived some day, it would merely cover the wave length region above 1550 Å, thus being insufficient for the reflecting properties here striven for.

Fig. 7. The same applies to the metal tin which is known to particularly increase its electrical conductivity with decreasing temperature.

Fig. 8 at last shows the reflecting power of a pure electron gas not degenerated, in thermal equilibrium, under the influence of the mutual collisions of the electrons, which solely collide still here.

The effective collisional cross-sections of the mutual collisions of electrons being of similar magnitude like the cross-sections of the collisions of electrons against ions in a plasma, this picture of the reflecting power of pure electron gas shows a certain similarity to the corresponding picture Fig. 5 of the singly ionized plasma. The dotted branches of the curves lie already within the region of densities so high, at comparatively moderate temperatures, that degeneration of the electron gas will result therefrom. This electron gas thus will not be governed any more by the laws of BOLTZMANN statistics, but by those of FERMI statistics. The effective collisional cross sections of mutual electron collisions will thus subside to the magnitude of the classical electron cross section.

Fig. 9 correspondingly shows the reflective power of pure, degenerate, electron gas in thermal equilibrium, under the influence of the mutual electron collisions occurring here very rarely.

This reflective power thus fully satisfies the demands made by weapons beams as well as by photonic propulsion, meaning that nature apparently does not throw any fundamental obstacles in our way, neither with a view to supply, nor with a view to directing intensive photonic beams.

Reference

1. E. SÄNGER, IXth International Astronautical Congress, Amsterdam 1958, Proceedings, p. 817. Wien: Springer, 1959.

The Three-Body Problem Earth–Moon–Spaceship[1]

By

W. Gröbner[2] and F. Cap[2]

With reference to the method discussed on the occasion of the IXth International Astronautical Congress, Amsterdam, the solution of the astronomical n-body problem using LIE series is discussed and the known algebraic integrals (conservation of momentum, conservation of angular momentum, conservation of energy) are reproduced. In order to prepare the solution of the three-body problem, the two-body problem is solved by using the new method. Furthermore, the anomaly λ is introduced as an independent variable in order to obtain a parametric representation $r(t)$, $\varphi(t)$. In doing this, a generalization of KEPLER's equation was found. The solutions of the two-body problem are discussed. Next, the solution of the plane three-body problem is given in which case a decomposition of the LIE operator is useful. After a thorough discussion of the initial data, the closed solution of the three-body problem is given, and two different methods for numerical computation are given in such detail that immediate programming is possible. The voluminous calculations with respect to the spatial three-body problem are not described in extenso due to lack of space and time.

Zusammenfassung

Das Dreikörperproblem Erde-Mond-Weltraumschiff. Unter Bezugnahme auf die beim IX. Internationalen Astronautischen Kongreß in Amsterdam besprochene Methode wird zunächst die Lösung des astronomischen n-Körperproblems mit Hilfe LIEscher Reihen besprochen, und die bekannten zehn algebraischen Integrale (Schwerpunktsatz, Drehimpulssatz, Energiesatz) werden reproduziert. Als Vorbereitung für die Lösung des Dreikörperproblems wird dann das Zweikörperproblem mit Hilfe der neuen Methode gelöst. Um eine Parameterdarstellung $r(t)$, $\varphi(t)$ der Bahnen zu bekommen, wird die Anomalie λ als unabhängige Variable eingeführt, wobei sich eine Verallgemeinerung der KEPLERschen Gleichung ergibt. Die so gewonnenen Lösungen des Zweikörperproblems werden diskutiert. Anschließend wird die Lösung des ebenen Dreikörperproblems in Angriff genommen, wobei sich eine Aufspaltung des LIE-Operators als sehr günstig erweist. Nach einer ausführlichen Diskussion der Anfangsdaten wird die geschlossene Lösung des Dreikörperproblems angeführt, und zwei verschiedene numerische Rechenverfahren werden so detailliert angegeben, daß eine sofortige Programmierung möglich ist. Die sehr umfangreichen Rechnungen zum räumlichen Dreikörperproblem werden infolge Platz- und Zeitmangels nicht in extenso wiedergegeben.

Résumé

Le problème des trois corps Terre-Lune-Astronef. S'appuyant sur la méthode présentée au 9ème Congrès à Amsterdam, la solution du problème des n-corps, utilisant des séries de LIE, est discutée et les intégrales connues (conservation de la quantité

---

[1] Abbreviated version. For a full account of these investigations see Astronaut. Acta **5**, 287—312 (1959).

[2] Innsbruck University, Innsbruck, Austria.

de mouvement et de son moment, conservation de l'énergie) sont reproduites. Préparant la solution du problème des trois corps, le problème des deux corps est résolu par la nouvelle méthode. L'anomalie λ est introduite comme variable indépendante dans le but d'obtenir une représentation paramétrique. Ceci conduit à une généralisation de l'équation de KEPLER. Les solutions du problème des deux corps sont discutées. Dans la solution du problème plan des trois corps une décomposition de l'opérateur de LIE est utile. Après discussion complète des donnèes intiales, la solution est donnée sous forme finie. Deux méthodes de calcul numérique sont données avec de tels détails que leur programmation immédiate est possible. Les calculs volumineux du problème général des trois corps ne sont pas décrits in extenso, faute d'espace et de temps.

On the Flight Path of a Hypervelocity Glider Boosted by Rockets[1,2]

By

Angelo Miele[3]

Abstract — Zusammenfassung — Résumé

On the Flight Path of a Hypervelocity Glider Boosted by Rockets. The problems arising from the analysis of the flight paths of hypervelocity vehicles have received considerable attention in recent years. Historically, Sänger and Bredt were among the first to realize the military importance of hypervelocity vehicles. Eggers, Allen, and Neice, on the other hand, produced a simple and interesting comparative analysis of the performance of ballistic missiles, skip-type vehicles, and glide-type vehicles.

The engineering assumptions used by Eggers, Allen, and Neice with glide-type vehicles are essentially equivalent to imagining that the aircraft flies along the so called equilibrium path[4] under the following circumstances: 1. all the propellant mass is burned instantaneously; 2. the kinetic energy thus impressed to the vehicle is gradually expended along a coasting trajectory flown at the maximum value of the lift to drag ratio.

The above hypotheses may convey an impression of arbitrariness to a superficial observer. Yet, the resultant trajectory is endowed with an important property; it is the *extremal path* (i.e., a solution of the Euler equations) associated with the following problem: "Given the initial mass and the final mass, the initial velocity and the final velocity, find the *angle of attack program* and *the burning program* which maximize the range of a hypervelocity vehicle."

A discussion of the above problem of the Calculus of Variations and an analytical proof of the properties associated with the path studied by Eggers and his associates is offered in the present article.

Über die Flugbahn eines raketengetriebenen Hyperschallgleiters. Die aus der Untersuchungsmethode der Flugbahn von Hyperschallgleitern sich ergebenden Probleme haben in den letzten Jahren beträchtliche Beachtung gefunden. Vom historischen Standpunkt aus waren Sänger und Bredt unter den ersten, welche die militärische Bedeutung von Hyperschallgleitern erkannt haben. Andererseits fanden Eggers, Allen und Neice eine einfache und interessante vergleichende Untersuchungsmethode

---

[1] Abbreviated version. For a full account of these investigations see Astronaut. Acta **5**, 367—379 (1959).

[2] This research was supported by the United States Army through the Army Ballistic Missile Agency under Contract No. DA-33-008-ORD-1664.

[3] Formerly, Professor of Aeronautical Engineering, Purdue University, Lafayette, Indiana, U.S.A.; at present, Director of Astrodynamics and Flight Mechanics, Boeing Scientific Research Laboratories, Seattle, Washington, U.S.A.

[4] Equilibrium trajectory is, by definition, a flight path along which the weight is balanced by the aerodynamic lift plus that portion of the centrifugal force which is due to the curvature of the Earth.

der Leistung von ballistischen Geschossen sowie Flugkörpern vom Wellenflug-Typ und Gleiter-Typ.

Die von Eggers und Mitarbeitern für Gleitflugzeuge benützten technischen Annahmen sind im wesentlichen gleichbedeutend mit der Vorstellung, daß der Flugkörper entlang der sogenannten Gleichgewichtsbahn[1] unter folgenden Bedingungen fliegt: 1. Die gesamte Treibstoffmasse wird momentan verbrannt; 2. die dem Flugkörper so erteilte kinetische Energie wird allmählich entlang einer Bahn verbraucht, die antriebslos beim Maximalwert des Verhältnisses von Auftrieb zu Widerstand durchflogen wird.

Diese Hypothesen könnten bei einem oberflächlichen Beobachter den Eindruck der Willkürlichkeit erwecken. In der Tat ist die resultierende Flugbahn mit einer wichtigen Eigenschaft ausgestattet: Sie ist die extremale Bahn (d. h. eine Lösung der Eulerschen Gleichungen), in Verbindung mit folgendem Problem: „Wenn die Anfangs- und die Endmasse sowie die Anfangs- und Endgeschwindigkeit gegeben sind, sind diejenigen Programme für den Angriffswinkel und für die Durchführung der Verbrennung zu finden, welche die größte Reichweite eines Hyperschallgleiters bewirken."

Die vorliegende Arbeit enthält eine Diskussion des obigen Problems der Variationsrechnung und einen analytischen Beweis derjenigen Eigenschaften, die mit der von Eggers und Mitarbeitern untersuchten Bahn verbunden sind.

La trajectoire d'un planeur hypersonique à fusées d'appoint. L'analyse des trajectoires d'engins hypersoniques a reçu une grande attention ces dernières années. Sänger et Bredt ont été parmi les premiers à reconnaître l'importance militaire des véhicules hypersoniques. Eggers, Allen et Neice ont comparé d'une façon simple et intéressante les performances d'engins balistiques, planeurs et véhicules à "ressaut" (skip-type).

Les hypothèses de travail de Eggers, Allen et Neice pour les planeurs reviennent essentiellement à supposer que l'appareil suit une trajectoire d'équilibre (poids = = portance + force centrifuge due à la courbure de la terre) dans les conditions suivantes: a) la consommation des ergols est instantanée, b) l'énergie cinétique impartie est graduellement dépensée le long d'une trajectoire libre à finesse aérodynamique maximum. Cette trajectoire jouit d'une propriété importante: c'est une extrémale du problème suivant: "Les masses initiale et finale étant données ainsi que les vitesses terminales, programmer l'angle d'attaque et la consommation en vue de rendre le rayon d'action maximum".

L'article discute ce problème du calcul des variations et donne les preuves analytiques des propriétés de la trajectoire.

Fundamental Hypotheses and Equations of Motion

A high speed vehicle operating over a spherical Earth in a resisting medium is considered. It is assumed that: (a) the thrust is tangent to the flight path; (b) the trajectory is contained in a plane passing through the center of the Earth and inclined at a small angle with respect to the local horizon; (c) the component of the weight on the tangent to the flight path is negligible with respect to the drag; (d) both the Coriolis acceleration and that part of the centripetal acceleration which is due to the time rate of change of the inclination of the velocity with respect to the local horizon are neglected in the equation of motion on the normal to the flight path; (e) the mass flow of the rocket engine is bounded between an upper value and a lower value.

---

[1] Eine Gleichgewichtsbahn wird als eine Flugbahn definiert, entlang deren das Gewicht durch den aerodynamischen Auftrieb plus jenen Teil der Zentrifugalkraft ausbalanciert wird, der aus der Krümmung der Erde folgt.

Thus, the equations of motion are written as follows:

$$\dot{X} - V = 0 \tag{1}$$

$$\dot{h} - V\theta = 0 \tag{2}$$

$$\dot{V} + \frac{D(h, V, L) - V_e\beta}{m} = 0 \tag{3}$$

$$L - mg_0\left(1 - \frac{V^2}{R_0 g_0}\right) = 0 \tag{4}$$

$$\dot{m} + \beta = 0 \tag{5}$$

where

$$0 \leqslant \beta \leqslant \beta_{max}. \tag{6}$$

In the above equations X is the range, h the altitude, V the velocity, m the mass, D the drag, L the lift, θ the inclination of the flight path with respect to the local horizon, β the mass flow of the rocket engine, g_0 the acceleration of gravity at sea level, and R_0 the radius of the Earth.

The Mayer Problem

In the class of functions $X(t)$, $h(t)$, $V(t)$, $L(t)$, $\theta(t)$, $\beta(t)$, and $m(t)$ which are consistent with eqs. (1) to (5), with the inequality (6), and with certain prescribed end-conditions, the MAYER problem seeks that particular set such that the range is a maximum.

Since the unknown functions are seven in number, while the constraining equations are five, the variational problem admits 2 degrees of freedom. This is logical, in view of the possibility of controlling both the angle of attack program and the burning program.

After writing the EULER equations and making use of the transversality condition and the LEGENDRE-CLEBSCH condition, the following conclusions are reached:

1. *Angle of Attack Program.* For each instantaneous velocity, the flight altitude is to be adjusted in such a way that the overall drag is a minimum.

2. *Burning Program.* The extremal arc can only include sub-arcs of two kinds in the order indicated below: (a) sub-arcs flown with maximum engine output; (b) sub-arcs flown by coasting. It is mathematically proved that no variable thrust sub-arc may appear in the composition of the extremal arc.

Remarks

It is to be emphasized that the present conclusions are limited to the cruising portion of the flight path. Should the complete trajectory from take-off to landing be considered, some modification would probably appear in the optimum flight program.

It is also stressed that the simplified equations used in the present article are applicable to the class of glide paths, but not to the class of skip paths. Thus, the present analysis points out that the conventional treatment of the gliding flight is optimal with respect to the class of glide paths, but not necessarily with respect to the larger class of glide and skip paths.

References

1. E. SÄNGER and J. BREDT, A Rocket Drive for Long Range Bombers. Bureau of Aeronautics, Navy Department, Translation CGD-32, 1944.
2. A. J. EGGERS, JR., H. J. ALLEN and S. E. NEICE, A Comparative Analysis of the Performance of Long Range Hypervelocity Vehicles. NACA T.N. 4046, October, 1957.
3. A. MIELE, General Variational Theory of the Flight Paths of Rocket-Powered Aircraft, Missiles and Satellite Carriers. Astronaut. Acta 4, 264 (1958).
4. G. A. BLISS, Lectures on the Calculus of Variations. Chicago: University Press, 1946.
5. A. MIELE and J. O. CAPPELLARI, JR., Some Variational Solutions to Rocket Trajectories over a Spherical Earth. Purdue University, School of Aeronautical Engineering, Report No. A-58-9, November, 1958.
6. T. EDELBAUM, Personal Communication, February, 1959.
7. A. MIELE, Variational Approach to Problems of Hypervelocity Flight. Purdue University, School of Aeronautical Engineering, Report No. A-59-7, 1959.

On the Numerical Computation of Free Trajectories of a Lunar Space Vehicle[1]

By

J. M. J. Kooy[2]

One of the possibilities for the smaller countries to contribute something to the advance of astronautics, is to carry out astronautical computations. By the kindness of "Bull Nederland", Amsterdam, it has become possible also to contribute to this work in the Netherlands, by providing the occasional support of a powerful electric computer.

As a beginning of these investigations, a threedimensional orbit has been computed according to which the vehicle approaches the moon up to about 11,000 km, thereby coming in a position at which the hidden lunar hemisphere can be observed, whereas, after passing through a loop, the vehicle again returns to terrestrial neighbourhood. The launching site has been chosen on the terrestrial equator, in order to take full profit of the rotation of the earth.

The "zero orbit", which the vehicle would describe if it were only subjected to the terrestrial attraction, is chosen in the plane of the equator. The motion of the sun and the moon with respect to a geocentric system of reference, not rotating with respect to the celestial sky, taking into account the mutual perturbations, has firstly been computed throughout a time lapse covering the travel time of the lunar rocket. The amount of the speed in the last burn-out point has been chosen just sufficient in order to bring the vehicle at lunar distance. The elliptical zero orbit and the start time from the last burn-out point have been determined in such a way that rocket and moon would meet at the apogee, if the lunar and solar attraction would not come into play. With the initial conditions of free flight in zero approximation so obtained, the corresponding real motion has been computed mechanically, taking into account the solar and lunar perturbations. Then, by suitable variation of the initial position and speed as to magnitude and direction, soon a satisfactory orbit was obtained. Thereby it appeared that a variation of $1^0/_{00}$ of the magnitude of the initial speed renders the orbit quite unsuitable.

In order to meet this difficulty, in a following investigation, with a launching site on northern latitude (of Cape Canaveral) a hyperbolic zero orbit has been chosen and the effects of the application of a retro rocket in the vicinity of the moon have been explored, assuming the retro thrust still in instantaneous negative speed direction and constant numerical values throughout the short braking periods. In this way, orbits have been obtained, hitting the moon, and also circumnavigating the moon with subsequent return to the terrestrial surface. In all cases, the lunar and solar perturbations have been taken into account.

---

[1] Abbreviated version. For a full account of these investigations see J. M. J. Kooy and J. Berghuis, Astronaut. Acta 6, 115—143 (1960).

[2] Lector K.M.A., St. Ignatiusstraat 99a, Breda, Holland.

Further, also the possibility of obtaining a moon satellite has been investigated and a corresponding orbit has been computed. It appeared that the KEPLER motion of the satellite so obtained (with respect to a selenocentric system of reference, not rotating with respect to the celestial sky) is strongly disturbed by the combined terrestrial and solar perturbations.

Zusammenfassung

Numerische Untersuchung der freien Bewegung einer Mondrakete. Durch die von „Bull Nederland", Amsterdam, freundlich angebotene, gelegentliche Hilfe einer elektronischen Rechenmaschine ergab sich die Möglichkeit, einige numerische Bahnberechnungen für Mondraketen durchzuführen. Als erster Fall wurde eine dreidimensionale Bahn berechnet, wobei das Raumfahrzeug sich dem Mond bis auf etwa 11.000 km nähert und dabei in eine Lage kommt, von wo aus die Rückseite des Mondes beobachtet werden kann. Nach Durchlaufen einer Schleife kommt das Fahrzeug wieder in Erdnähe. Bei diesem Fall wurde der Startplatz am Äquator angenommen und beim letzten Brennschluß eine sogenannte ballistische Anfangsgeschwindigkeit der freien Fahrt angewendet, die gerade ausreicht, um das Raumfahrzeug auf Mondentfernung zu bringen.

Bei einer weiteren Untersuchung wurde der Startplatz in nördlicher Breite (von Cape Canaveral) gewählt und beim letzten Brennschluß eine hyperbolische Geschwindigkeit angenommen. Die Auswirkung einer Bremsrakete in der Nähe des Mondes wurde studiert. Entsprechende Bahnen, nach Abbremsen, um den Mond zu treffen und auch um den Mond mit folgender Rückkehr zur Erdoberfläche zu umfahren, wurden berechnet. Auch wurde die Möglichkeit untersucht, durch geeignetes Bremsen in Mondnähe einen Mondsatelliten zu erhalten, und eine entsprechende Bahn wurde berechnet.

Bei allen Untersuchungen wurden auch die Störungen der Sonne berücksichtigt. Es stellte sich heraus, daß die KEPLER-Bewegung des Mondsatelliten in Beziehung auf ein selenozentrisches Koordinatensystem sehr stark durch die kombinierte Auswirkung von Erde und Sonne gestört wird.

Résumé

Investigation numérique du mouvement gravitant d'un véhicule d'espace lunaire. Par la générosité de la S.A. "Bull Nederland", Amsterdam, il nous est devenu possible à contribuer à des recherches astronautiques numériques. Comme introduction à ces investigations théoriques une orbite trois-dimensionelle est déterminée selon laquelle le véhicule approchera la lune jusqu'à 11.000 km., dans une position de laquelle l'hémisphère postérieur de la lune pourrait être observé. Ensuite le véhicule, toujours en gravitant, retourne au voisinage de la terre. Il paraît que la précision nécessaire de la valeur de la vitesse initiale du mouvement gravitant est $1^0/_{00}$.

Dans un deuxième exemple la valeur de la vitesse intiale est plus grande que la vitesse locale hyperbolique et l'application d'une fusée de frein dans le voisinage de la lune est le sujet d'une investigation numérique. En appliquant une telle fusée de frein, des orbites circumnavigant la lune ont été obtenues, le véhicule aussi retournant à la surface de la terre.

Aussi la possibilité d'obtenir un satellite de la lune est investigée et pour ceci une orbite a été calculée. Il paraît que le mouvement Keplérien d'un tel satellite artificiel lunaire est fortement troublé par les perturbations combinées de la terre et du soleil.

Application of Magnetohydrodynamics to Astronautics

By

Arthur Kantrowitz[1]

Abstract — Zusammenfassung — Résumé

Application of Magnetohydrodynamics to Astronautics. Magnetohydrodynamics has two presently visible applications to space flight. First, it seems quite likely that Magnetohydrodynamics methods of electrical propulsion will be very useful in the provision of exhaust jets in the specific impulse range 1500—5000 seconds. At higher specific impulses the competitive position between ion rockets and Magnetohydrodynamics is still to be determined. A discussion of various Magnetohydrodynamics devices to produce thrust in various specific impulse ranges will be given.

Second, Magnetohydrodynamics plays a dominant role in the dynamics of interplanetary gas masses. Thus, the dynamics of interplanetary gas clouds will be important because these gas masses will transport the highly energetic particles which are dangerous to human beings. Experimental reproduction of the fast shock waves present in the interplanetary plasma will be discussed. Some preliminary laboratory results obtained with shock waves in the velocity range of 50 centimetres per microsecond will be given.

Anwendung der Magnetohydrodynamik auf die Astronautik. Die Magnetohydrodynamik hat zwei gegenwärtig erkennbare Anwendungsmöglichkeiten für die Weltraumfahrt. Erstens scheint es durchaus wahrscheinlich, daß magnetohydrodynamische Methoden des elektrischen Antriebs sehr nützlich für die Erzeugung von Ausstoßstrahlen im Bereich spezifischer Impulse von 1500 bis 5000 Sekunden sein werden. Für größere spezifische Impulse muß der relative Wert der Ionenraketen und der magnetohydrodynamischen Systeme noch ermittelt werden. Eine Erörterung verschiedener magnetohydrodynamischer Techniken zur Erzeugung des Schubs in verschiedenen Bereichen des spezifischen Impulses wird gegeben.

Zweitens spielt die Magnetohydrodynamik eine beherrschende Rolle in der Dynamik der interplanetarischen Gasmassen. So wird sich die Dynamik der interplanetarischen Gasnebel als bedeutungsvoll erweisen, weil diese Gasmassen die Partikel hoher Energie transportieren, die für den Menschen gefährlich sind. Es wird die experimentelle Reproduktion der schnellen Stoßwellen diskutiert, die im interplanetarischen Plasma vorhanden sind. Einige vorläufige Laboratoriumsergebnisse werden angegeben, die mit Stoßwellen im Geschwindigkeitsbereich von 50 cm/μsec erhalten wurden.

Application de la magnétohydrodynamique en astronautique. La magnétohydrodynamique a, à ce jour, deux applications évidentes au vol spatial: Premièrement, il semble très probable que les méthodes magnétohydrodynamiques de propulsion électrique seront des plus utiles pour l'obtention de jets avec impulsions spécifiques de l'ordre de 1500 à 5000 secondes. Pour des impulsions spécifiques plus élevées, la valeur relative des fusées ioniques et des procédés magnétohydrodynamiques doit

---

[1] Avco-Everett Research Laboratory, Everett, Massachusetts, U.S.A.

encore être déterminée. Une description de plusieurs techniques magnétohydrodynamiques permettant d'obtenir des poussées avec différentes valeurs d'impulsion, sera
donnée.

Deuxièmement, la magnétohydrodynamique joue un rôle dominant dans la dynamique des masses gazeuses interplanétaires. Par exemple, la dynamique des nuages
gazeux interplanétaires sera importante parce que ces masses gazeuses transportent
des particules à haute énergie dangereuses pour le genre humain. La reproduction
expérimentale des ondes de choc très rapides se trouvant dans le plasma interplanétaire, sera discutée. Quelques résultats préliminaires obtenus en laboratoire grâce à
des ondes de choc se mouvant à des vitesses de l'ordre de 50 cm. par microseconde
seront donnés.

Observation Satellites:
Problems, Possibilities and Prospects

By

Amrom H. Katz[1]

Abstract — Zusammenfassung — Résumé

Observation Satellites: Problems, Possibilities and Prospects. The emergence of special-purpose orbiting vehicles has focused attention on the observation satellite as a useful tool. Earth satellites of this type could be employed for both detailed and broad inspection, for mapping and geodetic purposes, for astronomical photography free of the image-degrading and narrow-band-pass effects of the Earth's atmosphere, for studies in the several Earth sciences and for meteorological observation.

Restricting this argument to satellites that are instrumented to make essentially pictorial or photographic observations of the Earth, its clouds, or celestial objects, the author discusses the kinds of information that can be obtained and the problems involved, the ways of returning this information from the satellite, and the purposes which may be served by such information.

During the past one hundred years, photography has developed greatly as a practical tool for high-altitude observation, chiefly using aircraft as platforms. However, at the altitudes contemplated for satellites (100 to 10,000 miles, or more) the geometry of ground coverage is somewhat complicated by the fact that a flat Earth can no longer be assumed. Observation can be conducted with a number of sensors operating in different portions of the electromagnetic spectrum. The emphasis in the present paper is on photographic film systems with physical recovery of data.

An important parameter in describing the performance of observation satellite systems is "ground resolution". The author discusses the general concept of photographic resolution in relation to scale number, observing that if, in achieving a given ground resolution, one can trade between these two parameters, he should trade in the direction of lower resolution and smaller scale number. He defines four levels of observational detail in terms of ground resolution. These could be thought of as constituting an "observation sieve" through which (for example), for a given amount of film, smaller and smaller areas are scrutinized with increasing detail.

The shutterless continuous-strip camera and the panoramic camera are described and illustrated, and three rules-of-thumb for designing such camera systems are postulated. Each of the two methods of returning information to Earth from a satellite—telemetering by video link and physical recovery of film—is shown to have certain advantages for certain applications. Telemetering, for example, provides data quickly but at a low delivery rate; physical recovery introduces a large delay, but after recovery the production rate is tremendous.

The optical and photographic problems involved in long-distance observation from a 24-hour satellite or from the moon are discussed and shown to be not inconsiderable.

Observation of celestial bodies by means of satellites holds many exciting possibilities. The Earth sciences, too, stand to benefit in significant ways. The application of wide-angle photography to weather observation from satellites appears to be especially interesting.

---

[1] Engineering Division, The RAND Corporation, 2400 Colorado Avenue, Santa Monica, California, U.S.A.

On the international scene, it is clear that observation satellites could play a role in inspection systems and in systems to provide warning against surprise attack.

The author concludes with a discussion of families of satellites employing various known booster combinations. Performance possibilities afforded by the use of payloads several times larger than those now available are considered.

Beobachtungssatelliten: Probleme, Möglichkeiten und Aussichten. Das Auftauchen von Weltraumschiffen, die einem bestimmten Zweck dienen, hat die Aufmerksamkeit auf den Beobachtungssatelliten als ein nützliches Hilfsmittel gelenkt. Erdsatelliten von diesem Typ könnten für einige kartographische und geodätische Zwecke und für astronomische Photographie angewendet werden. Sie müßten aber frei von bildverschlechternder und bildbeschränkender Wirkung der Erdatmosphäre sein und ebenso für Studien in den verschiedenen Erdwissenschaften und für meteorologische Beobachtungen verwendet werden können.

Der Autor beschränkt seine Darlegung auf Satelliten, die mit Instrumenten versehen sind, um mit ihnen hauptsächlich filmische und photographische Beobachtungen der Erde, ihrer Wolken oder himmlischer Objekte anzustellen; er erörtert die Arten dieser Informationen, die eingezogen werden können, und die darin enthaltenen Probleme, ebenso die Wege, auf denen man diese Informationen von dem Satelliten zurücksenden kann, und die Zwecke, denen diese Informationen dienen können.

Während der letzten hundert Jahre hat sich die Photographie außerordentlich als ein praktisches Mittel für Beobachtungen aus großer Höhe, hauptsächlich vom Flugzeug aus, entwickelt. Bei den Höhen jedoch, die für Satelliten in Betracht kommen (100 bis 10000 Meilen oder mehr), ist die Geometrie der Erde etwas durch die Tatsache kompliziert, daß die Erde nicht mehr als flach angenommen werden kann. Die Beobachtung kann mit Hilfe einer Anzahl empfindlicher Strahlen durchgeführt werden, die in verschiedenen Bereichen des elektromagnetischen Spektrums wirken. Das Hauptgewicht in der vorliegenden Abhandlung liegt auf dem photographischen Filmsystem, um diese Angaben physikalisch aufzufangen.

Ein wichtiger Teil bei der Beschreibung der Leistung des Beobachtungssatelliten ist die „ground resolution". Der Autor diskutiert die allgemeine Auffassung von der photographischen Auflösung im Verhältnis zur Maßstabzahl und stellt dabei fest, daß man in Richtung auf die niedrigere Auflösung und die kleinere Maßstabzahl hinarbeiten soll, wenn man — um eine gegebene „ground resolution" zu erhalten — zwischen diesen beiden Parametern arbeiten soll. Er definiert dabei 4 Stufen von Beobachtungsangaben auf Grund dieser „ground resolution". Diese könnten dafür gedacht sein, daß sie ein Beobachtungssystem darstellen, durch das z. B. für eine gegebene Länge des Films kleine und kleinere Felder mit zunehmenden Einzelangaben ganz genau untersucht werden.

Die verschlußlose, ununterbrochene Streifenkamera und die panoramische Kamera werden beschrieben und illustriert und drei Faustregeln für den Entwurf solch eines Kamerasystems angegeben. Jede der beiden Methoden, wie man Informationen von einem Satelliten-Telemeter durch eine Fernsehverbindung oder physische Bergung des Films zur Erde zurücksenden kann, wird aufgezeigt, um gewisse Vorteile für bestimmte Anwendungen zu haben. Über das Telemetersystem z. B. werden Angaben schnell weitergeleitet, aber mit einer langsamen Lieferungsrate; die physische Bergung hat eine große Verzögerung zur Folge, aber nach dem Empfang ist die Leistung äußerst hoch. Die optischen und photographischen Probleme, welche Weitsichtbeobachtungen von einem 24-Stunden-Satelliten oder vom Mond einschließen, werden besprochen und es wird gezeigt, daß sie nicht unbeträchtlich sind.

Beobachtungen von Himmelskörpern mittels Satelliten haben viele erregende Möglichkeiten. Auch die Erdwissenschaften sind ein eindringlicher Beweis für diese Vorteile. Besonders interessant scheint die Anwendung der Weitwinkelphotographie für die Wetterbeobachtung von Satelliten aus zu sein.

Auf internationaler Ebene ist es klar, daß Beobachtungssatelliten eine Rolle im Inspektionssystem spielen können, und ebenso Systeme, die Warnung vor einem Überraschungsangriff geben.

Der Autor schließt mit einer Diskussion über die verschiedenen Satellitengruppen, die verschiedene bekannte Hilfsmotorkombinationen anwenden. Ihre Leistungsmöglichkeiten, die durch den Gebrauch von Nutzlasten geschaffen werden, werden um ein Mehrfaches größer erachtet als jene, die heute zur Verfügung stehen.

Satellites d'observation: problèmes, possibilités et perspectives. L'apparition de véhicules orbitaux à missions spécialisées a mis l'accent sur le rôle utile du satellite d'observation. Ce type de satellite pourrait servir aux fins d'inspections d'ensembles ou de détails, en cartographie et géodésie, pour la photographie astronomique libérée des effets de filtrage de l'atmosphère, pour l'observation météorologique et en général pour faire progresser plusieurs des sciences de la Terre.

L'auteur se limite aux satellites instrumentés dans le but essentiel d'effectuer des observations photographiques ou des prises de vues de la Terre, de ses nuages ou des objets célestes. Il discute les types d'information possibles et les problèmes qu'ils posent, les moyens de recevoir cette information du satellite et les usages qui peuvent en être faits.

Durant les cent dernières années, la photographie s'est développée jusqu'à devenir un instrument pratique pour l'observation de haute altitude, spécialement à bord des avions. Aux altitudes envisagées pour les satellites (de 160 à 16 000 km ou plus) la courbure de la Terre ne peut plus être ignorée, ce qui complique la géométrie de la restitution. L'observation peut être conduite avec un certain nombre d'éléments sensibles à des intervalles différents du spectre électromagnétique. Dans cet article l'accent est mis sur l'utilisation de film photographique avec restitution physique des donnés.

Le "pouvoir séparateur" est un paramètre important du système d'observation à bord du satellite. L'auteur discute le concept général du pouvoir séparateur en relation avec l'échelle de l'observation et pour autant que ces paramètres puissent être variés, marque sa préférence pour un pouvoir séparateur plus faible associé à une plus petite échelle, quand il s'agit d'obtenir une séparation donnée des détails au sol. Il définit quatre niveaux d'observation pour ces détails; ils constituent une sorte de filtre à travers lequel on peut par exemple analyser des surfaces de plus en plus petites de façon de plus en plus détaillée.

La caméra à déroulement continu sans obturateur et la caméra panoramique sont décrites et illustrées. Trois règles simples pour leur conception sont avancées. Des deux méthodes de retransmission des informations recueillies par le satellite — liaison télémétrique en video ou récupération des films—chacune a ses avantages particuliers. La première transmet sans délai mais à un rythme lent; pour la seconde, une fois le long délai de récupération passé, le dépouillement est extrêmement rapide.

Les problèmes d'optique et de photographie soulevés par l'observation à longue distance à partir d'un satellite stationnaire ou de la lune ne sont pas négligeables.

L'observation des corps célestes promet d'être fertile en résultats scientifiques. Les sciences de la Terre profiteront aussi de ce nouvel instrument. L'application de la photographie à grande ouverture angulaire paraît spécialement intéressante pour la météorologie.

Du point de vue de la situation internationale, il est certain que les satellites d'observation pourraient jouer un rôle dans les dispositifs d'inspection et d'avertissement contre une attaque par surprise.

L'auteur conclut par une discussion des performances d'une famille de satellites utilisant diverses combinaisons de fusées de lancement connues. Les possibilités offertes par des charges utiles plusieurs fois supérieures à celles actuellement réalisables sont considérées.

Libration of an Artificial Earth Satellite

By

V. V. Beletsky[1]

Abstract — Zusammenfassung — Résumé

Libration of an Artificial Earth Satellite. The paper deals with the determination, by the methods of LYAPUNOV and CHETAYEV, of conditions for the existence and stability of the relative equilibrium of an orbiting solid body in a Newtonian central force field, and with the examination of the oscillations of the body near the position of relative equilibrium. The investigated problem is an idealisation of movements actually existing in the solar system, i. e. the movement of the moon relative to earth, as well as of possible movements of artificial earth satellites.

The conclusions drawn from the examination of the problem are that "the movement of an artificial earth satellite around the centre of mass will be of the libration type provided the stability conditions are observed and provided the initial angular rotation speeds near the mass centre are small compared with the angular rotation velocity of the mass centre. However, if the initial angular velocities are comparatively great, as in the first Soviet sputniks, the movement of the satellite will be a component of the unperturbed EULER-POINCEAU movement near the kinetic moment vector and the secular precessional movement of the kinetic moment vector itself". Some minor factors of perturbation would affect the actual libration of an artificial satellite, but "investigations show that, if the basic stability conditions are fulfilled, the libration in the presence of these perturbation factors will differ little from the undisturbed libration".

Libration eines künstlichen Erdsatelliten. Die Arbeit befaßt sich mit Hilfe der Methoden von LYAPUNOW und TSCHETAJEW mit der Bestimmung der Bedingungen für die Existenz und Stabilität des relativen Gleichgewichtes eines in einem Newtonschen Zentralkraftfeld umlaufenden festen Körpers; ferner mit der Prüfung der Oszillationen des Körpers in der Nähe des Ortes des relativen Gleichgewichtes. Das untersuchte Problem ist eine Idealisierung von im Sonnensystem tatsächlich vorkommenden Bewegungen, z. B. der Bewegung des Mondes relativ zur Erde ebenso wie möglicher Bewegungen künstlicher Erdsatelliten.

Die Schlußfolgerungen sind nachstehende: Die Bewegung eines künstlichen Erdsatelliten um das Massenzentrum werden von der Art einer Libration sein, vorausgesetzt, daß die Stabilitätsbedingungen eingehalten werden und die anfänglichen Rotationswinkelgeschwindigkeiten in der Nähe des Massenzentrums klein sind im Vergleich zur Rotationswinkelgeschwindigkeit des Massenzentrums. Wenn jedoch die anfänglichen Winkelgeschwindigkeiten verhältnismäßig groß sind, wie dies bei den ersten der sowjetischen Sputniki der Fall war, wird die Bewegung des Satelliten eine Komponente der ungestörten EULER-POINCEAU-Bewegung nahe dem kinetischen Momentenvektor und der säkularen Präzessionsbewegung des kinetischen Momentenvektors selbst sein. Einige geringere Störungsfaktoren würden die tatsächliche Libration eines künstlichen Satelliten beeinflussen, doch zeigen die Untersuchungen des Verfassers, daß bei Erfüllung der grundlegenden Stabilitätsbedingungen die Libration beim Vorhandensein dieser Störungsfaktoren nur wenig von der ungestörten Libration abweichen wird.

---

[1] STEKLOV Institute of Mathematics of the Academy of Sciences of the U.S.S.R., MOSCOW, U.S.S.R.

Librations d'un satellite artificiel de la Terre. Il s'agit de la détermination par la méthode de Liapounov et Chetaev des conditions d'existence et de stabilité de l'équilibre relatif d'un solide en orbite dans un champ de force central Newtonien et de l'étude des oscillations autour de la position d'équilibre. Le problème analysé est une idéalisation des mouvements existants dans le système solaire, par exemple le mouvement de la lune relativement à la Terre et les mouvements possibles d'un satellite artificiel.

Les conclusions sont que le mouvement d'un satellite autour du centre de masse sera du type d'une libration si les conditions de stabilité sont observées et si la vitesse initiale de rotation au voisinage du centre de masse est petite comparée à la vitesse angulaire du centre de masse. Si cependant la vitesse initiale de rotation était relativement grande, comme dans le cas des premiers satellites soviétiques, le mouvement du satellite serait une combinaison d'un mouvement d'Euler-Poinceau non perturbé au voisinage de l'axe du moment cinétique et de la précession de l'axe cinétique lui-même. Des causes de perturbation secondaires affectent les librations réelles mais les recherches montrent que, si les conditions fondamentales de stabilité sont remplies, la libration en présence de ces perturbations ne diffère que peu d'une libration non perturbée.

Determination of Atmospheric Density at an Altitude of 430 km by the Sodium Vapour Diffusion Method

By

I. S. Shklovsky[1] and V. G. Kurt[2]

Abstract — Zusammenfassung — Résumé

Determination of Atmospheric Density at an Altitude of 430 km by the Sodium Vapour Diffusion Method. In order to confirm the recent air density observations derived from the deceleration of Soviet and American satellites in the altitude range of 270—750 km, the analysis of the diffusion of sodium vapour emitted from rockets is proposed. This method was originally suggested by BATES and several American experiments were made at comparatively low altitudes, viz.: 75—140 km.

The first experiment of this type in the upper atmosphere was made on the 19th September, 1958. When the Soviet rocket reached 430 km, at the apex of its trajectory, the programmed ignition of thermite caused the ejection of sodium vapour from two vapourisers containing 2 kg. of metallic sodium each. The sodium vapour cloud, lit by the rays of the sun above the earth shadow, was photographed and the photographs were analysed by the standard astrophysical method. The atmospheric temperature was estimated as more than 1600 °K on the basis of the assumption of atomic oxygen as the basic constituent.

The density figures obtained by this method, which can be used up to 500—600 km, were compared with the data from satellite deceleration and are shown to be between 10^{-14} and 10^{-15} g/cm³ at 430 km altitude.

Bestimmung der atmosphärischen Dichte in 430 km Höhe mit Hilfe der Diffusion von Natriumdampf. Um die jüngsten Dichtemessungen zu bestätigen, die sich aus der Bremsung der sowjetischen und amerikanischen Satelliten im Höhenbereich zwischen 270 und 750 km ableiten lassen, wird die Analyse der Diffusion von aus Raketen ausgestoßenem Natriumdampf vorgeschlagen. Diese Methode wurde ursprünglich von BATES angeregt und einige amerikanische Versuche wurden bei verhältnismäßig niedrigen Höhen, nämlich 75 bis 140 km, ausgeführt.

Der erste Versuch dieser Art in der Hohen Atmosphäre wurde am 19. September 1958 unternommen. Als die sowjetische Rakete am Gipfelpunkt ihrer Bahn die Höhe von 430 km erreicht hatte, bewirkte die programmierte Zündung von Thermit die Ausstoßung von Natriumdampf aus zwei Verdampfern, die jeder 2 kg metallisches Natrium enthielten. Die Natriumdampfwolke wurde von den Sonnenstrahlen oberhalb des Erdschattens beleuchtet und photographiert. Die photographischen Aufnahmen wurden nach den astrophysikalischen Standardmethoden analysiert. Aus dem angenommenen Grundbestandteil (atomarer Sauerstoff) wurde die Temperatur der Atmosphäre auf mehr als 1600 °K geschätzt.

---

[1] Institute for Atmospheric Physics of the Academy of Sciences of the U.S.S.R., Moscow, U.S.S.R.

[2] Junior Researcher, State Astronomical Institute (SHTERNBERG Institute), Moscow, U.S.S.R.

Die nach diesem Verfahren, das bis zu Höhen von 500 bis 600 km anwendbar ist, erhaltenen Dichtewerte wurden mit den Daten verglichen, die aus der Bremsung von Satelliten abgeleitet wurden. Sie liegen in einer Höhe von 430 km zwischen 10^{-14} und 10^{-15} g/cm³.

Détermination de la densité atmosphérique à l'altitude de 430 km par diffusion de vapeur de sodium. Pour confirmer les observations de densité déduites du freinage des satellites soviétiques et américains aux altitudes comprises entre 270 et 750 km., la méthode de diffusion d'une vapeur de sodium émise par une fusée est proposée. Elle a été originellement suggérée par Bates et plusieurs expériences américaines ont eu lieu à des altitudes relativement faibles (75 à 140 km.).

La première expérience de ce genre en haute atmosphère a eu lieu le 19 Septembre 1958. Quand la fusée soviétique a atteint son altitude maximum de 430 km. l'allumage programmé de thermite a provoqué l'éjection de vapeur de sodium par deux vaporisateurs contenant 2 kilos de sodium métallique chacun. Le nuage de vapeur illuminé par les rayons solaires au dessus de l'ombre terrestre a été photographié, puis analysé par les méthodes usuelles de l'astrophysique. La température a été déterminée comme étant supérieure à 1600 °K en admettant l'oxygène atomique comme constituant principal.

Les chiffres de densité obtenus, utilisables jusqu'à 500—600 km. ont été comparés à ceux déduits du freinage et sont compris entre 10^{-14} et 10^{-15} g/cm³ à 430 km. d'altitude.

Theorie der relativistischen n-Stufenrakete[1]

Von

M. Subotowicz[2]

(Eingegangen am 5. Juni 1959)

Zusammenfassung — Abstract — Résumé

Theorie der relativistischen n-Stufenrakete. Es wird gezeigt, daß die kinematischen Gleichungen der relativistischen Einstufen- und eindimensionalen Raketenbewegung im leeren, gravitationslosen Raum mit Hilfe aufeinanderfolgender LORENTZ-Transformationen erhalten werden können (Teil I).

Die Differentialgleichungen der relativistischen n-Stufenrakete werden abgeleitet. Diese Gleichungen beschreiben die Änderungen von Masse und Geschwindigkeit der Vielstufen-Rakete sukzessive in den Bezugssystemen Σ_0, Σ_1, ... Σ_{n-1}, die in der vorliegenden Arbeit definiert sind. Es wurde die Endgeschwindigkeit der relativistischen n-Stufenrakete relativ zum Inertialbezugssystem Σ_0 erhalten. Die Optimalisierungsmöglichkeiten der relativistischen n-Stufenrakete werden skizziert (Teil II).

Die Bewegungsgleichungen werden für den Fall konstanter lokaler Beschleunigung gelöst. Die Koordinaten der Rakete werden als Funktionen der Eigenzeit und der in dem Bezugssystem Σ_0 gemessenen Zeit ausgedrückt. Es werden auch die Transformationsgleichungen für n lokale, gleichförmig beschleunigte Bezugssysteme abgeleitet (Teil III).

Theory of the n-Step Relativistic Rocket. It is shown, that the kinematics equations of the relativistic 1-step and 1-dimensional rocket motion in empty, gravitationless space can be obtained by means of successive LORENTZ transformations (part I).

The differential equations of the n-stage relativistic rocket are derived. These equations describe the mass and velocity changes of the multistep rocket successively in the reference systems Σ_0, Σ_1, ..., Σ_{n-1}, defined in this paper. The final velocity of the n-step relativistic rocket relative to the inertial reference system, Σ_0 was obtained. The optimization possibilities of the n-step relativistic rocket are sketched (part II).

The equations of motion are solved for the case of constant local acceleration. The coordinates of the rocket are expressed as functions of the proper time and of the time measured in Σ_0. The transformation equations for n uniformly accelerated local reference systems are derived (part III).

Théorie relativiste de la fusée à n étages. La première partie montre comment les équations cinématiques de la fusée relativiste à un étage et une dimension peuvent être obtenues à l'aide de transformations de LORENTZ successives. L'équation différentielle de la fusée à n étages est dérivée. Elle décrit les changements de masse et de vitesse en terme de systèmes de référence successifs décrits dans le texte.

La vitesse terminale par rapport au système inertial Σ_0 est obtenue et les possibilités d'optimisation de la fusée à n étages sont esquissées.

---

[1] Ein wesentlicher Teil dieser Arbeit wurde am 23. April 1958 bei dem Treffen des Vorstandes der „Polskie Towarzystwo Astronautyczne" (P.T.A.) vorgelegt.

[2] Mitglied der P.T.A., Warszawa, ul. Narbutta 85, Politechnika, pok. 307.

Anstalt für Experimentelle Physik der M. Curie-Skłodowska-Universität in Lublin, Polen.

Les équations du mouvement sont résolues dans le cas d'une accélération locale constante. Les coordonnées de la fusée sont données en fonction du temps propre et du temps mesuré dans Σ_0. Les équations de transformation pour n systèmes de référence locaux uniformément accélérés sont dérivées.

Einleitung

Das Problem der relativistischen Theorie der Vielstufenrakete ist bisher noch nicht behandelt worden (siehe die Literaturangaben am Ende der vorliegenden Arbeit). Eine kurze Bemerkung über die Notwendigkeit von Vielstufenraketen bei in ferner Zukunft zu erwartenden Flügen zu den Fixsternen ist in der Arbeit von Peschka [8] enthalten, doch gibt es noch keine Theorie derartiger Raketen. Es scheint die Notwendigkeit für eine solche Theorie für den Fall von Ionenraketen mit großer Ausströmungsgeschwindigkeit zu bestehen — nicht nur für das Problem des Sternenfluges.

Peschka [8] und Kooy [5] zeigten, daß Flüge zu den Sternen durch Anwendung des Reaktionsprinzips nur verwirklicht werden könnten, wenn die Massenverhältnisse ungeheuer groß würden. Unter solchen Bedingungen muß das Prinzip der Vielstufenraketen angewendet werden.

In der vorliegenden Arbeit wird angenommen, daß die Reise zu den Sternen mit Hilfe der Kernenergie mittels Ionen- oder Photonenraketen auszuführen ist.

Der gewaltige Energiebetrag, der zur Erreichung eines ausreichenden Schubes durch Ausströmung eines Ionen- oder Photonenstrahls erforderlich ist, bringt notwendigerweise den Transport riesiger Massen von Brennstoffen und Strahlsubstanzen an Bord der Sternenschiffe mit sich.

Wir nehmen an, daß die Geschwindigkeit der künftigen Vielstufenraketen zu den Sternen vergleichsweise nahe der Lichtgeschwindigkeit sein wird, und versuchen, die Bewegung dieser Raketen durch Anwendung der relativistischen Mechanik zu analysieren.

Wahrscheinlich wird man in der Zukunft *andere* Möglichkeiten des Fluges zu den Sternen finden, verschieden von denjenigen, die sich auf das Reaktionsprinzip (oder anders ausgedrückt, auf das Raketenprinzip) gründen. Beispielsweise sind einige Anregungen hinsichtlich der wichtigen Folgerungen aus einer besseren Erkenntnis der Gravitation in einer Veröffentlichung von Bondi [3] zu finden.

Symbole

Abschnitt I

J	Inertialbezugssystem
R	beschleunigtes, mit der Rakete verbundenes Bezugssystem
X, Y, Z, T	Koordinaten von J
x, y, z, T_0	Koordinaten von R
m_0	Ruhemasse der Rakete beim Start
m_k	Ruhemasse der Rakete nach Verbrauch des Brennstoffes
m	augenblickliche Ruhemasse der Rakete
$r = m_0/m_k$	Massenverhältnis
T	durch den Beobachter im Inertialsystem J gemessene Zeit
T_0	Eigenzeit der Astronauten, durch den Beobachter in R gemessen
$V = dX/dT$	Geschwindigkeit der Rakete, in J gemessen
u_0	Ausströmungsgeschwindigkeit relativ zur Rakete
b	Beschleunigung, in J gemessen
b_0	Beschleunigung, in R gemessen
$\varrho(T_0) = dm/dT_0$	Geschwindigkeit, mit der sich die Ruhemasse ändert
$\theta(T_0)$	Funktion, definiert in Gl. (2)

Abschnitte II und III

Σ_0 Inertial-Bezugssystem, verbunden mit den Fixsternen und dem Startort der Vielstufenrakete

Σ_1 Inertial-Bezugssystem, verbunden mit der ersten Stufe nach völligem Verbrauch des Brennstoffes und Erreichen der Maximalgeschwindigkeit v_1 durch die erste Stufe in Beziehung zu Σ_0

Σ_2 Inertial-Bezugssystem, verbunden mit der zweiten Stufe nach völligem Verbrauch des Brennstoffes und Erreichen der Maximalgeschwindigkeit v_2 durch die zweite Stufe in Beziehung auf Σ_1

Σ_{n-1} Inertial-Bezugssystem, verbunden mit der $(n-1)$ten Stufe nach völligem Verbrauch des Brennstoffes und Erreichen der Maximalgeschwindigkeit v_{n-1} durch die $(n-1)$te Stufe in Beziehung zu Σ_{n-2}

$m_1, \ldots, m_n$ augenblickliche Ruhemasse der 1., $\ldots$, $n$ten Raketenstufe in jedem Punkt der Bahn

$m_{o1}, \ldots, m_{on}$ Anfangs-Ruhemasse der 1., $\ldots$, $n$ten Stufe

$m_{k1}, \ldots, m_{kn}$ End-Ruhemasse der 1., $\ldots$, $n$ten Stufe

$dm_1, \ldots, dm_n$ Ruhemasse eines Elements der ausströmenden Substanz der 1., $\ldots$, $n$ten Stufe

$m_{e1}, \ldots, m_{en}$ Leer-Ruhemasse der 1., $\ldots$, $n$ten Stufe

$m_{p1}, \ldots, m_{pn}$ Treibstoff-Ruhemasse der 1., $\ldots$, $n$ten Stufe

m_L Ruhemasse der Nutzlast; Nutzlast der $n$ten Stufe

$\lambda_1 = m_{o2}/m_{o1}, \ldots, \lambda_n = m_{0(n+1)}/m_{on} = m_L/m_{on}$ Nutzlast-Parameter der 1., $\ldots$, $n$ten Stufe

$r_1 = m_{o1}/(m_{o1} - m_{p1}), \ldots, r_n = m_{on}/(m_{on} - m_{pn})$ Ideale Massen-Parameter der 1., $\ldots$, $n$ten Stufe

$\varepsilon_1 = m_{e1}/m_{o1}, \ldots, \varepsilon_n = m_{en}/m_{on}$ Konstruktionsparameter der 1., $\ldots$, $n$ten Stufe

$$\Lambda = \prod_{i=1}^{n} \lambda_i = m_L/m_{o1} \qquad \text{Parameter der totalen Nutzlast}$$

$x_1, \ldots, x_n$ x-Achsen-Koordinaten in $\Sigma_0, \ldots, \Sigma_{n-1}$

$y_1, \ldots, y_n$ y-Achsen-Koordinaten in $\Sigma_0, \ldots, \Sigma_{n-1}$

$z_1, \ldots, z_n$ z-Achsen-Koordinaten in $\Sigma_0, \ldots, \Sigma_{n-1}$

$X_1, \ldots, X_n$ Weg der 1., $\ldots$, $n$ten Stufe vom Startort der 1. Stufe, gemessen relativ zu Σ_0

$s_1, \ldots, s_n$ Weg der 1., $\ldots$, $n$ten Stufe vom Beginn der Tätigkeit des Triebwerkes dieser Stufe an, gemessen in Σ_0

$\tau_{o1}, \ldots, \tau_{on}$ Eigenzeit, gemessen in der betreffenden Stufe durch die Astronauten vom Beginn der Tätigkeit des Motors dieser Stufe an

$\tau_{ok1}, \ldots \tau_{okn}$,   Eigenzeit, gemessen in der betreffenden Stufe durch die Astronauten vom Beginn bis zum Ende des Brennstoffverbrauches in dieser Stufe

$\tau_1, \ldots, \tau_n$ zugehörige Zeit, gemessen in $\Sigma_0, \ldots, \Sigma_{n-1}$ für die entsprechende Stufe

$T_1, \ldots T_n$ Flugzeit der 1., $\ldots$, $n$ten Stufe relativ zu Σ_0 vom Start der 1. Stufe an, gemessen in Σ_0

$T_{o1}, \ldots, T_{on}$ Flugzeit der 1., $\ldots$, $n$ten Stufe, gemessen auf der Rakete der 1., $\ldots$, $n$ten Stufe durch die Astronauten relativ zu Σ_0 vom Startort der 1. Stufe an.

t_n Flugzeit der $n$ten Stufe vom Beginn der Tätigkeit des Triebwerks dieser Stufe an, gemessen relativ zu Σ_0

$t_{on} = \tau_{on}$ Flugzeit der $n$ten Stufe vom Beginn der Tätigkeit des Triebwerks dieser Stufe, gemessen durch die Astronauten

$v_1, \ldots, v_n$ Geschwindigkeit der 1., $\ldots$, $n$ten Stufe relativ zu $\Sigma_0, \ldots, \Sigma_{n-1}$

$V_1, \ldots, V_n$ Geschwindigkeit der 1., $\ldots$, $n$ten Stufe relativ zu Σ_0

$v_1', \ldots, v_n'$ Geschwindigkeit des Massenelements der ausströmenden Substanz der 1., $\ldots$, $n$ten Stufe relativ zu $\Sigma_0, \ldots, \Sigma_{n-1}$

u_0 — Ausströmungsgeschwindigkeit relativ zu der betreffenden Rakete, für alle Stufen gleich

$u_1, \ldots, u_n$ — Ausströmungsgeschwindigkeit der 1., $\ldots$, $n$ten Stufe relativ zu Σ_0

b_0 — lokale Beschleunigung, durch die Astronauten in der Rakete relativ zum Bezugssystem gemessen, das mit dieser Rakete verbunden ist

$b_1, \ldots, b_n$ — Beschleunigung der 1., $\ldots$, $n$ten Stufe relativ zu Σ_0

$R_n = r_1 \cdot r_2 \ldots r_n; \quad \bar{R}_n = \ln R_n$

$\varrho_1 (\tau_{o1}) = dm/d\tau_{o1}, \ldots, \varrho_n (\tau_{on}) = dm/d\tau_{on}$ — Geschwindigkeit, mit der sich die Masse der 1., $\ldots$, $n$ten Stufe ändert

$\theta_1 (\tau_{o1}), \ldots, \theta_n (\tau_{on})$ — Funktionen für die 1., $\ldots$, $n$te Stufe, definiert in den Gln. $(19_1), \ldots, (19_n)$

Θ_n — Funktion, definiert in Gl. (20) für die n-Stufenrakete

I. Aufeinanderfolgende Lorentz-Transformationen und die Kinematik der ersten Stufe

Aufeinanderfolgende Lorentz-Transformationen der Relativitätstheorie und die Kinematik der relativistischen Rakete

Das Problem der Bewegung der relativistischen n-Stufenrakete mit willkürlicher Beschleunigung der verschiedenen Stufen und die Analyse der für diese Bewegung charakteristischen Parameter in verschiedenen Bezugssystemen, insbesondere in einem Inertialsystem, ist das typische Problem der Relativitätstheorie. Das Ziel ist, die aufeinanderfolgenden Koordinatentransformationen zu finden, welche die Bewegung in einer Serie von n verschiedenen, beschleunigten (allgemein) Bezugssystemen beschreiben.

Geeignete Transformationsgleichungen gibt MøLLER [7] in den Kapiteln IV und VIII seines Buches "The Theory of Relativity" an. Diese Gleichungen sollten in einer für die vorliegende Arbeit erforderlichen Form angegeben werden, um die Bewegung der relativistischen Stufenrakete zu analysieren.

Wir bezeichnen die Koordinaten eines Punktes in einem Inertial-Bezugssystem J mit (X, Y, Z, T), während die Koordinaten dieses Punktes in einem beschleunigten System R (x, y, z, T_0) sind. Bei Beginn der Bewegung lassen wir den Ursprung O des Systems R sich in der Richtung der X-Achse des Systems J bewegen. Die Transformationsgleichungen, welche die Variablen der Systeme R und J miteinander verbinden, sind:

$$X = c \int_0^{T_0} \sin \mathrm{hyp}\, \theta \, d T_0 + x \cdot \cos \mathrm{hyp}\, \theta, \quad Y = y, \quad Z = z$$

$$T = \int_0^{T_0} \cos \mathrm{hyp}\, \theta \, d T_0 + \frac{x}{c} \sin \mathrm{hyp}\, \theta \tag{1}$$

wo θ eine willkürliche Funktion von T_0 ist, die für $T_0 = 0$ gleich null ist. Die Form von $\theta = \theta (T_0)$ wird durch die Konstruktion der Rakete, die Art der Brennstoffe und den Typus der beschleunigten Bewegung des Bezugssystems R in Beziehung zu J bestimmt.

$$\theta (T_0) = \frac{1}{c} \int_0^{T_0} b_0 (T_0) \, d T_0 = -\frac{u_0}{c} \int_0^{T_0} \frac{\varrho (T_0)}{m} \, d T_0 = \frac{u_0}{c} \ln \frac{m_0}{m_k} = \frac{u_0}{c} \ln r, \tag{2}$$

wo $b_0(T_0)$ die in R gemessene Beschleunigung ist. Die Funktion $\varrho (T_0)$ zeigt die Geschwindigkeit, mit der sich die Ruhemasse der Rakete ändert: $\varrho(T_0)$ ist eine nichtpositive Funktion; m ist die augenblickliche Masse der Rakete; m_0 ist die Ruhemasse der Rakete zur Zeit $T_0 = 0$, m_k die Ruhemasse zur Zeit T_{0k} nach Verbrauch des Brennstoffes, $r = m_0/m_k$ das Massenverhältnis.

Aus den Gln. (1) und (2) kann man leicht die wohlbekannte Formel von ACKERET [1] für die Geschwindigkeit der Rakete im System J zu Zeit T_{0k} ableiten:

$$V = dX/dT = c \operatorname{tg} \operatorname{hyp} \theta = c \operatorname{tg} \operatorname{hyp}\left(\frac{u_0}{c} \ln r\right) = c\, \frac{r^{2\,u_0/c} - 1}{r^{2\,u_0/c} + 1}\,. \tag{3}$$

Aus (1) und (2) kann man (SUBOTOWICZ [13a]) alle Ergebnisse der Arbeiten von ACKERET [1], SHEPHERD [13], BADE [2], KRAUSE [6] und ebenso einige Resultate von KOOY [5] und SÄNGER [10, 11] gewinnen.

Setzt man in Gl. (1) $x = 0$, erhält man die integrierte Gleichung der Raketenbewegung in Parameterform, die erstmals von BADE [2] abgeleitet worden ist.

II. Die Ableitung und Lösung der Differentialgleichungen der relativistischen n-Stufenrakete

Differentialgleichungen der relativistischen Vielstufenrakete

Wir nehmen an, daß sich die Rakete mit willkürlicher, beschleunigter Bewegung in einem schwerefeldfreien Raum entlang der x-Achse bewegt (geradlinige Bahn). Die Bewegung jeder Stufe wird in einem zugehörigen Bezugssystem $\Sigma_0, \ldots, \Sigma_{n-1}$ behandelt. Dies gestattet die Benützung der Methode von BADE [2] zur Ableitung der Differentialgleichung für die relativistische Einstufenrakete auch für unseren Fall einer relativistischen n-Stufenrakete.

Das Erhaltungsgesetz des gesamten Vierer-Impulses für jede Stufe ist:

$$\Delta \Sigma m_1\, d\, x_1{}^\mu / d\tau_1 = 0 \tag{4_1}$$

$$\Delta \Sigma m_n\, d\, x_n{}^\mu / d\tau_n = 0\,. \tag{4_n}$$

Die Symbole werden hier in analoger Bedeutung wie in der Veröffentlichung von BADE gebraucht. Σ ist die Summierung über die Teilchen des Systems, $dx_1{}^\mu/d\tau_1, \ldots, dx_n{}^\mu/d\tau_n$ die Komponenten der Vierer-Geschwindigkeit eines Teilchens in der betreffenden Stufe, Δ symbolisiert die Differenz der Werte vor und nach Wechselwirkung unter den Teilchen.

Wenn die Geschwindigkeit der ausströmenden Teilchen relativ zur Rakete für alle Stufen dieselbe und gleich u_0 ist und ihre Richtungen entgegengesetzt zur Bewegung der Rakete sind, können wir diese Geschwindigkeiten relativ zu den zugehörigen Bezugssystemen $\Sigma_0, \ldots, \Sigma_{n-1}$ in folgender Weise schreiben:

$$v_1{}' = \frac{v_1 - u_0}{1 - \dfrac{v_1\,u_0}{c^2}}, \quad \ldots, \quad v_n{}' = \frac{v_n - u_0}{1 - \dfrac{v_n\,u_0}{c^2}}\,. \tag{$5_1, \ldots 5_n$}$$

Unter Benützung des Prinzips der Erhaltung des gesamten Vierer-Impulses des Systems können wir genau die Ableitung der Differentialgleichung der relativistischen n-Stufenrakete am Beispiel der Einstufenrakete zeigen (SUBOTOWICZ [13a]).

Aus (4_1) ergibt sich:

$$dm_1/m_1 + \frac{1}{u_0} \cdot \frac{dv_1}{(1 - v_1^2/c^2)} = 0\,. \tag{6_1}$$

Ändert man in analoger Weise nur die Indizes von 1 in 2, 3, $\ldots n$, so erhält man die Gleichungen der relativistischen Raketen für die 2., 3., $\ldots n$te Stufe.

Diese Gleichungen beschreiben die entsprechende Bewegung der relativistischen n-Stufenrakete der Reihe nach in den Systemen $\Sigma_1, \Sigma_2, \ldots, \Sigma_{n-1}$:

$$d m_2/m_2 + \frac{1}{u_0} \cdot \frac{d v_2}{(1 - v_2^2/c^2)} = 0 \qquad (6_2)$$

$$d m_n/m_n + \frac{1}{u_0} \cdot \frac{d v_n}{(1 - v_n^2/c^2)} = 0 . \qquad (6_n)$$

Geschwindigkeit der relativistischen n-Stufenrakete relativ zu Σ_0

In den Gleichungen $(6_1), \ldots, (6n)$ werden die Variablen getrennt und wir können sie leicht integrieren. Wir nehmen an:

bei $\tau_{o\,1} = 0$ gilt $m_1 = m_{o\,1}$; bei $\tau_{o1} = \tau_{o\,k1}$ ist $m_1 = m_{k\,1}$. Von der Zeit 0 bis $\tau_{o\,k1}$ ändert sich die in Σ_0 gemessene Geschwindigkeit v_1 von 0 bis v_{1k}.

Für die folgenden Stufen sind die folgenden analogen Bedingungen erfüllt:

bei $\tau_{o\,n} = 0$ ist $m = m_{o\,n}$; bei $\tau_{o\,n} = \tau_{o\,kn}$ ist $m_n = m_{kn}$. Von der Zeit 0 bis $\tau_{o\,k}$ ändert sich die Geschwindigkeit v_n, gemessen im Bezugssystem Σ_{n-1}, von 0 bis v_{nk}. $\qquad (a)$

Nach Integration von $(6_1), \ldots, (6_n)$ erhält man:

$$r_1 = \frac{m_{o\,1}}{m_{k\,1}} = \left(\frac{1 + v_{1k}/c}{1 - v_{1\,k}/c} \right)^{c/2\,u_0} \qquad (7_1)$$

$$r_2 = \frac{m_{o\,2}}{m_{k\,2}} = \left(\frac{1 + v_{2k}/c}{1 - v_{2\,k}/c} \right)^{c/2\,u_0} \qquad (7_2)$$

$$r_n = \frac{m_{on}}{m_{kn}} = \left[\left(1 + \frac{v_{nk}}{c} \right) \middle/ \left(1 - \frac{v_{nk}}{c} \right) \right]^{c/2\,u_0} . \qquad (7_n)$$

Aus $(7_1), \ldots, (7_n)$ ergibt sich:

$$v_{1k} = c\,(r_1^{\,2\,u_0/c} - 1)/(r_1^{\,2\,u_0/c} + 1) \qquad (8_1)$$

$$v_{2k} = c\,(r_2^{\,2\,u_0/c} - 1)/(r_2^{\,2\,u_0/c} + 1) \qquad (8_2)$$

$$v_{nk} = c\,(r_n^{\,2\,u_0/c} - 1)/(r_n^{\,2\,u_0/c} + 1) . \qquad (8_n)$$

Wir setzen:

$$\prod_{i=1}^{n} (m_{o\,i}/m_{k\,i}) = \prod_{i=1}^{n} r_i = R_n \qquad (9)$$

oder:

$$R_n = \prod_{i=1}^{n} [(1 + v_{i\,k}/c)/(1 - v_{i\,k}/c)]^{c/2\,u_0} . \qquad (10)$$

Es ist jedoch sehr wichtig, die augenblickliche Geschwindigkeit der gegebenen Stufe relativ zu Σ_0 kennen, dem Inertialsystem, das mit dem Startort der Vielstufenrakete verbunden ist.

Aus der Definition von $v_{1k}, \ldots, v_{nk}$, ferner von $V_1, \ldots, V_n$ und dem Additionstheorem der relativistischen Geschwindigkeiten erhalten wir:

$$V_1 = v_{1k} \qquad (11_1) , \qquad\qquad V_2 = \frac{V_1 + v_{2k}}{1 + V_1 \cdot v_{2k}/c^2} \qquad (11_2)$$

$$V_3 = \frac{V_2 + v_{3k}}{1 + V_2 \cdot v_{3k}/c^2} \quad (11_3) , \qquad \ldots , \qquad V_n = \frac{V_{n-1} + v_{nk}}{1 + V_{n-1} \cdot v_{nk}/c^2} . \qquad (11_n)$$

Wir können jetzt die in (11_1), (11_2), ..., (11_n) auftretenden Ausdrücke in folgender Weise transformieren:

$$\frac{1 + v_{1k}/c}{1 - v_{1k}/c} = \frac{1 + V_1/c}{1 - V_1/c} \quad (12_1), \qquad \frac{1 + v_{1k}/c}{1 - v_{1k}/c} \cdot \frac{1 + v_{2k}/c}{1 - v_{2k}/c} = \frac{1 + V_2/c}{1 - V_2/c} \qquad (12_2),$$

$$\ldots, \quad \prod_{i=1}^{n} \left(\frac{1 + v_{ik}/c}{1 - v_{ik}/c} \right) = \frac{1 + V_n/c}{1 - V_n/c}. \tag{12_n}$$

Nach Einsetzen von (12) in (10) ergibt sich:

$$R_n = \left(\frac{1 + V_n/c}{1 - V_n/c} \right)^{c/2\,u_0} \qquad (13) \quad \text{or} \quad V_n = \frac{R_n^{\,2\,u_0/c} - 1}{R_n^{\,2\,u_0/c} + 1}. \tag{13a}$$

Wenn $\overline{R}_n = \ln R_n$ (14), können wir (13a) in sehr einfacher Form schreiben:

$$V_n = c \cdot \text{tg hyp} \left(\overline{R}_n \cdot \frac{u_0}{c} \right). \tag{15}$$

Die Formeln (13, 13a und 15) sind die gesuchten Ausdrücke für das Massenverhältnis (13) oder die Endgeschwindigkeit (13a, 15) der relativistischen n-Stufenrakete relativ zum Inertial-Bezugssystem Σ_0, das mit dem Startort verbunden ist.

Ein anderer Vorgang zur Auffindung der Geschwindigkeit V_n

Wir setzen voraus, daß die Endgeschwindigkeiten jeder Stufe relativ zu Σ_0, bzw. Σ_1, ..., Σ_{n-1} gleich v_{1k} bzw. v_{2k}, ..., v_{nk} sind. Gemäß Gl. (3) haben wir dann:

$$V_1 = v_{1k} = c \cdot \frac{(r_1^{\,2\,u_0/c} - 1)}{(r_1^{\,2\,u_0/c} + 1)}. \tag{11_1ab}$$

$$v_{2k} = c \cdot \frac{(r_2^{\,2\,u_0/c} - 1)}{(r_2^{\,2\,u_0/c} + 1)}. \tag{11_2a}$$

Aus (11_2), $(11_1 ab)$, $(11_2 a)$ erhält man:

$$V_2 = c \cdot \left[\frac{(r_1 \cdot r_2)^{2\,u_0/c} - 1}{(r_1 \cdot r_2)^{2\,u_0/c} + 1} \right]. \tag{11_2 b}$$

In analoger Weise kann gezeigt werden, daß die Endgeschwindigkeit der $n$ten Stufe relativ zu Σ_0, mit V_n bezeichnet, ist:

$$V_n = c \cdot \left[\frac{(r_1 \cdot r_2 \ldots r_n)^{2\,u_0/c} - 1}{(r_1 \cdot r_2 \ldots r_n)^{2\,u_0/c} + 1} \right]. \tag{13a}$$

In den vorhergehenden Bemerkungen war angenommen worden, daß die Ausströmungsgeschwindigkeit für alle Stufen gleich u_0 war. Das bedeutet, daß die Triebwerke aller Stufen auf analoge Weise konstruiert sein müssen und in allen Triebwerken die gleiche Strahlmasse verwendet wird. Es könnten sowohl Ionen- wie Photonenraketen sein. Eine solche relativistische Stufenrakete könnte bloß *eine* Reaktionskammer besitzen, die der Reihe nach von allen Stufen der Vielstufenrakete benützt wird. Nach Verwendung der gesamten Strahlmasse können alle Vorrichtungen und Brennstoff- (oder Strahlsubstanzen-) Behälter entfernt werden, wenn sie für das Arbeiten der höheren Stufen der relativistischen Rakete unnötig geworden sind.

Wenn die Ausströmungsgeschwindigkeit, gemessen relativ zu der betreffenden

Stufe, verschieden ist, und zwar der Reihe nach gleich $u_{o\,1}$, $u_{o\,2}$, ..., $u_{o\,n}$, dann ist V_n gegeben durch:

$$V_n = c \cdot \frac{(r_1^{u_{o\,1}} \cdot r_2^{u_{o\,2}} \dots r_n^{u_{o\,n}})^{2/c} - 1}{(r_2^{u_{o\,1}} \cdot r_2^{u_{o\,2}} \dots r_n^{u_{o\,n}})^{2/c} + 1} = c \cdot \mathrm{tg\ hyp}\left[\frac{1}{c} \cdot \ln\left(\prod_{i=1}^{n} r_i^{u_{o\,i}}\right)\right]. \quad (13\,\mathrm{b})$$

Die Funktion Θ_n der relativistischen n-Stufenrakete

Wenn wir den Ausdruck V_n in (15) für die relativistische n-Stufenrakete mit einem solchen Ausdruck für die Einstufenrakete (3) vergleichen, sehen wir, daß die Form der Funktion θ für die n-Stufenrakete, mit Θ_n bezeichnet, folgende ist:

$$\Theta_n = \frac{u_0}{c} \ln (r_1 \cdot r_2 \dots r_n) = \frac{u_0}{c} \bar{R}_n. \quad (16)$$

Hier ist Θ_n nicht als Funktion der Eigenzeit $\tau_{o\,1}$ ausgedrückt, sondern als Funktion der Massenverhältnisse r_1, ..., r_n oder R_n.

Wir zeigen jetzt, wie man diese Funktion erhält:

$$\Theta_n = \Theta_n (r_1, r_2, \dots, r_n) = \Theta_n (\theta_1, \theta_2, \dots, \theta_n) = \Theta_n (\tau_{o\,1}, \tau_{o\,2}, \dots, \tau_{o\,n}).$$

Aus Gl. (4) ergibt sich nach einigen Rechnungen (SUBOTOWICZ [13a]), daß eine Funktion $\theta_1 (\tau_{o\,1})$ existiert

$$\cos \mathrm{hyp}\ \theta_1 = d\tau_1/d\tau_{o\,1}, \quad \text{und} \quad c \cdot \sin \mathrm{hyp}\ \theta_1 = dx_1/d\tau_{o\,1}. \quad (17_1)$$

Wenn

$$v_1 = dx_1/d\tau_1 = c \cdot \mathrm{tg\ hyp}\ \theta_1, \quad (18_1)$$

erhalten wir unter Substitution von (18_1) in (16_1) nach einfachen Rechnungen:

$$\theta_1 (\tau_{o\,1}) = -\frac{u_0}{c} \cdot \int_0^{\tau_{o1}} \frac{\varrho_1 (\tau_{o\,1})}{m_1} \, d\tau_{o\,1}, \quad (19_1)$$

wo $\varrho_1 (\tau_{o\,1}) = dm/d\tau_{o\,1}$ eine nichtpositive Funktion wie in Gl. (2) ist. Auf analoge Art können wir erhalten:

$$\theta_2 (\tau_{o\,2}) = -\frac{u_0}{c} \cdot \int_0^{\tau_{o2}} \frac{\varrho_2 (\tau_{o\,2})}{m_2} \, d\tau_{o\,2}, \quad (19_2)$$

$$\theta_n (\tau_{o\,n}) = -\frac{u_0}{c} \cdot \int_0^{\tau_{o\,n}} \frac{\varrho_n (\tau_{o\,n})}{m_n} \, d\tau_{o\,n}. \quad (19_n)$$

Wir definieren eine neue Funktion Θ, die die relativistische n-Stufenrakete beschreibt, als Summe von θ_1, ..., θ_n aus $(19_1, \dots, 19_n)$:

$$\Theta_n = \sum_{i=1}^{n} \theta_i. \quad (20)$$

Wir nehmen an, daß die Bedingung (α) erfüllt ist. Dann können wir gemäß (20) und $(19_1, \dots, 19_n)$ schreiben:

$$\Theta_n = -\frac{u_0}{c} \sum_{i=1}^{n} \int_0^{\tau_{o\,k\,i}} \frac{\varrho (\tau_{o\,k\,i})}{m_i} \, d\tau_{o\,i} = -\frac{u_0}{c} \sum_{i=1}^{n} \int_{m_{o\,i}}^{m_{k\,i}} \frac{dm_i}{m_i} = \frac{u_0}{c} \sum_{i=1}^{n} \ln\left(\frac{m_{o\,i}}{m_{k\,i}}\right) =$$

$$= \frac{u_0}{c} \prod_{i=1}^{n} r_i = \frac{u_0}{c} \ln R_n = \frac{u_0}{c} \cdot \bar{R}. \quad (21)$$

Wenn man dann (21) in (15) einsetzt, läßt sich der Ausdruck für die Geschwindigkeit der relativistischen n-Stufenrakete V_n relativ zu Σ_0 in einer zu Gl. (3) analogen Form für die Einstufenrakete angeben:

$$V_n = c \cdot \operatorname{tg} \operatorname{hyp} \Theta_n \,. \tag{22}$$

Optimisierung der relativistischen n-Stufenrakete

Das hier benützte Kennzeichen der Optimisierung ist Minimalisierung der
Brutto-Ruhmasse für eine bestimmte verlangte Brennschlußgeschwindigkeit V_n
und Nutzlast m_L. Das bedeutet, daß $\Lambda^{-1} = $ Minimum. Wir nehmen an, daß die
Ausströmungsgeschwindigkeit u_0 in einem mit der Rakete verbundenen Bezugs-
system für alle Stufen gleich ist. Man findet leicht (VERTREGT [1956], SUBOTO-
WICZ [15, 16]) wie bei der „klassischen" Vielstufenrakete, daß in diesem Fall:

$$\lambda_1 = \lambda_2 = \ldots = \lambda_n = \lambda\,;\; r_1 = r_2 = \ldots = r_n = r\,;\; \varepsilon_1 = \varepsilon_2 = \ldots = \varepsilon_n = \varepsilon\,.$$

Dann ist

$$V_n = c \cdot \frac{R_n^{2\,u_0/c} - 1}{R_n^{2\,u_0/c} + 1} = c \cdot \frac{r^{2\,n\,u_0/c} - 1}{r^{2\,n\,u_0/c} + 1} \quad \text{oder}$$

$$r^{2\,n\,u_0/c} = \frac{1 + V_n/c}{1 - V_n/c} \,.$$

Wenn $(\lambda + \varepsilon) = \left(\dfrac{1 + V_n/c}{1 - V_n/c}\right)^{-c/2\,n\,u_0\,c}$ und $\lambda^{-n} = \Lambda^{-1}$, haben wir bei Optimisie-
rungsbedingungen:

$$\lambda^{-n} = \left[\left(\frac{1 + V_n/c}{1 - V_n/c}\right)^{-c/2\,n\,u_0} - \varepsilon\right]^{-n} = \Lambda^{-1} = \text{Minimum}\,. \tag{23a}$$

Die Parameter ε, V_n, u_0 wurden früher angegeben. Der optimale Wert von λ
für diese n-Stufenrakete wird mit λ_0 bezeichnet. Wir erhalten ihn durch gewöhn-
liches Aufsuchen der Extremwerte, unter Differenzieren, wobei n die unabhängige
Variable ist:

$$-\left[\left(\frac{1 + \frac{V_n}{c}}{1 - \frac{V_n}{c}}\right)^{-\frac{c}{2\,n\,u_0}} - \varepsilon\right]^{-(n+1)} \left\{\left[\left(\frac{1 + \frac{V_n}{c}}{1 - \frac{V_n}{c}}\right)^{-\frac{c}{2\,n\,u_0}} - \varepsilon\right] \cdot \right. \tag{23b}$$

$$\left. \cdot \ln\left[\left(\frac{1 + \frac{V_n}{c}}{1 - \frac{V_n}{c}}\right)^{-\frac{c}{2\,n\,u_0}} - \varepsilon\right] - \left(\frac{1 + \frac{V_n}{c}}{1 - \frac{V_n}{c}}\right)^{-\frac{c}{2\,n\,u_0}} \cdot \ln\left(\frac{1 + \frac{V_n}{c}}{1 - \frac{V_n}{c}}\right)^{-\frac{c}{2\,n\,u_0}}\right\} = 0\,.$$

Aus den Gln. (23b) und (23a) erhalten wir die transzendente Gleichung von
KRAUSE [6]:

$$\lambda \cdot \ln \lambda = (\lambda + \varepsilon) \cdot \ln (\lambda + \varepsilon)\,. \tag{23c}$$

Aus dieser Gleichung läßt sich leicht λ_0 finden (siehe KRAUSE [14], SUBOTO-
WICZ [15]). Nach Einsetzen von λ_0 in die Gl. (23a) und Auflösen nach n_0 — der
Optimalzahl der Stufen — erhalten wir schließlich:

$$n_0 = \frac{\dfrac{c}{2\,u_0} \cdot \ln\left(\dfrac{1 + V_n/c}{1 - V_n/c}\right)}{-\ln(\lambda_0 + \varepsilon)} \,. \tag{23}$$

Es besteht hier die theoretische Möglichkeit, die Rechnung für verschiedene Parameter in jeder Stufe, so wie es für die „klassische" n-Stufenrakete erfolgt ist (siehe Subotowicz [16]), auch für die relativistische Vielstufenrakete auszuführen.

III. Kinematik der relativistischen n-Stufenrakete

1. Die konstante lokale Beschleunigung ($b_0 =$ const.)

Aus Gl. (15) erhalten wir:

$$1 - (V_n/c)^2 = \cos \mathrm{hyp}^{-2} (\overline{R}_n \cdot u_0/c). \qquad (24)$$

Das Theorem der relativistischen Geschwindigkeitsaddition gibt für u_n, die Ausströmungsgeschwindigkeit relativ zu Σ_0:

$$u_n = \frac{V_n - u_0}{1 - \dfrac{u_0 \cdot V_n}{c_2}}. \qquad (25)$$

Für b_n erhält man aus (24):

$$b_n = b_0 \, [1 - (V_n/c)^2]^{3/2} = b_0 \cdot \cos \mathrm{hyp}^{-3} (\overline{R} u_0/c). \qquad (26)$$

Um die Flugzeit T_n zu bekommen, muß die Bewegungsgleichung der $n$ten Stufe (mit konstanter lokaler Beschleunigung) so geschrieben werden:

$$\frac{d}{dT_n} \left\{ V_n \cdot [1 - (V_n/c)^2]^{-1/2} \right\} = \dot{V}_n \cdot [1 - (V_n/c)^2]^{-3/2} = b_n \cdot [1 - (V_n/c)^2]^{-3/2} = \qquad (27)$$
$$= b_0 = \mathrm{const}.$$

Nach Integration $\displaystyle \int_0^{V_n} [1 - (V_n/c)^2]^{-3/2} \, dV = b_0 \cdot \int_0^{T_n} dT_n$

folgt:

$$T_n = V_n \cdot b_0^{-1} \cdot [1 - (V_n/c)^2]^{-1/2} = \frac{c}{b_0} \cdot \sin \mathrm{hyp} (\overline{R}_n \cdot u_0/c). \qquad (28)$$

Aus (28) findet man leicht:

$$V_n = b_0 \cdot T_n \cdot [1 + (b_0 T_n/c)^2]^{-1/2} = c \cdot \mathrm{tg} \, \mathrm{hyp} (\overline{R}_n \cdot u_0/c) \qquad (28\,\mathrm{a})$$

und

$$t_n = T_n - T_{n-1} = 2 \frac{c}{b_0} \sin \mathrm{hyp} \left(\frac{u_0}{c} \ln \sqrt{\frac{R_n}{R_{n-1}}} \right) \cdot \cos \mathrm{hyp} \left(\frac{u_0}{c} \cdot \ln \sqrt{R_n \cdot R_{n-1}} \right). \qquad (29)$$

Unter Verwendung von (28 a) kann man (26) als Funktion der Zeit T_n schreiben:

$$b_n = b_0 \cdot [1 + (b_0 T_n/c)^2]^{-2/3} = b_0 \cdot \cos \mathrm{hyp}^{-3} (\overline{R}_n \cdot u_0/c) \qquad (26\,\mathrm{a})$$

und (26 a) in Form einer Potenzreihe des Terms $(b_0 \cdot T_n/c)$ angeben. In der Zeit T_n während des beschleunigten Fluges durchfliegt die Rakete die Entfernung X_n, gegeben durch:

$$X_n = \int_0^{T_n} V_n dT_n = \frac{c^2}{b_0} \cdot \{[1 + (b_0 T_n/c)^2]^{1/2} - 1, =$$

$$= \frac{c^2}{b_0} \cdot [\cos \mathrm{hyp} (\overline{R} \cdot u_0/c) - 1] = \frac{2c^2}{b_0} \cdot \sin \mathrm{hyp}^2 (\overline{R}_n \cdot u_0/c). \, (30)$$

Für s_n erhalten wir aus (30)

$$s_n = X_n - X_{n-1} = \frac{c^2}{b_0} \cdot \{[1 + (b_0 \cdot T_n/c)^2]^{1/2} - [1 + (b_0 \cdot T_{n-1}/c)^2]^{1/2} =$$

$$= \frac{2c^2}{b_0} \cdot \sin \text{hyp}\left(\frac{u_0}{c}\sqrt{R_n \cdot R_{n-1}}\right) \cdot \sin \text{hyp}\left(\frac{u_0}{c}\sqrt{\frac{R_n}{R_{n-1}}}\right). \tag{31}$$

Aus (30) folgt auch die Zeit für die Bewältigung der Distanz X_n:

$$T_n = \frac{c}{b_0}\left[(1 + b_0 \cdot X_n/c)^2 - 1\right]^{1/2}. \tag{32}$$

Die Gln. (28a), (26a) und (30) liefern die Geschwindigkeit V_n, die Beschleunigung b_n und den Weg der n-Stufenrakete als Funktion der Zeit T_n (und der Massenverhältnisse R_n), gemessen relativ zu Σ_0. Die Abhängigkeit der Flugzeit T_n aller n-Stufen von den Massenverhältnissen R_n gibt die Gl. (28) an.

Die Flugzeit t_n und der Weg s_n der $n$ten Stufe vom Beginn der Tätigkeit des Raketentriebwerks dieser Stufe, gemessen relativ zu Σ_0 als Funktion der Massenverhältnisse R_n und R_{n-1}, geben die Ausdrücke (29) und (31). Gl. (32) stellt die Flugzeit T_n als Funktion der Distanz X_n dar.

Nun zeigen wir die Abhängigkeit von V_n und X_n von der Eigenzeit T_{on} der Astronauten. Das Element der Eigenzeit dT_0 und das Element der Zeit dT, die in Σ_0 gemessen werden, hängen durch folgende Beziehung miteinander zusammen:

$$dT_0 = [1 - (V/c)^2]^{1/2} dT. \tag{33}$$

Diese Beziehung kann für die n-Stufenrakete in folgender Form geschrieben werden:

$$dT_{on} = [1 - (V_n/c)^2]^{1/2} dT_n. \tag{33n}$$

Nach Integration von (33n) durch Verwendung von (28a) erhalten wir:

$$T_{on} = \int_0^{T_n} [1 - (V_n/c)^2]^{1/2} dT_n = \frac{c}{b_0} \cdot \int_0^{T_n} \frac{\frac{b_0}{c} \cdot dT_n}{[1 + (b_0 T_n/c)^2]^{1/2}} =$$

$$= \frac{c}{b_0} \cdot \ln\left\{\frac{b_0 T_n}{c} + \left[1 + \left(\frac{b_0 T_n}{c}\right)^2\right]^{1/2}\right\} = \frac{c}{b_0} \text{Ar} \sin \text{hyp} (b_0 T_n/c). \tag{34}$$

Wir kehren (34) um, um $T_n = T_n (T_{on})$ zu erhalten:

$$T_n = \frac{c}{b_0} \cdot \sin \text{hyp} (b_0 T_{on}/c). \tag{35}$$

Durch Verwendung von (35) können wir die Beziehung (33n) in geeigneterer Form schreiben:

$$dT_{on}/dT_n = [1 - (V_n/c)^2]^{1/2} = \cos \text{hyp}^{-1}\left(\frac{b_0}{c} \cdot T_{on}\right). \tag{36}$$

Nun kann man mit Hilfe von (36) die Beschleunigung b_n in (26), ebenso die Geschwindigkeit V_n und den Weg X_n als Funktion der Eigenzeit T_{on} der Astronauten angeben:

$$b_n = b_0 \cdot \cos \text{hyp}^{-3}\left(\frac{b_0 \cdot T_{on}}{c}\right) \tag{37}$$

und

$$V_n = b_0 \cdot \int_0^{T_{on}} \frac{\cos \text{hyp} (b_0 \cdot T_{on}/c) \, dT_{on}}{\cos \text{hyp}^3 (b_0 \cdot T_{on}/c)} = c \cdot \text{tg} \text{hyp}\left(\frac{b_0 T_{on}}{c}\right). \tag{38}$$

$$X_n = c \cdot \int_0^{T_{on}} \text{tg hyp}\left(\frac{b_0 \cdot T_{on}}{c}\right) \cdot \text{cos hyp}\left(\frac{b_0 T_{on}}{c}\right) dT_{on} =$$

$$= \frac{c^2}{b_0} [\text{cos hyp}\,(b_0 T_{on}/c) - 1] = 2\frac{c^2}{b_0} \cdot \text{sin hyp}^2\,(b_0 \cdot T_{on}/2c) \qquad (39)$$

und

$$s_n = X_n - X_{n-1} = \frac{c^2}{b_0}\left[\text{cos hyp}\,(b_0 T_{on}/c) - \text{cos hyp}\left(\frac{b_0 \cdot T_{0\,(n-1)}}{c}\right)\right] =$$

$$= \frac{2c^2}{b_0} \cdot \text{sin hyp}\left[\frac{b_0}{c}\sqrt{T_{on} \cdot T_{0\,(n-1)}}\right] \cdot \text{sin hyp}\left(\frac{b_0}{c}\sqrt{T_{on}/T_{0\,(n-1)}}\right). \qquad (40)$$

Aus (35) erhalten wir:

$$t_n = T_n - T_{n-1} = \frac{c}{b_0}\left[\text{sin hyp}\,(b_0 T_{on}/c) - \text{sin hyp}\left(\frac{b_0 \cdot T_{0\,(n-1)}}{c}\right)\right] =$$

$$= \frac{2c}{b_0} \cdot \text{sin hyp}\left(\frac{b_0}{c} \cdot \frac{T_{on} - T_{0\,(n-1)}}{2}\right) \cdot \text{cos hyp}\left(\frac{b_0}{c} \cdot \frac{T_{on} + T_{0\,(n-1)}}{2}\right). \qquad (41)$$

Die von den Astronauten gemessene Zeit T_{on}, die zur Erreichung der Geschwindigkeit V_n und der Entfernung X_n notwendig ist, erhalten wir nach Umkehrung der hyperbolischen Funktionen in den Gln. (38) und (39):

$$T_{on} = \frac{c}{b_0} \cdot \text{Ar tg hyp}\,(V_n/c) = \frac{2c}{b_0} \cdot \text{Ar sin hyp}\sqrt{\frac{b_0 \cdot X_n}{2 \cdot c^2}}. \qquad (42)$$

Die Abhängigkeit der von den Astronauten gemessenen Zeit T_{on} des Fluges der Vielstufenrakete und der „Erd"-Zeit T_n (eines Inertialsystem-Beobachters), und umgekehrt, äußert sich in den Beziehungen (34) und (35).

Die in Σ_0 gemessenen Größen, die Beschleunigung b_n, die Geschwindigkeit V_n und die von der Vielstufenrakete zurückgelegte Entfernung X_n als Funktion der „astronautischen" Zeit T_{on} sind in den Formeln enthalten, die leicht aus (1) und (2) abgeleitet werden können.

2. Die Form der Funktion Θ_n und der Transformationsgleichungen in Anwesenheit von „n" beschleunigten Systemen (konstante Lokalbeschleunigung relativ zu Σ_0)

Wenn wir die Ausdrücke (26) und (37), (15) und (38), (30) und (39) vergleichen, sieht man, daß gemäß (21)

$$b_0 T_{on}/c = \bar{R}_n \cdot u_0/c = \Theta_n. \qquad (43).$$

Aus (43) erhalten wir den interessierenden Wert für T_{on} als Funktion der Massenverhältnisse R_n, Ausströmungsgeschwindigkeit u_0 und Lokalbeschleunigung b_0:

$$T_{on} = \bar{R}_n \cdot u_0/b_0 \qquad (44)$$

oder

$$T_{on} = \frac{c}{b_0} \cdot \Theta_n. \qquad (44a)$$

Aus (35) und (28), (39) und (30) ersieht man, daß für den Fall einer konstanten Lokalbeschleunigung und eines geradlinigen Weges (hyperbolische Bewegung) sich für eine Vielstufenrakete in voller Analogie zu dem ähnlichen, für eine Einstufenrakete geltenden Ausdruck (1) folgende Koordinatentransformationen anschreiben lassen:

$$\left. \begin{aligned} X_n &= c \cdot \int\limits_0^{T_{on}} \sin \text{hyp}\, \Theta_n \, d\, T_{on} \\[2mm] T_n &= \int\limits_0^{T_{on}} \cos \text{hyp}\, \Theta_n \, d\, T_{on} \end{aligned} \right\}$$

und   (1n)

Statt θ in (1) für die Einstufenrakete müssen wir die Funktion Θ_n (21) für die n-Stufenrakete einführen, um die parametrische Form der Bewegungs-gleichungen in Σ_0 für eine relativistische Vielstufenrakete zu erhalten.

Literaturverzeichnis

A. Relativistische Raketenmechanik

1. J. Ackeret, Zur Theorie der Rakete. Helv. Phys. Acta **19**, 2 (1946).
1a. F. Cap, Relativitätstheorie und Astronautik. Proceedings, IXth International Astronautical Congress, Amsterdam 1958, Pt. I, p. 209. Wien: Springer, 1959.
2. W. L. Bade, Relativistic Rocket Theory. Amer. J. Physics **21**, 310 (1953).
3. H. Bondi, Negative Mass in General Relativity. Rev. Mod. Physics **29**, 423 (1957).
4. R. Esnault-Pelterie, L'Astronautique. Paris, 1930.
5. J. M. J. Kooy, On Relativistic Rocket Mechanics. Proceedings, VIIIth International Astronautical Congress, Barcelona 1957, p. 569. Wien: Springer, 1958; Astronaut. Acta **4**, 31 (1958).
6. H. G. L. Krause, Relativistische Raketenmechanik. Astronaut. Acta **2**, 30 (1956).
7. C. Møller, The Theory of Relativity, chapt. IV, VII. Oxford, 1952.
8. W. Peschka, Über die Überbrückung interstellarer Entfernungen. Astronaut. Acta **2**, 191 (1956).
9. E. Sänger, Die physikalischen Grundlagen der Strahlantriebstechnik. V.D.I. - Forschungsheft 437, 1953.
10. E. Sänger, Zur Theorie der Photonenraketen. 5. Heft der Mitteilungen aus dem Forschungsinstitut für Physik der Strahlantriebe. München, 1956.
11. E. Sänger, Die Erreichbarkeit der Fixsterne. Proceedings, VIIth International Astronautical Congress, Rome 1956. Roma: Associazione Italiana Razzi, 1956.
11a. E. Sänger, Zur Mechanik der Photonenstrahlantriebe. München, 1956. (Russische Übersetzung, Moskau, 1958.)
12. H. S. Seifert, M. M. Mills and M. Summerfield, Physics of Rockets. Amer. J. Physics **15**, 267 (1947).
13. L. R. Shepherd, Interstellar Flight. J. Brit. Interplan. Soc. **11**, 149 (1952).
13a. M. Subotowicz, Teoria relatywistycznej rakiety n-stopniowej, Technika Rakietowa. Materiały z II-ej Konferencji Techniki Rakietowej i Astronautyki w Warszawie, 1959 (in polnischer Sprache).

B. Klassische Mechanik der Stufenraketen

14. H. G. L. Krause, Allgemeine Theorie der Stufenrakete. Weltraumfahrt **4**, 52 (1953).
15. M. Subotowicz, Niektóre zagadnienia teorii rakiet wielostopniowych. Technika Rakietowa **1**, 173 (1957) (in polnischer Sprache).
16. M. Subotowicz, The Optimization of the n-Step Rocket with Different Parameters and Propellant Specific Impulses in Each Stage. Jet Propulsion **28**, 460 (1958).
und die in [14 bis 16] zitierten Arbeiten.

The Fluctuation of the Accelerations of Satellites and the Changing of the Upper Atmospheric Conditions

By

H. K. Paetzold[1] and H. Zschörner[2]

(With 6 Figures)

(Received August 13, 1959)

Abstract — Zusammenfassung — Résumé

The Fluctuation of the Accelerations of Satellites and the Changing of the Upper Atmospheric Conditions. Continuous radio observations of Sputnik III were made since its launching on 20 Mhz. The wave propagation of this frequency has been found sensibly influenced by the height of the satellite orbit, the time of day, the geographical latitude and the season. The acceleration of the satellite shows terrestrial and extraterrestrial influences. A latitude and season effect can be distinguished, and a dependency from day and night seems to exist. A pronounced variation of about 28 days is strongly correlated to the 20 cm solar radiation emission. Another more secular effect is shown markedly by Vanguard I whose acceleration increased by a factor of four in August 1958 and decreased gradually till August 1959. This effect is very weak below 250 kms. according to the acceleration of Sputnik III and Explorer IV. Contrarily the 28 days fluctuation increased much less to greater altitudes, so that it seems to be mainly located at 250 kms., while the secular effect occurs much higher above 300 kms.

Die Schwankung der Beschleunigung von Satelliten und die Veränderungen der Bedingungen in der hohen Atmosphäre. Kontinuierliche Radiobeobachtungen von Sputnik III wurden seit seinem Start ausgeführt. Danach wird die Wellenausbreitung auf 20 Mhz merklich beeinflußt von der Höhe der Satellitenbahn, von der Tageszeit, der geographischen Breite und der Jahreszeit. Die Beschleunigung des Satelliten zeigt terrestrische und extraterrestrische Einflüsse. Es besteht ein Breiten- und Jahreszeiten-effekt. Ferner scheint ein Einfluß von Tag und Nacht zu bestehen. Eine markante Variation mit einer Periode von 28 Tagen zeigt eine enge Korrelation zu den Schwankungen der solaren 20-cm-Radiostrahlung. Ein weiterer mehr säkularer Effekt wird bei Vanguard I beobachtet, bei dem die Beschleunigung im August 1958 um den Faktor 4 anstieg, um dann allmählich bis zum August 1959 wieder abzunehmen. Im Gegensatz dazu ist dieser Effekt unterhalb von 250 km nur schwach, wie die Beschleunigungen von Sputnik III und Explorer IV zeigen. Hingegen wachsen die 28-Tage-Variationen oberhalb von 250 km Höhe nur wenig mit der Höhe an, so daß sie in 250 km Höhe ihren Hauptsitz zu haben scheinen. Die säkulare Variation hat oberhalb von 300 km ihren Sitz.

Fluctuations des accélérations des satellites et modifications dans la haute atmosphère. Depuis son lancement des observations radioélectriques continues sur 20 Mhz ont été faites sur Sputnik III. La propagation à cette fréquence a été trouvée sensible à la

---

[1] Technische Hochschule, München, Bundesrepublik Deutschland.

[2] Max-Planck-Institut für Aeronomie, Institut für Stratosphären-Physik, (14b) Weissenau, Kreis Ravensburg, Bundesrepublik Deutschland.

hauteur de l'orbite, à l'heure d'observation, à la latitude géographique et à la saison. L'accélération du satellite révèle des influences terrestres et extra-terrestres. On peut distinguer un effet de latitude et de saison et, semble-t-il, un effet diurne-nocturne. Une variation prononcée d'une période de 28 jours est en correlation étroite avec l'émission solaire sur 20 cm. Une variation plus lente est très marquée par Vanguard I, dont l'accélération s'est accrue d'un facteur 4 en août 1958 pour décroître graduellement jusqu'en août 1959. Les accélérations de Sputnik III et Explorer IV montrent que cet effet est très faible au dessous de 250 km. La fluctuation d'une période de 28 jours ne s'accroît que faiblement avec l'altitude et semble avoir son siège principal à 250 km; tandis que l'effet à longue période doit avoir son siège bien au dessus de l'altitude de 300 km.

Since the launching of the first artificial satellite the orbits of the satellites have been observed at many places. We ourselves have observed continuously the passages of Sputnik III (1958 δ_2) from May 1958 with the aid of a modern direction finder which was developed by the firm Telefunken in Ulm. The direction of the wave front and the amplitude of the received signals can be measured at the front of a television tube. A movie-camera registers the pictures with a time relation of 0,1 sec.

I. Bearings on 20 Mcs

Since a homogeneous material has been obtained now for more than one year some details could be found about the influence of season and hour of day, of height of the orbit, of solar activity etc. on the wave propagation. In general

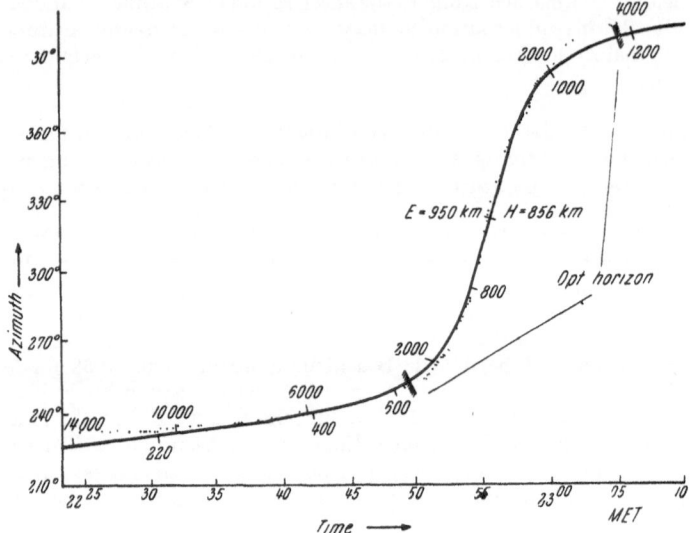

Fig. 1. Bearing on 20 MHz of 1958 δ_2

the distances varied very strongly at which the satellite could be received well and a defined wave front existed. At day time the conditions were worse than at night time. Periods of stronger solar activity were characterized by greater disturbances of the bearings in agreement with the ionospheric observations.

Sometimes the satellite could be received at very large distances as Figs. 1 and 2 demonstrate. The differences between the observed and the calculated directions are very small. So the radio waves must closely follow the great circle. Stronger deviations had often been found when the satellite passed the northern latitudes. In this case a considerable meridional gradient of electronic density must be supposed. Furthermore the satellite often could not be received when it stood to the north of the observation place though its height was 1000 km and its elevation amounted to 40° above the horizon. In this case the radio waves

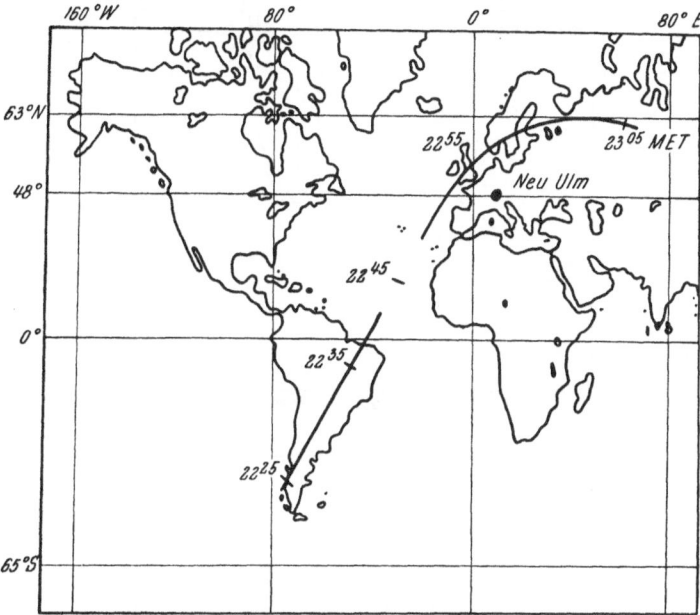

Fig. 2. Path o f 1958 δ_2 during the bearing in Fig.

must be reflected at the upper boundary of the F-layer into space due to a stronger decrease of the electronic density with the height than it has been observed in the average. In general the satellite signals at 20 MHz are not reflected into space so that a decrease of the electronic density by a factor of 2 is assumed for an interval of 200 km.

The accuracy of the time of passage has been found to be 2 sec. if several passages were observed to eliminate the ionospheric influences. This is sufficient to get the significant fluctuations of the acceleration of the satellite.

II. The Fluctuation of the Acceleration of the Satellite

It is well known that several observers early have seen that the acceleration of the different satellites varies markedly. The uncertainty whether these irregularities are caused by the special form of most satellites or not disappeared by the observations of Sputnik III (1958 δ_2) and its carrier rocket (1958 δ_1). Though the shape of the two bodies was totally different the irregularities occurred at the same time. In Fig. 3 the decrease $\dfrac{dP}{dn}$ of the revolution time P is given for the 1958 δ_2 our bearings with the direction finder have been used while the values for 1958 δ_1 are based on the observations of the German Moon-watch groups complemented by data of the Air Research Center, Bedford, Massachusetts.

It is important that the fluctuations of Vanguard I are in phase with those of Sputnik III as Fig. 4 demonstrates, though the perigee is very different for both satellites (Vanguard data from the Vanguard Computing Center, Washington). Therefore these fluctuations must be temporal but not local. The common time scale of these irregularities for different satellites is about 20—30 days.

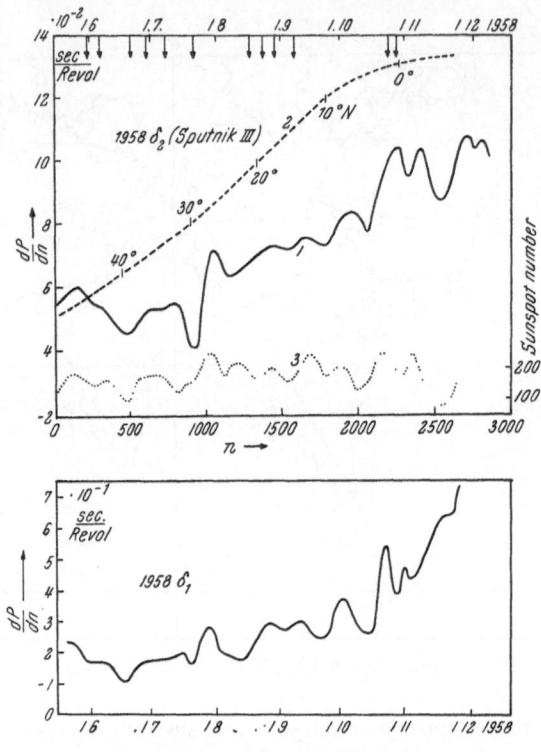

Fig. 3. Acceleration of Sputnik III and its rocket

Vanguard I furthermore shows a very strong change of the acceleration by a factor of 4 from the end of August 1958 till the end of October 1958. Then the effect slowly decreases again until in June 1959 the former low value of the acceleration of spring 1958 has been reached. The changes of the acceleration of Vanguard I amount to a factor 15 in extreme cases; Sputnik III only to the factor of 2.

III. The Variations of the Conditions in the Upper Atmosphere

The causes of the above fluctuations must be sought in variations of the air density, predominantly in the vicinity of perigee height. Other influences than electrical and magnetic forces seem to be too small. In first approximation the acceleration $\frac{dP}{dn}$ is proportional to the air density at the height of the perigee. Because the high level fluctuations are not very fast, it is probable that the upper atmosphere remains in the state of barometric equilibrium. This means that from the density variations the corresponding temperature changes can be derived.

For the analysis terrestrial and extraterrestrial influences must be distinguished. The first can be seen very clearly from the data of Sputnik III (Figs. 4 and 5). During May 1958 till December 1958 the perigee shifted from northern latitude to the equator. But the increase of the air drag was smaller than expected because of the oblateness of the earth (dotted line in Fig. 4 with the data of perigee latitude). This confirms a meridional gradient of density and temperature at 220 km height as already suggested by rocket ascents. In June 1959 the perigee of Sputnik III had reached a southern latitude of 65°. Due

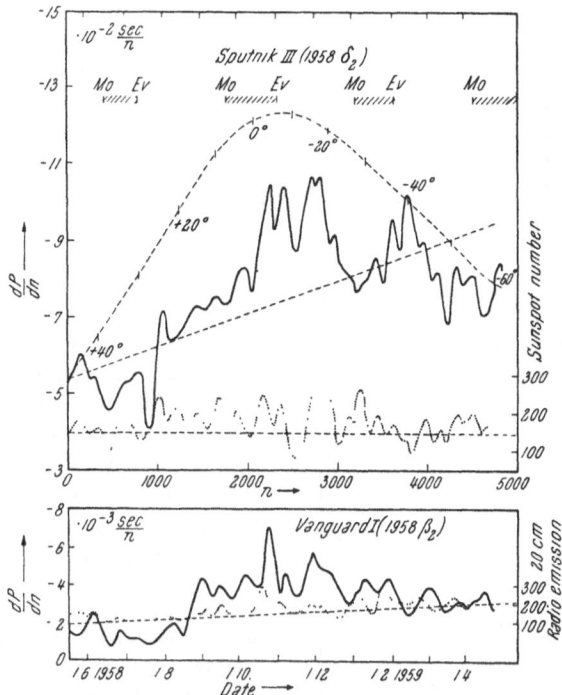

Fig. 4. Acceleration of 1958 δ_2 and 1958 β_2 and solar activity

to the slow sinking of the perigee one could expect an increase of the air drag by a factor of 2 from May 1958 to June 1959 (the inclined broken straight line in Fig. 4). In spite of this the acceleration has decreased nearly down to is initial value in May 1958 (Fig. 5). This suggested that the air density in 220 km height is smaller by a factor of 2 in winter time in higher latitudes. Finally a daily variation of the air density seems to occur with an amplitude of about 25 %. In Fig. 4 the dashed regions mark the night time of the perigee. It is well seen that the lowest values of density (and temperature) are reached at the end of the night and the highest values in the afternoon.

The extraterrestrial influence can be distinguished most clearly by the data of Vanguard I due to the smaller inclination of the orbit and its much faster perigee shifting. There seems to exist a close correlation between the faster fluctuation and the sunspot number which is also clearly shown by the variations of Sputnik III (Figs. 4 and 5). This correlation is still closer to the solar radio emission in the decimeter region. In Fig. 4 the dotted curve represents the intensity of the 1500 Mcs emission. The solar radiation responsible for the density fluctuation seems to originate from the corona condensations. It may be of importance that a small time lag seems to exist between fluctuations of the radiation

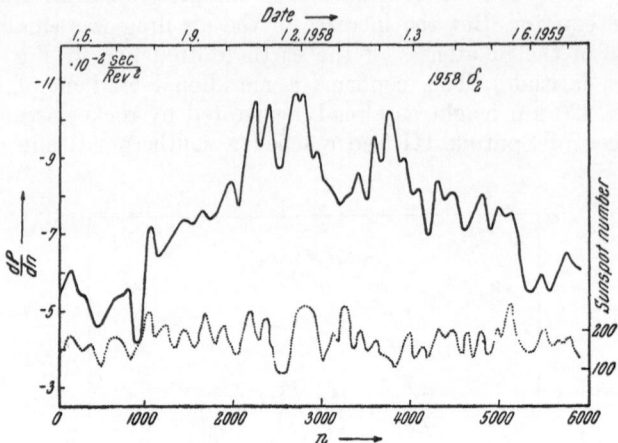

Fig. 5. Variable acceleration of Sputnik III

and the air density. There could not be found any correlations to single event of solar activity such as ionospheric and magnetic storms (arrow in Fig. 3).

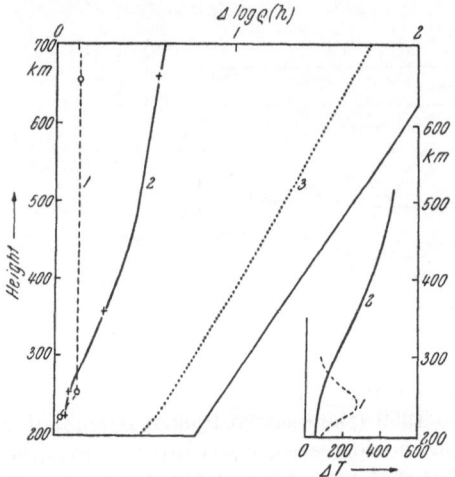

Fig. 6. Variation of air density

A crude estimation of the height of the variable heating is possible by observations of satellites with different perigee height. In Fig. 6 the dashed line *1* demonstrates the mean amplitude of the shorter fluctuation with about 27 days period for Sputnik I, Explorer IV and Vanguard I. The derived heating seems to take place mostly at an altitude of about 240 km. On the contrary the strong slow variation has been observed most markedly for Vanguard I, somewhat weaker for Explorer I and very weak for Sputnik III and Explorer IV (curve *2* of Fig. 6). The layer of strongest heating must be situated in higher altitudes of about 400 km.

According to the correlation of the upper air density to the solar activity one would expect that the density may be considerably smaller during a sunspot minimum. The difference between the ARDC-model 1956 and the satellite observations (curve *3* of Fig. 6) may be less than during the present sunspot maximum.

Telecomunicazioni spaziali e caratteristiche dei loro canali

A. Boni[1]

(Ricevuto il 27 agosto 1959)

Riassunto — Zusammenfassung — Abstract — Résumé

Telecomunicazioni spaziali e caratteristiche dei loro canali. I sistemi di telecomunicazione spaziale vengono classificati secondo le bande di onde elettromagnetiche impiegate, il tipo di servizio, terra-spazio o spazio, utilizzazione per telemetria, controllo, ecc.

I fattori determinanti il raggio d'azione della telecomunicazione vengono esaminati, in particolare il livello dei rumori, sia del radioricevitore (riducibile usando amplificatori di tipo MASER, MARVAR e simili), e delle sorgenti cosmiche. Questo disturbo, specialmente nella direzione di sorgenti intense, non può sempre essere evitato anche mediante antenne a fascio molto stretto. Gli estremi raggi d'azione si ottengono riducendo l'ampiezza della banda di modulazione, ad esempio fino a pochi cicli per secondo. Tuttavia la trasmissione dell'informazione diventa lentissima, la spesa unitaria di energia aumenta, e sono richiesti dispositivi di registrazione codificata. D'altronde l'informazione è per sè stessa differita.

A tali distanze è richiesta una elevata stabilità di frequenza, che potrà essere ottenuta mediante i così detti orologi atomici. Conseguentemente all'effetto Doppler, sarà necessario distinguere tra la frequenza assoluta e relativa dell'onda, in relazione anche alla rotazione terrestre. Saranno necessari dispositivi per la ricerca automatica della sintonia, l'agganciamento e la regolazione, specialmente durante l'accelerazione. Distanze stellari potranno presumibilmente raggiungersi con una catena di ripetitrici.

Nachrichtenübermittlungssysteme im Raum und Eigenschaften ihrer Kanäle.

Die Nachrichtenübermittlungssysteme im Raum werden je nach dem Band der verwendeten elektromagnetischen Wellen, nach der Art des Betriebes Erde-Raum oder Raum-Raum, nach der Benützung für Telemetrie, Fernsteuerung usw. klassifiziert.

Die die Reichweite der Nachrichtenübermittlung bestimmenden Faktoren werden untersucht, insbesondere die Störgeräuschschwelle des Radioempfängers (die durch Verstärker des Typs MASER, MARVAR u. ä. reduziert werden kann) wie auch die der kosmischen Störquellen. Letztere Störungen können speziell in der Richtung intensiver Störquellen auch mit Antennen mit sehr begrenztem Winkelbereich nicht immer vermieden werden. Die größte Reichweite erhält man, indem man die Breite des Modulationsbandes reduziert, beispielsweise bis auf wenige Schwingungen pro Sekunde. Dadurch wird die Nachrichtenübertragung sehr langsam, der Energieaufwand wächst, und es werden Einrichtungen mit verschlüsselter Registrierung erforderlich.

Bei den in Frage kommenden Entfernungen ist eine hohe Frequenzstabilität erforderlich, die mit Hilfe der sogenannten Atomuhren erhalten werden kann. Zufolge des Doppler-Effekts wird man zwischen der absoluten und relativen Wellenfrequenz auch in Beziehung auf die Erdrotation unterscheiden müssen. Vorrichtungen für automatische Abstimmung, Verstärkung und Regulation werden notwendig sein, besonders während der Anfahrt. Stellare Entfernungen werden vermutlich mit einer Relais-Kette überbrückt werden können.

---

[1] Via G. Sisco 7, Roma, Italia.

Telecommunication Systems in Space and Properties of their Channels. The tele-communication systems in space are classified according to the frequency band of the electromagnetic waves employed, according to the type of service such as earth-space, or space-space, or according to the purpose such as telemetry, remote control etc.

The factors determining the range of the telecommunication capability are investigated, especially the noise level of the radio receiver (which can be reduced by amplifiers of the MASER, MARVAR, and similar types), and also the cosmic sources of disturbing noise. Even with antennas of very narrow beam width such disturbances cannot be completely avoided. The maximum range is obtained by reducing the modulation band width, for instance down to a few cycles per second. This, however, makes the transmission of information very slow, increases the energy requirement per unit of information, and makes it necessary to resort to coded storage.

For the extreme distances here considered, it is necessary to provide high frequency stability which can be obtained with the so-called "atomic clocks.' Because of the Doppler effect, it will be necessary to differentiate between absolute and relative wave frequency, also with due regard to the earth rotation. Provision for automatic tuning, gain control, and regulation will be necessary, especially during acceleration. Stellar distances may have to be bridged by means of repeater chains.

Les transmissions d'information dans l'espace et les propriétés de leurs canaux. Les dispositifs de transmission d'information dans l'espace sont classés suivant les bandes électro-magnétiques utilisées, suivant la nature Terre-Espace ou Espace-Espace de la transmission et suivant le type d'utilisation: télémétrie, contrôle etc.

Les facteurs qui influencent le domaine de transmission sont étudiés. En particulier le bruit de fond des récepteurs, qui peut être réduit par l'emploi d'amplificateurs du type MASER, MARVAR etc., les sources de perturbation cosmiques qui ne peuvent toujours être évitées, même avec des antennes à faisceau angulaire très étroit. On obtient le plus large domaine par réduction de la largeur de la bande de modulation, en la réduisant par exemple à quelques Hertz. Cependant la rapidité de retransmission décroit, l'énergie nécessaire par unité d'information augmente, les signaux doivent être transmis en code.

L'éloignement requiert une augmentation de la stabilité de fréquence; elle peut s'obtenir à l'aide des "horloges atomiques". Par suite de l'effet Doppler il faut distinguer les fréquences absolues des fréquences relatives, aussi relativement à la rotation de la terre. Des installations d'accord automatique, de blocage et de régulation doivent être prévues. On peut envisager une chaine de relais pour couvrir les distances interstellaires.

I. Fattori della telecomunicazione spaziale

1. Sistemi di telecomunicazione spaziale

Le telecomunicazioni spaziali hanno come base delle loro possibilità la propagazione di onde elettromagnetiche, le quali, mediante la variazione dei loro parametri, possono diventare portatrici dei segnali.

Un sistema di comunicazione spaziale può essere classificato in primo luogo in relazione alla frequenza, o lunghezza, delle onde e.m. impiegate.

Una prima classe di sistemi è quella che impiega le così dette onde radio, od hertziane, la cui lunghezza si intende compresa nell'intervallo da 1 mm, o meno, fino a 1000 Km, e oltre, e quindi con frequenze da oltre 300 Giga Hz fino a 300 Hz, e meno.

Un secondo sistema è quello ottico, che impiega onde nella banda dello spettro visibile, 4000—7500 A, ed eventualmente le bande contigue, infrarosso (7500 A — 1 mm) e ultravioletto (4000—1800 A).

Onde di lunghezza più corta delle ultraviolette[1] non sembrano utilizzabili

---

[1] Raggi di SCHUMANN, 1800—2000 A; di LYMAN, 1200—500 A; raggi X, 500—0,5 A; raggi γ, <0,5 A.

allo stato attuale della tecnica, ma non si può escludere che lo saranno in futuro per speciali condizioni e progressivamente.

Considereremo nel seguito essenzialmente sistemi della prima classe, sistemi radio, in quanto questi consentono, ora, più agevolmente del sistema ottico di ottenere la trasmissione di una maggiore quantità di informazioni con minore energia e alle massime distanze.

Tuttavia il sistema ottico potrà trovare applicazione in qualche caso particolare[1], specie dove è richiesta la sola localizzazione del mezzo portante, e in tali casi i dispositivi di osservazione potranno eventualmente essere di tipo misto, telescopico-elettronico, accoppiati o no a registrazione fotografica.

Un secondo punto di vista per la classificazione dei sistemi di telecomunicazione spaziale considera i punti terminali, ed eventualmente gli intermedi, della linea di propagazione, così si hanno sistemi terra-spazio, spazio-spazio, terra-spazio-terra, e il sistema può essere semplice, per comunicazione unilaterale, o composto, per comunicazioni bilaterali. Specificazioni particolari riguardano il terminale nello spazio, ad esempio satelloide, sonda interplanetaria etc.

Un terzo punto di vista per la classificazione è diretto a specificare la particolare utilizzazione del sistema, ad esempio per scopi di telemisura, tracciamento, teleguida e controllo, per informazioni generali, e la trasmissione può essere automatica, indipendentemente dall'osservatore, o comandata dall'osservatore.

Il sistema può impiegare mezzi di registrazione per una trasmissione differita dell'informazione ed eventualmente con velocità cambiata.

Altri punti di vista per la classificazione potranno riguardare i particolari sistemi di modulazione impiegati, ad esempio del tipo a impulsi e relativi sottotipi, e altre speciali caratteristiche degli apparati di trasmissione, ricezione, e relative antenne, e dispositivi particolari.

La utilizzazione, che in un primo tempo può essere per solo fine scientifico, in un secondo tempo potrà essere a scopo industriale e commerciale, e queste formeranno branche separate e che potranno essere materia di specializzazione, ad esempio i satelliti terrestri potranno essere impiegati come ripetitori attivi o passivi per la telegrafia, telefonia e televisione [2].

2. I fattori determinanti il raggio d'azione del sistema

Il problema, che può essere considerato per primo, è l'espressione di:
$R =$ raggio d'azione della telecomunicazione, come funzione del gruppo di fattori, che lo determina.

Specifichiamo che l'esame del problema verrà limitato in relazione alle seguenti premesse, (a), (b):

(a) L'esame verrà condotto in prima approssimazione, per determinare essenzialmente l'ordine di grandezza delle variabili, trascurando la considerazione di effetti secondari e l'introduzione di coefficienti correttivi, che non siano troppo diversi dall'unità (quando siano applicati ad un prodotto di fattori).

---

[1] La segnalazione del mobile spaziale potrà ad esempio avvenire mediante il lampo nella reazione chimica di speciali sostanze, o mediante il lancio di fumogeni. Tali gas nello spazio vuoto, diventando rarefatti, possono raggiungere alte temperature per la irradiazione anche molto lontana da parte del sole, secondo la dimostrazione di FABRY [1]. Così la nuvola di gas diventa luminosa specie verificandosi condizioni di risonanza. Un esempio di queste applicazioni, nel caso dei lanci di missili lunari — Lunik dell'URRS, 2 Gennaio 1959 — riguarda la produzione di nuvole di vapore di sodio ai fini dell'avvistamento della rotta.

(b) Nel seguito ci riferiremo principalmente al caso di comunicazioni (del tipo telegrafico), che impiegano la trasmissione di impulsi di ampiezza costante, mentre la larghezza potrà assumere vari valori oppure restare costante, nel secondo caso la distanza tra gli impulsi potrà assumere vari valori.

I principali fattori, che intervengono a determinare il raggio d'azione R, sono i seguenti:

P_t = potenza, di cresta, del trasmettitore;

A_t, A_r = aree, efficaci, delle antenne, trasmettente e ricevente;

λ od f = lunghezza, o frequenza, dell'onda portante;

F_M = intervallo di frequenza dello spettro di modulazione;

$P_{N,1}$ = potenza media del disturbo, in ricezione, per 1 Hz di banda passante.

Specifichiamo tali fattori, e alcune formule ad essi relative, che saranno utilizzate in seguito. Verranno anche richiamate formule note, in particolare allo scopo di introdurre simboli, che saranno adottati negli ulteriori sviluppi.

La potenza di cresta P_t, massima potenza con cui il trasmettitore può alimentare l'antenna, è legata alla potenza media, P_m, dalla:

$$P_t = K_{Pm} P_m \qquad (1.2—1)$$

dove il fattore K_{Pm} è funzione della forma della modulazione.

Nel caso della emissione di treni di impulsi di ampiezza costante, e ciascuno di durata costante τ_p (come in alcuni tipi di modulazione, specie telegrafica e radar), il fattore K_{Pm}[1] è semplicemente il rapporto tra τ_p e l'intervallo medio di ripetizione dei successivi impulsi, $\bar{\tau}_r$:

$$K_{Pm} = \tau_p / \bar{\tau}_r. \qquad (1.2—2)$$

Per quanto riguarda le aree efficaci A_t, A_r delle antenne, esse possono convenientemente esprimersi come prodotto rispettivamente di aree convenzionali di riferimento, A_{tc}, A_{rc} per dei coefficienti, K_{At}, K_{Ar}, che considerano sia l'effetto dell'antenna in condizioni ideali di radiazione, sia le perdite che si verificano nell'attuazione pratica.

Con riferimento alle aree convenzionali A_{tc}, A_{rc} delle antenne, la convenzione per la loro misura è scelta in accordo al tipo di antenna; ad esempio nel caso di un sistema composto da un dipolo e da un riflettore parabolico, è usata come riferimento la superficie trasversale, che limita il riflettore, cioè:

$$A_c' = \pi D^2/4, \qquad (1.2—3)$$

nel caso di forma circolare, di diametro D.

Nel caso di una schiera di dipoli, distribuiti in una area rettangolare di dimensioni a e b, in larghezza ed altezza, si assume:

$$A_c'' = ab.$$

Con tale scelta per le aree convenzionali, i fattori correttivi K_{At}, K_{Ar} per ottenere le aree efficaci, restano in genere nell'intervallo 0,5—0,9.

Per le antenne riceventi, l'area A_r, è chiamata spesso area (efficace) di assorbimento.

Fattori connessi ai precedenti possono essere considerati, così, nel seguente n. I.3, introdurremo la velocità di informazione, collegata ad F_M, e, nel n. I.4, altre caratteristiche del sistema, collegate a P_{N1}, mentre ora specifichiamo per le antenne le caratteristiche G, guadagno e b, apertura del fascio.

---

[1] Nel caso dei sistemi radar il coefficiente K_{Pm} è detto "duty cycle".

Si definisce:

G = guadagno di una antenna direttiva = $\qquad$ (1.2—4)

= rapporto tra la densità del flusso di potenza irradiato in una particolare direzione e la densità del flusso di un radiatore isotropo, di pari potenza.

(Il flusso di potenza è proporzionale al vettore di POYNTING o al quadrato dell'intensità del campo.)

Il guadagno G verrà considerato in corrispondenza alla direzione di massima direttività, salvo diversa indicazione.

Si trova che G è collegato a λ ed all'area A_c, convenzionale, dell'antenna dalla formula:

$$G = K_G \cdot 4\pi A_c/\lambda^2, \qquad (1.2—5)$$

dove K_G è un fattore numerico, che tiene conto delle perdite, che si verificano nell'antenna, rispetto ad un radiatore ideale.

Nel caso di un dipolo con riflettore paraboloidico circolare, il fattore K_G è usualmente compreso nell'intervallo 0,50 — 0,65, mentre per una schiera di dipoli si ha $K_G = 0{,}80 — 0{,}95$[1].

Per quanto riguarda l'apertura del fascio, b, ci riferiremo al caso particolare del riflettore parabolico a sezione circolare, in cui la larghezza del fascio, tra i punti a metà potenza, è data da:

$$b \cong \lambda/D \qquad \text{(rad)}. \qquad (1.2—6)$$

Formule analoghe valgono per altri tipi di antenne, salvo l'introduzione di coefficienti correttivi. L'apertura si intende riferita al lobo maggiore del diagramma di direttività dell'antenna.

Nel seguito verranno prese in considerazione specialmente onde che poco risentono gli effetti di riflessione da parte della ionosfera, come sono le onde di frequenza superiore a circa 20 MHz, le quali anche subiscono una piccola rifrazione da parte della ionosfera, rifrazione della quale potrà essere tenuto conto per maggiore precisione nelle applicazioni relative al "tracciamento" dei veicoli spaziali.

3. Velocità di informazione e ampiezza della banda di modulazione

Mentre facciamo riferimento alla Appendice 1 per alcuni richiami sulla teoria della "informazione", riportiamo qui la formula di HARTLEY — TULLER — SHANNON, che esprime la capacità C, di trasmissione dell'informazione per un canale di comunicazione affetto da disturbi.

La detta formula per la capacità ha la seguente espressione, per un canale che ammette frequenze (di modulazione) nella banda $O \mapsto F_M$, ed è disturbato da un "disturbo termico bianco"[2] di potenza media N e quando la potenza media del segnale non supera P:

$$C = K_{UI} F_M \log (1 + P/N), \qquad (1.3—1)$$

---

[1] Effettivamente nel calcolo per l'antenna ricevente dell'area equivalente di assorbimento e del guadagno, occorre tener conto del tipo di polarizzazione dell'onda. Ad esempio nel caso di un'onda che non abbia un piano di polarizzazione fisso, ma con variazioni casuali, occorre introdurre un fattore correttivo dell'ordine di $^2/_3$ per l'area equivalente e per G_r, e ciò resterà sottinteso.

[2] E' chiamato così un disturbo la cui potenza media ha distribuzione uniforme in tutto lo spettro di frequenze considerato.

dove $K_{\sigma l}$ è un fattore, che assume il valore 1, se i logaritmi sono presi in base 2 e C è misurata in Bit/sec.

Un'altra formula è stata sviluppata da D. GABOR (cfr. Appendice 1), per il caso in cui sia stabilita la potenza massima del segnale invece che la potenza media, tuttavia le due formule non forniscono valori molto diversi quando il rapporto tra le potenze del segnale e del disturbo assume valori limitati, caso che sarà principalmente considerato in seguito.

Indicando con $\dot{I}$ la rapidità di trasmissione dell'informazione (misurata ad esempio in Bit/sec), la capacità C del canale è uguale al valore massimo che può assumere $\dot{I}$, con una adatta scelta della codificazione dei messaggi:

$$\dot{I}_{\max} = C \text{ (Bit/sec)} . \tag{1.3—2}$$

Osservando che non sempre sarà possibile applicare la codificazione ottima, useremo per la valutazione della $\dot{I}$, che è possibile trasmettere, un valore inferiore a quello fornito dalla (1.3—1). Precisamente, considerando che sarà in genere conveniente riferirsi per P/N al minimo valore, che tuttavia assicuri la intelligibilità (con eccezione per speciali tipi di trasmissione, ad esempio televisive), impiegheremo per il calcolo della velocità di informazione la semplice formula:

$$\dot{I} = K_{IF} F_M \text{ (Bit/sec)} \tag{1.3—3}$$

dove K_{IF} è un fattore, che resterà suggerito meglio dalla pratica.

Nel seguito basterà considerare per K_{IF} il suo ordine di grandezza, che può essere rappresentato dal valore 1 (Bit/ciclo).

Particolari formule per K_{IF} sono state sviluppate da M. H. NICHOLS e L. R. RAUCH [3] per vari sistemi di modulazione impiegati in radiotelemetria, brevemente indicati come sistemi AM-FM, PPM-AM, PAM-FM, etc.

4. Il rumore nel ricevitore

Il rumore, che si presenta all'uscita del ricevitore insieme col segnale amplificato, risulta dalla sovrapposizione di due gruppi di rumore, uno formato dai rumori che l'antenna raccoglie dall'ambiente esterno e immette all'entrata del ricevitore, il secondo gruppo formato dai rumori di origine interna al ricevitore stesso.

Una delle fonti di rumore è l'agitazione termica degli elettroni nei conduttori, detto "effetto JOHNSON" [4], dal primo che ne effettuò osservazioni sperimentali; la potenza del corrispondente rumore, detto brevemente "rumore termico" può essere valutata mediante una formula dovuta a NYQUIST [5, 6].

Una resistenza sia connessa all'entrata di un ricevitore; si suppone che la resistenza abbia un valore uguale all'impedenza di entrata e che abbia una temperatura assoluta T; la formula di NYQUIST fornisce per il valore, P_{NT}, del rumore all'entrata e relativamente a un valore B_F della banda di frequenze passante:

$$P_{NT} = k\,T\,B_F, \tag{1.4—1}$$

dove k è la costante di BOLTZMANN[1]:

$$k = 1{,}372 \cdot 10^{-23} \text{ Joule/°K} \cdot \text{ciclo} . \tag{1.4—2}$$

---

[1] Alcuni autori indicano per k il seguente valore, con riferimento a esperienze sul rumore termico: $k = 1{,}3805 \cdot 10^{-23}$ Joule/°K.

I rumori che l'antenna immette nel ricevitore possono essere essenzialmente distinti nei seguenti gruppi: rumori relativi all'emissione di radio onde da parte di sorgenti cosmiche [7], rumori per fluttuazioni nella ionosfera, rumori, detti "atmosferici", per scariche elettriche nell'atmosfera, effetti di interferenza da parte di altre radiotrasmissioni, "rumore termico" nella resistenza ohmica dell'antenna e nella linea di alimentazione del ricevitore.

Tra questi vari gruppi di rumore, quello prevalente è il rumore cosmico [8], nel caso di trasmissioni in una regione di frequenze 100 — 3000 MHz, caso che ci limiteremo a considerare nel seguito.

Tale rumore cosmico presenta, all'entrata del ricevitore, una potenza, P_{NC}, che (con riferimento alla resistenza di radiazione dell'antenna) può essere espressa dalla stessa formula (1.4—1), introducendo per T un adatto valore, T_{ac}, temperatura relativa ai rumori cosmici raccolti dall'antenna, quando è orientata in una data direzione:

$$P_{NC} = k\, T_{ac}\, B_F\,. \qquad (1.4—3)$$

Gli altri rumori che si manifestano all'uscita del ricevitore e di origine interna ad esso, sono dovuti a varie cause, agitazione termica nei conduttori, fluttuazioni nei tubi termoionici[1], nelle sorgenti di alimentazione, nella imperfezione dei contatti e così via.

All'uscita del ricevitore l'effetto globale dei rumori di origine interna è in gran parte dovuto al rumore nei primi stadi, amplificato attraverso il ricevitore, e può essere rappresentato schematicamente, per gli usi pratici, mediante l'introduzione di un così detto "fattore di rumore", μ, del ricevitore.

Tale fattore μ è definito come rapporto tra i valori che il rapporto segnale-disturbo assume all'entrata e all'uscita del ricevitore[2], cioè, indicando con P_S e P_N la potenza del segnale e del rumore in entrata e distinguendo con apici i valori in uscita, si ha:

$$\mu = (P_S/P_N)/(P_S'/P_N')\,. \qquad (1.4—4)$$

Questa si può scrivere:

$$\mu = (P_S/P_S')\, P_N'/P_N\,, \qquad (1.4—5)$$

cioè, indicando con G_s il guadagno del ricevitore (rapporto tra le potenze del segnale all'uscita ed entrata):

$$\mu = P_N'/G_s\, P_N\,. \qquad (1.4—6)$$

Nelle condizioni in cui si effettua la misura pratica di μ, il ricevitore è chiuso sopra una antenna fittizia, di impedenza uguale a quella di entrata del ricevitore ed avente la stessa temperatura ambientale, T_o, del ricevitore (nella misura si usa $T_o = 293°$); in queste condizioni[3] la potenza P_{No} del rumore in entrata è, secondo la (1.4—1):

---

[1] I rumori relativi ai tubi termoionici hanno come cause le fluttuazioni nell'emissione catodica e secondaria, i così detti "effetto mitraglia" (shot effect), originariamente studiato da Schottky, lo "effetto di scintillazione" (flicker effect), la disordinata ripartizione degli elettroni tra gli elettrodi e altri.

[2] Valori pratici di μ sono indicati nella letteratura tecnica, ad esempio [9] negli Atti del Comité Consultatif des Radiocommunications (C.C.I.R.); per radio ricevitori di buona qualità si ha $\mu = 3 \rightarrow 8$, nella regione 100 — 1000 MHz.

[3] Tali condizioni vengono attuate per la misura pratica di μ, misurando in un primo tempo la corrispondente potenza del rumore in uscita; in un secondo tempo si inserisce come segnale all'entrata il rumore prodotto da una sorgente, che viene regolato finchè il rumore in uscita si raddoppia; allora nella (1.4—5) si porra $P_S' = P_N'$, e si avrà $\mu = P_S/P_N$, dove P_N è dato dalla (1.4—7) e P_S si ha da una taratura della sorgente di rumore.

$$P_{No} = k \, T_o \, B_F \, . \tag{1.4—7}$$

Distingueremo nella potenza del rumore in uscita la parte P_{Ni}' di origine interna da quella inerente a P_{No}, cioè posto:

$$P_N' = P_{Ni}' + G_S \, P_{No} \tag{1.4—8}$$

si ha, sostituendo nella (1.4—6):

$$\mu = 1 + P_{Ni}'/G_S \, P_{No} \, . \tag{1.4—9}$$

Indicando con P_{Ni} la potenza del rumore di origine interna, riferita all'entrata del ricevitore, cioè posto:

$$P_{Ni} = P_{Ni}'/G_S \tag{1.4—10}$$

si può scrivere, secondo le (1.4—7, 9):

$$P_{Ni} = (\mu - 1) \, P_{No} = (\mu - 1) \, k \, T_o \, B_F \, . \tag{1.4—11}$$

Come si è detto precedentemente, ci limitiamo a considerare il caso che il rumore cosmico sia il principale rumore di origine esterna, perciò, tenendo conto del rumore di origine interna, il rumore totale riferito all'entrata del ricevitore, sarà, mediante le (1.4—3, 11):

$$P_N = P_{NO} + P_{Ni} = k \, B_F \, [T_{ac} + (\mu - 1) \, T_o] \, . \tag{1.4—12}$$

Indicando per brevità:

$$T_{er} = T_{ac} + (\mu - 1) \, T_o \, , \tag{1.4—13}$$

una temperatura equivalente per il calcolo, secondo la tipica formula (1.4—1) per il rumore termico, dell'effetto complessivo P_N del rumore di antenna e di quello interno del ricevitore, si avrà in definitiva:

$$P_N = k \, T_{er} \, F_M \, . \tag{1.4—14}$$

In questa abbiamo introdotto invece del simbolo B_F per la larghezza della banda passante[1], il simbolo F_M, già usato nei precedenti n. 1.2, 1.3.

Per ogni ciclo/sec di banda passante il rumore all'entrata sarà:

$$P_{N1} = k \, T_{er} \, . \tag{1.4—15}$$

Esprimendo, nelle (1.4—14, 15), k in Joule/°K, T_{er} in °K, P_N risulta in watt e P_{N1} in Watt/Hz.

5. Equazione tra le potenze emessa e ricevuta

Considereremo il caso della trasmissione chiamata "nello spazio libero".
Indichiamo con P_r la potenza del segnale all'ingresso del radioricevitore.

L'equazione che lega la potenza trasmessa, P_t, a quella ricevuta, P_r, con l'intervento di altri fattori, può essere ottenuta col noto procedimento, che è seguito nello studio dei sistemi di radiofari e, con qualche variante, per i sistemi radar [10], dove il cammino di trasmissione in andata è seguito da un secondo cammino di ritorno dopo riflessione sul bersaglio.

Ci riferiremo nel seguito essenzialmente all'equazione per la trasmissione ad una via.

---

[1] Un fattore correttivo potrà essere introdotto nella (1.4—14) per tener conto che, invece di F_M, dovrebbe essere considerata una "banda equivalente" per il calcolo del rumore, tenendo conto della curva di selettività dei vari circuiti del ricevitore.

Sia r la distanza tra il trasmettitore e il ricevitore.

La densità del flusso di potenza, che un radiatore isotropo produrrebbe a detta distanza r, è espressa da:

$$S_i = P_t / 4 \pi r^2 .$$ (1.5—1)

Nel caso di una antenna trasmettente di guadagno G_t, la densità del flusso diventa:

$$S = G_t S_i .$$ (1.5—2)

La potenza raccolta dall'antenna ricevente sarà:

$$P_r = A_r S .$$ (1.5—3)

In definitiva dalle (1.5—1, 2, 3) si ha per P_r la seguente espressione (una delle forme semplificate dell'equazione chiamata dei radiofari):

$$P_r = P_t G_t A_r / 4 \pi r^2.$$ (1.5—4)

Questa sarà assunta come potenza del segnale all'ingresso del ricevitore[1], trascurando in un primo esame le varie perdite tra antenna e ricevitore.

In questa formula potrebbe essere sostituita per G_t la sua espressione mediante A_t e λ^2, secondo la (1.2—5), oppure invece di A_r si potrebbe sostituire la sua espressione con G_r (e λ^2).

In particolare, al posto di A_r, area di assorbimento dell'antenna ricevente, considereremo un diametro equivalente D_r, definito da:

$$D_r^2 = (4/\pi) A_r .$$ (1.5—5)

Nel caso di antenna ricevente del tipo a riflettore paraboloidico, con sezione trasversale circolare di diametro D_{cr}, ammetteremo che il coefficiente riduttivo K_{Ar} (cfr. N. I.2), per il passaggio dall'area convenzionale A_{cr} alla efficace A_r, possa essere sostituito per approssimazione con 1:

$$K_{Ar} \cong 1,$$ (1.5—5′)

che equivale a:

$$D_r \cong D_{cr}.$$ (1.5—5″)

Per quanto riguarda G_t, di cui richiamiamo l'espressione (1.2—5):

$$G_t = K_{Gt} \cdot 4 \pi A_{et} / \lambda^2 ,$$ (1.5—6)

si ha, considerando al posto di A_{et} un diametro equivalente D_t, come nella (1.5—5):

$$G_t = K_{Gt} (\pi D_t / \lambda)^2 .$$ (1.5—7)

Facendo intervenire nella (1.5—4) i diametri D_t, D_r invece di G_t, A_r, si ha la formula espressiva:

$$P_r / P_t = K_t (D_t D_r / \lambda r)^2 ,$$ (1.5—8)

dove[2]:

---

[1] Un fattore correttivo potrà essere introdotto per tener conto della profondità di modulazione, che può essere raggiunta nei vari sistemi, con riguardo alla potenza di cresta del trasmettitore.

[2] Il fattore K_{Gt} assume il valore 0,96, e quindi $K_t \cong 0,6$ nel caso di un dipolo con riflettore paraboloidico circolare e in assenza di perdite; praticamente, con tale riflettore, come si è detto nel n. I.2, $K_{Gt} = 0,50 - 0,65$, quindi $K_t \cong 0,3 - 0,4$.

Una ulteriore riduzione potrà essere data al fattore K_t nella (1.5 — 8) per tener conto di altre perdite, finora trascurate (nell'attraversamento dell'atmosfera, nelle linee di alimentazione, ecc.).

$$K_t = (\pi/4)^2 \, K_{Gt} \, . \tag{1.5—9}$$

Nella (1.5—8) si è finora supposto che le antenne trasmettente e ricevente abbiano i loro assi di massima direttività orientati come la retta passante per il trasmettitore e il ricevitore; in caso diverso si dovrà sostituire nella (1.5—8) il coefficiente K_t con un altro, K'_t, del tipo:

$$K'_t = K_t \, \varphi_t \, \varphi_r \,, \tag{1.5—9'}$$

dove i coefficienti φ_t e φ_r tengono conto della riduzione di guadagno delle antenne, trasmettente e ricevente, per direzioni diverse da quella dei rispettivi massimi.

Le equazioni (1.5—4, 8) possono essere trasformate con l'impiego dei logaritmi, allo scopo di introdurre per le varie grandezze la misura in db (decibel), usata in pratica.

Al detto scopo, considerando per P_r un valore di riferimento, $P_r{}^*$, porremo:

$$P_{r(db)} = 10 \log_{10} (P_r/P_r{}^*) \tag{1.5—10}$$

ed analogamente per le aree (per i guadagni si assumerà $G_* = 1$).
Con riguardo ad r porremo:

$$r_{(db)} = 20 \log_{10} (r/r_*) \,, \tag{1.5—11}$$

ed analogamente per altre lunghezze, come D_t, D_r, λ.
Allora la (1.5—8) si scrive:

$$P_r = P_t + D_t + D_r - \lambda - r + K_{r1} \,, \text{(db)} \tag{1.5—12}$$

dove si è posto:

$$K_{r1} = 10 \log_{10} [K_t \, (P_t{}^*/P_r{}^*) \, (D_t{}^* \, D_r{}^*/\lambda_* \, r_*)^2] \, . \tag{1.5—13}$$

Assumendo per le potenze uno stesso riferimento, in particolare 1 Watt, e per le lunghezze pure uno stesso riferimento, ad esempio 1 m ($f_* = 300$ MHz, in corrispondenza a $\lambda_* = 1$ m), si ha semplicemente:

$$K_{r1} = 10 \log_{10} K_t \, . \tag{1.5—14}$$

Il valore di K_{r1}, con le dette unità, potrà variare praticamente nell'intervallo — 4 ⊣ — 5 db, e per valutazioni di massima si potrà assumere per K_{r1} il valore:

$$K'_{r1} = 0 \text{ db} \, . \tag{1.5—15}$$

Se invece si assume 10^6 Km $= 1$ MKm, come valore di riferimento per r e si lasciano invariati gli altri valori di riferimento (1 Watt per le potenze e 1 m per D_t, D_r, λ), si avrà per K_{r1} circa il valore:

$$K''_{r1} = \log_{10} (1/10^{18}) = - 180 \text{ db} \, . \tag{1.5—16}$$

I valori di P_t, D_t, D_r nelle (1.5—8, 12) sono da considerarsi come delle constanti caratteristiche per un dato sistema di stazioni trasmettente ricevente e ammettendo che venga assicurato automaticamente un costante reciproco orientamento delle antenne; allora P_r dipenderà solo da r, in un servizio mobile, e la (1.5—8) si potrà scrivere brevemente:

$$\boxed{P_r = P_t (D_{sa}/r)^2} \qquad (1.5\text{—}17)$$

dove D_{sa} è una costante del sistema:

$$D_{sa} = K_t (D_t D_r/\lambda) . \qquad (1.5\text{—}18)$$

Il fattore D_{sa}, che ha le dimensioni di una lunghezza, verrà chiamato "*diametro comparativo del sistema di antenne*"; esso, come mostra la (1.5—18), dipende dalla lunghezza d'onda di lavoro, λ, dai diametri equivalenti D_t e D_r, delle antenne ricevente e trasmettente, e dal mutuo orientamento di queste, e dal coefficiente numerico di rendimento, K_t, definito nella (1.5—9), e che dovrà essere sostituito con il coefficiente K'_t, indicato nella (1.5—9') se il mutuo orientamento delle antenne non è quello di massima direttività.

L'equazione (1.5—17), esprimendo le varie grandezze in decibel, diventa, con analogia alla (1.5—12):

$$P_r = P_t - r + D_{sa} + K_{r2}, \text{(db)} \qquad (1.5\text{—}19)$$

dove:

$$K_{r2} = 10 \log_{10} [(P_t*/P_r*) (D_{sa}*/r_*)^2] , \qquad (1.5\text{—}20)$$

indicando con asterisco gli elementi di riferimento.

Un modo di enunciare la (1.5—17) è il seguente: "*Le densità del flusso di potenza che due radiatori isotropi, di potenze P_r e P_t, producono alle rispettive distanze D_{sa} ed r, sono uguali.*"

6. Rapidità di informazione come funzione della distanza

Per la intelligibilità del segnale, la potenza P_r di esso in ricezione dovrà stare in un conveniente rapporto, K_{SN} rispetto alla potenza media del disturbo[1], riferita all'ingresso del ricevitore, e che è state indicata con P_N nel n. I.4:

$$P_r = K_{SN} P_N . \qquad (1.6\text{—}1)$$

Ricordiamo che P_N, secondo le (1.4—14, 15), si può esprimere come prodotto della larghezza della banda di modulazione, F_M, per la potenza specifica P_{N1} del rumore per 1 Hz di detta banda:

$$P_N = P_{N1} F_M . \qquad (1.6\text{—}2)$$

Dalle (1.6—1, 2):

$$P_r = K_{SN} P_{N1} F_M . \qquad (1.6\text{—}3)$$

Dal confronto di questa espressione di P_r con quella data nella (1.5—17), si ha, eliminando P_r:

$$\boxed{F_M r^2 = F_{SN} D_{sa}^2} \qquad (1.6\text{—}4)$$

dove si è posto:

$$F_{SN} = P_t / K_{SN} P_{N1} . \qquad (1.6\text{—}5)$$

---

[1] Per il rapporto segnale-disturbo, K_{SN}, sono ammessi anche valori prossimi a 1 in alcune applicazioni, come in radio astronomia e radar, di qualche unità in sistemi di telegrafia e di modulazione a impulsi, di alcune decine e più, nelle trasmissioni musicali e televisive.

Il fattore F_{SN}, che ha le dimensioni di una frequenza, potrà essere chiamato *"frequenza comparativa del sistema di comunicazione rispetto al rumore"*; nella (1.6—5) la potenza trasmessa, P_t viene uguagliata a K_{SN} volte la potenza del rumore relativa ad una banda di frequenze di larghezza F_{SN}.

La (1.6—4) mostra che, a parità di F_{SN} e D_{Sa}, l'ampiezza F_M della banda di modulazione deve essere ridotta come $1/r^2$, così alle massime distanze la F_M consentita diventa molto stretta.

Il problema si presenta diversamente in radio astronomia [8], dove per l'ascolto di sorgenti cosmiche deboli si cerca di aumentare la larghezza ΔF della banda del ricevitore; questo è in relazione col fatto che nei procedimenti impiegati, ad esempio sistema DICKE e sistema RYLE e VONBERG, la minima potenza rilevabile in ricezione è dell'ordine di grandezza relativa delle fluttuazioni di potenza del rumore, $\Delta P/P$, all'uscita del ricevitore, e queste fluttuazioni sono inversamente proporzionali alla radice quadrata del prodotto di ΔF per la costante di tempo τ_M dello strumento indicatore (scrivente) all'uscita del ricevitore. Invece nelle ricerche con echi radar dalla Luna e dai pianeti, sono impiegati ricevitori a banda stretta, associati a impulsi di durata relativamente lunga (0,01 sec) rispetto a quelli delle applicazioni comuni del radar (pochi $\mu \cdot$ sec).

Ricordiamo ora la (1.3—3), che lega F_M e la rapidità di informazione, $\dot{I}$:

$$\dot{I} = K_{IF} F_M , \qquad (1.6\text{—}6)$$

dove K_{IF} è un fattore, di cui abbiamo detto di considerare come ordine di grandezza il valore 1 Bit/ciclo.

Dalle (1.6—4, 6) si ottiene per la rapidità di informazione come funzione della distanza:

$$\boxed{\dot{I}\, r^2 = \dot{I}_{SN} D_{Sa}^{\ 2}} , \qquad (1.6\text{—}7)$$

dove si è indicato con $\dot{I}_{SN}$ il fattore:

$$\dot{I}_{SN} = K_{IF} F_{SN} = K_{IF}\, P_t / K_{SN}\, P_{N1} \qquad (1.6\text{—}8)$$

che chiameremo *"rapidità comparativa di informazione del sistema"*.

Ricordiamo che P_{N1} ha l'espressione, secondo le (1.4—13, 15):

$$P_{N1} = k\, T_{er} = k\, [T_{ac} + (\mu - 1)\, T_0] , \qquad (1.6\text{—}9)$$

che può essere valutata ad esempio in Joule/ciclo oppure in Watt/Hz.

Ricordiamo anche l'espressione (1.5—18) di D_{Sa} ("diametro comparativo del sistema di antenne"):

$$D_{Sa} = K_t\, (D_t\, D_r / \lambda) , \qquad (1.6\text{—}10)$$

dove, con riferimento alla (1.5—9′), il coefficiente K_t dovrà essere eventualmente sostituito con K'_t.

La (1.6—7) si può enunciare: *"Il prodotto della rapidità di informazione per il quadrato della distanza è una costante del sistema di comunicazione"*, ammettendo il sistema caratterizzato da valori costanti per i fattori $\dot{I}_{SN}$ e D_{Sa}, e in particolare per i termini P_t e P_{N1} di $\dot{I}_{SN}$, mentre con riguardo a D_{Sa}, si ammette un costante orientamento reciproco delle antenne.

Ammesso che la rapidità di informazione $\dot{I}$ si conservi costante entro un intervallo di tempo τ, il prodotto $\dot{I}_\tau$ rappresenta l'informazione totale, I_τ, trasmessa in detto periodo, allora moltiplicando per τ entrambi i membri della (1.6—7), questa diventa:

$$I_\tau r^2 = I_{SN, \tau} D_{Sa}{}^2, \qquad (1.6\text{—}11)$$

dove si è posto:

$$I_{SN, \tau} = \dot{I}_{SN} \tau = (K_{IF}/K_{SN} P_{N1}) P_t \tau. \qquad (1.6\text{—}12)$$

Le (1.6—11, 12) mostrano che: "L'informazione trasmessa entro il periodo τ è inversamente proporzionale al quadrato della distanza r, e direttamente proporzionale all'energia, $P_t \tau$. erogata entro detto periodo, a parità di altre condizioni", (cioè il flusso di informazione è proporzionale all'energia e la densità di detto flusso decresce come $1/r^2$).

II. Ordine di grandezza dei fattori per servizi tipici

1. I vari ordini di grandezza del raggio d'azione

Scelta una unità di misura per le lunghezze, U_l, ed una base numerica b_l, diremo che U_{ln} è una unità di ordine n per le lunghezze, se essa verifica la:

$$U_{ln} = b^n{}_l U_l. \qquad (2.1\text{—}1)$$

Ponendo $n = 0$, si osserva che U_l è di ordine 0:

$$U_l = U_{lo}. \qquad (2.1\text{—}2)$$

Le seguenti unità possono essere considerate, per una scelta di U_l:

— 1 m, oppure uno dei suoi multipli secondo una potenza intera di base 10, come 1 Km, 1 Mm, 1 Gm, 1 Tm[1]
— 1 u. a. = 1 unità astronomica = distanza media Terra-Sole = 149,6 MKm
— 1 Anno luce = 9,46 TKm = 63.300 u. a.
— 1 Parsec = $(360 \cdot 60^2/2 \pi)$ u. a. $\cong$ 206.265 u. a. = 3,258 anni luce = 30,84 TKm.

Nel seguito adotteremo: $U_l = 1$ Km, e $b_l = 100$; in tale modo il rapporto tra una unità di ordine n ed una di ordine $n - 1$, è, in decibel, secondo la (1.5—11):

$$(U_{ln}/U_{l, n-1})_{db} = 20 \log_{10} b_l = 40 \text{ db}. \qquad (2.1\text{—}3)$$

Diremo che una distanza r è dell'ordine di grandezza di U_{ln} quando la differenza non supera ± 20 db, cioè quando $r^2/U_{ln}{}^2$ è compreso tra $1/10$ e 10.

Una illustrazione di tali U_{ln} con riguardo alle comunicazione spaziali è fornita dalla Tab. 1.

2. Grandezza dei fattori caratteristici per vari tipi di servizi

Esamineremo gli ordini di grandezza dei vari fattori caratteristici del sistema di comunicazione spaziale ai vari ordini di distanza, giovandoci della espressione (1.6—4) della ampiezza di modulazione, F_M, come funzione della distanza.

Premettiamo alcune indicazioni sui diametri D_r, D_t, equivalenti delle antenne e sulla potenza specifica del rumore, P_{N1}.

Nel caso di antenne formate da un dipolo con riflettore paraboloidico limitato da una apertura circolare, il diametro di questa verrà assunto per D_r e D_t.

Nel caso di una antenna ricevente costituita da un dipolo risonante, si ha, per polarizzazione qualsiasi[2]:

---

[1] Ricordiamo che: G = giga = 10^9, T = tera = 10^{12}.

[2] L'area corrispondente di assorbimento è: $A_r = \lambda^2/4\pi$. Nel caso di polarizzazione parallela il precedente valore di A_r deve essere moltiplicato per 3/2, e allora D_r diventa uguale al valore di D_t indicato nella (2.2—2).

Tab. 1. *I vari ordini di distanze nelle comunicazioni spaziali*

$$n = \frac{1}{2} \log_{10} (U_{lm}/U_l).$$

$U_l = 1$ km, $b_l = 100$.

Ordine n	U_{lm}	r db	Intervallo	Distanza in tempo luce	Tipo di distanza	Osservazioni
—1	10 m	—40	1 ÷ 100 m		Interno-(Base)	Distanze quali a bordo di un veicolo spaziale; ricetrasmettitori per uso personale.
0	1 Km	0	100 m ÷ 10 Km	3,3 μsec	(Intorno)-Base	Distanze di esercizio nelle immediate prossimità di una base spaziale.
1	100 Km	40	10 ÷ 1.000 Km	330 μsec	(Base)-Sub-base	Distanze di comunicazione tra una base spaziale e sub-basi prossime.
2	10^4 Km	80	10^3 ÷ 10^5 Km	33 msec	(Base)-Satellite	Comunicazioni tra una base planetaria e prossimi satelliti.
3	1 MKm	120	10^5 ÷ 10^7 Km	1 sec-luce = 0,3 U_{l3}	lunare	Distanza media Terra-Luna = 0,3844 MKm.
4	100 MKm	160	1 ÷ 100 GKm	1 min-luce = 0,18 U_{l4}	solare e dei pianeti prossimi	Distanza dalla Terra al Sole, Marte, Venere, Asteroidi, e, all'estremo, Giove; 1 u. a. = 1,496 U_{l4}.
5	10 GKm	200	10^{11} ÷ 10^{13} Km	1 ora-luce = 0,108 U_{l5}	dei pianeti lontani	Distanza Sole—Plutone = 39,5 u. a. = 0,59 U_{l5}.
6	1 TKm	240	10 ÷ 1.000 TKm	1 Anno-luce = 9,46 U_{l6}	peristellare	Distanze presumibili per stazioni ripetitrici in comunicazioni stellari.
7	100 TKm	280		~10 Anni luce	stellare prossima	1 Parsec = 0,3084 U_{l7}; distanza minima stellare Sole—α Centauri = 1,3 Parsec = 4,24 Anni luce ≅ 0,4 U_{l7}.
8	10^4 TKm	320		~10^3 Anni luce	del sistema locale	
9	10^{18} Km	360		~10^5 Anni luce	delle galassie vicine	
10	10^{20} Km	400		~10^7 Anni luce	delle galassie lontane	

$$D_r = \lambda/\pi \cong 0.32\ \lambda. \tag{2.2—1}$$

Per un radiatore formato da un dipolo (così detto infinitesimo) il massimo guadagno è $G_{dt} = 3/2$, e in corrispondenza[1], dalla (1.5—7), con $K_{Gt} = 1$:

$$D_t = \sqrt{3/2}\ \lambda/\pi \cong 0.39\ \lambda. \tag{2.2—2}$$

Nel calcolo del fattore caratteristico D_{sa}, mediante la (1.6—10), assumeremo approssimativamente $K_t = 0.4$.

Per quanto riguarda P_{N1}, riferendoci alle (1.4—13, 15), stabiliamo prima un valore per T_{er}, assumendo a titolo indicativo:

$$T_{ac} = 10\ °K, \qquad T_0 = 293\ °K, \qquad \mu = 3.5. \tag{2.2—3, 3', 3''}$$

In corrispondenza si ha:

$$T_{er} = 10 + 2.5 \times 293 \cong 750°\ K, \tag{2.2—4}$$

pertanto:

$$P_{N1} = k\,T_{er} \cong 1 \cdot 10^{-20}\ \text{Watt/Hz}. \tag{2.2—5}$$

Indicheremo con E_{Fr} il comune valore dei due membri della (1.6—4):

$$E_{Fr} = F_M\,r^2 = F_{SN}\,D_{sa}{}^2\ (\text{m}^2\ \text{Hz}). \tag{2.2—6}$$

Potremo dare ad E_{Fr} il nome di "*estensione della telecomunicazione, in frequenza e raggio quadrato*".

La misura di E_{Fr} potrà essere data in m² Hz oppure in db:

$$E_{Fr\ (db)} = 10 \log_{10} F_M\,r^2_{(\text{m}^2\text{Hz})}. \tag{2.2—7}$$

Con riferimento ai vari ordini di distanze, n, considerati nella Tab. 1, abbiamo riportato nella Tab. 2 dei valori indicativi per i fattori caratteristici del sistema di comunicazione spaziale, in relazione a servizi caratterizzati da valori crescenti della "estensione", E_{Fr} (colonne 2 e 3).

Ricordiamo che, secondo la (2.1—3), valori interi consecutivi di n corrispondono a valori di r nel rapporto 100, cioè differenti di 40 db; la stessa differenza dovrebbe essere mantenuta tra i corrispondenti valori della estensione E_{Fr}, se un valore indipendente da n venisse dato alla frequenza di modulazione F_M, per caratterizzare un tipo di servizio.

Tuttavia, per non giungere a valori proibitivi nel dimensionamento del sistema di comunicazione ai più alti valori di n, si è ammessa una graduale restrizione per i valori della banda F_M, e in corrispondenza i valori adottati per E_{Fr}, ai vari ordini n, differiscono in media di $100/3 = 33$ db.

Allora, definito un "*grado di estensione del servizio*", $°E_{Fr}$ (colonna 1, Tab. 2):

$$°E_{Fr} = 3\ \% \ [E_{Fr(db)} - 100]\,, \tag{2.2—8}$$

la corrispondenza tra n ed $°E_{Fr}$ è espressa, per servizi telegrafici, da:

$$n = °E_{Fr}\,, \tag{2.2—9}$$

mentre per servizi televisivi:

$$n = °E_{Fr} - 1. \tag{2.2—10}$$

Si intende che le (2.2—9, 10) sono atte a indicare solo gli ordini di grandezza.

I valori di F_M, riportati in Tab. 2, colonna 13, specificano la maggiore lentezza del servizio al crescere di n (sigle: l = lento, ll = lentissimo).

---

[1] Nel caso di un radiatore mezza onda il guadagno è $G'_{dt} \cong 1.64$, invece di 3/2, e in corrispondenza si otterrebbe per D_t un valore un poco maggiore (del 4 %) di quello fornito dalla (2.2—2).

Tab. 2. *Valori dei fattori caratteristici per telecomunicazioni spaziali*

$$E_{Fr} = F_{SN} D^2_{Sa} = F_M r^2 \text{ (m}^2\text{ Hz) (2.2-6)} \qquad F_{SN} = P_t/K_{SN} P_{N1} \text{ (1.6-5)} \qquad P_{N1} = K T_{er} \text{ (1.4-15)}$$

$$E_{Fr} \text{ (db)} = 10 \log_{10} E_{Fr} \text{ (m}^2\text{ Hz) (2.2-7)} \qquad D_{Sa} = K_t (D_t D_t/\lambda) \text{ (1.5-18)} \qquad K_t = 0,4 \qquad P_{N1} = 10^{-20} \text{ W/Hz, per } T_{er} = 750° \text{ K}$$

TF = Telefonia   TG = Telegrafia   TV = Televisione

1 Grado del servizio °E_{Fr}	2 E_{Fr} m² Hz	3 E_{Fr} db	4 P_t	5 T_{er}	6 K_{SN}	7 F_{SN} Hz	8 λ m	9 F M Hz	10 D_t m	11 D_r m	12 D_{Sa}	13 F_M Hz	14 r km	15 n	16 Tipo di servizio	17 S
0	10^{10}	100	10 μW	750°K	250	$4 \cdot 10^{12}$	1	300	dip.	dip.	0,05	10^4	1	−1; 0	TF (interno ed intorno a una) Base	$3 \cdot 10^{-5}$
1	10^{14}	140	10 mW	750°K	25	$4 \cdot 10^{16}$	1	300	dip.	dip.	0,05	10^4 / 10^7	100 / 1	1 / −1; 0	TF / TV (Base) ⇄ Sub-base Base	$3 \cdot 10^{-5}$
2	10^{17}	170	100 mW	750°K	25	$4 \cdot 10^{17}$	1	300	dip.	3	0,5	10^3 / 10^7	10^4 / 100	2 / 1	TF / TV (Base) ⇄ Satellite Sub-base	$3 \cdot 10^{-6}$
3	10^{20}	200	100 mW	750°K	2,5	$4 \cdot 10^{18}$	0,30	1000	dip.	30	5	100 / 10^6	1 MKm / 10^4	3 / 2	TG / TV Lunare Satelliti (lenta)	10^{-7}
4	10^{24}	240	1 W	750°K	1	10^{20}	0,30	1000	0,7	100	100	100 / 10^6	100 MKm / 1 MKm	4 / 3	TG / TV Solare (pianeti vicini) Lunare	10^{-7}
5	10^{28}	280	100 W	750°K	1	10^{22}	0,10	3000	1,5	150	1000	100 / 10^6	10 GKm / 100 MKm	5 / 4	TG / TV pianeti lontani pianeti vicini	$3 \cdot 10^{-8}$
6	10^{31}	310	1 KW	75°K	1	10^{24}	0,10	3000	4	200	3000	10 / 10^5	1 TKm / 10 GKm	6 / 5	TG-l / TV-ll peristellare (l = lenta) pianeti lontani	$3 \cdot 10^{-9}$
7	10^{33}	330	10 KW	75°K	1	10^{25}	0,10	3000	10	250	10000	1	40 TKm	7	TG-ll —α Centaurus (ll = lentissima)	$3 \cdot 10^{-10}$

Stabiliti i valori di E_{F_r} e di F_M, i valori di r (e di n, colonne 14, 15) risultano dalla (2.2—6):

$$r^2 = E_{F_r}/F_M. \qquad (2.2—11)$$

Il valore di E_{F_r} viene raggiunto dimensionando i fattori F_{SN} e D_{Sa}^2, dei quali E_{F_r} è il prodotto, come indica la (2.2—6).

Per tale dimensionamento è stato usato il seguente procedimento.

In un primo tempo abbiamo scelto una serie di valori per la potenza trasmessa, P_t (colonna 4, Tab. 2); successivamente sono stati scelti dei valori (colonna 6) per il rapporto segnale-disturbo, K_{SN}, eseguendo qualche tentativo; mediante P_t, K_{SN} e il valore per P_{N1} indicato nella (2.2—5), si ottengono i valori di F_{SN} (colonna 7), applicando la (1.6—5):

$$F_{SN} = P_t/K_{SN}P_{N1}. \qquad (2.2—12)$$

In un secondo tempo si è provveduto alla scelta di serie di valori per la lunghezza d'onda, λ, e per i diametri D_t e D_r delle antenne (colonne 8, 10, 11), mediante i quali sono stati ottenuti i valori del "diametro comparativo del sistema di antenne", D_{Sa} (colonna 12), applicando la (1.6—10):

$$D_{Sa} = K_t (D_t D_r/\lambda), \qquad (2.2—13)$$

e ricordando le (2.2—1,2) per il caso di dipoli (per K_t è stato adottato il valore 0,4).

Mentre per $n = 0$ e 1, le antenne sono entrambe considerate dipoli, per $°E_{F_r} = 2$ e specie per $°E_{F_r} = 3$ (caso di trasmissioni di bande telegrafiche da distanze lunari e di trasmissioni televisive da veicoli satelliti), l'antenna ricevente è supposta un dipolo con riflettore parabolico.

Per i valori di $°E_{F_r}$ da 4 a 7, è supposto che entrambe le antenne siano dipoli con riflettori parabolici e orientate reciprocamente nella direzione di massima direttività (con controllo automatico), così le dimensioni del sistema restano limitate, in tali condizioni di massimo guadagno.

Per $°E_{F_r} \geqslant 3$, la potenza P_t indicata è quella per il trasmettitore della sonda spaziale, mentre D_r si riferisce al diametro dell'antenna della base di ascolto.

Nei casi di $°E_{F_r} = 3$ e 4, può essere domandata una guida della sonda mediante controllo dalla base, e allora i valori per D_t e D_r, delle colonne 10 e 11, sono da scambiarsi tra loro, inoltre i valori di P_t potranno anche essere aumentati, per maggiore sicurezza.

Per limitare la banda passante di rumore nel ricevitore, il sistema dovrà possedere un sufficiente grado di stabilità della frequenza portante, F; tale grado può essere indicato approssimativamente dal fattore (colonna 17, Tab. 2):

$$s = F_M/F. \qquad (2.2—14)$$

Si osservi che, mentre nella (2.2—4) è stato indicato il valore di 750 °K per T_{er}, un valore più basso, 75 °K (colonna 5, Tab. 2) è stato ammesso per i servizi chiamati 6 e 7.

Nei numeri seguenti verranno date alcune indicazioni sulle limitazioni relative ai rumori di origine cosmica, inoltre sulle variazioni di frequenza nel sistema di comunicazione e sull'impiego di dispositivi speciali.

Ora osserviamo che, mentre la (2.2—6) ha condotto all'introduzione del fattore caratteristico E_{F_r}, "estensione in frequenza e raggio quadrato", analogamente, ricordando la (1.6—7), può essere introdotto un fattore, E_{j_r}, *"estensione del sistema in rapidità di informazione e raggio quadrato"*:

$$E_{ir} = \dot{I} r^2 = \dot{I}_{SN} D_{Sa}{}^2 \quad \text{(m}^2 \text{ Bit/sec)}. \tag{2.2—15}$$

Tra questi due fattori, per le (1.3—3) e (1.6—8), esiste la relazione:

$$E_{ir} = K_{IF} E_{Fr}. \tag{2.2—16}$$

Avendo adottato approssimativamente per K_{IF} il valore 1 Bit/ciclo, i valori numerici di E_{Fr} e di E_{ir} coincidono, quando si esprima il primo fattore in m² Hz ed il secondo in m² Bit/sec.

III. Il rumore cosmico e limitazioni inerenti

Nell'esame del problema distingueremo varie parti, cominciando da alcune proprietà generali, che verranno poi applicate a casi particolari.

1. Emissione specifica di una sorgente di radiazione

L'emissione di radio onde da parte di corpi cosmici presenta caratteristiche, che possono essere paragonate con quelle dell'emissione del corpo nero.

Consideriamo un elemento di superficie $d\sigma_*$ del corpo emittente le radiazioni ed un cono elementare, di angolo solido $d\Omega_*$ ed uscente da $d\sigma_*$ (cioè con vertice che può essere confuso con $d\sigma_*$); sia θ_* l'angolo tra l'asse del cono e il vettore $\vec{n}_*$ della normale a $d\sigma_*$.

La potenza dP_* delle radiazioni emesse da $d\sigma_*$ entro detto angolo solido e in un intervallo di frequenze $F \frown F + dF$, ammette una espressione del tipo [11]:

$$dP_* = E_{\theta_*, F} \, d\sigma_* \, d\Omega_* \, dF, \tag{3.1—1}$$

dove $E_{\theta_*, F}$, funzione di F e θ_*, è la "emissione specifica" (per unità di superficie, di angolo solido e di banda di frequenze), misurata ad esempio in Watt/ m² · Hz · sterad.

Indicando con $E_{0,F}$ l'emissione specifica in direzione di $\vec{n}_*$, si dice che l'emissione segue la legge di LAMBERT se[1]:

$$E_{\theta_*, F} = E_{0,F} \cos \theta_*. \tag{3.1—2}$$

L'emissione dipende dalla temperatura del corpo, T (assoluta), e nel caso del corpo nero la legge di PLANCK fornisce:

$$E_{0,F} = (2 \, h \, F^3/c^2)/[\exp (h F/kT) — 1], \tag{3.1—3}$$

dove k è la costante di BOLTZMANN, già indicata nella (1.4—2), c è la velocità della luce ($= 299796$ Km/sec, nel vuoto) ed h è la costante di PLANCK ($= 6{,}547 \cdot 10^{-34}$ Joule · sec).

L'esponenziale, nella (3.1—31), può essere approssimato con una funzione lineare dell'argomento, se questo è abbastanza piccolo, ad esempio se:

$$h F/kT < 1/10, \tag{3.1—4}$$

e allora si ha la formula di RAYLEIGH-JEANS:

$$E_{0,F,R} = 2 \, k T/\lambda^2, \tag{3.1—5}$$

---

[1] In tale caso l'integrale di $E_{\theta_*, F}$ esteso a tutto un emisfero, ($\Omega_* = 2\pi$), chiamato "potere emissivo", è: $E_F = \pi E_{0,F}$.

avendo specificato $E_{0,F}$ con l'ulteriore indice R, con riguardo al particolare campo di valori di F e T in cui è considerata.

Precisamente, affinchè la (3.1—4) sia soddisfatta, si richiede, in base ai valori di h e K, che:

$$F < 2 \cdot 10^{10} \, T, \tag{3.1—6}$$

e ad esempio, per $T = 1°$, $F < 20000$ MHz (onde radio).

Nella (3.1—1) si può sostituire dF con $d\lambda$, purchè si introduca una funzione $E_{\theta_*, \lambda}$, tale che:

$$E_{\theta_*, \lambda} \, d\lambda = E_{\theta_*, F} \, dF, \tag{3.1—7}$$

la quale fornisce $E_{\theta_*, \lambda}$ mediante $E_{\theta_*, F}$, essendo:

$$dF/d\lambda = -F/\lambda. \tag{3.1—8}$$

Consideriamo ora un elemento $d\sigma_r$ di una superficie, che riceva le radiazioni emesse da $d\sigma^*$, e sia $d\sigma_{rn}$ la proiezione di $d\sigma_r$ in un piano perpendicolare alla congiungente i centri di $d\sigma_r$ e $d\sigma_*$.

Indicando con θ_r l'angolo che detta congiungente forma con la normale $\vec{n}_r$ a $d\sigma_r$, si avrà:

$$d\sigma_{rn} = \cos \theta_r \, d\sigma_r. \tag{3.1—9}$$

L'angolo solido $d\Omega_*$, prima considerato genericamente, sia in particolare l'angolo sotto il quale $d\sigma_r$ è visto da $d\sigma_*$, allora, indicando con r la distanza tra $d\sigma_r$ e $d\sigma_*$, risulta:

$$d\Omega_* = (1/r^2) \, d\sigma_{rn}. \tag{3.1—10}$$

Una espressione analoga vale per l'angolo solido $d\Omega_r$ sotto cui $d\sigma_*$ è visto da $d\sigma_r$:

$$d\Omega_r = (1/r^2) \, d\sigma_{*n}, \tag{3.1—11}$$

dove:

$$d\sigma_{*n} = \cos \theta_* \, d\sigma_*. \tag{3.1—12}$$

La potenza dP_r delle radiazioni, che $d\sigma_r$ riceve da $d\sigma_*$, con riguardo alla banda dF, non differisce dalla dP_*, indicata nella (3.1—1), così, sostituendo per $d\Omega_*$ la sua espressione (3.1—10):

$$dP_r = E_{\theta_*, F} (1/r^2) \, d\sigma_{rn} \, d\sigma_* \, dF. \tag{3.1—13}$$

Questa, se l'emissione verifica la legge di LAMBERT, può essere scritta, tenendo presenti le (3.1—11, 12):

$$dP_r = E_{0,F} \, d\sigma_{rn} \, d\Omega_r \, dF, \tag{3.1—14}$$

e simmetricamente la (3.1—1) si scrive in tale caso:

$$dP_* = E_{0,F} \, d\sigma_{*n} \, d\Omega_* \, dF. \tag{3.1—15}$$

La potenza ricevuta, dP_r, dipende, come mostra la (3.1—14), dall'angolo solido $d\Omega_r$ sotto cui è vista la sorgente, e non separatamente da r, $d\sigma_*$ e θ_* (così, un corpo che emetta radiazioni luminose come il corpo nero, non è visto diversamente, se esso è piatto o convesso).

2. Le equazioni della potenza dei rumori cosmici in ricezione

Applichiamo le (3.1—5, 14) per l'esame di alcuni aspetti della ricezione dei rumori cosmici.

Indicheremo allora in particolare con dP_c la potenza ricevuta delle radiazioni cosmiche, e con $dP_{c,1}$ la potenza relativa ad una banda di frequenze di larghezza unitaria, cioè porremo:

$$dP_c = dP_{c,1} \, dF, \qquad\qquad (3.2\text{—}1)$$

$$dP_{c,1} = E_{0,F} \, d\sigma_{rn} \, d\Omega_r, \qquad\qquad (3.2\text{—}2)$$

dove per $E_{0,F}$ potremo usare la particolare espressione (3.1—5), nel campo delle radiofrequenze.

Intenderemo precisamente in seguito che la potenza elementare $dP_{c,1}$, indicata nella (3.2—2), si riferisca alla potenza disponibile all'ingresso del ricevitore, allora l'elemento superficiale $d\sigma_{rn}$ in detta formula assume il significato di elemento della superficie di assorbimento dell'antenna ricevente, in conformità della definizione di questa superficie.

L'antenna ricevente sia costituita da un dipolo, sul quale vengano concentrate le onde raccolte da un riflettore paraboloidico circolare.

La superficie apparente del riflettore, per un osservatore nella direzione dell'asse del paraboloide, può essere assunta come superficie convenzionale di riferimento, A_{re}.

La superficie di assorbimento, che indicheremo con A_{rm}, in relazione a radiazioni provenienti secondo detto asse, si otterrà, come detto al n. I.2, moltiplicando A_{re} per un coefficiente correttivo, K_{Ar}, poco inferiore ad 1:

$$A_{rm} = K_{Ar} \pi D^2_{cr}/4, \qquad\qquad (3.2\text{—}3)$$

dove D_{cr} è il diametro del riflettore.

Il valore di K_{Ar} potrà essere ottenuto in corrispondenza ad una determinazione sperimentale di A_r, mediante la misura della potenza ricevuta da un trasmettitore abbastanza lontano (così che possa essere considerato come una sorgente puntiforme) e impiegando le (1.5—1, 2, 3), che legano la potenza trasmessa e ricevuta:

$$A_{rm} = (P_r/G_t \, P_t) \, 4 \pi \, r^2. \qquad\qquad (3.2\text{—}4)$$

Tale valore di A_r si riferisce, come detto, al caso che la sorgente (puntiforme) di radiazioni sia nella direzione dell'asse del paraboloide, secondo la quale il diagramma della direttività della ricezione (rilevata sperimentalmente per sorgenti puntiformi) presenta un valore massimo, G_{rm}, per il guadagno dell'antenna ricevente.

Per altre direzioni, individuate ad esempio da certe coordinate sferiche φ, ψ la superficie di assorbimento può essere definita mediante:

$$A_r (\varphi, \psi) = A_{rm} \, G_r (\varphi, \psi)/G_{rm}. \qquad\qquad (3.2\text{—}5)$$

Osservando che le dimensioni di A_r possono essere considerate come infinitesime rispetto alle grandi distanze tra il ricevitore e la sorgente di emissione, l'espressione (3.2—5) di A_r può essere sostituita al posto di $d\sigma_{rn}$ nella (3.2—2). Questa formula allora diventa:

$$dP_{c,1} = (A_{rm}/G_{rm}) \, E_{0,F} \, (\varphi, \psi) \, G_r \, (\varphi, \psi) \, d\Omega_r, \qquad\qquad (3.2\text{—}6)$$

dove $E_{o,F}(\varphi,\psi)$ indica l'emissione specifica delle sorgenti situate nella direzione dell'asse del $d\Omega_r$, avendo supposto una distribuzione continua di sorgenti.

La potenza delle radiazioni ricevute, entro una banda unitaria di frequenze, e da tutte le direzioni, P_{e1}, si otterrà integrando la (3.2—6) rispetto a $d\Omega_r$:

$$P_{e1} = (A_{rm}/G_m) \int E_{o,F}(\varphi,\psi)\, G_r(\varphi,\psi)\, d\Omega_r. \qquad (3.2—7)$$

Nel caso che le sorgenti abbiano una distribuzione uniforme, cioè $E_{o,F}$ sia (approssimativamente) costante entro la regione angolare dove $G_r(\varphi,\psi)$ conserva un sensibile valore, allora $E_{o,F}(\varphi,\psi)$ potrà essere portato fuori del segno integrale nella (3.2—7).

Con riferimento a tale caso, può essere definito un valore efficace dell'angolo solido di ricezione, Ω_r, mediante la:

$$\Omega_r\, G_{rm} = \int G_r(\varphi,\psi)\, d\Omega_r, \qquad (3.2—8)$$

cioè Ω_r è l'angolo entro il quale, se il guadagno restasse idealmente costante e uguale al massimo, G_{rm}, verrebbe ricevuta da sorgenti (uniformemente distribuite) di radiazioni una potenza uguale a quella, che viene ricevuta con guadagno variabile G da tutte le direzioni.

D'altra parte la scala per valutare G è scelta convenzionalmente in modo da soddisfare la condizione di normalizzazione:

$$\int G_r(\varphi,\psi)\, d\Omega_r = 4\pi, \qquad (3.2—9)$$

pertanto Ω_r risulta legato a G_{rm} dalla[1]:

$$\Omega_r = 4\pi/G_{rm}. \qquad (3.2—10)$$

Nel detto caso, di una distribuzione di sorgenti con valore uniforme di $E_{o,F}$, l'espressione (3.2—7) della potenza ricevuta, entro una banda unitaria, si riduce alla seguente, tenendo presenti le (3.2—8, 9, 10):

$$P_{e,1} = E_{o,F}\, A_{rm}\, \Omega_r. \qquad (3.2—11)$$

Se le sorgenti si estendono solo per una regione angolare $\Delta\Omega_r$, che sia piccola rispetto a Ω_r, allora l'espressione della potenza ricevuta, P'_{e1} si avrà dalla (3.2—11) sostituendo $\Delta\Omega_r$ al posto di Ω_r:

$$P'_{e1} = E_{o,F}\, A_{rm}\, \Delta\Omega_r. \qquad (3.2—12)$$

Nei casi in cui l'emissione specifica $E_{o,F}$ non può essere ammessa costante nella regione considerata per Ω_r, si potrà stabilire solo un valore medio, $\overline{E}_{o,F}$, che soddisfa la:

$$\overline{E}_{o,F} = \int E_{o,F}(\varphi,\psi)\, G_r(\varphi,\psi)\, d\Omega_r, \qquad (3.2—13)$$

tuttavia $\overline{E}_{o,F}$ dipenderà sia dalla distribuzione, $E_{o,F}(\varphi,\psi)$ sia da $G_r(\varphi,\psi)$, cioè dal tipo di antenna.

---

[1] Una relazione analoga alla (3.2—10) può essere scritta per il caso di una antenna trasmettente.

3. Potenza ricevuta nel caso di sorgenti cosmiche con distribuzione uniforme

Per il caso in cui la (3.2—11) sia applicabile (valore uniforme di $E_{0,F}$), il valore dell'angolo solido efficace, Ω_r, può essere ottenuto moltiplicando per un coefficiente correttivo, $K_{\Omega r}$, prossimo all'unità, il valore dell'angolo solido corrispondente all'apertura del fascio, b, che è stata indicata nella (1.2—6) per una antenna con riflettore paraboloidico e con riferimento ai punti a metà potenza del diagramma di direttività:

$$\Omega_r = K_{\Omega r} (\pi/4) (\lambda/D_{cr})^2 . \qquad (3.3—1)$$

Ricordando l'espressione (3.2—3) di A_{rm}, si avrà:

$$A_{rm} \Omega_r = K_{Er} \lambda^2/2 , \qquad (3.3—2)$$

dove si è posto:

$$K_{Er} = K_{\Omega r} K_{Ar} \pi^2/8 , \qquad (3.3—3)$$

coefficiente di valore prossimo ad 1.

La (3.3—2) mostra che il prodotto dell'area di assorbimento (massima) per l'angolo solido efficace non dipende dal diametro D_{cr} dell'antenna.

Sostituendo tale espressione di $A_{rm} \Omega_r$ nella (3.2—11), si avrà:

$$P_{c1} = K_{Er} E_{0,F} \lambda^2/2 . \qquad (3.3—4)$$

Adottando in particolare per $E_{0,F}$ l'espressione approssimata (3.1—5) di RAYLEIGH-JEANS, si avrà:

$$P_{c1} = K_{Er} k T , \qquad (3.3—5)$$

la quale mostra che la potenza delle radiazioni cosmiche, ricevuta entro una banda unitaria di frequenze, dipende esclusivamente dalla temperatura T delle sorgenti.

Nella (3.3—5) il fattore K_{Er}, che resta prossimo ad 1, è quasi indipendente dalle dimensioni dell'antenna ed anche il tipo di questa ha piccola influenza[1]:

La temperatura T è quella delle particelle cosmiche, generalmente allo stato gassoso, che emettono quelle particolari radiazioni.

Le misure dei rumori cosmici, effettuate esplorando il cielo con radiotelescopi, hanno indicato che T dipende dalla lunghezza d'onda, λ, e questo mostrerebbe che l'emissione proviene da componenti gassose disposte a varia profondità.

Per caratterizzare la potenza delle radio onde emesse da una sorgente cosmica, è stato introdotto nell'uso in radioastronomia il parametro "temperatura di emissione della sorgente", T_{ec}, che viene definita come la temperatura che dovrebbe avere il corpo nero per emettere radiazioni di potenza uguale a quella della sorgente, e nella stessa banda di frequenze.

Per il calcolo di tale T_{ec} può essere impiegata la (3.3—5) che fornisce:

$$T_{ec} = P_{c1}/K_{Er} k , \qquad (3.3—6)$$

nel caso di sorgenti diffuse in una regione angolare più ampia dell'angolo solido efficace dell'antenna.

Ricordiamo che nella (1.4—3) è stata introdotta una temperatura convenzionale T_{ac}, per il calcolo della potenza disponibile dei rumori cosmici all'ingresso del ricevitore, secondo la:

$$P_{c1} = k T_{ac} . \qquad (3.3—7)$$

---

[1] In pratica la presenza di lobi secondari del diagramma di direttività può influenzare le misure della P_{c1}.

La relazione tra T_{ee} e T_{ac}, che risulta dalle (3.3—6, 7), è:

$$T_{ac} = K_{Er} \, T_{ee} \, . \tag{3.3—8}$$

4. Espressioni del coefficiente K_{Er}

Il coefficiente K_{Er}, che è stato definito nella (3.3—3) mediante $K_{\Omega r}$ e K_{Ar}, può anche essere espresso mediante il coefficiente K_G, già indicato nella (1.2—5), che riportiamo:

$$A_c = (1/K_G) \, G \, \lambda^2/4\pi \, . \tag{3.4—1}$$

Questa può essere scritta come segue, per l'antenna ricevente di guadagno massimo G_{rm} ed area di assorbimento $A_{rm} = K_{Ar} \, A_c$:

$$A_{rm} = (K_{Ar}/K_G) \, G_{rm} \, \lambda^2/4\pi \, , \tag{3.4—2}$$

oppure, essendo $G_{rm} = 4\pi/\Omega_r$:

$$A_{rm} \, \Omega_r = (K_{Ar}/K_G) \, \lambda^2 \, . \tag{3.4—3}$$

Dal confronto con la (3.3—2) si ha un'altra espressione per K_{Er}:

$$K_{Er} = 2 \, K_{Ar}/K_G \, . \tag{3.4—4}$$

Inoltre, dalle (3.3—3) e (3.4—4), risulta:

$$K_G = 16/\pi^2 \, K_{\Omega r} \cong 1{,}62/K_{\Omega r} \, . \tag{3.4—5}$$

La (3.4—2) è usata rispettivamente coi valori $K_{Ar}/K_G = 1$ e $2/3$ nei casi di ricezione di onde con o senza un piano fisso di polarizzazione, ed in corrispondenza $K_{Er} = 2$ e $4/3$, mentre per $K_{Ar}/K_G = {}^1/_2$ si ha $K_{Er} = 1$, che fornisce $T_{ee} \cong T_{ac}$.

5. Indicazioni sui risultati delle misure dei rumori cosmici

L'emissione delle radio onde cosmiche presenta una componente di fondo, distribuita in tutto il cielo, che è poi costellato di sorgenti più intense, aventi estensione da una frazione ad alcune decine di minuti primi.

I valori della temperatura di emissione, T_{ee}, per la componente di fondo, variano da circa $100\,000°$ K, per lunghezze d'onda di circa 10 m, decrescendo fino a circa $1°$ K, per $\lambda = 10$ cm.

I diagrammi che rappresentano T_{ee} in funzione di λ in scale logaritmiche hanno forma circa rettilinea, cioè T_{ee} è proporzionale ad una potenza di λ, con esponente circa $2, 5$.

Secondo questa dipendenza di T da λ, la formula (3.1—5) di RAYLEIGH-JEANS fornisce per l'emissione specifica $E_{0,F}$ un valore che varia solo lentamente rispetto a λ (come $\lambda^{1/2}$); da tale legge si discostano però le sorgenti più intense.

Le sorgenti di estensione $> 20'$ si trovano in generale concentrate nel piano galattico e alcune, ad esempio in misure con $\lambda = 3{,}7$ m, mostrano densità di flusso, P'_{e1}/A_{rm} anche superiori a 10^{-24} Watt/m² Hz.

Poichè in corrispondenza a $20'$ si ha $\Delta \Omega_r \cong 3 \cdot 10^{-5}$ sterad, risulta dalla (3.2—12) per tale densità di flusso:

$$E_{0,F} = P'_{e,1}/A_{rm} \, \Delta \Omega_r \cong 3 \cdot 10^{-20} \text{ Watt/m² Hz} \cdot \text{sterad}, \tag{3.5—1}$$

e quindi, applicando la formula di RAYLEIGH-JEANS:

$$T_{ee} = (\lambda^2/2 \, k) \, E_{0,F} \cong (3{,}7)^2 \cdot 3 \cdot 10^{-20}/2 \cdot 1{,}372 \cdot 10^{-23} \cong 15.000° \text{ K} \, . \tag{3.5—2}$$

Come altro esempio, per $\lambda = 1,5$ m (200 M Hz), si trova al polo galattico una emissione di fondo $E_{0,F} \cong 5 \cdot 10^{-21}$ Watt/m² Hz sterad, a cui corrisponde, secondo la (3.5—2), $T_{ec} \cong 400°$ K.

Si presentano anche sorgenti che, pure avendo una limitata estensione angolare, sono molto intense, come la Cygnus A (19 N 4 A), Cassiopeia A (23 N 5 A), Virgo A (12 N 1 A), Taurus A (05 N 2 A).

Ad esempio la Cygnus A presenta rispettivamente per $\lambda = 0,1 — 1 — 10$ m le densità di flusso $10 — 30 — 400 \cdot 10^{-24}$ Watt/m² Hz, e tenendo presente che l'estensione della sorgente è di circa 1' (nelle varie direzioni), si hanno, in base alle 3.5—1,2) le rispettive temperature $4 \cdot 10^4$—10^7—10^9 °K, di cui l'ultima è più alta, con un fattore 10^5, della temperatura 10^4 °K, ordine di grandezza della temperatura superficiale delle stelle.

La elevatissima emissione di radio onde, che si manifesta per la Cygnus A, si presume associata con la collisione di due galassie, che si verifica in quel luogo.

6. Differenze tra il caso di sorgenti estese e quello di sorgenti puntiformi

Dal punto di vista del raggiungimento, in ricezione, di un sufficiente rapporto segnale-disturbo, possono essere distinti due casi: primo, di sorgenti aventi una estensione angolare superiore (o dello stesso ordine di grandezza) dell'angolo solido Ω_r efficace dell'antenna ricevente; secondo caso, che la sorgente abbia una estensione di valore trascurabile rispetto a Ω_r, cioè che la sorgente possa considerarsi puntiforme.

Nel primo caso, ammesso che il rumore cosmico sia prevalente rispetto a quello di origine interna al ricevitore, cioè, con riferimento alla (1.6—9),

$$T_{ac} \gg (\mu — 1) \, T_0, \qquad\qquad (3.6—1)$$

la potenza del rumore all'ingresso del ricevitore, come mostra la (3.3—5), non varia aumentando A_r cioè le dimensioni dell'antenna, (in quanto Ω_r si riduce inversamente).

D'altra parte, per una data densità di flusso prodotta dal trasmettitore, la potenza del segnale ricevuto è proporzionale all'area di assorbimento, A_{rm}, come indicano le (1.5—1, 2, 3, 4), quindi in detto caso, un aumento del rapporto segnale-disturbo potrà essere ottenuto aumentando A_{rm}, finchè possibile praticamente (ammettendo fissati la potenza e guadagno del trasmettitore, P_t e G_t; esempi sono stati dati nel precedente paragrafo, cfr. Tab. 2).

Consideriamo ora il secondo caso, di sorgenti puntiformi, o più in generale, aventi una estensione $\Delta \Omega_e$ trascurabile rispetto a Ω_r.

In questo caso, sia la potenza del segnale in ricezione, sia quella del rumore della sorgente cosmica sono proporzionali ad A_{rm}, così il rapporto segnale-disturbo non è influenzato da un aumento di A_{rm}, ma è determinato solo dal rapporto delle rispettive densità di flusso (sempre ammesso trascurabile il rumore interno del ricevitore).

7. Definizione di una "temperatura equivalente di emissione del trasmettitore"

Indicando con $S'_{c,1}$ la densità del flusso di potenza del rumore per 1 Hz di banda entrante nel ricevitore, si ha secondo la (3.2—12), per una sorgente cosmica del tipo (quasi) puntiforme:

$$S'_{c,1} = P'_{c,1}/A_{rm} = E_{0,F}\, \Delta\,\Omega_{r,c} \ (\mathrm{Watt/m^2\ Hz})\,, \qquad (3.7\text{—}1)$$

oppure, mediante la (3.1—5):

$$S'_{c,1} = 2\,k\,T_{ec}\, \Delta\,\Omega_{r,c}/\lambda^2\,. \qquad (3.7\text{—}2)$$

Indicheremo analogamente con $S_{t,1}$ la densità del flusso di potenza del segnale, prodotto dal trasmettitore. riferita a 1 Hz della sua banda di modulazione. F_M, cioè, ricordando le (1.5—1, 2):

$$S_{t,1} = P_t\,G_t/4\,\pi\,r^2\,F_M \ (\mathrm{Watt/m^2\ Hz})\,. \qquad (3.7\text{—}3)$$

In questa formula il fattore G_t può essere sostituito dalla sua espressione (1.2—5), che riscriviamo:

$$G_t = K_G \cdot 4\,\pi\,A_{et}/\lambda^2\,. \qquad (3.7\text{—}4)$$

Pertanto:

$$S_{t,1} = (P_t/F_M)\, \Delta\,\Omega_{r,t}/\lambda^2\,, \qquad (3.7\text{—}5)$$

dove si è posto:

$$\Delta\,\Omega_{r,t} = A_t/r^2\,, \qquad (3.7\text{—}6)$$

angolo solido sotto cui è vista dal ricevitore l'area equivalente dell'antenna trasmettente (rispetto al guadagno):

$$A_t = K_{Gt}\,A_{et}\,. \qquad (3.7\text{—}7)$$

Le espressioni (3.7—2. 5) delle densità di flusso, $S'_{c,1}$ ed $S_{t,1}$, del rumore e del segnale, hanno forme in parte simili, e che possono farsi coincidere, quando si definisce una *"temperatura equivalente di emissione del trasmettitore"*:

$$T_t = P_t/2\,k\,F_M\,, \qquad (3.7\text{—}8)$$

allora, $S_{t,1}$ si scrive:

$$S_{t,1} = 2\,k\,T_t\, \Delta\,\Omega_{r,t}/\lambda^2\,. \qquad (3.7\text{—}9)$$

Per la uguaglianza delle densità di flusso del rumore e del segnale, entro una stessa banda di frequenze, si richiede, secondo le (3.7—2,9):

$$T_t\, \Delta\,\Omega_{r,t} = T_{e,c}\, \Delta\,\Omega_{r,c}\,, \qquad (3.7\text{—}10)$$

oppure, indicando con A_{ec} la superficie apparente della sorgente cosmica e con r_c la sua distanza dal ricevitore:

$$T_t\,A_t/r^2 = T_{ec}\,A_{ec}/r_c^2\,. \qquad (3.7\text{—}11)$$

Se le distanze r ed r_c sono uguali (almeno come ordine di grandezza), l'uguaglianza delle potenze del segnale e del rumore richiede che il parametro:

$$L = T\,A\,, \qquad (3.7\text{—}12)$$

prodotto della temperatura per l'area emittente, che chiameremo "area-temperatura", dovrà assumere lo stesso valore per il trasmettitore e per la sorgente cosmica:

$$T_t\,A_t = T_{ec}\,A_{ec}\,. \qquad (3.7\text{—}13)$$

Calcoliamo ad esempio il valore del prodotto $T_{ec}\,A_{ec}$ per il Sole.

Nell'intervallo $\lambda = 0{,}01 - 0{,}1$ m la temperatura T_{ec} del Sole, in giorni tranquilli, ha andamento crescente da 30.000 a 50.000° K.

Nell'intervallo $\lambda = 0{,}1 - 1$ m la temperatura presenta una variazione maggiore, circa con legge: $T_{ec} = 10^6\,\lambda^{1,3}$, raggiungendo 10^6 °K per $\lambda = 1$ m, e conservando circa lo stesso valore nell'intervallo $\lambda = 1 - 10$ m.

Riferiamoci al caso $\lambda = 0,3$ m $(F = 1000$ MHz$)$, dove $T_{ec} \cong 200.000°$ K.

Le misure mostrano che T_{ec} non è costante nelle varie regioni del disco solare, e che anzi per $\lambda = 3 - 5$ m l'emissione si produce fino ad una distanza dal centro del Sole circa doppia del raggio della fotosfera, tuttavia, per il caso considerato $\lambda = 0,3$ m, il valore di T_{ec} resta circa costante sul disco solare e decresce rapidamente fuori di esso.

Calcoleremo allora A_{ec} in corrispondenza al raggio del Sole, $R_s = 697.000$ Km, e si ha: $A_{ec} = \pi R_s^2 \cong 1,53 \cdot 10^{18}$ m², quindi per il Sole, a $\lambda = 0,3$ m:

$$L_S = A_{ec} T_{ec} \cong 3,1 \cdot 10^{23}\,°\text{K} \cdot \text{m}^2 \,. \qquad (3.7—14)$$

Come si è detto con riguardo alla (3.7—12), il valore di $L_t = A_t T_t$ per il trasmettitore della sonda spaziale deve essere uguale ad L_S, quando la sonda passa in prossimità del Sole e si ammetta un rapporto segnale-rumore uguale ad 1.

Supponendo $A_t \cong 1$ m², che corrisponde secondo le (3.7—4, 7) a: $G_t = 4\,\pi/0,09 \cong 140 \cong 21,5$ db, il trasmettitore dovrà avere la temperatura equivalente di emissione $T_t = L_t/A_t = 3,1 \cdot 10^{23}$ °K, cioè, secondo la (3.7—8):

$$P_t/F_M = 2\,k\,T_t \cong 8,4\;\text{Watt/Hz} \,. \qquad (3.7—15)$$

8. Definizione del parametro "area-energia per ciclo"

Mentre riteniamo, che il parametro T_t, temperatura equivalente di emissione del trasmettitore, sia suggestivo, ed il parametro $L_t = T_t A_t$, di "area-temperatura" sia semplice, tuttavia i loro valori numerici, essendo risultati molti grandi nell'esempio considerato, potrà convenire l'uso del seguente parametro:

$$Q_t = 2\,k\,L = 2\,k\,A\,T \;\text{(m² joule/ciclo)} \qquad (3.8—1)$$

che chiameremo "area-energia per ciclo", del trasmettitore, o della sorgente cosmica in genere.

In particolare, per il trasmettitore, Q_t ammette l'espressione seguente, che unisce le principali caratteristiche A_t, P_t, F_M:

$$Q_t = A_t\,P_t/F_M \,. \qquad (3.8—2)$$

Mediante il definito parametro Q, la (3.7—13) si scrive:

$$Q_t = Q_e \,, \qquad (3.8—3)$$

uguaglianza richiesta quando si ammetta un rapporto segnale-rumore uguale ad 1 e le distanze del trasmettitore e della sorgente di rumore dal ricevitore si possano considerare approssimativamente uguali (si suppone inoltre che il ricevitore lasci passare la sola banda F_M).

Nell'esempio cui si riferisce la (3.7—15) si ha:

$$Q_t \cong 8,4 \text{ m}^2 \text{ joule/ciclo} \cong 8,4 \text{ m}^2 \text{ Watt/Hz} \,. \qquad (3.8—4)$$

Calcoliamo, con i dati di detto esempio, e per un ricevitore situato sulla Terra, la densità di flusso $S'_{e,1}$, mediante la (3.7—2); lo stesso valore si otterrà in questo caso per la densità $S_{t,1}$, espressa dalla (3.7—3).

Essendo la distanza Terra-Sole: $r_{eS} = 1,496 \cdot 10^{11}$ m, si avrà prima:

$$\Delta \Omega_{r,e} = A_{ec}/r_{eS}^2 \cong 6,83 \cdot 10^{-5} \text{ sterad} \,, \qquad (3.8—5)$$

(come pure si otterrebbe partendo dal diametro apparente del Sole, medio tra quello all'afelio, 31' 30" e quello al perielio, 32' 35").

L'emissione specifica del Sole, per i valori considerati di T_{ec} e λ, risulta:

$$E_{0,F} = 2\ k\ T_{ec}/\lambda^2 = 2,74 \cdot 10^{-23} \times 200.000/(0,3)^2 \cong$$
$$\cong 6,09 \cdot 10^{-17}\ \text{Watt/m}^2\ \text{Hz} \cdot \text{sterad}, \qquad (3.8\text{—}6)$$

pertanto:

$$S'_{e,1} = E_{0,F}\, \Delta\, \Omega_{r,e} \cong 4,2 \cdot 10^{-21}\ \text{Watt/m}^2\ \text{Hz} . \qquad (3.8\text{—}7)$$

Come si è detto, un valore uguale si trova per $S_{t,1}$, avendo assunto:

$$T_t/T_{ec} = A_{ec}/A_t = 3,1 \cdot 10^{23} , \qquad (3.8\text{—}8)$$

rapporto uguale anche a $\Delta\, \Omega_{r,e}/\Delta\, \Omega_{r,t}$, nell'ipotesi $r = r_{ec}$.

Aumentando l'area di assorbimento dell'antenna ricevente, A_{rm}, si riduce Ω_r, ma un miglioramento del rapporto segnale-disturbo si ottiene nelle descritte condizioni solo quando l'ampiezza del fascio, corrispondente a Ω_r, diventa inferiore al diametro apparente del Sole ($\sim 32'$, per un osservatore sulla Terra).

Come altro esempio, cerchiamo il valore minimo che può ricevere il parametro Q_t (area-energia per ciclo) per il trasmettitore di una sonda spaziale in prossimità del pianeta Marte, che allora potrà essere considerato come la sorgente principale di rumore.

Si ammetterà ancora soddisfatta una formula del tipo della (3.6—1), modificata in modo da tener conto che Marte è da considerare (generalmente) come sorgente puntiforme per un ricevitore sulla Terra.

Tenendo presente che il semiasse maggiore dell'orbita di Marte è: $a_{rM} = 1,524$ u. a., la distanza minima Terra-Marte è dell'ordine: $r_{T,M} \cong 0,524$ u. a. $\cong 78$ MKm.

In relazione al raggio di Marte, $R_M \cong 3450$ Km, si ha la superficie apparente: $A_{eM} = \pi\, R^2_M \cong 37,4 \cdot 10^{12}$ m², alla quale corrisponde come angolo solido massimo sotto cui Marte è visto dalla Terra: $\Omega_{rM} = A_{eM}/r^2_{T,M} \cong 6,14 \cdot 10^{-9}$ sterad, e in corrispondenza l'apertura del fascio è: $b_M \cong 18''$.

La temperatura media della superficie di Marte è: $T_{eM} \cong 250\ °\text{K}$, quindi il parametro area-temperatura assume il valore: $L_M = A_{eM} T_{eM} \cong 9,35 \cdot 10^{15}$ m² °K, ed il parametro area-energia per ciclo risulta: $Q_M = 2\ k\ L_M = 0,25 \cdot 10^{-6}$ m² Watt/Hz.

9. Limitazioni portate dai rumori solari nel controllo di una sonda spaziale

Consideriamo il caso che una sonda spaziale debba ricevere dalla Terra dei radio segnali di comando.

Se la sonda spaziale è per pianeti lontani, ad esempio per Plutone, che dista 39,5 u.a. dal Sole, l'angolo parallattico sotto il quale dalla sonda sono visti la Terra e il Sole, sarà inferiore a:

$$b_{Pl} \cong 1/39,5 \cong 0,025\ \text{rad}. \qquad (3.9\text{—}1)$$

L'angolo solido corrispondente ad un fascio di apertura b_{Pl} è:

$$\Omega_{Pl} \cong \pi\, b_{Pl}^2/4 \cong 5,0 \cdot 10^{-4}\ \text{sterad} . \qquad (3.9\text{—}2)$$

Per una antenna paraboloidica circolare di angolo Ω_{Pl}, il guadagno corrispondente ad Ω_{Pl} è:

$$G_{Pl} = 4\,\pi/\Omega_{Pl} \cong 25.000 . \qquad (3.9\text{—}3)$$

Per $\lambda = 0,3$ m, l'area di assorbimento dell'antenna ricevente, a bordo della sonda, dovrà essere, assumendo $K_{Er} = 1$ nella (3.3—2):

$$A_{rm} = K_{Er}\, \lambda^2/2\, \Omega_r \cong 0,09 \cdot 10^3 \cong 90\ \text{m}^2 . \qquad (3.9\text{—}4)$$

Presumibilmente antenne di minori dimensioni verranno adottate per la sonda, perciò a dette distanze e specie a distanze superiori, il trasmettitore terrestre dovrà soddisfare la (3.8—3) per un rapporto tra segnale e rumore solare uguale ad 1, cioè ricordando la (3.8—4), la "area-energia per ciclo" dovrà almeno avere il valore: $Q_t = A_t P_t/F_M = 8{,}4 \ m^2 \ \text{Watt/Hz}$ in corrispondenza al considerato $\lambda = 0{,}3$ m.

Un analogo esame potrà essere fatto per altri valori di λ e del rapporto segnale-rumore.

10. Condizioni di confronto tra una sorgente puntiforme e una diffusa

Nelle (3.2—11,12) abbiamo già indicato le potenze $P_{e,1}$ e $P'_{e,1}$, che, entro una banda unitaria di frequenze, vengono ricevute rispettivamente nel caso di una sorgente estesa uniformemente a tutto l'angolo solido Ω_r dell'antenna ricevente e nel caso di una sorgente, che abbracci solo un piccolo angolo $\varDelta \Omega_r$, e che diremo sorgente quasi-puntiforme.

Un confronto tra $P_{e,1}$ e $P'_{e,1}$ si avrà dal rapporto tra le (3.2—11,12), allora si elimina $A_{r,m}$:

$$P_{e,1}/P'_{e,1} = (E_{0,F}/E'_{0,F}) \ (\Omega_r/\varDelta \Omega_r) \ , \tag{3.10—1}$$

avendo indicato con $E'_{0,F}$ l'emissione specifica delle sorgenti quasi-puntiformi.

Affinchè risulti:

$$P_{e,1}/P'_{e,1} = 1 \ , \tag{3.10—2}$$

si richiede:

$$\varDelta \Omega_r = \Omega_r \ E_{0,F}/E'_{0,F} \ . \tag{3.10—3}$$

In questa il rapporto tra le emissioni specifiche può essere sostituito da quello tra le temperature di emissione, ricordando l'espressione (3.1—5) di RAYLEIGH-JEANS:

$$\varDelta \Omega_r = \Omega_r \ T_{ec}/T'_{ec} \ , \tag{3.10—4}$$

nella quale la lunghezza d'onda, λ, non figura esplicitamente.

La (3.10—4) è applicabile anche se la sorgente quasi-puntiforme che viene considerata è il radio-trasmettitore, allora si sostituirà T_{ec}' con T_{ec}, secondo la (3.7—8); così la (3.10—4) appare come una estensione della (3.7—10).

Applichiamo la (3.10—4) ad esempio per valutare a quale distanza da un radioricevitore l'emissione del Sole e quella di fondo del cielo sono ricevute con uguale potenza.

Faremo il confronto in corrispondenza di $\lambda = 0{,}1$ m, dove, in conformità dei dati di III.5 e III.7, la temperatura di emissione del Sole e quella di fondo del cielo sono rispettivamente 50.000 e 1° K, cioè:

$$T'_{ec}/T_{ec} = 50.000 \ . \tag{3.10—5}$$

Supporremo una antenna ricevente paraboloidica di diametro $D_r = 25$ m. Ricordando la (1.2—5):

$$G_{rm} = K_{Gr} \cdot 4 \pi A_e/\lambda^2 \ , \tag{3.10—6}$$

si ha, assumendo $K_{Gr} = 0{,}65$ ed $A_e = \pi D_r^2/4$:

$$G_{rm} = K_{Gr} (\pi D_r/\lambda)^2 \cong 40 \cdot 10^4 \ . \tag{3.10—7}$$

L'ampiezza efficace del fascio, b_r, è legata a $\Omega_r \cong \pi b_r^2/4$, cioè a $G_{rm} = 4\pi/\Omega_r$:

$$b_r \cong 4/G_{rm}^{1/2} \ , \tag{3.10—8}$$

quindi per la (3.10—7):

$$b_r \cong 4/632 \cong 0{,}0063 \text{ rad} \cong 22' \, . \qquad (3.10—9)$$

Assumendo per la parallasse media del Sole dalla Terra il valore 32', l'angolo b_r, indicato nella (3.10—9), è uguale alla parallasse del Sole da una distanza (in unità astronomiche):

$$r_s' = 32/22 \cong 1{,}5 \text{ u.a.} \qquad (3.10—10)$$

Per ottenere che, secondo la (3.10—4), il rapporto $\Omega_r/\Delta\,\Omega_r$ sia uguale al valore 50.000, indicato nella (3.10—5), la distanza del radioricevitore dal Sole dovrà essere:

$$r_s = r_s' \, (50.000)^{1/2} \cong 330 \text{ u. a.} \qquad (3.10—11)$$

A parità di altre condizioni il valore di r_s varia proporzionalmente a D_r, così si riduce a 33 u.a. per $D_r = 2{,}5$ m, e diventa:

$$r_s'' \cong 3300 \text{ u.a.} \qquad (3.10—12)$$

per il valore $D_r = 250$ m, quale indicato in Tab. 2 per il servizio di comunicazioni spaziali chiamato di "Grado 7".

Il raggio d'azione di tale servizio è: $r_{\alpha c} \cong 40 \, T$ km (cfr. Tab. 2), cioè (essendo 1 u. a. $= 149{,}6$ Mkm):

$$r_{\alpha c} \cong 270.000 \text{ u. a.} \qquad (3.10—13)$$

Assumendo che la stella α-Centaurus abbia una radio emissione dello stesso ordine di grandezza di quella del Sole, la ricezione di essa da una distanza $r_{\alpha c}$, impiegando $D_r = 250$ m appare nel seguente rapporto di potenza, rispetto a quella di fondo del cielo, relativa a un angolo di apertura $b_r'' = 22/10 = 2'$, 2:

$$(r_s''/r_{\alpha c})^2 = 1/6700 \, , \qquad (3.10—14)$$

cioè la seconda è predominante.

La potenza, $P''_{c,1}$, del rumore cosmico ricevuto in tali condizioni, può essere calcolata con la (3.3—5), ponendo $K_{Er} = 1$ e $T = 1°$ K:

$$P''_{c,1} = k \cdot 1° = 1{,}372 \cdot 10^{-23} \text{ Watt/Hz} \, . \qquad (3.10—15)$$

Con i dati del detto esempio di Tab. 2 ($P_t = 10$ KW, $D_{Sa} = 10.000$ m, $F_M = 1$ Hz) la potenza $P_{r,1}$ del segnale, per 1 Hz di banda passante, è, applicando la (1.5—17):

$$P_{r,1} = P_r/F_M = P_t \, (D_{Sa}/r)^2/F_M \cong 62 \cdot 10^{-23} \text{ Watt/Hz} \, , \qquad (3.10—16)$$

la stessa che si avrebbe per un rumore cosmico di fondo alla temperatura $T_{ec} = 62/1{,}372 \cong 45°$ K (Nella Tab. 2 è indicata $T_{er} = 75$ °K, e in corrispondenza, dalla $E_{Fr} = F_M \, r^2 = 10^{33}$ m² Hz, si ha più precisamente: $r \cong 32 \, T$ Km, invece di $r_{\alpha c} \cong 40 \, T$ Km).

In detto esempio la temperatura equivalente di emissione del trasmettitore è:

$$T_t = P_t/2 \, k \, F_M = 3{,}6 \cdot 10^{26} \text{ °K} \, , \qquad (3.10—17)$$

e la "area-energia per ciclo":

$$Q_t = A_t \, P_t/F_M \cong 500.000 \text{ m² Watt/Hz} \, , \qquad (3.10—18)$$

essendo: $A_t = K_a \pi \, D_t^2/4 \cong 50$ m².

La densità di flusso dovuta al rumore cosmico può essere espressa analogamente alle (3.7—1,2):

$$S_{e1} = P_{e1}/A_{rm} = 2{,}74 \cdot 10^{-28} \text{ Watt/m}^2 \text{ Hz}, \qquad (3.10\text{---}19)$$

essendo: $A_{rm} \cong 50.000$ m², secondo la (3.2—3), con $K_{Ar} = 1$.

Si osservi che $S_{e,1}$ è legata ad una corrispondente Q_e dalla:

$$Q_e = S_{e,1} \, \lambda^2 \, r^2 \, . \qquad (3.10\text{---}20)$$

Altre relazioni possono essere stabilite tra i parametri definiti in questo paragrafo, come Q, T_1, ecc., ed i parametri definiti nel paragrafo II.2, come E_{Fr}, ecc.

IV. Interazioni tra il moto e la telecomunicazione

1. I parametri cinematici che influenzano la telecomunicazione

Il trasmettitore ed il ricevitore del sistema di telecomunicazione spaziale hanno generalmente un moto relativo uno rispetto all'altro.

I parametri caratteristici di tale moto, sia le distanze, sia i vettori di velocità, e accelerazione, hanno influenza su altri parametri caratteristici della radiocomunicazione.

Nel seguito daremo alcune indicazioni sui problemi che si presentano.

2. La distanza reciproca e il tempo di ritardo della comunicazione

Le radio onde emesse dal trasmettitore giungono al ricevitore con un tempo τ di ritardo, data da:

$$\tau = r/c, \qquad (4.2\text{---}1)$$

dove r è la distanza reciproca e c la velocità della luce.

Il tempo τ ha come ordine di grandezza 1 millisec nelle comunicazioni coi satelliti artificiali terrestri, 1 sec nelle comunicazioni Terra-Luna, da alcuni minuti ad alcune decine di minuti per i pianeti prossimi, alcune ore per i pianeti lontani e diventa di 4,2 anni per la stella più prossima, α-Centaurus (ad 1,3 parsec).

Il valore di τ deve essere tenuto in conto nel tracciamento delle orbite dei veicoli spaziali, in funzione del tempo variabile t.

Indicando con $\Delta s \, (\tau)$ lo spostamento del veicolo spaziale sull'orbita, si ha approssimativamente, per piccoli valori di τ:

$$\Delta s(\tau) \cong v_r \, \tau. \qquad (4.2\text{---}2)$$

Ad esempio, per un satellite terrestre, con $r = 3000$ Km e v_r avente come ordine di grandezza 10 Km/sec, si ha Δs dell'ordine di 100 m.

Per una sonda lunare, con $v_r \cong 3$ Km/sec, il valore di Δs è ~ 3 Km.

Per una sonda relativa al pianeta Marte, e che debba essere controllata dalla Terra, si ha τ dell'ordine di 10 min, e ΔS dell'ordine di 3000 Km, con riferimento ad una velocità relativa della sonda rispetto a Marte di 5 Km/sec.

La (4.2—2) fornisce solo una generica indicazione specie per i grandi valori di τ, quando v_r è soggetta a notevoli variazioni.

L'integrazione delle equazioni del moto in base ai valori dei parametri cinematici determinati in istanti precedenti può fornire dati di maggiore precisione sulla posizione in funzione di t.

Tuttavia la precisione ottenibile è da presumere che diventi sempre minore al crescere della distanza r, come uno studio analitico del problema, nei vari casi, potrebbe stabilire.

3. Il ritardo τ e la velocità di informazione

Diversi casi possono essere esaminati, secondo il sistema di comunicazioni, unilaterali e bilaterali, per segnali telegrafici, telemetrici in genere, televisivi, ecc.

Quando il tempo di ritardo τ è notevole, potrà convenire di effettuare la trasmissione dell'informazione con una velocità diversa, ed anche molto minore, della velocità con la quale l'informazione è stata ripresa dagli strumenti, e ciò inserendo una fase intermediaria di registrazione, ad esempio di tipo magnetico.

In tale modo potrà essere ridotta la potenza del trasmettitore della sonda, in relazione alla diminuzione dell'ampiezza del canale occorrente per trasmettere l'informazione: tuttavia l'energia occorrente, prodotto della potenza per l'intervallo di tempo, resterà legata alla informazione totale.

4. Frequenza assoluta e relativa dell'onda portante

Il sistema trasmettente, Σ_T, emetta radio onde con una frequenza F_T; per un sistema ricevente, Σ_R, che abbia un moto relativo M_{Σ_T, Σ_R} rispetto a Σ_T, la frequenza delle onde propagantisi è rilevata con un valore differente, F_R, secondo il noto "effetto Doppler".

Viene considerata principalmente la propagazione nello spazio libero, tralasciando di esaminare le correzioni dovute alla propagazione nella ionosfera (d'altronde piccole, per frequenze > 100 MHz, e rilevabili solo con precisi metodi di zero).

Siano definiti due punti, C_R e C_T, come centri dei due sistemi Σ_R e Σ_T e sia $\overrightarrow{v_{TR}}$ il vettore della velocità relativa di C_R rispetto a Σ_T[1].

Indicando con $v_{TR,\varrho}$ il componente di $\overrightarrow{v_{TR}}$ secondo la congiungente $C_R C_T$ (positiva nel caso si avvicinamento), si ha:

$$F_R = F_T \left(1 + v_{TR,\varrho}/c\right) . \qquad (4.4-1)$$

Questa formula dovrebbe essere corretta per tener conto, tra altri effetti secondari:

a) del moto di Σ_T, come pure di Σ_R, intorno ai rispettivi centri (come ad esempio si verifica per alcuni veicoli satelliti);

b) di effetti relativistici, specie al crescere del rapporto v_{TR}/c.

Indichiamo con $\varDelta F_{TR}$ la differenza assoluta delle frequenze:

$$\varDelta F_{TR} = F_R - F_T = (v_{TR,\varrho}/c)\, F_T . \qquad (4.4-2)$$

La variazione specifica delle frequenze (rapportata ad F_T), che indicheremo con $\varepsilon_{F,TR}$:

$$\varepsilon_{F,TR} = \varDelta F_{TR}/F_T = v_{TR,\varrho}/c , \qquad (4.4-3)$$

dipende sola da $v_{TR,\varrho}$.

La valutazione di una velocità v, in particolare di $v_{TR,\varrho}$ può essere fatta in vari modi, ad esempio in decibel rispettto ad n ceurto livello, oppure, come illustrato in Tab. 3, introducendo un "grado di v", che indicheremo con $^\circ v$, o più brevemente con m, e definito da:

$$v = c \cdot 10^{m-7} . \qquad (4.4-4)$$

---

[1] Tali centri C_R e C_T potranno essere stabiliti secondo i vari casi nei centri di gravità dei rispettivi sistemi, o negli estremi delle linee di alimentazione delle antenne (trascurando la irradiazione di tali linee), cioè, nel caso di riflettori parabolici, in prossimità del loro fuoco.

Tab. 3. *Valori tipici della velocità*

$$m = 7 + \log_{10}(v/c)$$

$m = {}^\circ v$	$\varepsilon = v/c$	v km/sec	Esempi tipici
1	10^{-6}	0,3	Velocità, nella rotazione terrestre, per un punto circa a latitudine 45°
2	10^{-5}	3	Velocità di un veicolo satellite o di una sonda lunare
3	10^{-4}	30	Velocità di rivoluzione della Terra intorno al Sole (velocità di una sonda planetaria lenta)
4	10^{-3}	300	Velocità di una sonda planetaria veloce
5	10^{-2}	3000	Velocità di una sonda stellare lenta
6	10^{-1}	30000	Velocità di una sonda stellare veloce

Il grado $m = 1$ corrisponde ad una velocità di 1080 Km/ora, che è poco inferiore ad 1 Mach (nelle ordinarie condizioni), così m è poco inferiore al $\log_{10}$ della velocità espressa in Mach.

Nella Tab. 4 riportiamo dei valori indicativi dello scostamento di frequenza ΔF_{TR}, secondo la (4.4—2), per alcuni valori del grado m di $v_{TR, \varrho}$ e della frequenza portante F_T.

Tab. 4. *Valori tipici dello scostamento di frequenza,* ΔF_{TR}

$v_{TR, \varrho}$	0,3	3	30 Km/sec
$m = {}^\circ v_{TR, \varrho}$	1	2	3
$\varepsilon = v_{TR, \varrho}/c$	10^{-6}	10^{-5}	10^{-4}
$F_T = 100$ MHz $(\lambda = 3\,\text{m})$	100 Hz	1 KHz	10 KHz
$F_T = 10^3$ MHz $(\lambda = 0,3\,\text{m})$	1 KHz	10 KHz	100 KHz
$F_T = 10^4$ MHz $(\lambda = 3\,\text{cm})$	10 KHz	100 KHz	1 MHz

Nella considerazione del moto relativo del sistema ricevente, Σ_R, rispetto al sistema trasmettente, Σ_T, potrà convenire di considerare tale moto come risultante di più moti componenti, ad esempio come nel seguente caso, abbastanza generale, dal quale possono derivarsi altri casi particolari.

Il sistema trasmettente, Σ_T, sia installato sopra un veicolo spaziale, S_T, che abbia un moto di rivoluzione, come un satellite, intorno ad un pianeta, P_i; il sistema ricevente, Σ_R, sia a sua volta installato sopra un veicolo spaziale S_R, che abbia un moto di rivoluzione intorno ad un altro pianeta, P_j.

Indichiamo con T_{01} e T_{02} due terne di assi di riferimento, di origini O_1 e O_2; T_{02} presenti rispetto a T_{01} un moto rigido M_{12}. Le velocità, $\vec{v}_{1)Q}$ e $\vec{v}_{2)Q}$, che un punto mobile Q presenta rispetto a T_{01} e T_{02}, sono legate dal noto "principio dei moti relativi":

$$\vec{v}_{1)Q} = \vec{v}_{1)Q^2} + \vec{v}_{2)Q}, \tag{4.4—5}$$

dove $\vec{v}_{1)Q^2}$, chiamata "velocità di trascinamento" è la velocità rispetto a T_{01} di quel particolare punto, Q^2, immaginato solidale a T_{02}, e che all'istante considerato coincide con Q.

Se in particolare il moto M_{12} si riduce ad una traslazione, che indicheremo con M'_{12}, allora qualsiasi punto di T_{02} avrà rispetto a T_{01} la stessa velocità, uguale in particolare a quella $\vec{v}_{1)02}$ dell'origine O_2, e la (4.4—5) si ridurrà a:

$$\vec{v}_{1)Q} = \vec{v}_{1)02} + \vec{v}_{2)Q}. \tag{4.4—6}$$

Indicando in generale con T^* una terna che abbia orientamento invariabile rispetto alle stelle fisse, il moto di una terna T^*_{02} rispetto ad una T^*_{01} è del tipo M'_{12} ed è applicabile la (4.4—6).

Nell'esempio preso in esame, considereremo delle terne del tipo T^*, con origini nei baricentri dei veicoli spaziali, dei pianeti e del sole, che indicheremo con G_{ST}, G_{SR}, G_{Pi}, G_{Pj}, G_{SO}.

La velocità (di trascinamento) di un punto qualsiasi $Q^{(ST)}$, della terna T^*_{GST}, rispetto alla T^*_{GSO}, è data da:

$$\vec{v}_{GSO)Q^{(ST)}} = \vec{v}_{GSO)GPi} + \vec{v}_{GPi)GST}. \tag{4.4—7}$$

Ammetteremo che il moto del sistema trasmettente Σ_T rispetto alla T^*_{GST} sia approssimativamente rappresentato mediante la sola velocità, $\vec{v}_{GST)QTK}$, di un conveniente punto Q_{TK} di Σ_T; la velocità di questo punto rispetto alla T^*_{GSO}, che indicheremo brevemente con: $\vec{v}_{GSO)\Sigma T}$, sarà secondo la (4.4—6):

$$\vec{v}_{GSO)\Sigma T} = \vec{v}_{GSO)Q^{(ST)}} + \vec{v}_{GST)QTK}. \tag{4.4—8}$$

Il contributo della $\vec{v}_{GST)QTK}$ nella (4.4—8) può essere in genere trascurato, se il moto di Σ_T intorno al punto G_{ST} si effettua lentamente, tanto più nei casi in cui l'orientamento di Σ_T rispetto alle stelle fisse è mantenuto costante da adatti dispositivi giroscopici e da eventuali dispositivi di puntamento a cellule foto-elettriche o con ricevitori di radio stelle. Tuttavia la considerazione della $\vec{v}_{GST)QTK}$ potrà intervenire in particolari studi della radio propagazione.

Equazioni analoghe alle (4.4—7,8) possono essere scritte per il sistema ricevente, Σ_R, sostituendo in dette equazioni, i simboli R e P_j al posto di T e P_i.

La velocità relativa del sistema ricevente rispetto al trasmettente, che indicheremo brevemente con $\vec{v}_{TR}$, si otterrà dalla differenza:

$$\vec{v}_{TR} = \vec{v}_{GSO)\Sigma R} - \vec{v}_{GSO)\Sigma T}, \tag{4.4—9}$$

cioè, trascurando i termini $\vec{v}_{GST)QTK}$ e $\vec{v}_{GSR)QRK}$:

$$\vec{v}_{TR} \cong [\vec{v}_{GSO)GPj} + \vec{v}_{GPj)GSR}] - [\vec{v}_{GSO)GPi} + \vec{v}_{GPi)GST}]. \tag{4.4—10}$$

A questa si può dare anche la forma:

$$\vec{v}_{TR} \cong [\vec{v}_{GSO)\,GPj} - \vec{v}_{GSO)\,GPi}] + [\vec{v}_{GPj)\,GSR} - \vec{v}_{GPi)\,GST}], \qquad (4.4\text{—}11)$$

nella quale è messa in evidenza, entro la prima parentesi, la velocità relativa dei due pianeti, e nella seconda parentesi, la velocità relativa dei due veicoli spaziali, quando si facessero coincidere i centri di P_i e P_j.

Nella equazione (4.4—10) invece sono messi in evidenza nella prima parentesi i termini relativi al sistema ricevente e nella seconda parentesi quelli relativi al sistema trasmettente, e tale forma di equazione suggerisce che, nel caso di più stazioni di ascolto, una stazione centrale fornisca i valori della seconda parentesi, mentre le singole stazioni interessate all'ascolto, calcolino la rispettiva prima parentesi.

Per determinare, secondo la (4.4—1), la frequenza di ricezione F_R, data quella di trasmissione, F_T, si dovrà poi calcolare la componente, $v_{TR,\varrho}$, di $\vec{v}_{TR}$, secondo la congiungente $C_R C_T$, ciò che potrà essere fatto ad esempio in base alla conoscenza dei coseni direttori di $\vec{v}_{TR}$ e $C_R C_T$ rispetto alla T^*_{GSO}.

Potrà essere conveniente di calcolare $v_{TR,\varrho}$ come somma delle componenti secondo $C_R C_T$ dei vettori che compariscono a secondo membro della (4.4—10) o (4.4—11). Nel caso di grandi valori della distanza del trasmettitore dal ricevitore, la direzione di $C_R C_T$ resta quasi costante durante notevoli intervalli di tempo.

Ad esempio, se il ricevitore si trova sulla superficie terrestre ed il trasmettitore è a bordo di una sonda planetaria, il contributo portato alla $v_{TR,\varrho}$ dalla rotazione terrestre è calcolabile con una formula approssimata di semplice forma.

Indicando con v_{TR,ϱ,T_e} tale componente, si trova per questa la seguente espressione:

$$v_{TR,\varrho,T_e} = w_{T_e} R_{T_e} \cos \varphi_R \cos \varphi_{T_0} \operatorname{sen} (\lambda_{T_0} - \lambda_R), \qquad (4.4\text{—}12)$$

dove: w_{T_e} ed R_{T_e} sono la velocità angolare ed il raggio della Terra, φ_R e λ_R sono la latitudine e longitudine del ricevitore sulla superficie terrestre, φ_{T_0} e λ_{T_0} sono le analoghe per il punto, T_o, che diremo sub-trasmettitore, punto in cui la superficie terrestre è intersecata dalla retta che unisce il centro della Terra col trasmettitore.

Analoghe considerazioni possono essere svolte per tener conto del moto di rivoluzione degli indicati pianeti, P_i e P_j, intorno al Sole.

Così, se i pianeti P_i e P_j sono rispettivamente Marte e la Terra, supposto per approssimazione che le orbite stiano in uno stesso piano ed abbiano forma circolare, di raggi a_{Ma} ed a_{Te}, e indicando con v_{Ma} e v_{Te} le velocità di rivoluzione intorno al Sole, il contributo alla $v_{TR,\varrho}$ portato da tali moti è espresso da:

$$v_{TR,\varrho,riv} = v_{Ma} \operatorname{sen} \gamma_{Ma} - v_{Te} \operatorname{sen} \gamma_{Te} \qquad (4.4\text{—}13)$$

dove γ_{Ma} e γ_{Te} sono gli angoli che la retta Terra-Marte forma rispettivamente con le rette Sole-Marte e Sole-Terra.

La (4.4—13) si può anche scrivere:

$$v_{TR,\varrho,riv} = [1 - (a_{Te}/a_{Ma})^{3/2}] v_{Te} \operatorname{sen} \gamma_{Te} \qquad (4.4\text{—}14)$$

tenendo presente che, per le leggi di KEPLERO, risulta $v_{Ma}/v_{Te} = (a_{Te}/a_{Ma})^{1/2}$.

I valori di $v_{TR,\varrho,riv}$ possono essere riportati in effemeridi, in funzione del tempo.

In base ai valori riportati nelle Tab. 3 e 4 potrà essere valutato l'ordine di grandezza degli effetti sulla variazione di frequenza, che vengono portati da componenti come nelle (4.4—12,14).

All'uso delle denominazioni: "frequenza assoluta" e "frequenza relativa", potrà essere preferito quello di "frequenza di trasmissione" e "frequenza di ricezione".

5. Effetto dell'accelerazione del moto relativo sulle variazioni di frequenza

Questo effetto consiste in una variazione della differenza ΔF_{TR}, tra le frequenze di trasmissione e ricezione, con una rapidità che, a parità di altre condizioni, è proporzionale alla grandezza dell'accelerazione nel moto relativo del ricevitore rispetto al trasmettitore.

L'effetto deve essere tenuto presente, ad esempio, nella progettazione dei dispositivi destinati a rendere automatica la sintonia tra ricevitore e trasmettitore, specie nei casi in cui si operi con strette bande di modulazione.

Si ricorda che, secondo il teorema di CORIOLIS, l'accelerazione assoluta è somma delle accelerazioni relativa, di trascinamento e complementare.

V. Problemi particolari delle telecomunicazioni spaziali

Nei numeri precedenti sono stati esaminati alcuni aspetti generali dei problemi delle telecomunicazioni spaziali. Per ogni particolare tipo di servizio e di utilizzazione, sorgono poi numerosi problemi che interessano i vari campi speciali.

Accenneremo nel seguito ad alcuni di questi problemi, senza però entrare in dettagli.

1. Dispositivi di registrazione a bordo del veicolo spaziale

Tali dispositivi sono destinati a ottenere che l'informazione venga trasmessa differita nel tempo, eventualmente con velocità diversa, ed eventualmente solo in seguito ad un segnale di comando.

I dispositivi potranno usare i sistemi di registrazione magnetica, su nastro, filo, tamburo, disco, od altri tipi di memorie, ad esempio a ferriti.

Potrà trovare applicazioni anche la registrazione fotografica, su adatti film.

2. Dispositivi per un campione di frequenza

Alle distanze estreme di comunicazione, come è stato illustrato, l'ampiezza della banda di modulazione, che è praticamente possibile, diventa molto stretta, e ciò richiede l'attuazione di sistemi di trasmissione e ricezione a frequenza notevolmente stabile.

A distanze oltre i pianeti esterni, e specie stellari, la stabilità richiesta, come indicato in Tab. 2, entra nell'intervallo di una unità rispetto a 10^8 fino a 10^{10}.

Precisioni di frequenza di questo ordine di grandezza superano le possibilità dei sistemi di controllo con quarzo piezoelettrico, e si dovrà pensare ai cosidetti "orologi atomici", che impiegano frequenze inerenti a fenomeni molecolari; il peso di tali dispositivi può essere contenuto entro poche decine di Kg.

Il campione di frequenza, ad alta precisione, installato sulla sonda spaziale ed uno analogo, installato a Terra, permetterranno, mediante l'impiego di radio-trasmettitori e ricevitori, il confronto con adatti dispositivi tra le frequenze ricevute e quelle locali, quindi di valutare la velocità relativa di allontanamento o avvicinamento, in relazione all'effetto Doppler.

Dispositivi inerziali a bordo della sonda consentiranno, mediante integratori, l'indicazione della velocità propria, e in unione coi precedenti dispositivi di confronto per l'effetto Doppler, potranno facilitare le correzioni e la taratura a intervalli di tali integratori e inoltre l'agganciamento automatico di frequenza, per i ricevitori, specie alle grandi distanze, dove la banda di modulazione consentita diventa stretta.

Si osserva che variazioni di frequenza che abbiano andamento regolare, ad esempio per variazioni di temperatura a bordo o variazioni di velocità sull'orbita, possono in parte essere previste per estrapolazione e compensate, mentre invece variazioni irregolari della frequenza ricevuta, ad esempio per variazioni negli effetti della ionosfera, non sono suscettibili di semplice compensazione entro brevi intervalli di tempo.

3. Temperatura del sistema

La intensità dei rumori nel ricevitore dipende dalla temperatura assoluta alla quale si trovano i componenti di esso, specie nei primi stadi.

Dispositivi di raffreddamento, ammessi per alcuni componenti del ricevitore, potranno trovare soluzioni semplificate a bordo dei veicoli spaziali, con uno schermaggio da radiazioni termiche esterne.

La riduzione dei rumori potrà essere spinta fino in prossimità dei valori teorici, usando speciali amplificatori di tipo recente, come i "Maser", "Marvar", fondati su effetti elettronici nello stato solido, oppure altri tipi, ad esempio gli "amplificatori parametrici".

4. Orientamento dei dispositivi della sonda

Alcuni dispositivi della sonda richiedono per un corretto funzionamento che i loro assi vengano disposti secondo particolari direzioni, così i dispositivi per la ripresa fotografica e televisiva, i dispositivi radar per la misura di distanze, e le stesse antenne radio, di trasmissione e ricezione, per ridurre al minimo la potenza richiesta.

La sonda, all'istante in cui cessa la propulsione da parte dei razzi vettori, conserva una forza viva di rotazione, che è causa di moti di tipo precessionale della sonda intorno al suo centro di gravità.

Tali moti possono in parte venire compensati con l'azione di razzi di controllo a bordo della sonda, che potranno però agire solo per brevi intervalli di tempo e d'altra parte la sonda nel suo moto è soggetta a campi di forza perturbatori.

Assicurare a bordo una base giroscopica, che conservi orientamento costante rispetto alle stelle fisse, richiede una spesa di energia, e anche limitando le perdite per attrito nelle sospensioni cardaniche non sarà facile rifornire tale energia con semplici pile solari, d'altronde non utilizzabili in alcuni casi.

Il problema dell'orientamento della sonda si presenta così di particolare difficoltà, ed una sua soluzione soddisfacente richiederà ancora molti progressi, indispensabili d'altra parte per vari impieghi delle sonde.

5. Automazione delle sonde spaziali

Le sonde spaziali potranno essere dotate di un sistema più o meno complesso di strumenti misuratori delle condizioni fisiche esterne e interne, di apparati radio riceventi e trasmettenti, di sorgenti di potenza, di registratori, calcolatori e vari.

Gli strumenti misuratori delle grandezze fisiche e relativi trasduttori dovranno essere particolarmente progettati in relazione alle condizioni che si presentano a bordo del satellite, come accelerazioni, variazioni termiche, modi di trasmissione delle indicazioni, intervalli di misura e inoltre dovranno soddisfare contemporaneamente a requisiti di minimo peso, sicurezza di funzionamento e precisione, assoluta e relativa.

In relazione alle indicazioni da trasmettere verrà adottato l'uno o l'altro dei vari sistemi di radiotelemetria [3], e il particolare tipo di codificazione dei segnali (che potrà anche essere di tipo complesso, in modo da ottenere un significato denso per ogni Bit di informazione).

Il programma di operazioni che dovrà essere svolto dalla sonda diventa gradualmente sempre più complesso, quando sono richieste correzioni automatiche dell'orientamento, regolazioni della traiettoria, eventuali fasi di decelerazione per l'atterraggio, emissione dei segnali a intervalli di tempo prestabiliti oppure con radiocomando, registrazioni e altre operazioni.

Una sonda con un programma largo di operazioni appare allora sempre più chiaramente come una macchina dotata di un grado di automazione notevole, e potranno distinguersi per così dire organi di tipo "cervello", "memoria", "sensoriali", come in una macchina calcolatrice-operatrice a programma.

L'organo calcolatore, specie per sonde destinate alle massime distanze, dovrà essere in grado di affrontare una varietà di situazioni, che potranno richiedere l'introduzione nel calcolatore di organi dotati di una speciale logica.

Resta in questo modo indicata la strada di sviluppo delle sonde con una automazione crescente e l'importanza che avranno i relativi studi e attuazioni.

6. Sorgente di energia a bordo delle sonde

La missione che viene assegnata alla sonda può essere di breve o lunga durata, e in relazione potranno essere provviste a bordo della sonda sorgenti di energia a breve o lento esaurimento ed a ricarica, nel primo tipo potendo classificarsi le pile elettriche ed i gruppi elettrogeni.

I sistemi a ricarica, come un sistema di accumulatori alimentato da pile a effetto fotoelettrico, potranno essere convenienti per piccole potenze e quando la sonda non debba allontanarsi troppo dal Sole, poichè allora il suo rendimento tenderebbe ad annullarsi.

Sonde destinate in futuro per distanze stellari dovranno disporre di notevole potenza e per lunghi periodi, diventando elevato il rapporto energia per ogni unità di informazione ottenuta; riteniamo che lo scopo non potrà essere raggiunto che dotando la sonda di un generatore elettro-nucleare, il che implica una sonda di tonnellaggio non indifferente, che d'altronde si associa alla complessità degli impianti di cui dovranno essere dotate tali sonde.

Così una sonda per distanze estreme viene prevista come un complesso automa, con organi calcolatori, sensori e registri, ed alimentato a energia nucleare, resistente alle più diverse azioni fisiche che potrà incontrare nei mezzi attraversati.

VI. Conclusioni

E' confortevole rilevare, secondo i dati di Tab. 2, che i principali problemi di una telecomunicazione spaziale, fino a distanze stellari, si prevedono di possibile soluzione, con un adeguato impiego delle risorse che la tecnica attuale dei sistemi di comunicazione già ha raggiunto, e con qualche progresso, che si prevede potrà essere ottenuto.

I problemi inerenti alle comunicazioni spaziali sono numerosi, e molto lavoro verrà richiesto per giungere ad efficaci apparecchiature, secondo le varie utilizzazioni.

Terminiamo formulando la previsione che già prima della fine di questo secolo la prima sonda stellare potrebbe essere lanciata verso la binaria "α-Centauri", o la semplice "Proxima Centauri", mentre nell'intervallo degli anni 2000—3000 verrà proseguita mediante sonde l'esplorazione delle stelle prossime, entro una decina di anni luce, al nostro Sole.

Appendice 1
Velocità di informazione e ampiezza della banda di modulazione

Nella teoria delle comunicazioni il canale, nel quale si effettua la trasmissione dei segnali, è detto di tipo discreto o continuo, secondo che i segnali siano distinti in un numero finito di livelli di intensità, oppure siano rappresentabili mediante funzioni continue del tempo (esempi tipici rispettivamente le trasmissioni telegrafiche e musicali).

I segnali, fisici, sono associati secondo un certo codice, ai simboli dei vari messaggi, che, in detta teoria, si dicono generati da una "sorgente di informazioni", discreta o continua.

Per la misura dell'informazione che, da un punto di vista statistico, si considera fornita in media dalla scelta di un messaggio, viene impiegata la seguente formula, nel caso di una sorgente discreta e indicando con p_i la probabilità dello i-esimo messaggio:

$$I = K_{UI} \sum_i p_i \log (1/p_i) . \qquad (A. 1{-}1)$$

Il fattore K_{UI} dipende dalle unità di misura adottate per l'informazione e dalla base scelta per i logaritmi.

Una unità spesso usata è il Bit (binary information unit), che attribuisce il valore 1 all'informazione inerente alla scelta tra due messaggi, equiprobabili (ad esempio 0 od 1, "sì" o "no").

Quando l'informazione è misurata in Bit e sono considerati nella (A. 1—1) i logaritmi in base 2, il fattore K_{UI} si riduce a 1.

Se i messaggi sono formati da successioni di M segni (almeno in media), l'informazione specifica, per ciascun segno, è:

$$I_1 = I/M . \qquad \text{(Bit/segno)} \quad (A. 1{-}2)$$

Può essere anche considerata l'informazione specifica rispetto al tempo, o rapidità di informazione, $\dot{I}$; indicando con $\dot{M}$ il numero di simboli generati per secondo dalla sorgente, si ha:

$$\dot{I} = I_1 \dot{M} . \qquad \text{(Bit/sec)} \quad (A. 1{-}3)$$

Nel caso che i diversi segni, in numero di L, abbiano tutti la stessa probabilità e indipendente dalla scelta dei segni precedenti nella successione, allora i messaggi hanno tutti una stessa probabilità:

$$p = 1/L^M, \qquad (A. 1{-}4)$$

e l'informazione della sorgente è:

$$I = M \log_2 L . \qquad \text{(Bit/messaggio)} \quad (A. 1{-}5)$$

In questo caso l'informazione specifica è: $I_1 = \log_2 L$ (Bit/segno), uguale al numero di cifre (binarie) occorrenti per esprimere il numero (decimale) L in base 2 (ad esempio, se $L = 32$, allora $I_1 = 5$).

Le precedenti formule sono suscettibili di estensione al caso di sorgenti continue di informazione.

Si dimostra che, indicando con $n(q)$ il numero dei messaggi, di lunghezza N, che occorre prendere per ottenere una probabilità totale q, quando i messaggi sono stati disposti nell'ordine di probabilità decrescenti, si ha:

$$I_1 = \lim_{N \to \infty} (1/M) \log_2 n(q) \ . \qquad \text{(Bit/segno)} \quad \text{(A. 1—6)}$$

Una espressione, simile nella forma alla precedente, è impiegata per definire la capacità, C, di un canale discreto:

$$C = \lim_{T = \infty} (1/T) \log_2 n(T) \ , \qquad \text{(Bit/sec)} \quad \text{(A. 1—7)}$$

dove $n(T)$ è il numero delle diverse serie di segnali permesse, considerando serie che abbiano ciascuna la stessa durata complessiva T, dei segnali.

Si dimostra che, per un canale privo di disturbi, la massima rapidità di trasmissione dell'informazione, che può essere raggiunta con una adatta codificazione, uguaglia C:

$$\dot{I}_{max} = C \ . \qquad \text{(Bit/sec)} \quad \text{(A. 1—8)}$$

Nel caso di un canale affetto da disturbi, l'informazione I della sorgente subisce nella trasmissione una perdita, che indicheremo con I_a, chiamata "equivocazione".

Il valore di I_a, che misura l'ambiguità del segnale ricevuto, può essere calcolato con la formula per la così detta informazione condizionale media, $I_{x,y}$, relativa al messaggio (variabile) X_i in entrata, quando è noto il messaggio (variabile) Y_i in uscita.

Detta formula è:

$$I_a = I_{x,y} = \sum_{i,j} p_{i,j} \log (1/p_{i,j}) \ , \qquad \text{(A. 1—9)}$$

essendo $p_{i,j}$ la probabilità condizionale che X sia uguale a X_i, quando è prefissato Y_i, mentre $p_{i,j}$ è la probabilità (non condizionata) dell'evento congiunto X_i e Y_i.

Nel caso di un canale con disturbi, la capacità di trasmettere informazioni è definita, analogamente alla (A. 1—8), da:

$$C = (\dot{I} - \dot{I}_a)_{max} \ , \qquad \text{(Bit/sec)} \quad \text{(A. 1—10)}$$

essendo il massimo ricercato rispetto a tutte le sorgenti, che possono essere usate come entrata nel canale.

Si dimostra che con una adatta codificazione, può essere raggiunta la rapidità di informazione C della (A. 1—10), con errori arbitrariamente piccoli.

Le precedenti formule sono suscettibili di estensione al caso di sorgenti continue di informazione e canali continui.

Diremo che la funzione $f(t)$ del tempo è limitata alla banda $0 \vdash F_M$, quando risulta dalla combinazione lineare di termini sinusoidali di frequenza non eccedente F_M; in tale caso $f(t)$ è determinata assegnando le sue ordinate per una serie di valori di t intervallati di $1/2 \ F_M$ (entro l'intervallo T di tempo).

Si dimostra, formula di HARTLEY — TULLER — SHANNON [12, 13], che la capacità di un canale, che ammette frequenze (di modulazione) nella banda

$0 \multimap F_M$, ed è disturbato da un "disturbo termico bianco" (cfr. n. I.3) di potenza media N e quando la potenza media del segnale non supera P, è data da:

$$C = F_M \log_2 (1 + P/N). \qquad\qquad \text{(Bit/sec)} \quad \text{(A. 1—11)}$$

Una formula modificata per la capacità di un canale in presenza di disturbi è stata sviluppata da D. Gabor [14], tenendo conto del ruolo dei battimenti, quando è fissata la potenza massima dei segnali, P_M, invece della potenza media.

La formula di Gabor per la capacità è:

$$C_G = k_{UI} F_M \log_2 [(1/2) + (1/2) (1 + 2 P_M/P_2)^{1/2}], \qquad (A. 1—12)$$

dove P_2 è la potenza del disturbo.

Per valori piccoli del rapporto segnale-disturbo, le formule (A. 1—11, 12) portano allo stesso risultato, mentre invece per grandi valori di tale rapporto la capacità secondo la (A. 1—12) cresce più lentamente di quanto indica la formula di Shannon.

Per intervalli di tempo abbastanza grandi, l'effetto dei battimenti può essere trascurato.

Bibliografia

1. G. Armellini, I fondamenti scientifici dell'astrofisica, XIV-345. (Cfr. Cap. 2, n. 15, p. 45.) Milano: Hoepli, 1953.
2. R. P. Haviland, The Communication Satellite. Proceedings of the VIIIth International Astronautical Congress, Barcelona 1957, p. 543. Wien: Springer, 1958.
3. M. H. Nichols and L. R. Rauch, Radiotelemetry, IX-461. New York: J. Wiley, 1956.
4. J. B. Johnson, Thermal Agitation of Electricity in Conductors. Phys. Rev. **32**, 97 (1928).
5. H. Nyquist, Thermal Agitation of Electric Charge in Conductors. Phys. Rev. **32**, 110 (1928).
6. J. L. Lawson and G. E. Uhlenbeck, Threshold Signals. Radiation Laboratory Series, n. 24. New York: McGraw-Hill, 1950.
7. P. Grivet et A. Blaquière, Le bruit de fond, X-495. Paris: Masson et Cie., 1958.
8. R. Hanbury Brown and A. C. B. Lovell, The Exploration of Space by Radio, XII-207. London: Chapman and Hall Ltd., 1957.
9. C.C.I.R. (Comité Consultatif International des Radiocommunications), Documents de la VIIe Assemblée Plenière, Londres 1953, Vol. I, Specie pp. 96-105.
10. Donald G. Fink, Radar Engineering, XII-644. New York: McGraw-Hill, 1947.
11. H. Bouasse, Théorie de l'emission, XXIV-439. Paris: Libraire Delagrave, 1925.
12. C. E. Shannon and W. Weaver, The Mathematical Theory of Communication, p. 117. (Specialmente pp. 67-70.) Urbana, Ill.: University of Illinois Press, 1949.
13. L. Brillouin, Science and Information Theory, XVII-320. (Specialmente pp. 247-257.) New York: Academic Press, 1956.
14. D. Gabor, Philos. Mag. (7) **41**, 1161 (1950).

Fuel Requirements for Inter-Orbital Transfer of a Rocket

By

R. N. A. Plimmer[1]

(With 13 Figures)

(*Received July 21, 1959*)

Abstract — Zusammenfassung — Résumé

Fuel Requirements for Inter-Orbital Transfer of a Rocket. This paper is concerned with coplanar orbit-to-orbit transfers and in particular with minimal fuel transfers. LAWDEN's equations governing an optimal fuel transfer are known to have analytic solutions for certain special cases but can only be solved in general using numerical methods. Optimal transfers between circular orbits (HOHMANN transfers), between an elliptical orbit and a circular orbit, between elliptical orbits having a common direction of major axes and the same eccentricity and, finally, between elliptical orbits which are identical except for the direction of their major axes are all discussed in detail, the field of all possible initial and final terminal orbits being considered. A comparison of the fuel requirements is made between these optimal transfers and certain other modes of non-optimal transfer.

Treibstofferfordernisse für Raketenübergänge zwischen Planetenbahnen. Die vorliegende Arbeit befaßt sich mit in derselben Ebene liegenden Übergängen zwischen Planetenbahnen, im besonderen solchen mit kleinstem Treibstoffverbrauch. Es ist bekannt, daß LAWDENS Gleichungen für Übergänge mit optimalem Treibstoffverbrauch analytische Lösungen für bestimmte Spezialfälle haben, jedoch in allgemeiner Form nur durch Anwendung numerischer Methoden lösbar sind. Eingehend erörtert werden optimale Übergänge zwischen Kreisbahnen (HOHMANN-Übergänge), zwischen einer elliptischen und einer kreisförmigen Bahn, zwischen elliptischen Bahnen mit gleichgerichteten großen Achsen und derselben Exzentrizität und schließlich zwischen elliptischen Bahnen, die mit Ausnahme der Richtung ihrer großen Achsen identisch sind. Dabei wird der Bereich aller möglichen Anfangs- und schließlichen Endbahnen in Betracht gezogen. Es werden die Treibstofferfordernisse zwischen diesen optimalen Übergängen und bestimmten anderen Arten des nicht-optimalen Überganges verglichen.

Les exigences en carburant pour le transfert inter-orbitial d'une fusée. Cet article concerne principalement les transferts coplanaires d'orbite à orbite et en particulier les transferts à consommation minimum. Les équations de LAWDEN gouvernant un transfert optimum admettent des solutions analytiques dans certains cas spéciaux mais ne peuvent être résolues en général que par des méthodes numériques. Les transferts optimums entre des orbites circulaires (transferts d'après HOHMANN), entre une orbite circulaire et une orbite elliptique, entre orbites elliptiques ayant une direction commune des axes principaux et la même excentricité et finalement entre des orbites elliptiques identiques excepté pour la direction de leurs axes

---

[1] Guided Weapons Department, Royal Aircraft Establishment, South Farnborough, Hants., United Kingdom.

principaux sont tous discutés en détail; les champs de tous les orbites initiales et finales étant considérés. La comparaison est faite entre ces transferts optimums et certains autres modes de transferts non-optimums.

I. Introduction

This paper presents a preliminary investigation into the fuel requirements for the optimal transfer of a rocket between two terminal orbits described under a central inverse square law field of attraction, the terminal orbits both being described in the same sense. Such trajectories are likely to be of importance in adjusting the orbit of an earth satellite to correct initial guidance errors, or in changing the orbit in order to make maximum use of the scientific payload. Transfer orbits are also of interest when it is desired to bring a number of rockets into coincidence — as, for example, in the build up of a space platform — and, of course, for interplanetary travel. The word rocket will refer to any vehicle which is powered by a jet motor. Only coplanar orbits will be considered in any detail since the more general equations of transfer for the corresponding three dimensional problem have yet to be deduced, although this problem will be dealt with in a later paper.

In 1925 HOHMANN [1] made an analysis of the problem of transferring a rocket from one circular orbit into another coplanar with the first and about the same attracting body. His results showed that the most economical fuel transfer path is along an elliptical orbit tangential at its apses to both circular orbits; fuel is expended rapidly at each apse in order to provide an impulsive thrust sufficient to carry the rocket into or out of the terminal orbits, the rocket being in a state of free fall between these impulses. Such a mode of transfer now bears his name.

In more recent years, LAWDEN [2, 3] has obtained the equations to be satisfied for the more general situation where the terminal orbits are coplanar ellipses. The form of optimal transfer is similar to a HOHMANN transfer in that impulses are applied at the terminal orbits, the rocket being solely under the influence of the attracting centre between these impulses. Although it is not possible, in general, to solve these equations analytically and recourse must be made to numerical methods, for many interesting cases a relatively simple form of solution may be obtained. It is with these solutions that we shall be concerned in this paper.

II. Fundamental Equations of Minimal Fuel Transfer for Coplanar Orbits

For ease of reference the notation used here corresponds to that used by LAWDEN in [2]. The reader is referred to this paper for the actual derivation of the equations occurring in this section. Using polar co-ordinates (r, θ) with respect to the centre of attraction and in the plane of the two terminal orbits, any elliptic orbit in this plane may be expressed in the form

$$u = \alpha^2 + \alpha\beta \cos(\theta + \gamma) \tag{1}$$

where $u = 1/r$ and α, β, γ are constants determining the orbit; we shall refer to this orbit as the orbit (α, β, γ). The reciprocal of α^2 is the semi-latus rectum l of the ellipse and β/α is equal to the eccentricity e. γ is the anti-clockwise perigee angle measured from the arbitrary reference line corresponding to $\theta = 0$ (Fig. 1). The semi-major axis, a, of the ellipse is equal to $1/\alpha^2 (1 - e^2)$.

In general, if $(\alpha_1, \beta_1, \gamma_1)$ and $(\alpha_2, \beta_2, \gamma_2)$ are the two terminal orbits, they will not intersect and the optimum mode of transfer is achieved by the application of a number of impulsive thrusts (at least two). The elements (α, β, γ) of the orbit will vary during the application of the thrust in a manner depending upon the magnitude and direction of the thrust and also the point of application on the orbit. They are related by the two equations

$$\alpha \beta \cos (\theta + \gamma) = u - \alpha^2 \qquad (2)$$

$$\alpha \beta \sin (\theta + \gamma) = (u - A \alpha) \tan \varphi \qquad (3)$$

where u, θ, A, φ are constants having the following physical significance:

(i) $(1/u, \theta)$ is the point at which the impulse is applied.

(ii) φ is the angle (Fig. 1) between the direction of the applied thrust and the normal to the radial vector at the point $(1/u, \theta)$.

(iii) If μu^2 is the attraction per unit mass, then $\mu^{1/2} A \sin \varphi$ is the component of rocket velocity in the direction normal to the thrust which, of course, remains constant during the short period of the thrust.

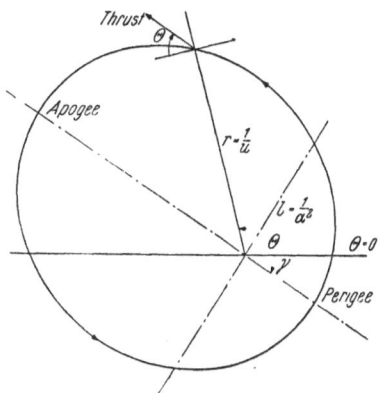

Fig. 1. Notation for elliptic orbit (α, β, γ) and thrust direction φ

It is seen that these four constants specify any general thrust on an orbit.

For transfer by two impulses, let $(u_1, \theta_1, A_1, \varphi_1)$ specify the first impulse on the orbit $(\alpha_1, \beta_1, \gamma_1)$ and $(u_2, \theta_2, A_2, \varphi_2)$ the second impulse on the terminal orbit $(\alpha_2, \beta_2, \gamma_2)$. During the period between these two impulses the rocket will orbit in free flight along an ellipse (α, β, γ). Because of the relations (2) and (3), the following boundary conditions must therefore be satisfied for such a general transfer

$$\alpha_1 \beta_1 \cos (\theta_1 + \gamma_1) = u_1 - \alpha_1^2 \qquad (4)$$
$$\alpha_1 \beta_1 \sin (\theta_1 + \gamma_1) = (u_1 - A_1 \alpha_1) \tan \varphi_1 \qquad (5)$$
$$\alpha \beta \cos (\theta_1 + \gamma) = u_1 - \alpha^2 \qquad (6)$$
$$\alpha \beta \sin (\theta_1 + \gamma) = (u_1 - A_1 \alpha) \tan \varphi_1 \qquad (7)$$
$$\alpha \beta \cos (\theta_2 + \gamma) = u_2 - \alpha^2 \qquad (8)$$
$$\alpha \beta \sin (\theta_2 + \gamma) = (u_2 - A_2 \alpha) \tan \varphi_2 \qquad (9)$$
$$\alpha_2 \beta_2 \cos (\theta_2 + \gamma_2) = u_2 - \alpha_2^2 \qquad (10)$$
$$\alpha_2 \beta_2 \sin (\theta_2 + \gamma_2) = (u_2 - A_2 \alpha_2) \tan \varphi_2. \qquad (11)$$

Using the methods of the calculus of variations in order to minimise the fuel expenditure, [2] determines the WEIERSTRASS-ERDMANN corner conditions which must be satisfied at the discontinuities in the elements (α, β, γ) produced by the impulses. These conditions state that the expressions

$$\left(\frac{u + \alpha^2}{A \alpha} + 1 \right) \cos \varphi,$$

$$\left(1 + \frac{\alpha}{A} \right) (u^2 - \alpha^2) \cos \varphi + (u - A \alpha) \tan \varphi \sin \varphi \qquad (12)$$

and

$$\left(\frac{u}{A} - A \right) \sin \varphi$$

are continuous from the end of the first impulse to the beginning of the second impulse. Hence optimal transfer is affected if the equations

$$\left(\frac{u_1 + \alpha^2}{A_1 \alpha} + 1\right) \cos \varphi_1 = \left(\frac{u_2 + \alpha^2}{A_2 \alpha} + 1\right) \cos \varphi_2, \tag{13}$$

$$\left(1 + \frac{\alpha}{A_1}\right)(u_1 - \alpha^2) \cos \varphi_1 + (u_1 - A_1 \alpha) \tan \varphi_1 \sin \varphi_1 =$$

$$= \left(1 + \frac{\alpha}{A_2}\right)(u_2 - \alpha^2) \cos \varphi_2 + (u_2 - A_2 \alpha) \tan \varphi_2 \sin \varphi_2 \tag{14}$$

and

$$\left(\frac{u_1}{A_1} - A_1\right) \sin \varphi_1 = \left(\frac{u_2}{A_2} - A_2\right) \sin \varphi_2 \tag{15}$$

are satisfied. The eleven eqs. (4) to (11) and (13) to (15) constitute the LAWDEN equations for optimal transfer via two impulses and are sufficient to determine the eleven unknowns (α, β, γ), $(u_1, \theta_1, A_1, \varphi_1)$ and $(u_2, \theta_2, A_2, \varphi_2)$.

The principles outlined above may easily be extended to cover any number of impulses merely by application of the boundary conditions (2) and (3) and the continuity eqs. (12) between impulses. For example, for transfer via three impulses, eighteen equations may be written down sufficient for the determination of the eighteen unknowns specifying the three impulses and the two intermediate transfer paths. This solution, although giving a mathematically stationary value for the fuel expenditure, does not necessarily represent an absolute minimum. No more than two impulse transfers will be considered in this paper since it is believed, although not established, that two impulse transfer leads to the absolute minimum fuel expenditure for coplanar orbits.

For any impulse (u, θ, A, φ) the amount of fuel used is determined by the equation

$$\mu^{-1/2} c \log \frac{m}{m'} = u \left|\left(\frac{1}{\alpha} - \frac{1}{\alpha'}\right) \sec \varphi\right|$$

where m is the mass of the rocket before the impulse, m' its mass afterwards and c is the jet velocity relative to the rocket (assumed constant). Having solved the equations of transfer it is then an easy matter to find the total fuel used by substitution in this equation for each of the impulses. More explicitly, for a two impulse transfer, the fuel expenditure is given by

$$\mu^{-1/2} c \log \frac{m}{m'} = u_1 \left|\left(\frac{1}{\alpha_1} - \frac{1}{\alpha}\right) \sec \varphi_1\right| + u_2 \left|\left(\frac{1}{\alpha_2} - \frac{1}{\alpha}\right) \sec \varphi_2\right| \tag{16}$$

$$= \alpha_1 F, \text{ say,}$$

where m' is now the final mass of the rocket after the second impulse. We will call F, which is obviously dimensionless and independent of the scale of the configuration, the fuel factor coefficient.

For purposes of fuel comparison later, it is useful to have some quantity which characterizes the terminal orbits. The most obvious choice is the total (gravitational) energy of an orbit. Now if E is the total energy per unit mass, then

$$E = -\frac{\mu}{2a},$$

or, in terms of α and the eccentricity of the orbit

$$E = -\frac{1}{2} \mu \alpha^2 (1 - e^2).$$

Consequently, if the ratio of the major axes of the two terminal orbits is n^2, i.e. $a_1 = n^2 a_2$, then

$$E_2 = n^2 E_1.$$

Also,

$$\alpha_1 F = \frac{F}{\sqrt{a_1 (1 - e_1{}^2)}} = F \sqrt{\frac{-2 E_1}{\mu (1 - e_1{}^2)}} = \frac{F}{n} \sqrt{\frac{-2 E_2}{\mu (1 - e_1{}^2)}}$$

$$= N^{12}(e_1) \sqrt{\frac{-2 E_1}{\mu}} = N^{21}(e_1) \sqrt{\frac{-2 E_2}{\mu}}$$

where

$$N^{12}(e_1) = \frac{F}{\sqrt{1 - e_1{}^2}}$$

and

$$N^{21}(e_1) = \frac{F}{n \sqrt{1 - e_1{}^2}}$$

so that

$$n N^{21}(e_1) = N^{12}(e_1).$$

The quantities $N^{12}(e_1)$ and $N^{21}(e_1)$ will be called the normalized fuel factor coefficients since they give a measure of the fuel required to effect transfer from an initial orbit of known total gravitational energy per unit mass E_0, say. For example, if transfer is from orbit 1 to orbit 2, then E_1 will be equal to E_0 and

$$c \log \frac{m}{m'} = N^{12}(e_1) \sqrt{-2 E_0}$$

whereas for transfer from orbit 2 to orbit 1, E_2 will be equal to E_0 and then

$$c \log \frac{m}{m'} = N^{21}(e_1) \sqrt{-2 E_0}.$$

If the argument of $N(e_1)$ is omitted it will be understood that orbit 1 is circular, i.e. $e_1 = 0$, and then $n N^{21} = N^{12} = F$. It is clear from the definition of F and N that both these quantities increase monotonically with increasing fuel expenditure.

In certain circumstances it is possible for the two terminal orbits to intersect. If this is the case, they will intersect in two points and then a single impulse of the correct magnitude and direction at either of these points is obviously sufficient to achieve a transfer. Although this mode of transfer again represents a mathematically stationary value for the fuel expenditure, it does not, as will be evident later, correspond to an absolute optimal mode. For this particular mode of transfer the fuel requirement [3] is given by

$$\mu^{-1/2} c \log \frac{m}{m'} = \sqrt{\beta_1{}^2 + \beta_2{}^2 - 2 \beta_1 \beta_2 \cos (\gamma_1 - \gamma_2) - (\alpha_1 - \alpha_2)^2 \left(1 + \frac{2 u}{\alpha_1 \alpha_2}\right)}$$

$$= \alpha_1 F_s, \tag{17}$$

the points of intersection of the terminal ellipses being obtained from the relationship

$$u = \alpha_1{}^2 + \alpha_1 \beta_1 \cos (\theta + \gamma_1) = \alpha_2{}^2 + \alpha_2 \beta_2 \cos (\theta + \gamma_2). \tag{18}$$

In the following sections, the normalised fuel factor coefficients are determined for various terminal orbit configurations.

III. Hohmann Transfers

This mode of transfer refers to the optimal trajectory between two concentric circles. The terminal orbits are thus described by $r = 1/\alpha_1{}^2$, $\beta_1 = 0$ and $r = 1/\alpha_2{}^2$, $\beta_2 = 0$, γ_1 and γ_2 being indeterminate. Let $\alpha_2 = n\,\alpha_1$ where $n < 1$ so that suffix 1 denotes the inner circle. Obviously at least two impulses are required to effect transfer. In the case of two impulses, eqs. (5) and (11) become

$$(\alpha_1 - A_1)\tan\varphi_1 = 0 \tag{19}$$
$$(\alpha_2 - A_2)\tan\varphi_2 = 0. \tag{20}$$

The possible solutions $A_1 = \alpha_1$ or $A_2 = \alpha_2$ of eqs. (19) and (20) lead to contradictions with eqs. (4) to (15) and consequently we must have $\varphi_1 = \varphi_2 = 0$ for transfer from circle 1 to 2. The alternative solution $\varphi_1 = \varphi_2 = \pi$ would correspond to transfer from circle 2 to circle 1. Eq. (15) is now satisfied and eqs. (13) and (14) merely serve to determine the constants A_1 and A_2. The remaining eqs. (4) to (11) give the relations

$$u_1 = \alpha_1{}^2, \qquad\qquad \theta_1 + \gamma = 0$$
$$u_2 = \alpha_2{}^2 = n^2\alpha_1{}^2, \qquad \theta_2 + \gamma = \pi$$

and

$$\alpha^2 = u_1 - \alpha\,\beta$$
$$\alpha^2 = u_2 + \alpha\,\beta$$

so that

$$2\,\alpha^2 = \alpha_1{}^2 + \alpha_2{}^2 = \alpha_1{}^2\,(1 + n^2) \tag{21}$$

with $\alpha_1 > \alpha > \alpha_2$. The eccentricity of the intermediate ellipse is

$$e' = \frac{\beta}{\alpha} = \frac{u_1 - u_2}{u_1 + u_2} = \frac{1 - n^2}{1 + n^2}$$

and it may be deduced that the constants A_1 and A_2 are given by

$$A_1 = -\,\alpha_1\frac{\sqrt{2\,(1 + n^2)^3}}{1 + 3\,n^2}, \qquad A_2 = -\,\alpha_1\frac{\sqrt{2\,(1 + n^2)^3}}{3 + n^2}.$$

This trajectory, the HOHMANN trajectory, thus corresponds to any ellipse which is cotangential to the two circles, tangential impulses being applied at each of its apses in order for the rocket to acquire the necessary velocity increments. It now follows by use of eqs. (16) and (21) that fuel expended for a HOHMANN transfer is given by

$$\mu^{-1/2}c\log\frac{m}{m'} = \alpha_1{}^2\left(\frac{1}{\alpha} - \frac{1}{\alpha_1}\right) + \alpha_2{}^2\left(\frac{1}{\alpha_2} - \frac{1}{\alpha}\right)$$
$$= \alpha_1\left[(n - 1) + (1 - n^2)\sqrt{\frac{2}{1 + n^2}}\right] \qquad (n \leqslant 1) \tag{22}$$

and the fuel factor coefficient is

$$F_H = n - 1 + (1 - n^2)\sqrt{\frac{2}{1 + n^2}}.$$

The corresponding expression for the fuel expenditure when $n > 1$ may be deduced from eq. (22) by noting that for transfer from circle 2 to circle 1, the transfer orbit would be along the same path as that for transfer from circle 1 to 2 if the velocities of the rocket in the transfer and terminal orbits were reversed in direction, the physical difference being that the direction of the impulses at the apses would be reversed in order to obtain a decrease in velocity, although the fuel used would be the same. Mathematically this corresponds to the transformation of α_1, α_2, and n in eq. (22) being replaced by α_2, α_1 and $1/n$ respectively. Hence for $n > 1$ the fuel requirement is given by

$$\mu^{-1/2} c \log \frac{m}{m'} = \alpha_2 F_H \left(\frac{1}{n}\right) = n \alpha_1 F_H \left(\frac{1}{n}\right)$$

$$= \alpha_1 \left[(n^2 - 1) \sqrt{\frac{2}{1 + n^2} + 1 - n} \right] \qquad (n \geqslant 1). \qquad (23)$$

The normalized fuel factor coefficient, $N_H{}^{12} = F_H$, for a HOHMANN transfer is shown plotted as a function of n^2 in Fig. 2. Two general points of interest are evident from this graph. First of all, it is seen that as the final circular orbit is increased in size $(n < 1)$ more fuel is required to effect optimal transfer until this orbit is about ten times as large as the initial orbit, after which the fuel requirement decreases. This is due to the fact that, although the fuel expenditure at the first impulse always increases as the final orbit is made larger, the fuel required at the second impulse in order to bring the rocket into the final orbit eventually reaches a maximum value after which it decreases. Secondly, to effect transfer from a given circular orbit to another circular orbit k times as small requires a greater expenditure of fuel than to effect transfer to a circular orbit k times as large.

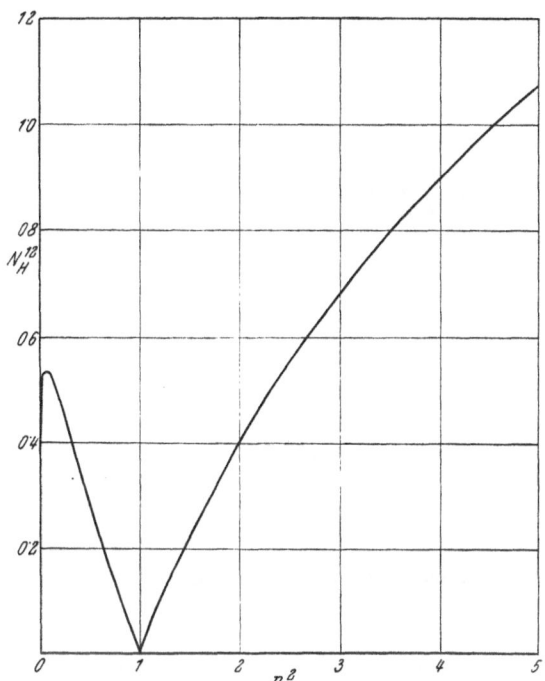

Fig. 2.   Normalized fuel factor coefficient for a HOHMANN transfer

IV. Transfer between Circular and Elliptical Orbits

After the HOHMANN transfers, the next logical configuration to investigate is that of transfer by means of two impulses between a circle $r = 1/\alpha_1{}^2$, $\beta_1 = 0$, and an ellipse $(\alpha_2, \beta_2, 0)$ of eccentricity e, there being no lack of generality in taking γ_2 zero. As before, we will denote the ratio of the semi-major axes by n^2 so that

$$a_1 = n^2 a_2$$

or, in terms of the α's,

$$n^2 \alpha_1{}^2 = \alpha_2{}^2 (1 - e^2). \qquad (24)$$

There are now several possible configurations to consider depending upon the relative magnitudes of n and the eccentricity e. These may be divided conveniently into the following three groups:

(i) $n^2 < 1 - e$. In this case the ellipse lies entirely outside the circle, its perigee touching the circle when $n^2 = 1 - e$.

(ii) $1 - e < n^2 < 1 + e$. The ellipse intersects the circle in two points, its apogee touching the circle when $n^2 = 1 + e$.

(iii) $1 + e < n^2$. For this configuration the circle lies entirely outside the ellipse.

These configurations are illustrated in Fig. 3.

Since $\beta_1 = 0$, eq. (5) implies that

$$(\alpha_1 - A_1) \tan \varphi_1 = 0$$

regardless of the eccentricity of the ellipse. The solution $A_1 = \alpha_1$ cannot be allowed because we know that for $e = 0$ (HOHMANN transfer), this leads to contradiction with the other equations. Hence we must again accept the solution $\varphi_1 = 0$ or π i.e., a tangential impulse must be applied on the circular orbit. With either of these values for φ_1, eq. (15) then states that

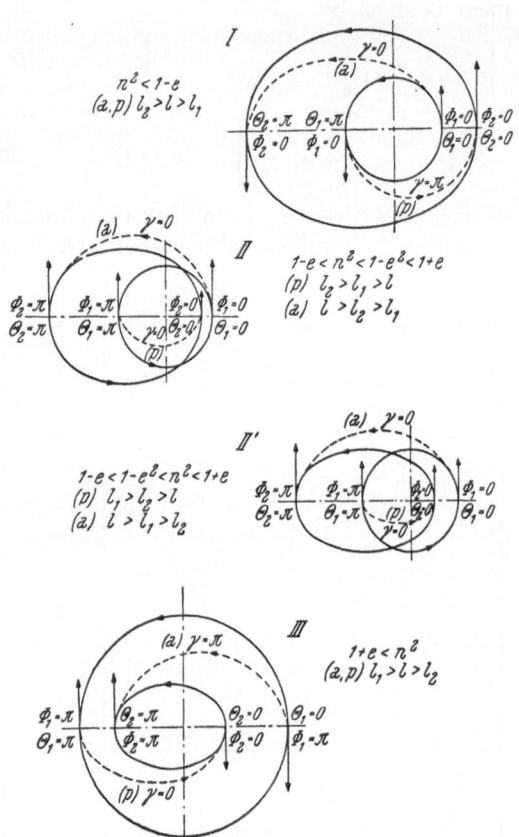

Fig. 3. Various configurations for transfer from circle to ellipse

$$(u_2 - A_2{}^2) \sin \varphi_2 = 0$$

and for a similar reason to the above, namely that $A_2{}^2 \neq u_2$ even for $e = 0$, the only allowable solutions are $\varphi_2 = 0$ or π. Eqs. (4) to (11) now imply that

$$\left.\begin{aligned}
u_1 &= \alpha_1{}^2 \\
\theta_2 &= 0 \text{ or } \pi, & u_2 &= \alpha_2{}^2 (1 \pm e) \\
\theta_1 + \gamma &= 0 \text{ or } \pi, & u_1 &= \alpha^2 \pm \alpha \beta \\
\theta_2 + \gamma &= 0 \text{ or } \pi, & u_2 &= \alpha^2 \pm \alpha \beta,
\end{aligned}\right\} \quad (25)$$

the signs to be attached to the various equations depending upon the configuration of the terminal orbits and the mode of transfer (two modes are possible for any configuration). To be more specific, consider the particular situation when the elliptic orbit lies outside the circular orbit i.e., $n^2 < 1 - e$ and transfer is from the circle to the ellipse. This is the situation corresponding to (i) in Fig. 3. The two appropriate solutions of eqs. (25) which are consistent with this configuration are:

$$
\left.
\begin{aligned}
(p)\ & \varphi_1 = \varphi_2 = 0 \\
& \theta_1 = \pi, \qquad \theta_2 = 0, \qquad \gamma = \pi \\
& u_1 = \alpha_1{}^2, \qquad u_2 = \alpha_2{}^2(1+e) = \frac{n^2 \alpha_1{}^2}{1-e} \\
& u_1 = \alpha^2 + \alpha\beta, \ u_2 = \alpha^2 - \alpha\beta
\end{aligned}
\right\} \quad (26)
$$

so that

$$
\alpha^2 = \frac{1}{2}(u_1 + u_2) = \frac{\alpha_1{}^2(1-e+n^2)}{2(1-e)},
$$

and

$$
\left.
\begin{aligned}
(a)\ & \varphi_1 = \varphi_2 = 0 \\
& \theta_1 = 0, \qquad \theta_2 = \pi, \qquad \gamma = 0 \\
& u_1 = \alpha_1{}^2, \qquad u_2 = \alpha_2{}^2(1-e) = \frac{n^2 \alpha_1{}^2}{1+e} \\
& u_1 = \alpha^2 + \alpha\beta, \qquad u_2 = \alpha^2 - \alpha\beta
\end{aligned}
\right\} \quad (27)
$$

so that

$$
\alpha^2 = \frac{1}{2}(u_1 + u_2) = \frac{\alpha_1{}^2(1+e+n^2)}{2(1+e)}.
$$

Solution (26) shows that the transfer ellipse is tangential at the perigee of the terminal ellipse, whereas (27) corresponds to a transfer ellipse tangential at apogee, both modes being tangential to the initial circular orbit. For both situations it is easily deduced that

$$
\frac{1}{\alpha_2} > \frac{1}{\alpha} > \frac{1}{\alpha_1} \qquad (n^2 < 1 - e).
$$

This is also immediately apparent from the geometry of the system when one recalls that the semi-latus rectum $l = 1/\alpha^2$. Substitution of the solutions (26) and (27) into eq. (16) gives that the corresponding fuel factor coefficients are $(e_1 = 0)$

$$
\left.
\begin{aligned}
F_p = N_p{}^{12} &= \frac{1-e-n^2}{1-e}\sqrt{\frac{2(1-e)}{1-e+n^2}} - 1 + n\sqrt{\frac{1+e}{1-e}} \\[2mm]
F_a = N_a{}^{12} &= \frac{1+e-n^2}{1+e}\sqrt{\frac{2(1+e)}{1+e+n^2}} - 1 + n\sqrt{\frac{1-e}{1+e}}
\end{aligned}
\right\} \quad
\begin{aligned} (n^2 < 1 - e) \end{aligned} \quad (28)
$$

The modes of transfer corresponding to the other configurations are all shown in Fig. 3 and it is now quite straightforward to form the correct solutions to eqs. (25). The distinction made in Fig. 3 (ii), due to the relative positioning of n^2 in the inequality $1 - e < 1 - e^2 < 1 + e$, effects the relative size of the latera recta of the terminal orbits but does not modify the equations for the fuel consumption. The fuel factor coefficients for the various cases are summarized below:

$$
\left.
\begin{aligned}
F_p = N_p{}^{12} &= -\sqrt{\frac{2(1-e+n^2)}{1-e}} + 1 + n\sqrt{\frac{1+e}{1-e}} \\[2mm]
F_a = N_a{}^{12} &= \sqrt{\frac{2(1+e+n^2)}{1+e}} - 1 - n\sqrt{\frac{1-e}{1+e}}
\end{aligned}
\right\} \quad
\begin{aligned} (1 - e < n^2 < 1 + e) \end{aligned}
$$

$$
(29)
$$

$$F_p = N_p{}^{12} = -\frac{(1 - e - n^2)}{1 - e}\sqrt{\frac{2(1 - e)}{1 - e + n^2}} + 1 - n\sqrt{\frac{1 + e}{1 - e}}$$

$$(1 + e < n^2)$$

$$F_a = N_a{}^{12} = -\frac{(1 + e - n^2)}{1 + e}\sqrt{\frac{2(1 + e)}{1 + e + n^2}} + 1 - n\sqrt{\frac{1 - e}{1 + e}}$$

$$(30)$$

For the particular case when the terminal orbits touch, the above equations reduce to the following simple forms;

$$F_a = F_p = \sqrt{1 + e} - 1$$

$$(n^2 = 1 - e)$$

$$(31)$$

$$F_a = F_p = 1 - \sqrt{1 - e}$$

$$(n^2 = 1 + e)$$

$$(32)$$

and only one impulse is necessary to effect transfer.

When the terminal orbits intersect, case (ii), transfer may be effected by the application of a single impulse at either point of intersection corresponding to the solution

$$u = \alpha_1{}^2 = \alpha_2{}^2 (1 + e \cos\theta),$$
$$\beta_1 = 0,$$
$$\gamma_1 = \gamma_2 = 0,$$
$$n^2 \alpha_1{}^2 = \alpha_2{}^2 (1 - e^2),$$
$$\beta_2 = e \alpha_2$$

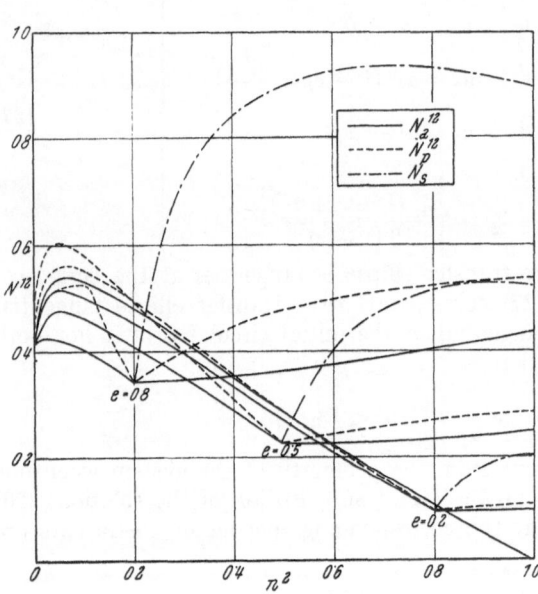

Fig. 4. Normalized fuel factor coefficients for transfer from a circle to an ellipse

of eq. (18). The fuel factor coefficient for single impulse transfer, eq. (17), then reduces to

$$F_s = \sqrt{3 - n^2 - \frac{2\sqrt{1 - e^2}}{n}}. \qquad (33)$$

Now it is not immediately apparent from inspection of eqs. (28), (29), (30) and (33) which mode of transfer is the most economical for any particular configuration. This may be determined in the following manner. For illustration, let us consider eqs. (30). Expansion of these equations in terms of the eccentricity leads to the equations

$$F_a = F_H + e\,n\left[1 - \sqrt{\frac{2}{1 + n^2}\frac{n(n^2 + 3)}{2(1 + n^2)}}\right] + 0\,(e^2)$$

and

$$F_p = F_H - e\,n \left[1 - \sqrt{\frac{2}{1 + n^2} \frac{n\,(n^2 + 3)}{2\,(1 + n^2)}} \right] + 0\,(e^2).$$

Since it may be shown that the expression in brackets is always positive $(n > 0)$, being zero only for $n = 1$, it follows that

$$F_a > F_p$$

for small values of e. But, equating F_a and F_p, leads to the result that $F_a = F_p$ only when $n^2 = 1 + e$, $1 - e$ or 0; or $e = 0$ or 1. Consequently, we must conclude that

$$F_a > F_p \qquad (n^2 > 1 + e)$$

$$(34)$$

for all values of e and n^2 satisfying the inequality $n^2 > 1 + e$. Similar reasoning for eqs. (28), (29) and (33) gives that

$$F_a < F_p < F_s$$
$$(1 - e < n^2 < 1 + e)$$

$$(35)$$

and

$$F_a < F_p \qquad (n^2 < 1 - e).$$

$$(36)$$

Eq. (35) shows immediately that it is always wasteful of fuel if transfer is made by use of a single impulse rather than by either of the other two modes of transfer.

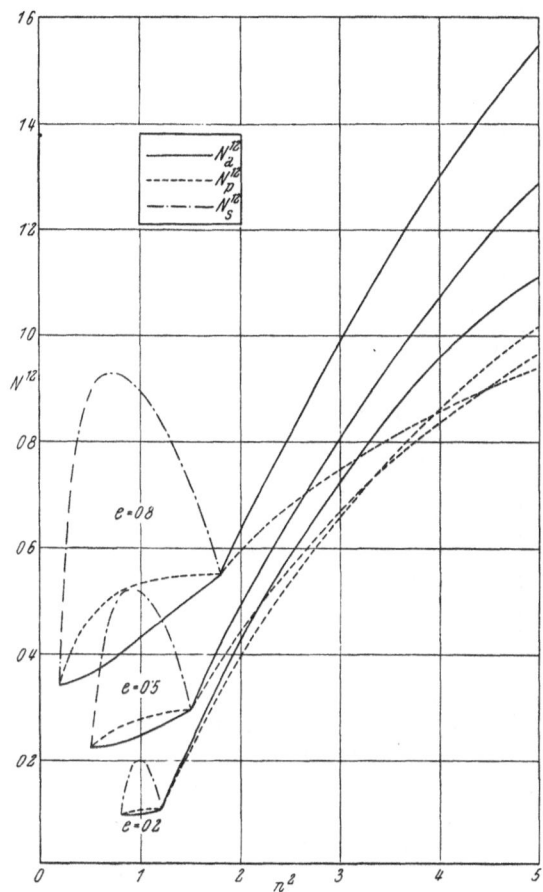

Fig. 5. Normalized fuel factor coefficients for transfer from a circle to an ellipse

For the converse situation, when transfer is to be made from the elliptical orbit to the circular orbit, it is clear from Fig. 3 that the geometry of the intermediate transfer orbits would be the same as previously for any particular configuration. Physically, the impulses would be applied in the reverse directions but of the same magnitudes and points of application. Since the elliptical orbit will now have the reference energy E_0 per unit mass, the normalized fuel factor coefficient will be N^{21}, $(e_1 = 0)$, which we know may be determined from the equations

$$n\,N^{21} = N^{12} = F.$$

$$(37)$$

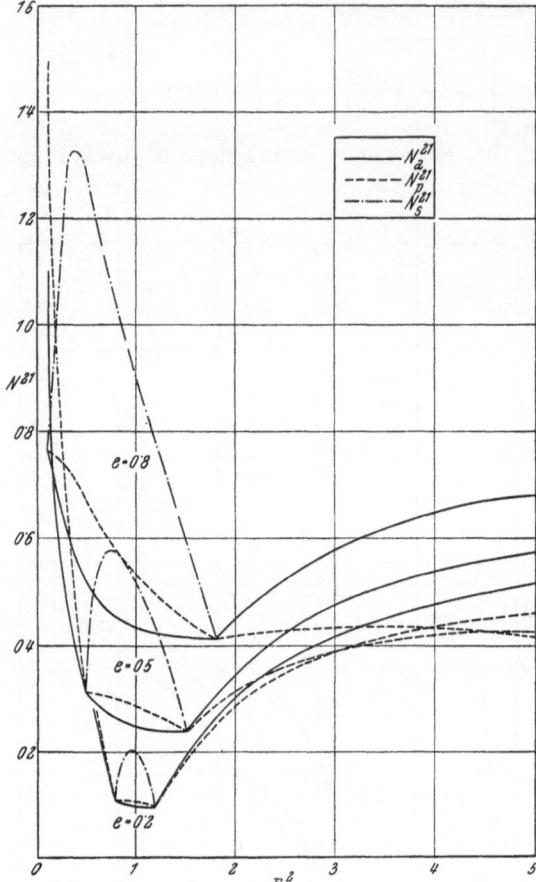

Fig. 6. Normalized fuel factor coefficients for transfer from an ellipse to a circle. Note that suffix 1 refers to the circular orbit

The inequalities (34), (35) and (36) will still hold, of course, for either N^{12} or N^{21} replacing the F's so that the most economical mode of transfer for any terminal configuration can be immediately specified.

Figs. 4 and 5 show the normalized fuel factors coefficients for transfer from a circular to an elliptical orbit, whilst Fig. 6 shows those corresponding to transfer from an elliptical to a circular orbit. (For $e = 0$, $N_a = N_p = N_s$ and the curves reduce to those corresponding to a HOHMANN transfer.) In general, these curves show that for transfer via two impulses, a wrong choice of transfer path (corresponding to a tangential impulse at the wrong apse of the elliptical terminal orbit) could lead to a considerable wastage of fuel, particularly when the terminal elliptic orbit is highly eccentric, and that transfer via a single impulse is even more wasteful of fuel.

V. Transfer between Two Elliptical Orbits Having a Common Direction of Major Axes

This configuration constitutes a slight generalization of that considered in section IV, the circular orbit being replaced by an elliptical orbit and $\gamma_1 - \gamma_2$ again being equal to 0 or π.

Consider eqs. (5), (7), (9) and (11). If one imposes the restriction that the impulses are tangential to the terminal orbits, i.e. φ_1, φ_2 equal to 0 or π, then these equations imply that

$$\left. \begin{array}{ll} \theta_1 + \gamma_1 = 0 \text{ or } \pi, & \theta_1 + \gamma = 0 \text{ or } \pi \\ \theta_2 + \gamma_2 = 0 \text{ or } \pi, & \theta_2 + \gamma = 0 \text{ or } \pi \end{array} \right\} \tag{38}$$

and, consequently,

$$\left. \begin{array}{l} \theta_1 - \theta_2 = 0 \text{ or } \pi \\ \gamma_1 - \gamma = 0 \text{ or } \pi \\ \gamma_2 - \gamma = 0 \text{ or } \pi \\ \gamma_1 - \gamma_2 = 0 \text{ or } \pi. \end{array} \right\} \tag{39}$$

Eq. (15) is automatically satisfied and eqs. (13) and (14) simply determine the constants A_1 and A_2 for optimum transfer. The physical interpretation of eqs. (39) is that transfer by means of two impulses which are tangential to the terminal orbits is only possible if the terminal orbits and the transfer orbits all have a common direction of major axes. The remaining boundary conditions, eqs. (4), (6), (8) and (10), then imply that the transfer orbits are tangential at one apse of each terminal orbit, the impulses being applied at these apses. Now, the solutions for $\beta_1 = 0$, $\beta_2 \neq 0$ (or $\beta_2 = 0$, $\beta_1 \neq 0$) are φ_1, $\varphi_2 = 0$ or π (section IV), and since it is to be expected that the solutions to the LAWDEN equations are continuous with respect to any variation of the terminal orbits, and in particular the parameter β_1 it follows (since for β_1 and $\beta_2 \neq 0$, then $\varphi_1, \varphi_2 = 0$ or π are solutions for $\gamma_1 - \gamma_2 = 0$ or π) that the converse of the above statement is true, namely that $\gamma_1 - \gamma_2 = 0$ or π implies that φ_1 and φ_2 are 0 or π for optimal transfer. Presumably this can also be deduced directly by algebraic manipulation of the LAWDEN equations but the analysis is not immediately evident.

Since we now know that the transfer ellipses are tangential at the apses of the terminal orbits if these have the same major axes, there is no difficulty in determining the corresponding fuel factor coefficients. In the following section, a particular case of this configuration is considered, namely that of optimal transfer between orbits having the same eccentricities and major axes.

VI. Transfer between Isoeccentric Ellipses Having a Common Direction of Major Axes

Two cases will be considered:

(i)   when the two major axes are in the same sense $(\gamma_1 = \gamma_2)$ and

(ii) when the two major axes are in the opposite sense $(\gamma_1 - \gamma_2 = \pi)$.

Let the eccentricity of the two terminal ellipses be denoted by e and, for the present, let n be less than unity so that the suffix 1 refers to the smaller ellipse. Also we will take $\gamma_1 = 0$, as this merely fixes the position of the major axes, and consider transfer from 1 to 2.

For case (i), the appropriate solutions are readily deduced to be:

(i pa) $\gamma_1 = \gamma_2 = \gamma = \varphi_1 = \varphi_2 = \theta_1 = 0, \qquad \theta_2 = \pi$

$$u_1 = \alpha_1{}^2 (1 + e), \qquad u_2 = n^2 \alpha_1{}^2 (1 - e), \qquad 2\alpha^2 = u_1 + u_2, \qquad l_1 < l < l_2$$

for transfer from the perigee of 1 to the apogee of 2, and

(i ap) $\gamma_1 = \gamma_2 = \varphi_1 = \varphi_2 = \theta_2 = 0, \qquad \theta_1 = \pi$

$$\gamma = 0 \ (u_1 < u_2), \qquad \gamma = \pi \ (u_1 > u_2)$$

$$u_1 = \alpha_1{}^2 (1 - e), \qquad u_2 = n^2 \alpha_1{}^2 (1 + e), \qquad 2\alpha^2 = u_1 + u_2, \qquad l_1 < l < l_2$$

when the transfer is from the apogee of 1 to the perigee of 2.

Substituting these solutions into eq. (16) shows that the corresponding fuel factor coefficients are

$$\left. \begin{array}{l} F_{pa} = \sqrt{2}\, \dfrac{(1 + e) - n^2 (1 - e)}{\sqrt{(1 + e) + n^2 (1 - e)}} - (1 + e) + n (1 - e) \\[1.5em] \qquad\qquad\qquad\qquad\qquad\qquad\qquad\qquad (n < 1) \\[1em] F_{ap} = \sqrt{2}\, \dfrac{(1 - e) - n^2 (1 + e)}{\sqrt{(1 - e) + n^2 (1 + e)}} - (1 - e) + n (1 + e). \end{array} \right\} \quad (40)$$

Also, using the arguments of section IV, it may be deduced that

$$F_{pa} < F_{ap} \qquad (n < 1) \tag{41}$$

for any given values of e and n.

Analysis of case (ii) leads to the solutions:

(ii aa) $\gamma_1 = \theta_2 = 0,$ $\qquad \gamma_2 = \gamma = \theta_1 = \pi$

$u_1 = \alpha_1^2 (1 - e),$ $\qquad u_2 = n^2 \alpha_1^2 (1 - e),$ $\qquad 2\alpha^2 = u_1 + u_2$

with

$$\varphi_1 = 0, \qquad \varphi_2 = \pi, \qquad l > l_2 > l_1 \qquad \text{for} \qquad \frac{1 - e}{1 + e} < n^2 < 1$$

and

$$\varphi_1 = 0, \qquad \varphi_2 = 0, \qquad l_2 > l > l_1 \qquad \text{for} \qquad n^2 < \frac{1 - e}{1 + e} < 1$$

for transfer between the apogees of the terminal ellipses and

(ii pp) $\gamma_1 = \gamma = \theta_1 = 0,$ $\qquad \gamma_2 = \theta_2 = \pi$

$u_1 = \alpha_1^2 (1 + e),$ $\qquad u_2 = n^2 \alpha_1^2 (1 + e),$ $\qquad 2\alpha^2 = u_1 + u_2$

with

$$\varphi_1 = \pi, \qquad \varphi_2 = 0, \qquad l_2 > l_1 > l \qquad \text{for} \qquad \frac{1 - e}{1 + e} < n^2 < 1$$

and

$$\varphi_1 = 0, \qquad \varphi_2 = \pi, \qquad l_2 > l > l_1 \qquad \text{for} \qquad n^2 < \frac{1 - e}{1 + e} < 1$$

for transfer between the perigees of the terminal ellipses.

For $n^2 = (1 - e)/(1 + e)$, the two terminal ellipses touch and transfer is effected directly by a single impulse at the point of contact. The fuel factor coefficients for these two modes of transfer are:

$$
\begin{aligned}
F_{aa} &= \sqrt{2(1 - e)(1 + n^2)} - (1 - e)(1 + n) & \left(\frac{1 - e}{1 + e} < n^2 < 1\right) \\
&= (1 - n^2)\sqrt{\frac{2(1 - e)}{1 + n^2}} - (1 - e)(1 - n) & \left(n^2 < \frac{1 - e}{1 + e}\right) \\
&= -1 + e + \sqrt{1 - e^2} & \left(n^2 = \frac{1 - e}{1 + e}\right)
\end{aligned}
$$

and

$$
\begin{aligned}
F_{pp} &= (1 + n)(1 + e) - \sqrt{2(1 + e)(1 + n^2)} & \left(\frac{1 - e}{1 + e} < n^2 < 1\right) \\
&= (1 - n^2)\sqrt{\frac{2(1 + e)}{1 + n^2}} - (1 + e)(1 - n) & \left(n^2 < \frac{1 - e}{1 + e}\right) \\
&= -1 + e + \sqrt{1 - e^2} & \left(n^2 = \frac{1 - e}{1 + e}\right)
\end{aligned}
\tag{42}
$$

For those situations satisfying the inequality $1 > n^2 > (1 - e)/(1 + e)$, the two terminal orbits intersect and consequently transfer may be achieved by the application of a single impulsive thrust at the points of intersection. Using eq. (18), the points of intersection are given by

$$u = \alpha_1^2 (1 + e \cos \theta) = n^2 \alpha_1^2 (1 - e \cos \theta)$$

i.e.

$$u = \frac{2\alpha_1^2 n^2}{1 + n^2} \quad \text{and} \quad \cos\theta = -\frac{(1 - n^2)}{e(1 + n^2)}.$$

Substituting these solutions into eq. (17) leads to the expression

$$F_s = \sqrt{e^2(1 + n)^2 - (1 - n)^2\left(1 + \frac{4n}{1 + n^2}\right)} \qquad \left(\frac{1 - e}{1 + e} < n^2 < 1\right) \qquad (43)$$

for the fuel factor coefficient for transfer via a single impulse applied at either of the two points of intersection.

Finally, using eqs. (42) and (43), it may be deduced that the following inequalities are satisfied:

$$F_{aa} < F_{pp} < F_s \qquad \left(\frac{1 - e}{1 + e} < n^2 < 1\right)$$

and (44)

$$F_{aa} > F_{pp} \qquad \left(n^2 < \frac{1 - e}{1 + e}\right).$$

Now, since the terminal orbits have identical shapes, it is clear [using the arguments leading to eq. (23)] that the fuel factor expressions for $n > 1$ may be deduced from those for $n < 1$ if one makes the mathematical transformations

$$1 \to 2, \qquad 2 \to 1, \qquad n \to \frac{1}{n}$$

and also $a \to p$, $p \to a$ for $\gamma_1 = \gamma_2 = 0$. Making this transformation, we may obtain the additional expressions

$$F_{ap} = \sqrt{2}\,\frac{n^2(1 + e) - (1 - e)}{\sqrt{n^2(1 + e) + (1 - e)}} - n(1 + e) + (1 - e)$$

$$\left.\begin{array}{c}(n > 1)\end{array}\right\} \quad (45)$$

$$F_{pa} = \sqrt{2}\,\frac{n^2(1 - e) - (1 + e)}{\sqrt{n^2(1 - e) + (1 + e)}} - n(1 - e) + (1 + e)$$

$$F_{aa} = \sqrt{2(1 - e)(1 + n^2)} - (1 - e)(1 + n) \qquad \left(\frac{1 - e}{1 + e} < n^2 < \frac{1 + e}{1 - e}\right)$$

$$= (n^2 - 1)\sqrt{\frac{2(1 - e)}{1 + n^2}} - (1 - e)(n - 1) \qquad \left(n^2 > \frac{1 + e}{1 - e}\right)$$

$$= 1 + e - \sqrt{1 - e^2} \qquad \left(n^2 = \frac{1 + e}{1 - e}\right)$$

$$F_{pp} = (1 + n)(1 + e) - \sqrt{2(1 + e)(1 + n^2)} \qquad \left(\frac{1 - e}{1 + e} < n^2 < \frac{1 + e}{1 - e}\right) \quad (46)$$

$$= (n^2 - 1)\sqrt{\frac{2(1 + e)}{1 + n^2}} - (1 + e)(n - 1) \qquad \left(n^2 > \frac{1 + e}{1 - e}\right)$$

$$= 1 + e - \sqrt{1 - e^2} \qquad \left(n^2 = \frac{1 + e}{1 - e}\right)$$

$$F_s = \sqrt{e^2(1 + n)^2 - (1 - n)^2\left(1 + \frac{4n}{1 + n^2}\right)} \qquad \left(\frac{1 - e}{1 + e} < n^2 < \frac{1 + e}{1 - e}\right)$$

and the inequalities

$$F_{ap} < F_{pa} \qquad (n > 1)$$

$$F_{aa} < F_{pp} < F_s \qquad \left(\frac{1-e}{1+e} < n^2 < \frac{1+e}{1-e} \right) \Bigg\} \qquad (47)$$

$$F_{aa} > F_{pp} \qquad \left(n^2 > \frac{1+e}{1-e} \right)$$

The normalized fuel factor coefficients are finally obtained using the equation

$$N^{12}(e) = F/\sqrt{1 - e^2}$$

and, clearly, the inequalities (41), (44) and (47) are all satisfied if the corresponding $N^{12}(e)$'s replace the F's. Thus, for any particular configuration of the terminal ellipses, we may immediately state which of the possible modes of transfer leads to the absolute minimum expenditure of fuel.

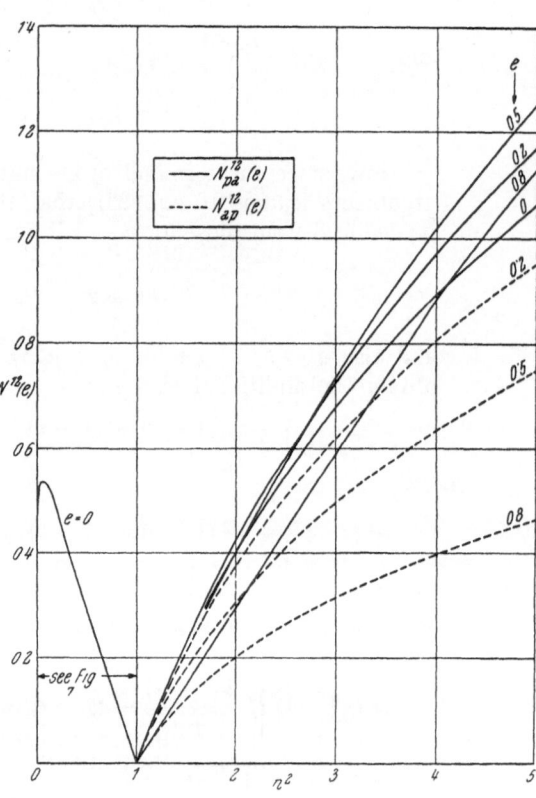

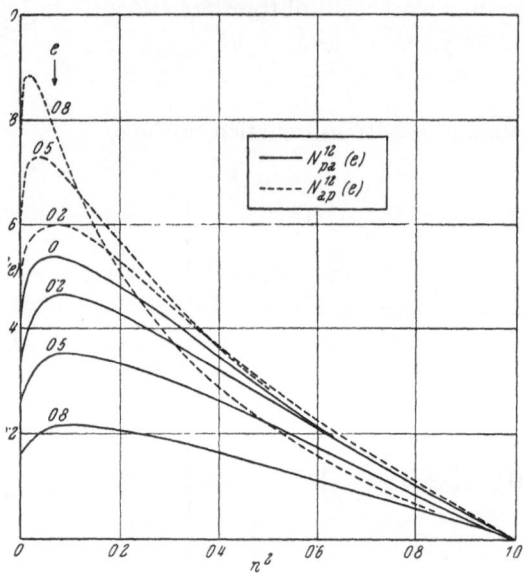

Fig. 7. Normalized fuel factor coefficients for transfer between isoeccentric ellipses $(\gamma_1 - \gamma_2 = 0)$

Fig. 8. Normalized fuel factor coefficients for transfer between isoeccentric ellipses $(\gamma_1 - \gamma_2 = 0)$

The normalized fuel factor coefficients for the various configurations and modes of transfer are shown in Figs. 7 and 8 for $\gamma_1 - \gamma_2 = 0$ and in Figs. 9 and 10 for $\gamma_1 - \gamma_2 = \pi$. Again it is evident that for two impulse transfer a considerable wastage of fuel is possible if the wrong mode of transfer is employed, especially for highly eccentric terminal orbits. Single impulse transfer, when it is possible, is even more wasteful of fuel.

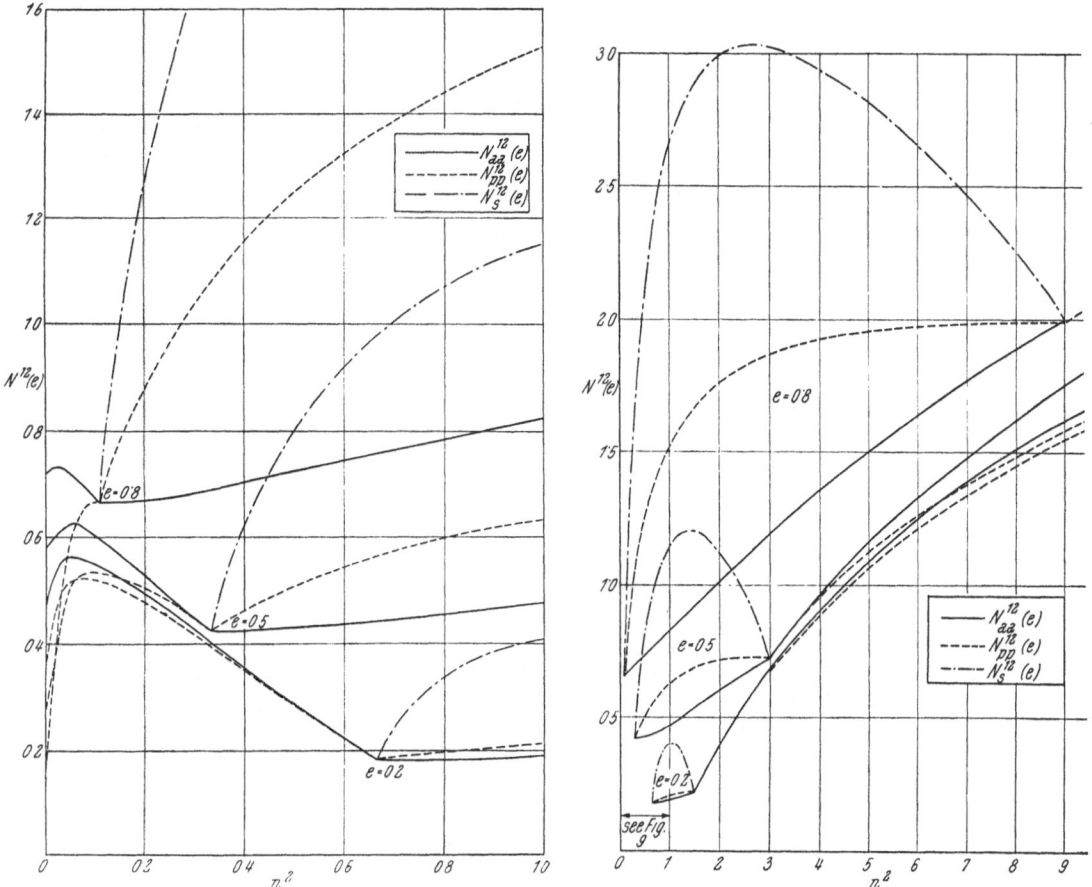

Fig. 9. Normalized fuel factor coefficients for transfer between isoeccentric ellipses $(\gamma_1 - \gamma_2 - \pi)$

Fig. 10. Normalized fuel factor coefficients for transfer between isoeccentric ellipses $(\gamma_1 - \gamma_2 = \pi)$

VII. Fuel Expenditure to Effect Rotation of an Elliptical Orbit

All the configurations considered so far have virtually had their major axes along the same line. As a consequence of this the optimal fuel trajectories have been realised by the application of impulses which are tangential to the terminal orbits and, mathematically, this has lead to relatively simple analytic solutions. Another configuration which is of interest and which allows some simplification in the mathematics, although not giving an analytic solution, is that corresponding to optimal transfer between the ellipses $(\alpha_1, \beta_1, \pm \gamma_1)$. Physically this corresponds to the rotation of an elliptical orbit through an angle $2\gamma_1$. The solution to this problem has been given by LAWDEN in [3] and only the relevant results of the analysis will be given here (with a slight change in notation).

Because of the symmetry of the terminal ellipses about the line $\theta = 0$, it is to be expected that the optimum two impulse solution will also show symmetry about this line (Fig. 11). In fact, it is shown that if an impulse characterized by (u, θ, A, φ) is applied on one of the terminal orbits in order to bring the rocket into the transfer orbit, then the impulse $(u, -\theta, A, \pi - \varphi)$ must be applied in order to return the rocket onto the other terminal orbit, the transfer orbit being $(\alpha, \beta, 0)$. Introducing the quantities

$$\left.\begin{array}{cc} x = \dfrac{u}{\alpha^2} & z = \dfrac{\alpha}{\alpha_1} \\[2mm] P = e_1 \cos \gamma_1, & Q = e_1 \sin \gamma_1, \end{array}\right\} \tag{48}$$

where $e_1 = \beta_1/\alpha_1$ is the eccentricity of either of the terminal orbits, then it is proved that, for optimal (absolute) transfer, the equations

$$\left.\begin{array}{c} x z^2 = 1 - \dfrac{P \sqrt{1 - x^2} + Q \sqrt{x (2 x + 1)}}{\sqrt{1 + x + x^2}} \\[4mm] \text{and} \\[2mm] z \left(\dfrac{x z}{1 + x} + 1 \right) = \sqrt{\dfrac{2 x + 1}{x (1 + x) (1 - x^3)}} \, [P \sqrt{x (2 x + 1)} - Q \sqrt{1 - x^2}] \end{array}\right\} \tag{49}$$

must both be satisfied; the thrust directions then being given by

$$\tan^2 \varphi = \frac{x (1 - x)}{(x + 1) (2 x + 1)} . \tag{50}$$

Also, θ may be determined from the equations

$$\sin \theta : \cos \theta : 1 = \sqrt{x (2 x + 1)} : - \sqrt{1 - x^2} \sqrt{1 + x + x^2}. \tag{51}$$

Although eqs. (49) do not admit an analytical solution for x and z, it is quite straightforward to find a numerical solution for any (known) values of P and Q. By plotting z as a function of x from each equation, an approximate solution may be found graphically. A more accurate solution may then be obtained by a process of linear interpolation.

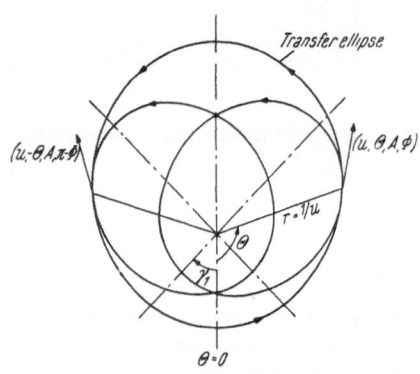

Fig. 11. Transfer orbit for minimal fuel expenditure between the elliptical orbits $(\alpha_1, \beta_1, \pm \gamma_1)$

Having determined x and z the fuel factor coefficient may be determined as follows. From eq. (16) and the symmetry of the solution, it is clear that

$$F_{\gamma 1} = \frac{2 u}{\alpha_1} \left| \sec \varphi \left(\frac{1}{\alpha_1} - \frac{1}{\alpha} \right) \right| ,$$

the same amount of fuel being expended at each impulse. But, from eqs. (48)

$$u = x \alpha^2 = x z^2 \alpha_1^2, \qquad \alpha = z \alpha_1$$

and consequently

$$F_{\gamma 1} = 2 x z^2 \left| \sec \varphi \left(1 - \frac{1}{z} \right) \right| \tag{52}$$

where φ is given by eq. (50). The normalized fuel factor coefficient is then

$$N_{\gamma 1}(e_1) = F_{\gamma 1} / \sqrt{1 - e_1^2}.$$

Table I shows the solutions obtained for terminal orbits of eccentricities 0.2, 0.5 and 0.8, the corresponding normalized fuel factor coefficients being plotted as a function of γ_1 in Fig. 12.

Table I. $(\alpha_1 = 1)$

e_1	$\gamma_1{}^0$	x	z	u	φ^0	θ^0	$N_{\gamma 1}$
0.2	45	0.9865	0.9259	0.8457	2.65	95.5	0.1382
0.2	90	1.0	0.8944	0.8	0	90	0.1928
0.5	22.5	0.8470	0.8870	0.6661	9.14	109.4	0.1981
0.5	45	0.9115	0.7907	0.5698	6.96	104.4	0.3514
0.5	67.5	0.9745	0.7284	0.5167	3.73	97.5	0.4457
0.5	90	1.0	0.7071	0.5	0	90	0.4783
0.8	45	0.7405	0.5716	0.2419	11.91	116	0.6173
0.8	90	1.0	0.4472	0.2	0	90	0.8240

Again, since the terminal orbits for this configuration always intersect in two points, transfer is possible by means of a single impulse at either of the points of intersection, namely $\theta = 0$, $u = \alpha_1{}^2 (1 + e_1 \cos \gamma_1)$ or $\theta = \pi, u = \alpha_1{}^2 (1 - e_1 \cos \gamma_1)$. For either of these points of intersection, it is readily deduced from eq. (17) that the fuel factor coefficient for single impulse transfer is

$$F_s = 2\, e_1 \left| \sin \gamma_1 \right|,$$

the normalized fuel factor coefficient being

$$N_s = 2\, \frac{e_1 \left| \sin \gamma_1 \right|}{\sqrt{1 - e_1{}^2}}.$$

(53)

Fig. 12 also shows N_s plotted as a function of γ_1 for $e_1 = 0.2$, 0.5 and 0.8. The curves for N_s and $N_{y 1}$ are symmetrical about the line $\gamma_1 = \pi/2$, since the configurations corresponding to γ_1 and $\pi - \gamma_1$ are identical.

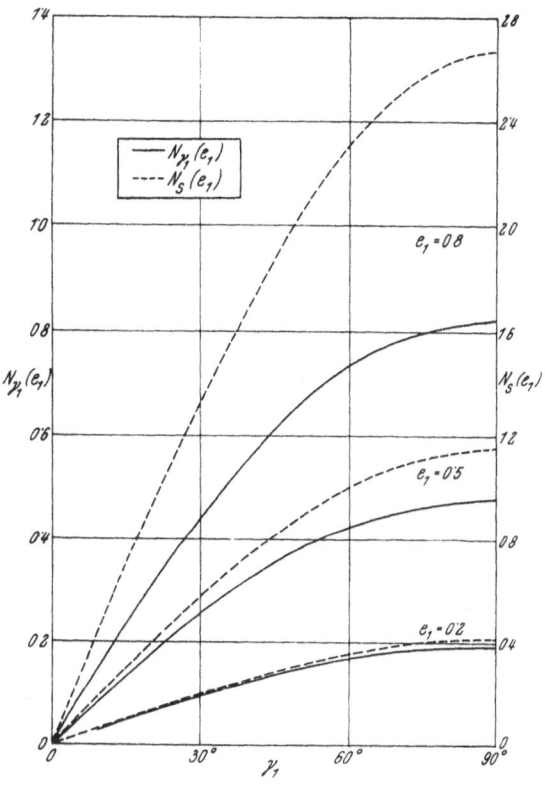

Fig. 12. Normalized fuel factor coefficients for rotation of an ellipse

Again it is seen that transfer by means of a single impulse is wasteful of fuel. Also, for either mode of transfer, the greater the eccentricity of the terminal orbits the more fuel is expended for a given rotation and for terminal orbits of fixed eccentricity, the larger the rotation the greater the fuel expended.

VIII. Fuel Expenditure for Earth Satellite Navigation

Although N gives a relative measure of the fuel expended, or mass reduction of the rocket, for a particular mode of transfer between the terminal orbits, the actual amount of fuel consumed will depend upon the gravitational energy of the rocket in its initial orbit and the jet velocity c of the rocket motor. Now the equation relating these quantities is

$$c \log m/m' = N \sqrt{-2 E_0},$$

E_0 being the energy per unit mass of the initial orbit. Consequently, if the percentage mass reduction is R, then

$$R = \frac{m - m'}{m} \times 100\%$$
$$= 100 \left(1 - e^{-\lambda N}\right)\%$$
$$(54)$$

where

$$\lambda = \sqrt{-2 E_0}/c$$
$$(55)$$

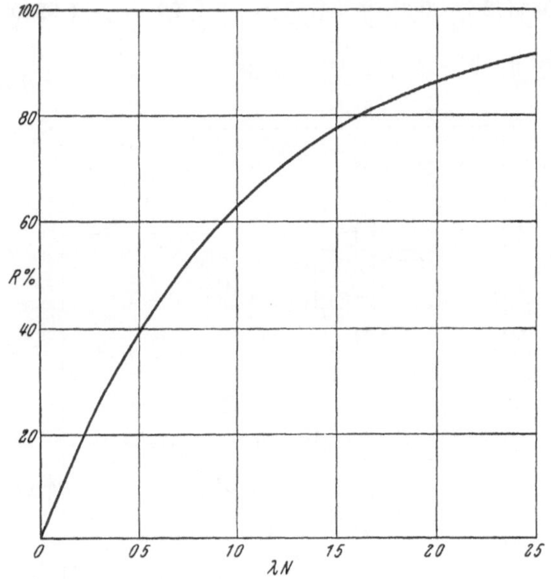

For reference purposes, R is shown plotted as a function of λN in Fig. 13.

Let us now consider the fuel requirements for the inter-orbital transfer of an earth satellite assuming a jet velocity of 10,000 ft./sec. The force constant μ_e for the earth is approximately 1.41×10^{16} ft.^{3}/sec^2. Thus for a satellite orbit having a semi-major axis of 5,000 miles, the total energy per unit mass is

Fig. 13. Percentage mass reduction plotted as a function of λN

$$E = - \frac{\mu_e}{2a} = - \frac{1.41 \times 10^{16}}{5280 \times 10^4} \simeq - 2.8 \times 10^8 \ (\text{ft./sec.})^2.$$

We will assume that the satellite has roughly this amount of energy/unit mass in its initial orbit. The corresponding value for λ is then determined from (55) to be approximately 2.5. Consequently, knowing the appropriate value of N for any particular configuration of the terminal orbits and the mode of transfer, the mass reduction for an earth satellite can be obtained immediately, from Fig. 13.

For example, if the initial orbit is circular and the final orbit has eccentricity e (Fig. 5), when both orbits have the same total energy per unit mass (i.e. both have the same semi-major axis, or, $n = 1$), then for $e = 0.2$, the mass reduction is 22% for either of the two stationary modes of transfer ($N = 0.10$) and 39% for transfer via a single impulse ($N = 0.2$). As the eccentricity e increases so does the percentage mass reduction for all three modes of transfer; for $e = 0.8$ the mass reduction is 65% for optimal transfer via two impulses ($N = 0.43$), 73% for transfer via the other stationary two impulse transfers ($N = 0.53$) and

90% for a single impulse transfer ($N = 0.90$). The latter mode of transfer would be extremely wasteful of fuel.

Consider now the effect of transferring the satellite to a similar but larger orbit (Figs. 7 and 8). Increasing the size of the orbit to even twice that of the original orbit ($n^2 = \frac{1}{2}$), regardless of its eccentricity, never requires more than 50% mass reduction ($N < 0.3$) for optimal transfer, the mass reduction decreasing as the eccentricity increases and the final orbit decreases in size. For a given increase, or decrease, in the size of the initial orbit, the HOHMANN transfers, corresponding to circular terminal orbits, require the most fuel.

Finally, let us consider the fuel requirements to effect rotation of a satellite orbit. Fig. 12 shows that the fuel expenditure, for optimal two impulse transfer or single impulse transfer, increases as the angle of rotation and eccentricity of the orbit increase, reaching a maximum value when the major axis has been reversed in direction ($\gamma_1 = 90^0$). For complete reversal of an orbit of eccentricity 0.2, the mass reduction is 39% ($N = 0.2$) for two impulse transfer and 63% for single impulse transfer ($N = 0.4$), whereas for the corresponding situation when $e_1 = 0.8$, the optimal two impulse mode of transfer causes an 87% mass reduction ($N = 0.82$) compared with 98% for a single impulse transfer ($N = 2.68$). On the other hand, if we now study Fig. 9, which corresponds to transfer from an elliptical orbit to a similar larger orbit rotated through 180^0, it will be seen that the values of N for optimal two impulse transfer are almost independent of n when the two terminal orbits intersect. Consequently, in this case, we may conclude that besides reversing the orientation of an orbit we may at the same time increase its size without any appreciable change in fuel expenditure. For example, from Fig. 9, if we reverse the orientation and double the size ($n^2 = \frac{1}{2}$) of an orbit of eccentricity 0.5, the corresponding mass reduction is 65% ($N = 0.43$) compared with a mass reduction of 70% ($N = 0.48$) for reversal without change in size of the initial orbit. Continuing in this manner, we may immediately deduce the minimal fuel expenditure to effect transfer of a satellite between any of the terminal orbit configurations considered in the text.

It is worth noting that the fuel expenditure will vary considerably for different jet velocities of the rocket. For example, if the jet velocity was only 5,000 ft./sec. instead of the assumed 10,000 ft./sec., the corresponding value of λ would be doubled [eq. (55)]. For the orbital transfers considered above this means that λ would now be 5.0 instead of 2.5 and the fuel expenditure would be increased; an original mass reduction of 10% ($\lambda N = 0.1$) would become a mass reduction of 18% ($\lambda N = 0.2$), and a 40% reduction ($\lambda N = 0.5$) would become 63% ($\lambda N = 1.0$). On the other hand, increasing the jet velocity of the rocket will lead to a saving of fuel. A jet velocity of 50,000 ft./sec. instead of the assumed 10,000 ft./sec., will reduce the value of λ to 0.5. Thus a previous mass reduction of 10% ($\lambda N = 0.1$) will now be only 2% ($\lambda N = 0.02$), whilst a 40% mass reduction ($\lambda N = 0.5$) will be replaced by a 10% reduction in mass ($\lambda N = 0.10$).

IX. Concluding Remarks

Although the results obtained in this paper enable one to calculate the necessary fuel expenditure of a rocket for transfer between a large selection of initial and final orbits, a great deal of further research still remains to be done. Two limitations of the paper are immediately apparent. In the first place, only coplanar terminal orbits have been considered and, secondly, only special cases of these have been analysed.

For the situation when the terminal orbits are coplanar, the equations governing an optimal fuel transfer, using any number of impulses, are known (LAWDEN equations). These equations must be solved, in general, using numerical analysis. Since the solutions are known for certain special configurations, it should not be too difficult to compute the solutions for other configurations by starting with a known solution and gradually changing the parameters specifying the terminal orbits. Also it would be extremely useful if one could determine the optimum number of thrusts to effect optimal transfer, it being suspected that this number is two.

For the more general problem of transfer between non-coplanar orbits, the equations which determine the optimal mode of transfer have yet to be deduced. Besides the quantities α, β, γ, two more parameters are now necessary to specify an elliptical orbit, one determining the inclination of the orbital plane to a reference plane and the other determining the line of intersection of these planes with respect to an arbitrary fixed line in the reference plane. This problem is now being studied by the author and it is intended to publish the results in a later paper.

Two further problems which require investigation are:

(a) the determination of the optimal fuel transfer trajectories when the transit time is specified or limited, and

(b) the determination of optimal time transfer orbits when the fuel expenditure is limited, specified or unlimited.

Both these problems would undoubtedly be extremely difficult because of the complex manner in which the time variable enters into the equations of motion. For the situation where the terminal orbits are circular, MOECKEL has considered the problem of reducing the transit time with the least excess energy [4].

References

1. W. HOHMANN, Die Erreichbarkeit der Himmelskörper. Munich: R. Oldenbourg, 1925.
2. D. F. LAWDEN, Inter-Orbital Transfer of a Rocket. Ann. Rep. Brit. Interplan. Soc., p. 321—333 (1952).
3. D. F. LAWDEN, The Determination of Minimal Orbits. J. Brit. Interplan. Soc. 11, 216 (1952).
4. W. E. MOECKEL, Interplanetary Trajectories with Excess Energy. Proceedings of the IXth International Congress, Amsterdam 1958, p. 96. Wien: Springer, 1959.

Methods of Analysing Observations on Satellites

By

G. V. Groves[1] and M. J. Davies[1]

(With 4 Figures)

(*Received July 21, 1959*)

Abstract — Zusammenfassung — Résumé

Methods of Analysing Observations on Satellites. A theory is developed for the precise determination of the elements of a satellite orbit from observational data. Account is taken of such effects as atmospheric refraction, aberration, the finite speed of light and the difference between geocentric and geographical latitude to meet the needs of analysing observations of accuracy 1 second of arc and 1/3 millisec of time. When analysing observations over more than a fraction of a revolution account needs to be taken of the effects of harmonics in the earth's gravitational field and of the drag effects of the atmosphere. These effects are discussed and the theory is formulated to enable the relevant parameters and coefficients to be determined.

Methoden zur Analyse von Satellitenbeobachtungen. Es wird eine Methode für die genaue Bestimmung der Elemente einer Satellitenbahn aus Beobachtungsdaten entwickelt. Dabei werden solche Effekte wie die atmosphärische Lichtbrechung, die Aberration, die endliche Lichtgeschwindigkeit und der Unterschied zwischen der geozentrischen und der geographischen Breite in Rechnung gesetzt, um die Erfordernisse zu erfüllen, wie sie zur Analyse von Beobachtungen mit einer Genauigkeit von 1 Bogensekunde und 1/3 Millisekunde (Zeit) unerläßlich sind. Bei der Analyse von Beobachtungen über mehr als einen Bruchteil eines Umlaufes müssen die Harmonieeffekte im Erdschwerefeld und die Effekte des Widerstandes der Atmosphäre berücksichtigt werden. Diese Einflüsse werden erörtert und eine Theorie formuliert, damit man die diesbezüglichen Parameter und Koeffizienten bestimmen kann.

Méthodes d'analyse des observations sur les satellites. Une théorie est établie pour la détermination précise des éléments de l'orbite d'un satellite à partir des données d'observation. Il est tenu compte des effets de la réfraction atmosphérique, de l'aberration, de la vitesse finie de la lumière et de la différence entre latitude géocentrique et géographique en vue d'obtenir une précision d'analyse d'une seconde d'arc et un tiers de milliseconde. Pour les observations étendues sur une portion appréciable de trajectoire il faut encore tenir compte des harmoniques du champ de gravitation et de la résistance aérodynamique. Ces effets sont discutés et la théorie susceptible d'en déduire les paramètres et coefficients désirés est formulée.

I. Introduction

For the precise tracking of satellites, we turn to optical methods for the greatest accuracy. So far, observations have been confined to periods when the

---

[1] Department of Physics, University College, London, United Kingdom.

satellite is illuminated by the sun against a darkened sky. Selfluminous or flashing satellites may be launched in the future and this would offer greater opportunities for observation, and also increase the possibility of simultaneous observations by means of which points of the trajectory could be triangulated directly [14].

Optical observations have been made chiefly by either visual or photographic means. Visual observations have been made extensively by Moonwatch teams, viewing through instruments of about 10 degrees field and recording a satellite passage to at best 0.1 second of time and 0.1 degree in angle. Observations down to the eighth or ninth magnitude have frequently been made by this method, which has been the main source of optical data on faint satellites.

Photographic methods offer greater accuracy in position determination than visual methods. Kinetheodolites have been used giving an angular accuracy of about 15 seconds of arc [10]. On account of the high angular velocity of a satellite, photographic recording is limited to only the brighter satellites, unless the equipment has special tracking facilities. This is the feature of the BAKER-NUNN camera in use at twelve tracking stations in various parts of the world. It has an $f/1$ lens system, and has now recorded the smallest of all satellites, the Vanguard I sphere.

Photo-electric devices offer a further method of detecting satellites with prospects of high sensitivity and high accuracy [15]. The equipment consists of a camera with a photo-multiplier behind the focal plane and looking at a screen mounted in the focal plane. The screen is opaque apart from a number of straight-line slits, and when the satellite image crosses one of these slits a signal is generated. The data derived from the equipment therefore consists of a series of time signals.

The present paper deals with the theory for analysing the time signals from photo-electric devices, and the theory is developed for deriving the orbital elements and other unknown parameters from such data and from photographic data. The development of the theory to embrace photo-electric data is an extension to that given by KING-HELE [7] and MERSON [10], which has been the basis for analysing the kinetheodolite observations of the R.A.E.

II. Determination of Satellite Direction by Photo-Electric Detector

Let $O\,x\,y$ be co-ordinate axes in the focal plane of the lens, and let the equation of the i-th slit $(i = 1, \ldots N)$ be

$$p_i\,x + q_i\,y + r_i = 0. \tag{1}$$

The analysis may be carried out by taking the path of the satellite image as

$$x = x_0 + x_1\,(t - t_0) + \ldots + x_p\,(t - t_0)^p$$
$$y = y_0 + y_1\,(t - t_0) + \ldots + y_p\,(t - t_0)^p \tag{2}$$

where t is the time at which the image is at (x, y). Then, from (1) and (2)

$$p_i\,x_0 + p_i\,(t_i - t_0)\,x_1 + \ldots + q_i\,y_0 + q_i\,(t_i - t_0)\,y_1 + \ldots + r_i = 0, \tag{3}$$

for $i = 1, \ldots N$, where t_i is the time at which the image crosses the i-th slit. If the number of slits is greater or equal to $2\,(p + 1)$, (3) can be solved for $x_0 \ldots x_p$ and $y_0 \ldots y_p$ by the method of least squares.

The minimum number of slits with which a determination is possible is readily seen to be four. The path of the satellite image is then determined as a straight line described at uniform speed. A suitable slit pattern would be a W, orientated with respect to the satellite's path so that all four lines are crossed. In the work carried out with this type of detector at University College London [15], the patterns used have been in the form of an M, giving angular coverages of up to 10^0.

On account of the limited accuracy of predictions, fields of view of several degrees are necessary, and the choice of $p = 1$ in (2) may not allow the full accuracy of the instrument, say $1''$ to $10''$, to be achieved. By increasing the number of slits and increasing p correspondingly in (2), the path may be obtained more accurately *over the whole field of view*.

As the field of view of an optical instrument can cover only a small fraction of a revolution of an orbit, observations need to be combined from different sightings, and the determination of the path over the whole field of view at a single sighting then adds little more than over a fraction of the field of view. A slit pattern is therefore proposed in which an M, covering the whole field of view, is combined with a multiplex of small M's, covering a small angular width as shown in Fig. 1. The multiplex consists of erect and inverted M's, which are here geometrically similar to the large M, although this need not necessarily be the case. If the path crosses all four lines of the large M, it will cross four lines of the multiplex. The path is first determined using the signals from the large M, and its points of intersection with the multiplex can then be calculated, so determining the particular four lines of the multiplex crossed. The path can then be determined more accurately for the central region of the field of view.

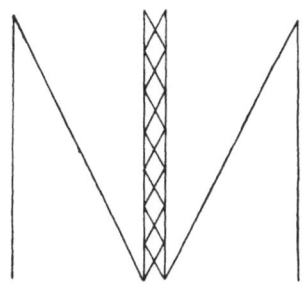

Fig. 1. Possible slit pattern for photoelectric detector

The conversion from (x, y) plate co-ordinates to the R.A. α and declination δ (or azimuth a and zenith distance z) has been dealt with elsewhere [2]. We have

$$\cot \delta\, e^{ia} = [(a_1 + i\, a_2)\, x + (b_1 + i\, b_2)\, y + (c_1 + i\, c_2)]/(a_0\, x + b_0\, y + c_0) \tag{4}$$

$$\tan z\, e^{-ia} = [(\alpha_1 + i\, \alpha_2)\, x + (\beta_1 + i\, \beta_2)\, y + (\gamma_1 + i\, \gamma_2)]/(\alpha_0\, x + \beta_0\, y + \gamma_0), \tag{5}$$

where the a's, b's, c's and the α's, β's, γ's are two sets of 'plate constants', which are related by

$$[\alpha_1\ \alpha_2\ \alpha_0] = [a_1\ a_2\ a_0]\, \Theta\, \Phi \quad \text{etc.,} \tag{6}$$

where

$$\Theta = \begin{bmatrix} -\cos\theta' & \sin\theta' & 0 \\ -\sin\theta' & -\cos\theta' & 0 \\ 0 & 0 & 1 \end{bmatrix} \qquad \Phi = \begin{bmatrix} \sin\phi & 0 & -\cos\phi \\ 0 & 1 & 0 \\ \cos\phi & 0 & \sin\phi \end{bmatrix} \tag{7}$$

θ' is L.S.T. at the instrument at the time of observation, and ϕ is the latitude.

Two methods are available for finding the plate constants, according to the design of the instrument. Where accurate mountings and scales are provided, the plate constants can be calculated directly from the azimuth and zenith distance of the optical axis, etc. [Groves and Owen [2], eqs. (76)]. Alternatively, the stars may be used as reference points from which the orientation of the instrument, and hence the plate constants, may be found. In practice, a photographic plate needs to be substituted for the photo-electric detector so that the slit pattern can be photographed with a surrounding star field. This method is tedious, and several sightings per satellite transit would not be easy to arrange. The use of an accurately-mounted instrument is to be preferred, with the star method reserved for occasional calibrations.

III. Axes, Notation and Geometrical Relationships of the Instantaneous Ellipse

The orbit of a satellite is in general a complicated three-dimensional curve, which approximates over a small region to an ellipse. It is therefore convenient to describe the orbit in terms of the orbital elements of an elliptic orbit, as these are then slowly varying quantities.

The orbital elements at any time are defined with respect to the ellipse which the satellite would describe if all perturbing forces were removed from that instant onwards. The perturbing forces include atmospheric drag and the forces arising from the departure of the gravitational field from that of a point mass, equal to the earth's mass and situated at the mass centre of the earth. This ellipse is referred to as the instantaneous ellipse at time t. The orbital elements will be taken to be a the semi major axis, e the eccentricity, i the inclination of the orbital plane to the equator, Ω the right ascension of the ascending node N, ω the angular distance of perigee from N in the direction of motion and χ the mean anomaly at time $t = 0$.

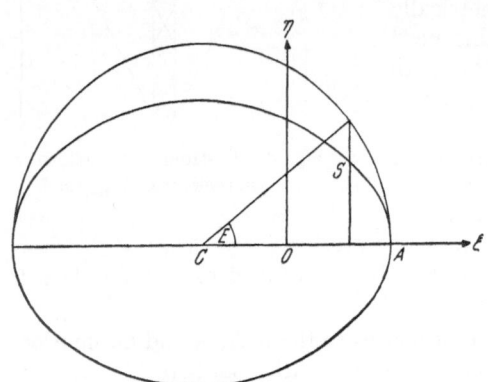

Fig. 2. Axes in orbital plane. O is mass-centre of earth. $CO = ae$

Let $O\,\xi\,\eta$ be axes in the plane of the ellipse, where O is the mass-centre of the earth, then the co-ordinates of the satellite S are, Fig. 2,

$$\xi = a \cos E - ae \tag{8}$$

$$\eta = b \sin E, \tag{9}$$

where E is the eccentric anomaly at time t.

Let $O\,x\,y\,z$ be celestial axes, where $O\,x$ is in the direction of the First Point of Aries, and $O\,z$ is towards the north celestial pole. If (l_1, m_1, n_1) are the direction cosines of $O\,A$ (the major axis of the instantaneous ellipse), then (Fig. 3)

$$l_1 = \cos \Omega \cos \omega - \sin \Omega \sin \omega \cos i \tag{10}$$

$$m_1 = \sin \Omega \cos \omega + \cos \Omega \sin \omega \cos i \tag{11}$$

$$n_1 = \sin \omega \sin i. \tag{12}$$

If $O\,B$ is at right angles to $O\,A$ in the sense shown in Fig. 3 and has direction cosines (l_2, m_2, n_2), then

$$l_2 = -\cos\Omega \sin\omega -$$
$$\qquad - \sin\Omega \cos\omega \cos i$$
$$\tag{13}$$

$$m_2 = -\sin\Omega \sin\omega +$$
$$\qquad + \cos\Omega \cos\omega \cos i$$
$$\tag{14}$$

$$n_2 = \quad \cos\omega \sin i. \tag{15}$$

The co-ordinates of S referred to $O\,x\,y\,z$ are therefore

$$\begin{bmatrix} x \\ y \\ z \end{bmatrix} = \begin{bmatrix} l_1\,\xi + l_2\,\eta \\ m_1\,\xi + m_2\,\eta \\ n_1\,\xi + n_2\,\eta \end{bmatrix} =$$

$$= S' \begin{bmatrix} \cos E \\ \sin E \\ 1 \end{bmatrix} \tag{16}$$

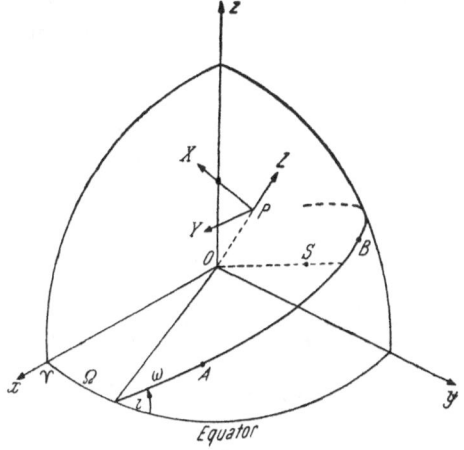

Fig. 3. Celestial axes $O\,x\,y\,z$ and ground axes $P\,X\,Y\,Z$. $O\,A$ is in the direction of perigee, and $A\,O\,B$ is a right angle. S = satellite. N = ascending node. P = observation post

by (8) and (9), where

$$S = \begin{bmatrix} a\,l_1 & a\,m_1 & a\,n_1 \\ b\,l_2 & b\,m_2 & b\,n_2 \\ -a\,e\,l_1 & -a\,e\,m_1 & -a\,e\,n_1 \end{bmatrix} \tag{17}$$

and

$$b = a\,(1 - e^2)^{1/2}. \tag{18}$$

Let P be an observation post on the earth's surface, then the co-ordinates of P are

$$[x'\ y'\ z'] = R\ [\cos\phi'\cos\theta'\quad \cos\phi'\sin\theta'\quad \sin\phi'] \tag{19}$$

where $R = O\,P$, ϕ' is the geocentric latitude of P and θ' is the L.S.T. at P.

Let $P\,Z$ be taken to the zenith and $P\,X$ to the north, then the co-ordinates of S referred to $P\,X\,Y\,Z$ are

$$[X\,Y\,Z] \equiv \varrho\,[\lambda\mu\,\nu] = [x - x'\quad y - y'\quad z - z']\,\Theta\,\Phi \tag{20}$$

where $\varrho = P\,S$, and λ, μ, ν are the direction cosines of $P\,S$ referred to $P\,X\,Y\,Z$. Hence by (16), (19) and (7)

$$[X\,Y\,Z] \equiv \varrho\,[\lambda\mu\,\nu] = [\cos E\quad \sin E\quad 1]\,S\,\Theta\,\Phi + R\,[\sin\varepsilon\quad 0\quad -\cos\varepsilon], \tag{21}$$

where

$$\varepsilon = \phi - \phi'. \tag{22}$$

IV. Aberration and Refraction Effects

Let t' be the time at which the satellite is observed, then on account of the finite speed of light, the position of the satellite is that for time

$$t = t' - \varrho/c \tag{23}$$

where c is the speed of light. For $\varrho = 3000$ km., the time difference amounts to 10 millisecs.

Aberration depends on the ratio of the speed of the satellite with respect to P relative to that of the speed of light. This may amount to 1 part in 40,000 giving rise to angular displacements of $5''$.

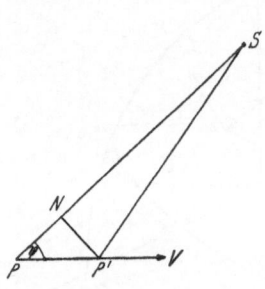

In Fig. 4, V is the velocity of P relative to S, and P moves to P' in the time interval taken by the light to travel from S to P'. $P'S$ is the apparent direction of S with direction cosines λ'', μ'', ν'' and PS is the true direction (apart from atmospheric refraction) with direction cosines λ', μ', ν'. If we define

$$[\varDelta\lambda\,\varDelta\mu\,\varDelta\nu] = [\lambda''\,\mu''\,\nu''] - [\lambda'\,\mu'\,\nu'], \quad (24)$$

then $[\varDelta\lambda\,\varDelta\mu\,\varDelta\nu]$ is in the direction $P'N$, and from the triangle of velocities $P\,P'\,N$,

Fig. 4. PS is the true direction of S, $P'S$ is the apparent direction due to aberration

$$c\,[\varDelta\lambda\,\varDelta\mu\,\varDelta\nu] = V\cos\psi\,[\lambda'\,\mu'\,\nu'] - V \quad (25)$$

where ψ is $\angle NPP'$, and V is the magnitude of

$$V = -\,[\dot{X}\,\dot{Y}\,\dot{Z}]. \quad (26)$$

Hence

$$c\,[\varDelta\lambda\,\varDelta\mu\,\varDelta\nu] = [\dot{X}\,\dot{Y}\,\dot{Z}] - (\lambda'\,\dot{X} + \mu'\,\dot{Y} + \nu'\,\dot{Z})\,[\lambda'\,\mu'\,\nu']. \quad (27)$$

From (21)

$$[\dot{X}\,\dot{Y}\,\dot{Z}] = \dot{E}\,[-\sin E\ \cos E\ 0]\,\boldsymbol{S\Theta\Phi} + [\cos E\ \sin E\ 1]\,\boldsymbol{S\Theta_1\Phi} + O\,(\dot{a},\dot{e},\ldots) \quad (28)$$

where

$$\boldsymbol{\Theta_1} = \theta' \begin{bmatrix} \sin\theta' & \cos\theta' & 0 \\ -\cos\theta' & \sin\theta' & 0 \\ 0 & 0 & 0 \end{bmatrix}. \quad (29)$$

θ' is the rate of change of L.S.T. with G.M.T., i.e. $1.002\,737\,8 + \varDelta\dot{\alpha}$ where $\varDelta\alpha$ is the nutation in R.A. The greatest value of $\varDelta\dot{\alpha}$ is approximately 1×10^{-7}, and so $\varDelta\dot{\alpha}$ may be neglected to this order of accuracy. In (28), the terms arising from the rates of change of the orbital elements have been neglected.

E is related to t by Kepler's equation

$$E - e\sin E - (n\,t + \chi) = 0, \quad (30)$$

where

$$n = 2\pi/T = \mu_0^{1/2}\,a^{-3/2} \quad (31)$$

$$\mu_0 = G\,M_0. \quad (32)$$

T is the period of revolution for the instantaneous ellipse, G is the gravitational constant and M_0 the mass of the earth. Hence

$$(1 - e\cos E)\,\dot{E} = n + 0\,(\dot{a},\dot{e}\ldots). \quad (33)$$

Atmospheric refraction displaces the satellite towards the zenith by amount $58''\tan z$, and so generally exceeds the aberration corrections. The apparent direction cosines λ', μ', ν' can be shown [2] to be related to the true direction cosines λ, μ, ν by

$$[\lambda \mu \nu] = (1 + k) [\lambda' \mu' \nu'] - k [0\ 0\ 1/\nu]$$ (34)

where k is the refraction constant ($= 0.000\ 28$ approximately for objects at infinity). (34) is an approximate relationship obtained by expanding in powers of k, but is accurate to $1''$ for $z < 68^0$ and to $1'$ for $z < 84^0$.

Table I. *Amounts by which the Refraction Correction is Less than the Astronomical Correction on Account of the Satellite's Proximity to the Observer*

Zenith distance, z	Height Z (km)						
	250	500	750	1000	2000	3000	4000
45^0	$2''.0$	$1''.0$	$0''.7$	$0''.5$	$0''.3$	$0''.2$	$0''.1$
75^0	$8''.9$	$5''.1$	$3''.7$	$3''.0$	$1''.8$	$1''.3$	$1''.1$

Even though the heights of satellites lie well above the appreciably refracting atmosphere, k is not independent of satellite height as might at first be supposed. For objects in the vicinity of the earth, there is a geometrical effect which alters k slightly from the value for an infinitely distant object. The theory has been developed by one of us (G.V.G.), and an account is in preparation. Table I shows the amount by which the refraction correction is less than the astronomical refraction for a selection of satellite heights and zenith distances.

V. Fundamental Relationship between Orbital Elements and Observational Data

The problem is now considered of relating a single observation on a satellite orbit to the orbital elements. The observation may be in the form of the R.A. and declination (or azimuth and zenith distance) of the satellite at a given time; or as the (x, y) plate co-ordinates from which α and δ (or a and z) can be found by (4) [or (5)]. If (x_0, y_0) are the plate co-ordinates of the satellite image at time t_0, then we may consider t_0 as the time at which the satellite would cross the slit $x = x_0$ and also the slit $y = y_0$. Thus in all cases, the positional data is expressible as the time of crossing certain straight-line slits, even though these slits may not have a physical existence.

Let λ'', μ'', ν'' be the apparent direction of S referred to axes $P\ X\ Y\ Z$ (Fig. 3). This direction departs from the true direction on account of aberration and refraction. The position of the satellite is that for time t, whereas the observer is positioned for time t' ($= t + \varrho/c$), on account of the finite speed of light. In terms of the apparent azimuth a'' and zenith distance z''

$$\lambda'' = \sin z'' \cos a'' \qquad \mu'' = - \sin z'' \sin a'' \qquad \nu'' = \cos z''.$$ (35)

From (5) and (35), the corresponding plate co-ordinates are

$$x'' = \frac{A_1 \lambda'' + A_2 \mu'' + A_0 \nu''}{C_1 \lambda'' + C_2 \mu'' + C_0 \nu''} \qquad y'' = \frac{B_1 \lambda'' + B_2 \mu'' + B_0 \nu''}{C_1 \lambda'' + C_2 \mu'' + C_0 \nu''}$$ (36)

where A_0, A_1 etc. are the cofactors of α_0, α_1 etc. in

$$\begin{bmatrix} \alpha_0 & \beta_0 & \gamma_0 \\ \alpha_1 & \beta_1 & \gamma_1 \\ \alpha_2 & \beta_2 & \gamma_2 \end{bmatrix}$$

i.e. $A_0 = \beta_1 \gamma_2 - \beta_2 \gamma_1$, $A_1 = \beta_2 \gamma_0 - \beta_0 \gamma_2$ etc.
If the satellite image lies on the slit

$$p\,x + q\,y + r = 0, \tag{37}$$

then λ'', μ'', ν'' satisfy

$$[\lambda'' \, \mu'' \, \nu''] \, p' = 0, \tag{38}$$

where

$$p' = \begin{bmatrix} A_1 & B_1 & C_1 \\ A_2 & B_2 & C_2 \\ A_0 & B_0 & C_0 \end{bmatrix} \begin{bmatrix} p \\ q \\ r \end{bmatrix}. \tag{39}$$

By (24) and (34)

$$(1+k)\,[\lambda'' \ \mu'' \ \nu''] = [\lambda \ \mu \ \nu] + [\varDelta\lambda \ \varDelta\mu \ \varDelta\nu] + k\,[0 \ 0 \ 1/\nu]. \tag{40}$$

Hence by (37)

$$\{[\lambda \ \mu \ \nu] + [\varDelta\lambda \ \varDelta\mu \ \varDelta\nu] + k\,[0 \ 0 \ 1/\nu]\}\,p' = 0. \tag{41}$$

By (21), (41) becomes

$$\{[\cos E \ \ \sin E \ \ 1]\,S\,\Theta\,\Phi + R\,[\sin\varepsilon \ \ 0 \ \ -\cos\varepsilon] + \delta\}\,p' = 0, \tag{42}$$

where

$$\delta = \varrho\,[\varDelta\lambda \ \varDelta\mu \ \varDelta\nu] + k\varrho\,[0 \ 0 \ 1/\nu]. \tag{43}$$

(42) is the fundamental relationship between the orbital elements of the instantaneous ellipse of a satellite at time t, which crosses a given slit at L.S.T. θ'. The effects of aberration and refraction are included in the small term δ.

When the observational data is provided as z'' and a'', and hence by (35) as λ'', μ'', ν'' we can put

$$\begin{bmatrix} A_1 & B_1 & C_1 \\ A_2 & B_2 & C_2 \\ A_0 & B_0 & C_0 \end{bmatrix} = \begin{bmatrix} 1 & 0 & 0 \\ 0 & 1 & 0 \\ 0 & 0 & 1 \end{bmatrix} \tag{44}$$

and the corresponding co-ordinates of the image are, by (35),

$$x'' = \lambda''/\nu'' \qquad y'' = \mu''/\nu''. \tag{45}$$

By (37), (39) and (44), the corresponding values for p can be taken as $[\nu''\ 0 - \lambda'']$ and $[0\ \nu'' - \mu'']$.

When the observational data is provided as the R.A. α'' and declination δ'',

$$[\lambda'' \, \mu'' \, \nu''] = [l'' \, m'' \, n'']\,\Theta\,\Phi, \tag{46}$$

where

$$l'' = \cos\delta'' \cos\alpha'' \qquad m = \cos\delta'' \sin\alpha'' \qquad n'' = \sin\delta'' \tag{47}$$

VI. Determination of Orbital Elements: General Theory

The theory is developed on the assumption that the orbital elements have been expressed as explicit functions of time

$$a = a\,(\{\alpha\};t) \text{ etc.} \tag{48}$$

Where $\{\alpha\}$ denotes a set of parameters whose determination is required from observations on the times t_i' at which the satellite crosses a number of slits. The slits need not necessarily be in the focal plane of the same instrument nor need all instruments necessarily be at the same observing point. It will be assumed that the values of R and ε are known for each observing post with adequate accuracy. On present knowledge R is likely to be correct to 100 m. When the accuracy of the observations and of the analysis is better than this amount, R and ε will need to be treated as unknowns.

We denote the unknowns by α_j, and choose them to minimize the differences $t_{i_1}' - t_i'$, where t_{i_1}' is the value of t_i' satisfying (30) and (42). In practice, the various instruments used may have different degrees of accuracy. In combining observations it is then necessary to associate with the differences $t_{i_1}' - t_i'$, the weight

$$w_i \propto 1/\sigma\, t_i', \tag{49}$$

where $\sigma\, t_i'$ is the standard deviation of t_i'. α_j are chosen to minimize

$$R^2 = \sum_i (t_{i_1}' - t_i')^2\, w_i^2. \tag{50}$$

The method of analysis has to be carried out by successive approximation starting with a set of approximately known values α_{j0}. Let t_{i0}' to the values of t_i' satisfying (30) and (42) with the unknowns given their appropriate values α_{j0}. Then

$$t_{i_1}' = t_{i0}' + \sum_j (\partial t_i'/\partial\alpha_j)_0\, \delta\alpha_j \tag{51}$$

to the first order in

$$\delta\alpha_j = \alpha_j - \alpha_{j0}. \tag{52}$$

From (50)

$$R^2 = \sum_i \left[\tau_i + \sum_j (\partial t_i'/\partial\alpha_j)_0\, \delta\alpha_j \right]^2 w_i^2, \tag{53}$$

where

$$\tau_i = t_{i0}' - t_i'. \tag{54}$$

The minimization of (53) with respect to α_j and hence to $\delta\alpha_j$ gives

$$0 = \sum_i \left[\tau_i + \sum_j (\partial t_i'/\partial\alpha_j)_0\, \delta\alpha_j \right] (\partial t_i'/\partial\alpha_j)_0\, w_i^2$$

i.e. in matrix notation

$$A'\, A\, \delta\alpha' + A'\, \tau' = 0, \tag{55}$$

where

$$A = \begin{bmatrix} w_1\,(\partial t_1'/\partial\alpha_1)_0 & w_1\,(\partial t_1'/\partial\alpha_2)_0 \cdots \\ w_2\,(\partial t_2'/\partial\alpha_1)_0 & w_2\,(\partial t_2'/\partial\alpha_2)_0 \cdots \\ \vdots & \vdots \end{bmatrix} \tag{56}$$

$$\delta\alpha = [\delta\alpha_1 \ldots] \tag{57}$$

$$\tau = [\tau_1 \ldots]. \tag{58}$$

(55) has the solution

$$\delta\alpha = -\tau A (A' A)^{-1}. \tag{59}$$

Improved values for α_j can then be found and the procedure repeated until convergence has been obtained.

A valuable feature of the least-squares method of analysis is that it enables the standard deviations of the unknowns to be estimated. It can be shown that $\sigma \alpha_j$, the standard deviation of α_j, can be estimated by

$$\sigma \alpha_j = \sigma (\delta \alpha_j) = (a_j R^2/N)^{1/2}, \tag{60}$$

where a_j is the j-th diagonal element of $(A' A)^{-1}$ and N is the number of degrees of freedom, i.e. number of observations minus number of unknowns. R^2 can be calculated from (53), which reduces to $\tau \tau'$ when $\delta \alpha_j$ have converged to zero.

To calculate t_0' from (42) and (30), a method of successive approximation may be used in which t_0' is given the observed value t' as an approximate value. S and Θ can then be evaluated approximately and E_0 calculated from (42), which has the form

$$k_1 \cos E + k_2 \sin E + k_3 = 0 \tag{61}$$

where k_1, k_2, k_3 are known. t_0 can then be found from (30), and t_0' from (23) and the process repeated. (A subsidiary loop in the calculation would enable the aberration and refraction effects to be included.) Convergence will be rapid for a near earth satellite as E changes about 15 times more rapidly then θ'.

VII. Evaluation of Partial Derivatives in A

(56) can be written

$$A = B\, C, \tag{62}$$

where

$$B = \begin{bmatrix} w_1\,(\partial t_1'/\partial\beta_1)_0 & w_1\,(\partial t_1'/\partial\beta_2)_0 \cdots \\ w_2\,(\partial t_2'/\partial\beta_1)_0 & w_2\,(\partial t_1'/\partial\beta_2)_0 \cdots \\ \vdots & \vdots \end{bmatrix} \tag{63}$$

$$C = \begin{bmatrix} (\partial\beta_1/\partial\alpha_1)_0 & (\partial\beta_1/\partial\alpha_2)_0 \cdots \\ (\partial\beta_2/\partial\alpha_1)_0 & (\partial\beta_2/\partial\alpha_2)_0 \cdots \\ \vdots & \vdots \end{bmatrix} \tag{64}$$

and $\beta_1 \ldots \beta_6$ denote $a, e, i, \Omega, \omega, \chi$.

Let (42) and (30) be expressed as

$$f\,(E, \beta_k, t') = 0 \tag{65}$$

$$g\,(E, \beta_k, t') = 0 \tag{66}$$

then

$$\frac{\partial t'}{\partial \beta_k} = -\frac{\partial (f, g)/\partial (\beta_k, E)}{\partial (f, g)/\partial (t', E)} . \tag{67}$$

From (42), neglecting the variation in δ,

$$\partial f/\partial \beta_k = [\cos E \quad \sin E \quad 1]\, S_k\, \Theta\, \Phi\, p', \tag{68}$$

where by (17) and (18)

$$S_1 \equiv \partial S/\partial a = \begin{bmatrix} l_1 & m_1 & n_1 \\ (1-e^2)^{1/2}\, l_2 & (1-e^2)^{1/2}\, m_2 & (1-e^2)^{1/2}\, n_2 \\ -e\, l_1 & -e\, m_1 & -e\, n_1 \end{bmatrix} \tag{69}$$

$$S_2 \equiv \partial S/\partial e = \begin{bmatrix} 0 & 0 & 0 \\ -\,ae\,(1-e^2)^{-1/2}\,l_2 & -\,ae\,(1-e^2)^{-1/2}\,m_2 & -\,ae\,(1-e^2)^{-1/2}\,n_2 \\ -\,a\,l_1 & -\,a\,m_1 & -\,a\,n_1 \end{bmatrix}$$

$$(70)$$

and by (10) to (15)

$$S_3 \equiv \partial S/\partial i = \begin{bmatrix} a\sin\Omega\sin\omega\sin i & -\,a\cos\Omega\sin\omega\sin i & a\sin\omega\cos i \\ b\sin\Omega\cos\omega\sin i & -\,b\cos\Omega\cos\omega\sin i & b\cos\omega\cos i \\ -\,ae\sin\Omega\sin\omega\sin i & ae\cos\Omega\sin\omega\sin i & -\,ae\sin\omega\cos i \end{bmatrix}$$

$$(71)$$

$$S_4 \equiv \partial S/\partial \Omega = \begin{bmatrix} -\,a\,m_1 & a\,l_1 & 0 \\ -\,b\,m_2 & b\,l_2 & 0 \\ a\,e\,m_1 & -\,a\,e\,l_1 & 0 \end{bmatrix}$$

$$(72)$$

$$S_5 \equiv \partial S/\partial \omega = \begin{bmatrix} a\,l_2 & a\,m_2 & a\,n_2 \\ -\,b\,l_1 & -\,b\,n_1 & -\,b\,n_1 \\ -\,a\,e\,l_2 & -\,a\,e\,m_2 & -\,a\,e\,n_2 \end{bmatrix}$$

$$(73)$$

$$S_6 \equiv \partial S/\partial \chi = 0.$$

$$(74)$$

Again from (42)

$$\partial f/\partial E = [-\sin E \quad \cos E \quad 0]\, S\,\Theta\,\Phi\,p' \tag{75}$$

$$\partial f/\partial t' = [\quad \cos E \quad \sin E \quad 1]\, S\,\Theta_1\,\Phi\,p', \tag{76}$$

where Θ is defined by (29).

From (30)

$$\partial g/\partial \beta_1 \equiv \partial g/\partial a = -\,(dn/da)\,t = \frac{3}{2}\,(n/a)\,t \tag{77}$$

$$\partial g/\partial \beta_2 \equiv \partial g/\partial e = -\sin E \tag{78}$$

$$\partial g/\partial \beta_3 \equiv \partial g/\partial i = 0 \tag{79}$$

$$\partial g/\partial \beta_4 \equiv \partial g/\partial \Omega = 0 \tag{80}$$

$$\partial g/\partial \beta_5 \equiv \partial g/\partial \omega = 0 \tag{81}$$

$$\partial g/\partial \beta_6 \equiv \partial g/\partial \chi = -\,1 \tag{82}$$

$$\partial g/\partial E = 1 - e\cos E \tag{83}$$

and

$$\partial g/\partial t' = -\,n \tag{84}$$

by (23), on neglecting the variation in ϱ/C.

The evaluation of the elements $\partial \beta_k/\partial \alpha_j$ of C depends on the functional forms (48), and this will be discussed in the following sections. The number of parameters included in these functions will need to be increased with the length of the time interval over which observations are analysed. For a sufficiently short time interval, the orbital elements may be taken as constant, and C then becomes a unit matrix. This case might be applicable when the observations are confined to a single transit. After one or two revolutions of a satellite, the changes in some of the elements (e.g. Ω) become significant even to a fairly low degree of accuracy, and account would need to be taken of this in the analysis.

VIII. Variation of the Orbital Elements with Time

In the previous section, a general analysis was presented, wherein the orbital elements were considered as arbitrary functions of time. The precise functional forms of the elements can only be found by integration of the equations of motion for those elements. In this way even if an exact solution is not possible, at least the functionals best suited to be assumed for the elements may be derived. If this is not done or if the equations of motion are intractable, then one may assume some polynomial form in time for the elements.

$$a = P_a(t) \tag{85}$$

and the essential parameters in these forms derived. The simplest method is to consider merely the secular variations, i.e. to assume all elements to vary linearly with time

$$a = a_0 + a_1 t. \tag{86}$$

Thus each element will have two degrees of freedom, and such an analysis would be identical to that assuming constant elements, except that twelve unknowns would be involved instead of six. The two simpler methods have the disadvantage that although they may be adequate for the problem, they conceal the physics of the variations by the introduction of arbitrary parameters. Following is an examination of the causes of the variation in the elements, and an analysis of data to account for linearly varying elements only.

The causes of the introduction of a secular variation are twofold. Firstly the air drag perturbation in a dissipative force which is continually whittling away the energy of the satellite and is the eventual cause of its descent to earth. Secondly the departure of the Earth's potential from a central force field, due to the irregular shape of the Earth, causes elemental changes resulting in an alteration in form of the orbit, with no loss in energy. These will be dealt with in detail.

The effect of the contact of the satellite with the top of the atmosphere is a drag on the satellite of magnitude $- K \varrho v^2$ where K is some constant, ϱ is air density and v is its velocity. This force exerted mainly at perigee, when the satellite is passing through the denser layers of air, causes a reduction of height at the succeeding apogee, accompanied by an increase in the orbital velocity, and a resulting decrease in the orbital period. A study of the behaviour of the orbital period, a relatively easily measured parameter, can thus lead to information about the air density at altitudes of perigee. This effect has been studied with some success by many authors (King-Hele and Leslie [6]; King-Hele [5]; Nonweiler [11]). The total effect of this drag force is seen to be a reduction in the major axis of the orbit, and a reduction in the eccentricity, orbital period, and mean height above ground level. A secondary effect of the atmosphere is caused by the rotation of this atmosphere with the earth. This rotation of the tenuous atmosphere at the heights involved suffices to produce a force orthogonal to the satellite trajectory. This force perturbes the ideally elliptic orbit and tends to alter the inclination of the orbital plane to the equatorial plane. (The same effect is produced by the earth's oblateness but is of a degree of magnitude smaller, and can thus be neglected in this connection.) This effect has been studied in detail and comparisons have been given between the theoretical estimates of di/dt and the observations on Sputnik II (1957/β) (Plimmer [12]; Cook [1]).

The departure of the shape of the Earth from a perfect sphere introduces zonal harmonics to the geopotential, the most important being the second and fourth harmonics. The geopotential function is usually written

$$U(r) = \frac{\mu}{r}\left[1 + J\left(\frac{R}{r}\right)^2\left(\frac{1}{3} - \sin\phi\right) + \frac{D}{35}\left(\frac{R}{r}\right)^4(35\sin^4\phi - 30\sin^2\phi + 3)\right] \quad (87)$$

where R is the equatorial radius of the Earth, ϕ is the geocentric latitude of r, and J and D are constants which characterise the geopotential. According to JEFFREYS [4] these are

$$J = 1.637 \times 10^{-3}; \qquad D = 10.6 \times 10^{-6}.$$

However much more accurate determinations of these constants are now available (LECAR et al [9]; KING-HELE and MERSON [7]). It is to be noticed that recent investigations have shown that it may not be sufficient to consider only the second and fourth harmonics in the geopotential when an accuracy of one second of arc and millisecond timing is envisaged. One such estimate of the coefficient of the third harmonic is (WHIPPLE [13])

$$A_3 = 3.0 \times 10^{41}\, cm^6/sec^2$$

in JEFFREYS notation, where the geopotential is given by

$$U(r) = \sum_1^\infty \frac{A_n}{r^{n+1}} S_n(\phi).$$

It is also probable that the longitudinal harmonics may not be negligible, and that harmonics of the type $S_n(\phi, \lambda)$ will have to be introduced. The effect of this potential function is to cause secular and periodic changes in the orbital elements. However by far the most important variations caused are those in Ω, the longitude of the ascending node, and in ω the argument of perigee. It is found that the line of nodes may regress at about 3^0 per day while the line of apsides may progress at about 4^0 per day. A third small perturbation in the orbit is in r, the instantaneous distance from the Earth's centre. It will be found that r will depart from the ideally elliptic orbit by some 2 n.m. due to the satellite tending to following the equigeopotential surface. However this is a periodic variation and will not be considered here.

It will be noticed that the effects on the orbital elements of these two main perturbing factors, the air drag and the non-central geopotential, cause quite distinct secular changes in the elements. The air drag affects mainly a, e, i, and T the orbital period, while the non-central geopotential affects mainly Ω and w. This fortunate behaviour facilities independent study of the effects.

In the special case where the secular variations only are considered, the matrix A may be constructed by evaluating the matrix B from (67) and using the matrix C in the form:

$$C = \begin{bmatrix} 1 & t & 0 & 0 & \dots & 0 & 0 \\ 0 & 0 & 1 & t & \dots & 0 & 0 \\ \cdot & \cdot & \cdot & \cdot & & \cdot & \cdot \\ \cdot & \cdot & \cdot & \cdot & & \cdot & \cdot \\ \cdot & \cdot & \cdot & \cdot & & \cdot & \cdot \\ 0 & 0 & \cdot & \cdot & \dots & 1 & t \end{bmatrix}.$$

References

1. G. E. Cook, R.A.E. Farnborough, Tech. Memo G.W. 351, 1959.
2 G. V. Groves and G. Owen (to be published).
3. K. G. Henize, Sky and Telescope 16, 108 (1957).
4. H. Jeffreys, The Earth. Cambridge: University Press, 1952.
5. D. G. King-Hele, Nature 183, 1224 (1959).
6. D. G. King-Hele and D. C. M. Leslie, Nature 181, 1761 (1958).
7. D. G. King-Hele and R. H. Merson, J. Brit. Interplan. Soc. 16, 446 (1958).
8. D. G. King-Hele and R. H. Merson, Nature 183, 881 (1958).
9. M. Lecar, J. Sorenson, and A. Eckels, J. Geophysic. Res. 64, 209 (1959).
10. R. H. Merson, Astronaut. Acta 5, 26 (1959).
11. T. R. F. Nonweiler, J. Brit. Interplan. Soc. 17, 14 (1959).
12. R. N. A. Plimmer, R.A.E. Farnborough, Tech. Note G.W. 504, 1959.
13. F. L. Whipple, Harvard Coll. Obs. Announcement Card No. 1420, 1958.
14. C. A. Whitney and G. Veis, Smithsonian Inst. Astrophysic. Obs., Special Report No. 19, 1958.
15. A. P. Willmore, Nature 182, 1008 (1958).

Second Colloquium on The Law of Outer Space
London 1959
Proceedings

Edited by

Andrew G. Haley

General Counsel, International Astronautical Federation
Washington/D.C., U.S.A.

and

Dr. Welf Heinrich Prince of Hanover

Frankfurt a. M., Germany

Approx. 180 pages. 6¾" × 9½". Wien: Springer-Verlag. 1960

Contents

Anfuso, V. L. (Washington/D.C.): Is Space the Way to Peace and Abundance? — **Beresford,** S. M. (Washington/D.C.): The Future of National Sovereignty. — **Binet,** H. T. P. (Montreal): Toward Solving the Space Sovereignty Problem. — **Bún,** Th. P. (São Paulo): The Impact of Spaceflight on World Economy. — **Cooper,** J. C. (Princeton/N.J.): Flight above the Air-Space. — **Feldman,** G. J. (New York/N.Y.): The Report of the United Nations Legal Committee on the Peaceful Uses of Outer Space: a Provisional Appraisal. — **Fonseca,** E. (Lisbon): Dynamical Limitation of the Freedom of Space. — **Furfey,** P. H. (Washington/D.C.): The Behavioral Sciences in the Space Age. — **Galloway,** E. (Washington/D.C.): The United Nations Ad Hoc Committee on the Peaceful Uses of Outer Space. Accomplishments and Implications for Legal Problems. — **Goedhuis**, D. (London): The Question of Freedom of Innocent Passage of Space Vehicles of One State through the Space above the Territory of Another State which is not Outer Space. — **Haley,** A. G. (Washington/D.C.): Space Exploration — the Problems of Today, Tomorrow and in the Future. — **Hanover,** Welf Heinrich Prince of (Frankfurt): Circle of Thoughts. — **Javitch,** R. A. (Montreal): Some Rules Regulating Earth-to-Earth, Space-to-Earth and Earth-to-Space Missiles and Interplanetary Vehicles. — **Kucherov,** S. (Washington/D.C.): Legal Problems of Outer Space. U.S.A. and Soviet Viewpoints. — **Lall,** S. S. (New Delhi): Space Exploration — Some Legal and Political Aspects. — **Machowski**, J. (New York/N.Y.): The Legal Status of Unmanned Space Vehicles. — **Meyer,** A. (Cologne): Some Problems Relating to Space Law. — **Pépin,** E. (Montreal): Proposals for the Future Work of the Permanent Legal Committee of the International Astronautical Federation. — **Rauchhaupt,** Fr. W. von (Heidelberg): World Space Law. The Basic Principles for its Codification. — **Rivoire,** J. (Paris): How to Introduce the Law into the Space? — **Rode-Verschoor,** I. H. Ph. de (Utrecht): The Influence of the Exploration of Outer Space on Mankind. — **Seara-Vazquez,** M. (Paris): The Functional Regulation of the Extra-Atmospheric Space. — **Smirnoff**, M. S. (Belgrade): The Role of I.A.F. in the Elaboration of the Norms of Future Space Law. — **Valladão,** H. (Rio de Janeiro): The Law of Interplanetary Space. — **Yeager,** Ph. B. (Washington/D.C.): Space and Cogno-Politics: a Third Force in World Affairs.

First Colloquium on the Law of Outer Space. The Hague 1958.

Proceedings. Edited by **Andrew G. Haley,** President, International Astronautical Federation, Washington, D.C., U.S.A., and Dr. **Welf Heinrich Prince of Hanover,** Frankfurt a.M., Germany. V, 126 pages. 6¾" × 9½". Wien: Springer-Verlag. 1959.

S 114.—, DM 19.—, sfr. 19.50, $ 4.50

Members of societies joined to the I.A.F. and subscribers of „Astronautica Acta" can obtain the Proceedings at a reduced price with a reduction of 20 per cent:

S 91.20, DM 15.20, sfr. 15.60, $ 3.60